Design Technology for Heterogeneous Embedded Systems

Gabriela Nicolescu • Ian O'Connor •
Christian Piguet
Editors

Design Technology for Heterogeneous Embedded Systems

Springer

Editors
Prof. Gabriela Nicolescu
Department of Computer Engineering
Ecole Polytechnique Montreal
2500 Chemin de Polytechnique Montreal
Montreal, Québec
Canada H3T 1J4
gabriela.nicolescu@polymtl.ca

Prof. Ian O'Connor
CNRS UMR 5270
Lyon Institute of Nanotechnology
Ecole Centrale de Lyon
av. Guy de Collongue 36
Bâtiment F7
69134 Ecully
France
ian.oconnor@ec-lyon.fr

Prof. Christian Piguet
Integrated and Wireless Systems Division
Centre Suisse d'Electronique et de
Microtechnique (CSEM)
Jaquet-Drotz 1
2000 Neuchâtel
Switzerland
christian.piguet@csem.ch

ISBN 978-94-007-9496-2 ISBN 978-94-007-1125-9 (eBook)
DOI 10.1007/978-94-007-1125-9
Springer Dordrecht Heidelberg London New York

Softcover reprint of the hardcover 1st edition 2012

Cover design: VTeX UAB, Lithuania

Printed on acid-free paper

Springer is part of Springer Science+Business Media (www.springer.com)

Contents

List of Contributors

El Mostapha Aboulhamid Department of Computer Science and Operations Research, University of Montreal, 2920 Chemin de la Tour Montreal, Montreal, Canada H3T 1J4

Gabriel Marchesan Almeida LIRMM, UMR 5506, CNRS–Université Montpellier 2, 161 rue Ada, 34095 Montpellier Cedex 5, France

Amara Amara Institut Superieur d'Electronique de Paris (ISEP), 21 rue d'Assas, 75270 Paris, France

Costin Anghel Institut Superieur d'Electronique de Paris (ISEP), 21 rue d'Assas, 75270 Paris, France, costin.anghel@isep.fr

David Atienza Embedded Systems Laboratory (ESL), Faculty of Engineering, Ecole Polytechnique Fédérale de Lausanne (EPFL), Lausanne, Switzerland, david.atienza@epfl.ch

José L. Ayala Embedded Systems Laboratory (ESL), Faculty of Engineering, Ecole Polytechnique Fédérale de Lausanne (EPFL), Lausanne, Switzerland, jose.ayala@epfl.ch;
Department of Computer Architecture (DACYA), School of Computer Science, Complutense University of Madrid (UCM), Madrid, Spain, jayala@fdi.ucm.es

Nadine Azemard LIRMM, UMR 5506, CNRS–Université Montpellier 2, 161 rue Ada, 34095 Montpellier Cedex 5, France

Cécile Belleudy University of Nice-Sophia Antipolis, LEAT, CNRS, Bat. 4, 250 rue Albert Einstein, 06560 Valbonne, France, belleudy@unice.fr

Marc Belleville CEA Leti, MINATEC, Grenoble, France

Pascal Benoit LIRMM, UMR 5506, CNRS–Université Montpellier 2, 161 rue Ada, 34095 Montpellier Cedex 5, France, Pascal.Benoit@lirmm.fr

Sébastien Bilavarn University of Nice-Sophia Antipolis, LEAT, CNRS, Bat. 4, 250 rue Albert Einstein, 06560 Valbonne, France, bilavarn@unice.fr

Dominique Borrione TIMA Laboratory, 46 Avenue Félix Viallet, 38031 Grenoble Cedex, France, Dominique.Borrione@imag.fr

Youcef Bouchebaba STMicroelectronics, 16 Fitzgerald Rd, Ottawa, ON, K2H 8R6, Canada, youcef.bouchebaba@st.com

Olivier Brousse LIRMM, UMR 5506, CNRS–Université Montpellier 2, 161 rue Ada, 34095 Montpellier Cedex 5, France

Jean-Frédéric Christmann CEA-LETI, MINATEC, 17 avenue des Martyrs, 38054 Grenoble Cedex 9, France

Fabien Clermidy CEA Leti, MINATEC, Grenoble, France

Cyril Condemine CEA-LETI, MINATEC, 17 avenue des Martyrs, 38054 Grenoble Cedex 9, France, cyril.condemine@cea.fr

Wim Dehaene ESAT–MICAS, Katholieke Universiteit Leuven, Leuven, Belgium

Jean-Luc Dekeyser INRIA Lille Nord Europe–LIFL–USTL–CNRS, Parc Scientifique de la Haute Borne, Park Plaza–Batiment A, 40 avenue Halley, 59650 Villeneuve d'Ascq, France, jean-luc.dekeyser@lifl.fr

Thibault Delavallée Université catholique de Louvain (UCL), Louvain-la-Neuve, Belgium, thibault.delavallee@uclouvain.be

Damien Deleruyelle IM2NP, Institut Matériaux Microélectronique Nanosciences de Provence, UMR CNRS 6242, Aix-Marseille Université, IMT Technopôle de Château Gombert, 13451 Marseille Cedex 20, France

Alain Fourmigue Department of Computer Engineering, Ecole Polytechnique Montreal, 2500 Chemin de Polytechnique Montreal, Montreal, Canada H3T 1J4, Alain.fourmigue@polymtl.ca

Abdoulaye Gamatié INRIA Lille Nord Europe–LIFL–USTL–CNRS, Parc Scientifique de la Haute Borne, Park Plaza–Batiment A, 40 avenue Halley, 59650 Villeneuve d'Ascq, France, abdoulaye.gamatie@lifl.fr

Patrice Gerin TIMA Laboratory, CNRS/Grenoble-INP/UJF, Grenoble, France, Patrice.Gerin@imag.fr

Luiza Gheorghe Department of Computer Engineering, Ecole Polytechnique Montreal, 2500 Chemin de Polytechnique Montreal, Montreal, Canada H3T 1J4

Olivier Ginez IM2NP, Institut Matériaux Microélectronique Nanosciences de Provence, UMR CNRS 6242, Aix-Marseille Université, IMT Technopôle de Château Gombert, 13451 Marseille Cedex 20, France

Bruno Girodias Department of Computer Engineering, Ecole Polytechnique Montreal, 2500 Chemin de Polytechnique Montreal, Montreal, Canada H3T 1J4

Benoit Gosselin Université Laval, Quebec, Canada, benoit.gosselin@gel.ulaval.ca

Christoph Grimm Institute of Computer Technology, Vienna University of Technology, Gußhausstraße 27-29/E384, 1040 Wien, Austria, grimm@ict.tuwien.ac.at

S. Gyger CSEM, Neuchâtel, Switzerland

Jan Haase Institute of Computer Technology, Vienna University of Technology, Gußhausstraße 27-29/E384, 1040 Wien, Austria, haase@ict.tuwien.ac.at

Mian Muhammad Hamayun TIMA Laboratory, CNRS/Grenoble-INP/UJF, Grenoble, France, Mian-Muhammad.Hamayun@imag.fr

Da He C-LAB, Paderborn University, Paderborn, Germany

Yannick Hervé Université de Strasbourg, Strasbourg, France, yrv@wanadoo.fr; Simfonia SARL, Strasbourg, France

L. Labrak CNRS UMR 5270, Lyon Institute of Nanotechnology, Ecole Centrale de Lyon, av. Guy de Collongue 36, Bâtiment F7, 69134 Ecully, France

Walid Lafi CEA-LETI, MINATEC, 17 avenue des Martyrs, 38054 Grenoble Cedex 9, France

Didier Lattard CEA-LETI, MINATEC, 17 avenue des Martyrs, 38054 Grenoble Cedex 9, France, didier.lattard@cea.fr

Yusuf Leblebici Microelectronic Systems Laboratory (LSM), Faculty of Engineering, EPFL, Lausanne, Switzerland, yusuf.leblebici@epfl.ch

Yves Leduc Advanced System Technology, Wireless Terminal Business Unit, Texas Instruments, Villeneuve-Loubet, France, y-leduc@ti.com

Jean-Didier Legat Université catholique de Louvain (UCL), Louvain-la-Neuve, Belgium, jean-didier.legat@uclouvain.be

Arnaud Legendre Simfonia SARL, Strasbourg, France

Igor Loiselle Université catholique de Louvain (UCL), Louvain-la-Neuve, Belgium, igor.loiselle@uclouvain.be

Philippe Manet Université catholique de Louvain (UCL), Louvain-la-Neuve, Belgium, philippe.manet@uclouvain.be

J.-M. Masgonty CSEM, Neuchâtel, Switzerland

Philippe Maurine LIRMM, UMR 5506, CNRS–Université Montpellier 2, 161 rue Ada, 34095 Montpellier Cedex 5, France

Samy Meftali INRIA Lille Nord Europe–LIFL–USTL–CNRS, Parc Scientifique de la Haute Borne, Park Plaza–Batiment A, 40 avenue Halley, 59650 Villeneuve d'Ascq, France, samy.meftali@lifl.fr

Nathalie Messina Advanced System Technology, Wireless Terminal Business Unit, Texas Instruments, Villeneuve-Loubet, France

Fabian Mischkalla C-LAB, Paderborn University, Paderborn, Germany

M. Morgan CSEM, Neuchâtel, Switzerland

Katell Morin-Allory TIMA Laboratory, 46 Avenue Félix Viallet, 38031 Grenoble Cedex, France, Katell.Morin-Allory@imag.fr

Wolfgang Mueller C-LAB, Paderborn University, Paderborn, Germany

Christophe Muller IM2NP, Institut Matériaux Microélectronique Nanosciences de Provence, UMR CNRS 6242, Aix-Marseille Université, IMT Technopôle de Château Gombert, 13451 Marseille Cedex 20, France, christophe.muller@im2np.fr

J.-L. Nagel CSEM, Neuchâtel, Switzerland

Gabriela Nicolescu Department of Computer Engineering, Ecole Polytechnique Montreal, 2500 Chemin de Polytechnique Montreal, Montreal, Québec, Canada H3T 1J4, gabriela.nicolescu@polymtl.ca

I. O'Connor CNRS UMR 5270, Lyon Institute of Nanotechnology, Ecole Centrale de Lyon, av. Guy de Collongue 36, Bâtiment F7, 69134 Ecully, France, ian.oconnor@ec-lyon.fr

Yann Oddos TIMA Laboratory, 46 Avenue Félix Viallet, 38031 Grenoble Cedex, France, Yann.Oddos@imag.fr

Frédéric Pétrot TIMA Laboratory, CNRS/Grenoble-INP/UJF, Grenoble, France, Frederic.Petrot@imag.fr

Pierre Paulin STMicroelectronics, 16 Fitzgerald Rd, Ottawa, ON, K2H 8R6, Canada

V. Peiris CSEM, Neuchâtel, Switzerland

C. Piguet Integrated and Wireless Systems Division, Centre Suisse d'Electronique et de Microtechnique (CSEM), Jaquet-Drotz 1, 2000 Neuchâtel, Switzerland, christian.piguet@csem.ch

Diego Puschini LIRMM, UMR 5506, CNRS–Université Montpellier 2, 161 rue Ada, 34095 Montpellier Cedex 5, France; CEA Leti, MINATEC, Grenoble, France

Imran Rafiq Quadri INRIA Lille Nord Europe–LIFL–USTL–CNRS, Parc Scientifique de la Haute Borne, Park Plaza–Batiment A, 40 avenue Halley, 59650 Villeneuve d'Ascq, France, imran.quadri@lifl.fr

Bettina Rebaud LIRMM, UMR 5506, CNRS–Université Montpellier 2, 161 rue Ada, 34095 Montpellier Cedex 5, France; CEA Leti, MINATEC, Grenoble, France

Michel Robert LIRMM, UMR 5506, CNRS–Université Montpellier 2, 161 rue Ada, 34095 Montpellier Cedex 5, France

Bertrand Rousseau Université catholique de Louvain (UCL), Louvain-la-Neuve, Belgium, bertrand.rousseau@uclouvain.be

D. Séverac CSEM, Neuchâtel, Switzerland

Gilles Sassatelli LIRMM, UMR 5506, CNRS–Université Montpellier 2, 161 rue Ada, 34095 Montpellier Cedex 5, France

Mohamad Sawan École Polytechnique de Montréal, Montreal, Canada

Arvind Sridhar Embedded Systems Laboratory (ESL), Faculty of Engineering, Ecole Polytechnique Fédérale de Lausanne (EPFL), Lausanne, Switzerland, arvind.sridhar@epfl.ch

Lionel Torres LIRMM, UMR 5506, CNRS–Université Montpellier 2, 161 rue Ada, 34095 Montpellier Cedex 5, France

Yves Vanderperren ESAT–MICAS, Katholieke Universiteit Leuven, Leuven, Belgium, yves.vanderperren@ieee.org

Guy Waltisperger CEA-LETI, MINATEC, 17 avenue des Martyrs, 38054 Grenoble Cedex 9, France

Jérôme Willemin CEA-LETI, MINATEC, 17 avenue des Martyrs, 38054 Grenoble Cedex 9, France

Chapter 1
Introduction

G. Nicolescu, I. O'Connor, and C. Piguet

Heterogeneous is an adjective, derived from Greek ("heteros", 'other' and "genos", 'kind'), used to describe an object or system that is composed of a number of items that are different from one another. It is the antonym of homogeneous, which signifies that the constituent parts of a system are of identical type. For example system heterogeneity in distributed systems refers to the existence, within the system, of different types of hardware and software, while data heterogeneity, in computing, refers to a mixing of data from two or more sources, often in two or more formats.

In fact, the term "heterogeneous" has many meanings, which originate to a large extent from the multiple fields of knowledge of the people that use the term. This book focuses on heterogeneity in the embedded systems and system on chip field, such that while we cannot pretend to establish an exhaustive list of the meanings of the term in this field, we can at least give our point of view.

The first and most obvious meaning is that more than one physical domain is involved in the functionality of the system. A domain represents a distinct part of the design space in which system components exist, and defines a set of common terminology, information on functionality and requirements for valid use. Indeed, a physical domain is based on the nature of the exchange of energy (i.e. power) used

G. Nicolescu
Department of Computer Engineering, Ecole Polytechnique Montreal, 2500 Chemin de Polytechnique Montreal, Montreal, Québec, Canada H3T 1J4
e-mail: gabriela.nicolescu@polymtl.ca

I. O'Connor (✉)
CNRS UMR 5270, Lyon Institute of Nanotechnology, Ecole Centrale de Lyon, av. Guy de Collongue 36, Bâtiment F7, 69134 Ecully, France
e-mail: ian.oconnor@ec-lyon.fr

C. Piguet
Integrated and Wireless Systems Division, Centre Suisse d'Electronique et de Microtechnique (CSEM), Jaquet-Drotz 1, 2000 Neuchâtel, Switzerland
e-mail: christian.piguet@csem.ch

G. Nicolescu et al. (eds.), *Design Technology for Heterogeneous Embedded Systems*,
DOI 10.1007/978-94-007-1125-9_1, © Springer Science+Business Media B.V. 2012

in hardware component functionality, and in communication between the components. The term “domain” here is synonymous with discipline and stems from the use of the term in the sense of “field of knowledge”. Some examples of physical domains are: electrical, mechanical, optical, thermal, hydraulic (using variations in the power of physical quantities such as charge, force, photon flux, temperature and volume flow as part of their functionality and/or as means of conveying information from one component to another).

Another meaning of “heterogeneous” concerns the target technology as the final fabric onto which all hardware components of the system will be deployed, and on which all component and system performance metrics ultimately depend. It can be defined as an association of raw materials (defining a set of material parameters), tools and process techniques (defining process parameters, including limits to physical geometries). This association leads to the technological fabrication process and the use of more than one basic material (silicon, III-V, organic, etc.) for the functional devices, whether by co-integration techniques (planar SoC[1] or stacked, i.e. 3D integrated circuits or 3DIC) or bonding (SiP[2]). When applied to SoC/SiP, it is implicit that one of the domains is electrical, and that one of the materials is silicon. From the technological viewpoint, many choices, issues and tradeoffs exist between the various packaging and integration techniques (above IC, SiP, heterogeneous integration, bulk, etc.).

While the general benefits of heterogeneous integration appear to be clear, this evolution represents a strong paradigm shift for the semiconductor industry. Moving towards diversification as a complement to the scaling trend that has lasted over 40 years is possible because the integration technology (or at least the individual technological steps) exists to do so. However, the capacity to translate system drivers into technology requirements (and consequently guidance for investment) to exploit such diversification is severely lacking. Such a role can only be fulfilled by a radical shift in design technology to address the new and vast problem of heterogeneous system design. However, the design technology must also remain compatible with standard “More Moore” flows, which are geared towards handling complexity both in terms of detail as device dimensions shrink (silicon complexity) and in terms of sheer scale of systems as the number of devices in the system grows (system complexity). Indeed, the micro-electronics industry, over the years and with its spectacular and unique evolution, has built its own specific design methods while focusing mainly on the management of complexity through the establishment of abstraction levels. Today, the emergence of device heterogeneity requires new approaches enabling the satisfactory design of heterogeneous embedded systems for the widespread deployment of such systems.

Other meanings of “heterogeneous” relate to the design process involved before the object exists physically (specification, synthesis, simulation, verification). In fact, the term was first commonly used to describe systems based on (digital) hardware and software, which can be more generally defined as system description

[1]System on Chip.

[2]System in Package.

using more than one level of abstraction—both for hardware and signal description. This is essentially driven by the need to handle the massive complexity of SoC/SiP by simplifying assumptions, and which in turn drives many of the requirements for modeling, design and simulation techniques of non-digital hardware. In fact in addition to the notion of physical domain, the use of abstract domain types in the design process is necessary to completely define the sphere of operation of a component. Multiple abstract domains exist and are either based on the intrinsic nature of the component functionality and communication, or on the way that this nature is described (usually in terms of reducing the description content to a strict minimum as a means of optimizing simulation time).

Specific branches of heterogeneity can also be identified, concerning the use of more than one behavioral domain (such as causal, synchronous, discrete, signal-flow, conservative), all of which find varying degrees of accuracy of data and of time. Various data domains exist, such as untyped objects or tokens, enumerated symbols, real values, integer values, logic values. The means of taking time into account also leads to various time-domains (e.g. continuous-time, discrete-time) and abstractions (cycle accurate, transaction accurate). While the temporal analysis of systems is necessary, it is not the only analysis domain where relevant information is to be found concerning performance metrics of (particularly continuous-time) components: analyses include time-domain, frequency-domain, static-domain, modulated time/frequency-domain. Generally, the heterogeneity of abstractions and models implies a heterogeneity of tools: a set of tools are required for a complete design flow.

Mastering heterogeneity in embedded systems by design technology is one of the most important challenges facing the semiconductor industry today and will be for several years to come. This book, compiled largely from a set of contributions from participants of past editions of the Winter School on Heterogeneous Embedded Systems Design Technology (FETCH), proposes a necessarily broad and holistic overview of design techniques used to tackle the various facets of heterogeneity in terms of technology and opportunities at the physical level, signal representations and different abstraction levels, architectures and components based on hardware and software, in all the main phases of design (modeling, validation with multiple models of computation, synthesis and optimization). It concentrates on the specific issues at the interfaces, and is divided into two main parts. The first part examines mainly theoretical issues and focuses on the modeling, validation and design techniques themselves. The second part illustrates the use of these methods in various design contexts at the forefront of new technology and architectural developments. The following sections summarize the contributions of each chapter.

1 Methods, Models and Tools

A model is a representation of a complex system and/or its environment. The system to be designed is a collection of communicating components that are organized through its architecture for a common purpose. The architecture defines the way in which the components are organized together to form a system. In order for each component to carry out a useful role within the system, it must be connectible in some way to other elements. Its functionality is thus defined by the conjunction of a particular behavior and an interface for communication. The functionality is manifest through the states of the component, which can evolve over time. The functionality of the component can be quantified by parameters and performance metrics.

Modeling is the first step of any design and/or analysis activity related to the architecture and function of a system. In general, a model is associated with an abstraction level and represents the set of properties specific to this level. The model enables a hierarchical design process where (i) the entire system and its environment are first represented at an abstract level, and (ii) model transformations refine this design and add detail as necessary to solve the design problem. The first part of this book will cover these various necessary phases: starting with the means of specifying design intent, the use of abstraction levels and refinement strategies are then described, ending with the two main activities in the design process, simulation (analysis) and design (synthesis).

1.1 Specifications

The design of embedded systems is not an isolated hardware design activity, and must take into account higher-level component descriptions, including software. The highest abstraction level for the description of a system is that used by the specifications.

In Chap. 2, the authors make the case that a general purpose modeling language (UML[3]) can be customized to the specific purpose of modeling electronic systems. The chapter covers recent advances of the UML language applied to SoC and hardware-related embedded systems design through several examples of specific UML profiles relevant to SoC design. Linking UML to existing hardware/software design languages and simulation environments is a key point to the discussion, and is illustrated through a concrete example of a UML profile for hardware/software co-modeling and code generation.

In Chap. 3, the authors show how it is possible to separate the key concerns of data and control to achieve an executable specification flow. The benefits of this technology-independent methodology are demonstrated through the validation and verification of complex systems accommodating a mixture of hardware and software, analog and digital, electronic and micromechanics components.

[3]Unified Modeling Language (http://www.uml.org).

1.2 Modeling, Abstraction and Reuse

Due to the sustained and exponential growth in complexity in embedded SoC architectures, exploration and performance estimation, particularly early in the design cycle, is becoming an increasingly difficult challenge. With the advent of the nanotechnology era, billions of transistors are available to form high-performance systems, and it is impossible to consider the transistor as the building block. It is widely acknowledged that to solve this problem, a continuum of model abstraction levels must be established during the course of the design of a system. A model at a higher abstraction level will hide certain characteristics of the object that it models, either in terms of performance metrics, or in terms of internal architecture, such that simulation at higher levels of abstraction can perform early validation of software. Models can be refined (i.e. made more accurate) by moving down the continuum of abstraction levels, implying that additional model characteristics must be defined. Through the selected design methodology (design space exploration, optimization, trial and error, etc.) performance metrics can be specified and/or architectural variants defined.

In Chap. 4, the authors propose a general autonomous integrated system model with functional, technological and structural scalability in mind. The model handles extensions to the scale of the system in distributed MPSoC systems, as well as to the dynamic nature of the system in reconfigurable and autonomous computing. The authors wrap up by considering more long term trends towards multi-agent bio-inspired approaches.

Chapter 5 covers techniques to achieve the goal of fast hardware/software system simulation on MPSoCs[4] for design choices and design validations. The authors argue that native simulation (rather than instruction-set simulation for example) is a promising solution to simulate software that can be linked to abstract hardware simulators for both functional and temporal hardware/software system validation.

Chapter 6 presents a high level component based approach for expressing system reconfigurability in SoC co-design. The authors firstly present a generic model for reactive control in SoC co-design. This allows the integration of control at different abstraction levels, in particular at the higher abstraction levels, as well as reconfigurability features. The work is validated through a case study using the UML MARTE[5] profile for the modeling and analysis of real-time embedded systems.

1.3 Simulation and Validation of Complex Systems

The refinement of models describing complex systems has to be validated, which is usually realized using simulation. For this it is necessary to define global executable models which require a model integration strategy. The strategy is based on

---

[4]Multi-Processor System on Chip.

[5]Modeling and Analysis of Real-Time and Embedded Systems (http://www.omgmarte.com).

the establishment of interfaces accommodating the various models types. Another possibility for validation is formal verification. This implies that all the models must be expressed using a given formalism, such that the set of properties can be verified.

The tradeoffs involved in validation are particularly evident in Chap. 7, which presents a hybrid platform composed of a simulation tool and a testbed environment to facilitate the design and improve test accuracy of new wireless protocols. The complexity of these protocols requires fast and accurate validation methodologies, and the authors combine the flexibility of simulation-based approaches with the speed and capability of testing designs in real life settings using testbed platforms.

Chapter 8 examines property-based verification, and its dynamic extension as applied to large systems that defeat formal verification methods. The authors describe a verification system in which temporal properties are automatically translated into synthesizable IPs,[6] while resulting monitors and generators are automatically connected to the design under verification.

1.4 Design, Optimization and Synthesis

Methodologies for the design of heterogeneous embedded systems using advanced technologies are critical to achieve, such that functionality can be reliably guaranteed, and such that performance can be optimized to one or more criteria. The establishment of methodologies is increasingly complex, since it has to cope with both very high-level system descriptions and low-level aspects related to technology and variability. Particular issues concern the formalization of global specifications and their continuous validation during design, hardware/software co-development, clarifying interaction between multiple concerns and physical domains. Tooling to support design methodologies also now inevitably at some stage uses numerical optimization. For hardware/software heterogeneity, a major challenge is the efficient mapping of software applications onto parallel hardware resources. This is a non-trivial problem because of the number of parameters to be considered for characterizing both the applications and the underlying platform architectures. For physical domain heterogeneity, the main issue is to build appropriate "divide and conquer" partitioning strategies while allowing the exploration of tradeoffs spanning several domains or abstraction levels.

Chapter 9 covers the major low-level issues (dynamic and static power consumption, temperature, technology variations, interconnect, reliability, yield), and their impact on high-level design, such as the design of multi-supply voltage, fault-tolerant, redundant or adaptive chip architectures. The authors illustrate their methodology through three heterogeneous multi-processor based systems: wireless sensor networks, vision sensors and mobile television.

In Chap. 10, the authors propose a new framework for the mapping of applications onto a many-core computing platform. This extensible framework allows

[6] Intellectual Property.

the exploration of several meta-heuristics, the addition of new objective functions with any number of architecture and application constraints. Experimental results demonstrate the relevance of using new meta-heuristics, and the power of the parallel framework implementation by significantly increasing the explored solution space.

The Functional Virtual Prototyping methodology is covered in Chap. 11, where a virtual prototype is defined as a model composed of multiple abstraction levels of a multi-domain system. The authors argue that this methodology can enable the formalization, exchange, and reuse of design knowledge, and that its widespread use could solve several issues in complex systems design, particularly for collaboration between multiple and geo-distributed design groups.

Chapter 12 addresses design complexity as related to multi-physics systems with a methodology based on hierarchical partitioning and multi-level optimization. The authors show how an optimization problem including cross-domain variables can be formulated to enable the exploration of trade-offs at an early design stage. The methodology is put into practice with an experimental framework, and is demonstrated with the optimization of the fill-factor and response speed in an active pixel sensor.

2 Design Contexts

From the miniaturization of existing systems (position sensors, labs on chip, etc.) to the creation of specific integrated functions (memory, RF tuning, energy, etc.), nanoscale and non-electronic devices are being integrated to create nanoelectronic and heterogeneous SoC/SiP and 3DICs. This approach will have a significant impact on several economic sectors and is driven by:

- the need for the miniaturization of existing systems to benefit from technological advances and improve performance at lower overall cost,
- the potential replacement of specific functions in SoC/SiP with nanoscale or non-electronic devices (nanoswitches, optical interconnect, magnetic memory, etc.),
- the advent of high-performance user interfaces (virtual surgical operations, games consoles, etc.),
- the rise of low-power mobile systems (communications and mobile computing) and wireless sensor networks for the measurement of phenomena inaccessible to single-sensor systems.

The second part of this book will cover design contexts from two points of view: firstly, how new technologies impact the design process, and secondly, how novel distributed autonomous sensor systems can be designed.

2.1 Designing with Emerging Technologies

New integration and fabrication technologies continually stretch the limits of design technology. 3D integration, consisting of stacking many chips vertically and connecting them together using Through Silicon Vias (TSVs) is a promising solution for heterogeneous systems, providing several benefits in terms of performance and cost. For example the first 3D processor (the 2008 Rochester Cube[7]) runs at 1.4 GHz and has abilities that the conventional planar chip cannot reach. However, the design process faced many difficulties because of the complexity of the design, such as ensuring synchronized operation of all of the layers and seamless inter-layer communication. 3D chips also demonstrate significant thermal issues (and consequently higher static power and lower reliability) due to the presence of processing units with a high power density, which are not homogeneously distributed in the stack.

The scaling of silicon technology and the integration of novel functional materials are also enabling exploration for alternative concepts for information processing, storage and communication in computing platforms. Adoption of such new technologies will only occur if significant improvement in the conventional compute figure of merit (number of operations per second·Watt·mm^3). As well as information technology, emerging technologies are also solving issues in biomedical applications, through the development of efficient low-power interface circuits between living objects and data gathering.

Chapter 13 analyzes both near-term and long-term technology alternatives for memory and logic. Scaling of conventional circuits is considered, as well as the development of novel logic circuits based on carbon nanotubes for reconfigurable circuits, nanowire crossbar matrices for memory, and graphene nanoribbons. Hybrid molecular-CMOS architectures, the most likely first step towards alternative architectures, are also discussed.

In Chap. 14, a new approach for the thermal control of 3D chips is discussed. The authors use both grid and non-uniform placement of TSVs as an effective mechanism for thermal balancing and control in 3D chips. A large part of the chapter is dedicated to the mathematical modeling of the material layers and TSVs, including a detailed calibration phase based on a real 5-tier 3D chip stack with several heaters and sensors to study the heat diffusion.

Chapter 15 covers 3D integration solutions for heterogeneous systems, with an overview of 3D manufacturing technologies and related concerns. An outlook to some potential applications is given, with particular focus on 3D MPSoC architectures for compute intensive systems.

In Chap. 16, the authors focus on alternative devices capable of switching between two distinct resistive states for non-volatile memories. In particular, the materials and their ability to withstand a downscaling of their critical dimensions are described. Particular attention is also given to the models describing the operation

[7] V.F. Pavlidis, E.G. Friedman, Three-Dimensional Integrated Circuit Design, Morgan Kaufman, 2009.

of such memory cells, and their implementation in electrical simulators to evaluate their robustness at the architectural level.

Chapter 17 presents dedicated circuit techniques and strategies to design and assemble dense embedded microsystems target towards bioelectrical signal recording applications. Efficient interface circuits to measure the weak bioelectrical signal from several cells in the cortical tissues are covered, and high-fidelity data-reduction strategies are demonstrated. Since power issues are predominant in in-vivo applications, particular attention is paid to on-chip power management schemes based on automatic biopotential detection, as well as low-power design techniques, ultra-low-power neural signal processing circuits, and dedicated implementation strategies enabling high multi-channel neural recording microsystem integration density.

2.2 *Designing Smart, Self-powered Radio Systems—Extreme Heterogeneity*

Perhaps the most extreme cases of heterogeneity in embedded systems today are in the field of sensor networks with wireless communication between nodes. Energy-autonomy is key to the deployment of such distributed sensor systems, since the sensor nodes have batteries or energy harvesters and must therefore be designed to consume very little power. Understanding at an early design stage the impact of design choices on power is a critical design step. Moreover, the energy harvesting system itself is a particularly heterogeneous system, including energy harvester, battery, antenna, sensors and electronic blocks. The main issues for this kind of system are the energy converter efficiency for small power transfer, the load consumption (RF,[8] sensors) in active and standby mode, and the embedded power management.

The application domains of such complex systems can be found in military, security, or high reliability systems such as aerospace and automotive. Meeting the joint constraints of specific requirements, multiple standards, low volume, high-reliability and cost mean that reconfigurable or reprogrammable hardware is the favored approach in this area.

Chapter 18 is focused on the integration of an energy- and data-driven platform for autonomous systems. The authors describe a global system description and specification, as well as the principles of three energy-harvesting techniques (mechanical vibrations, thermal flux, and solar radiation). The use of multiple sources for energy harvesting is shown to be feasible, when strategies for power management are considered with a focus on power path optimization.

In Chap. 19, the authors present a power model suited for multiprocessor power management based on the study of a video decoder application. The execution of this application on the multiprocessor, and the impact of the memory architecture on the energy cost, is analyzed in detail, as well as a power strategy suited to video processing based on the selection of operating points of frequency and voltage.

[8] Radiofrequency.

Chapter 20 covers the use of high-performance reconfigurable platforms in software-defined radio for multi-standard applications. Additional flexibility is explored through novel techniques of dynamic partial reconfiguration.

In Chap. 21, the authors propose an approach for complete system simulation and power estimation of wireless sensor networks, ultimately enabling sensor-node optimization at the architecture level. The authors argue that the accurate estimation of the power consumption of network nodes, requires both accurate and efficient modeling of the communication infrastructure and the architecture of the node. To achieve these goals, the developed simulation framework includes an Instruction Set Simulator and uses Transaction Level Modeling (TLM) with SystemC as the basis for simulation.

Acknowledgements We would like to take this opportunity to thank all the contributors to this book for having undertaken the writing of each chapter from the original winter school presentations and for their patience during the review process. We also wish to extend our appreciation to the team at Springer for their editorial guidance as well of course as for giving us the opportunity to compile this book together.

Part I
Methods, Models and Tools

Chapter 2
Extending UML for Electronic Systems Design: A Code Generation Perspective

Yves Vanderperren, Wolfgang Mueller, Da He, Fabian Mischkalla, and Wim Dehaene

1 Introduction

Larger scale designs, increased mask and design costs, 'first time right' requirements and shorter product development cycles motivate the application of innovative 'System on a Chip' (SoC) methodologies which tackle complex system design issues.[1] There is a noticeable need for design flows towards implementation starting from higher level modeling. The application of the Unified Modeling Language (UML) in the context of electronic systems has attracted growing interest in the recent years [16, 35], and several experiences from industrial and academic users have been reported [34, 58].

Following its introduction in 1995, UML has been widely accepted in software engineering and supported by a considerable number of Computer Aided Software Engineering (CASE) tools. Although UML has its roots in the software domain, the Object Management Group (OMG), the organization driving the UML standardization effort [44, 45], has turned the UML notation into a general-purpose modeling language which can be used for various application domains, ranging from business process to engineering modeling, mainly for documentation purposes. Besides the language complexity, the main drawback of such a broad target is the lack of

[1]While the term 'SoC' is commonly understood as the packaging of all the necessary electronic circuits and parts for a system on a *single* chip, we consider the term in larger sense here, and cover electronic systems irrespective of the underlying implementation technology. These systems, which might be multi-chip, involve several disciplines including specification, architecture exploration, analog and digital hardware design, the development of embedded software which may be running on top of a real-time operating system (RTOS), verification, etc.

Y. Vanderperren (✉) · W. Dehaene
ESAT–MICAS, Katholieke Universiteit Leuven, Leuven, Belgium
e-mail: yves.vanderperren@ieee.org

W. Mueller · D. He · F. Mischkalla
C-LAB, Paderborn University, Paderborn, Germany

G. Nicolescu et al. (eds.), *Design Technology for Heterogeneous Embedded Systems*,
DOI 10.1007/978-94-007-1125-9_2, © Springer Science+Business Media B.V. 2012

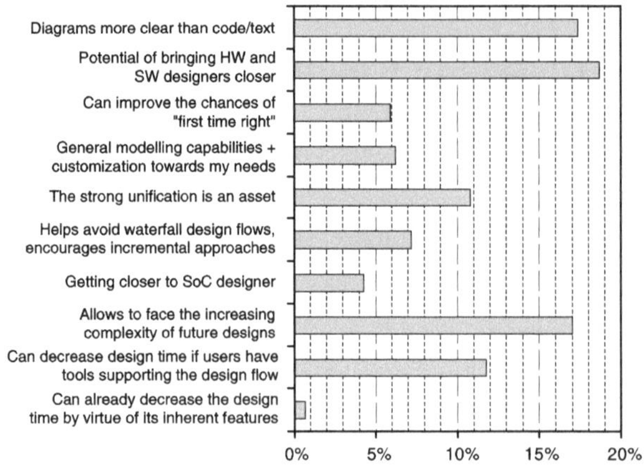

Fig. 2.1 Positive aspects of UML [62]

sufficient semantics, which constitutes the main obstacle for real engineering application. Therefore, application specific customizations of UML (UML profiles), such as the System Modeling Language (SysML) [38] and the UML Profile for SoC [42], are of increasing importance. The addition of precise semantics allows for the automatic generation of code skeleton, typically C++ or Java, from UML models.

In the domain of embedded systems, the complexity of embedded software doubled every 10 months in the last decades. Automotive software, for instance, may exceed several GBytes [22]. In this domain, complexity is now successfully managed by model-based development and testing methodologies based on MATLAB/Simulink with highly efficient C code generation. Unfortunately, the situation is more complex in electronic systems design than in the embedded software domain, as designers face a combination of various disciplines, the coexistence of multiple design languages, and several abstraction levels. Furthermore, multiprocessor architectures have become commonplace and require languages and tool support for parallel programming. In this multi-disciplinary context,

UML has great potential to unify hardware and software design flows. The possibility to bring designers of both domains closer and to improve the communication between them was recognized as a major advantage, as reported by surveys conducted during the UML-SoC Workshops at the Design Automation Conference (DAC) in 2006 (Fig. 2.1) and 2007 [63]. Additionally, UML is also perceived as a means to manage the increasing complexity of future designs and improve their specification. UML diagrams are expected to provide a clearer overview compared to text.

Significant issues remain, however, such as the perceived lack of maturity of tool support, the possible difficulty of acceptance by designers due to lack of knowledge, and the existence of different UML extensions applicable to SoC design but which are not necessarily compatible [62].

A detailed presentation of the UML is beyond the scope of this chapter and we assume that the reader has a basic knowledge of the language. The focus of this chapter is the concrete application of UML to SoC design, and follows the following structure. The next section introduces basic concepts of the UML extension mechanism, how to define a UML profile. Thereafter, we present some UML profiles relevant for UML for SoC and embedded systems design before we introduce one application in the context of SystemC/C++ co-simulation and -synthesis [27].

2 Extending UML

A stereotype is an extensibility mechanism of UML which allows users to define modeling elements derived from existing UML classifiers, such as classes and associations, and to customize these towards individual application domains [45]. Graphically, a stereotype is rendered as a name enclosed by ≪...≫. The readability and interpretation of models can be highly improved by using a limited number of well defined stereotypes. Additionally, stereotypes can add precise meanings to individual elements, enabling automatic code generation.

For instance, stereotypes corresponding to SystemC constructs can be defined, such as ≪*sc_module*≫, ≪*sc_clock*≫, ≪*sc_thread*≫, ≪*sc_method*≫ etc. The individual elements of a UML model can then be annotated with these stereotypes to indicate which SystemC construct they correspond to. The resulting UML model constitutes a first specification of a SystemC model, which can then be automatically generated. The stereotypes give to the UML elements the precise semantics from the target language (SystemC in this case).

As an example, Fig. 2.2 represents a Class Diagram with SystemC-oriented stereotypes. It corresponds to the simple bus example delivered with SystemC, with master, slave, and arbiter classes stereotyped as ≪*sc_module*≫ and connected to a bus. Modules are connected by a directed association with stereotype ≪*connect*≫. We introduce this stereotype as an abstraction for a port with associated interface where the flow points into the direction of the interface. An alternative and more detailed representation of the bus connection is provided by the explicit definition of the interface via a separate element with stereotype ≪*sc_interface*≫ (Fig. 2.3). Such examples illustrate how stereotypes add necessary interpretations to UML diagrams. A clear definition and structure of stereotypes is of utmost importance before applying UML for effective documentation and efficient code generation.

UML is defined on the basis of a metamodel, i.e., the UML language is itself described based on a model. This approach makes the language extremely flexible, since an application specific customization can be easily defined by the extension of that metamodel trough the definition of stereotypes. In theory, the principle of an application specific customization of UML through a so-called *UML profile* through stereotypes is simple. Considering a specific application domain, all unnecessary parts of the UML metamodel are stripped in a first step. In a second step, the resulting metamodel is extended. This mainly means the definition of a set of additional stereotypes and tagged values, i.e., stereotype attributes. In further steps, useful graphical icons/symbols, constraints, and semantic outlines are added.

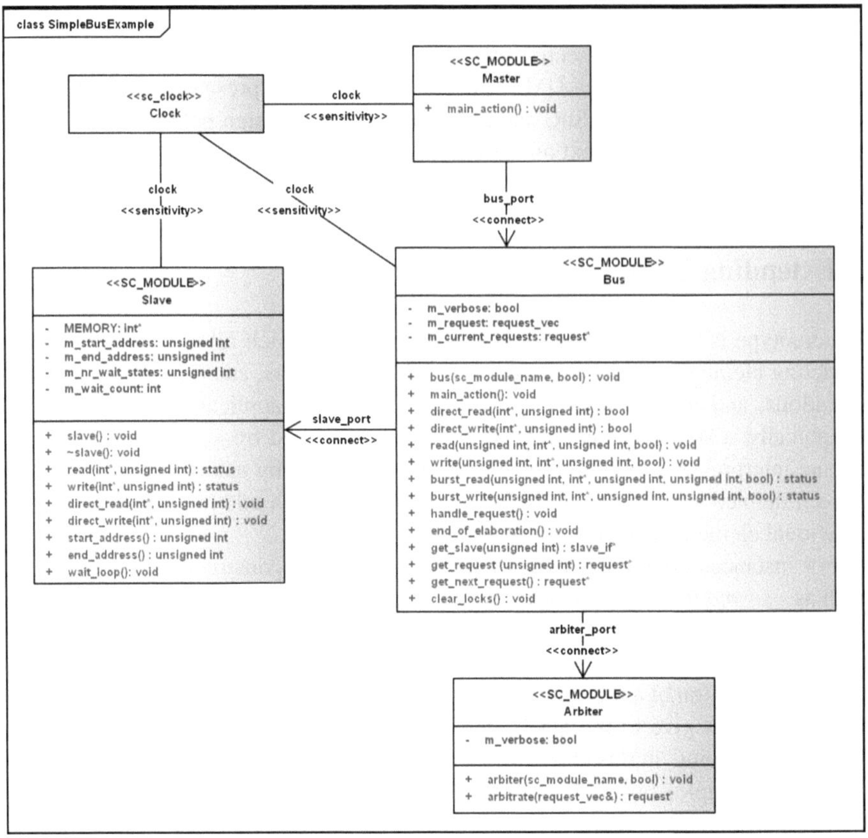

Fig. 2.2 UML simple bus class diagram

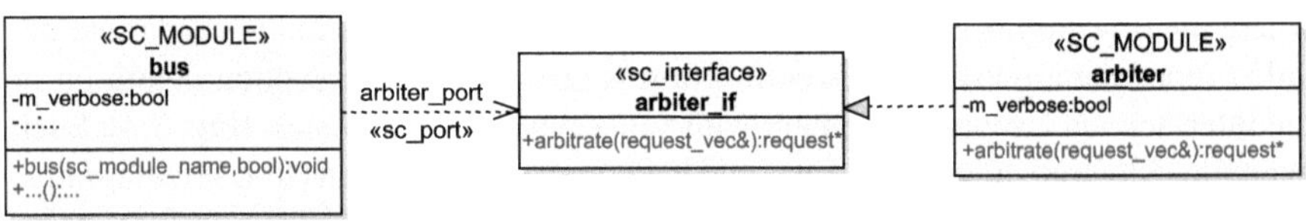

Fig. 2.3 UML arbiter interface

In practice, the first step is often skipped and the additional semantics weak, leaving room for several interpretations.

Definitions of stereotypes are often given in the form of a table. In most cases, an additional set of Class Diagrams is given, as depicted for example in Fig. 2.4, which shows an excerpt from the UML profile for SoC [42], which will be further discussed in Sect. 3.1. The extended UML metaclass *Port* is indicated by the keyword *«metaclass»*. For its definition, the stereotype *SoCPort* is specified with the keyword *«stereotype»* and linked with an extension relationship (solid line link with

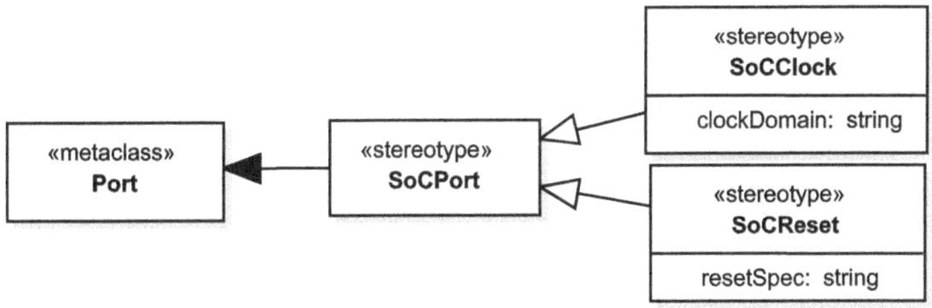

Fig. 2.4 UML stereotype definition

black head). The two other extensions, *SoCClock* and *SoCReset*, are simply specified with their tagged values as generalizations of *SoCPort*. After having defined those extensions, the stereotypes *«SoCPort»*, *«SoCClock»*, and *«SoCReset»* can be applied in Class Diagrams.

Several UML profiles are available as OMG standards and applicable to electronic and embedded systems modeling, such as the UML Testing Profile [39], the UML Profile for Modeling Quality of Service (QoS) and Fault Tolerance Characteristics and Mechanisms [40], the UML Profile for Schedulability, Performance and Time (SPT) [41], the UML Profile for Systems Engineering (which defines the SysML language) [38], the UML Profile for SoC [42], and MARTE (Modeling and Analysis of Real-Time Embedded Systems) [43].

The following sections will focus on the most important ones in the context of SoC design.

3 UML Extensions Applicable to SoC Design

3.1 UML Profile for SoC

The UML profile for SoC was initiated by CATS, Rational (now part of IBM), and Fujitsu in 2002. It is available as an OMG standard since August 2006 [42]. It targets mainly Transaction Level Modeling (TLM) SoC design and defines modeling concepts close to SystemC. Table 2.1 gives a summary of several stereotypes introduced in the profile and the UML metaclasses they extend.

The SoC profile introduces *Structure Diagrams* with special symbols for hierarchical modules, ports, and interfaces. The icons for ports and interfaces are similar to those introduced in [23]. Annex A and B of the profile provide more information on the equivalence between these constructs and SystemC concepts. Automatic SystemC code generation from UML models based on the SoC Profile is supported by tools from CATS [12] and the UML tool vendor ArtisanSW [19].

Table 2.1 Examples of stereotypes defined in the UML profile for SoC

SoC model element	Stereotype	UML metaclass
Module	SoCModule	Class
Process	SoCProcess	Operation
Data	Data	Class
Controller	Controller	Class
Protocol Interface	SoCInterface	Interface
Channel	SoCChannel	Class
Port	SoCPort	Port/Class
Connector	SoCConnector	Connector
Clock Port	SoCClock	Port
Reset Port	SoCReset	Port
Data Type	SoCDataType	Dependency

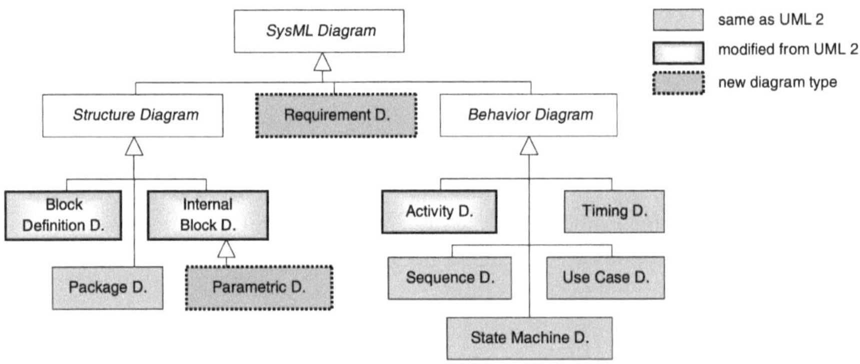

Fig. 2.5 Architecture of SysML

3.2 SysML

SysML is a UML profile which allows modeling systems from a domain neutral and Systems Engineering (SE) perspective [38]. It is the result of a joint initiative of OMG and the International Council on Systems Engineering (INCOSE). The focus of SE is the efficient design of complex systems which include a broad range of heterogeneous domains, including hardware and software. SysML provides opportunities to improve UML-based SoC development processes with the successful experiences from the SE discipline [59]. Strong similarities exist indeed between the methods used in the area of SE and complex SoC design, such as the need for precise requirements management, heterogeneous system specification and simulation, system validation and verification. The architecture of SysML is represented on Fig. 2.5. The main differences are summarized hereafter:

- **Structure:** SysML simplifies the UML diagrams used to represent the structural aspects of a system. It introduces the concept of *block*, a stereotyped class which describes a system as a structure of interconnected parts. A block provides a domain neutral modeling element that can be used to represent the structure of any kind of system, regardless of the nature of its components. In the context of a SoC, these components can be hardware or software based as well as analog or digital.

- **Behavior:** SysML provides several enhancements to Activity Diagrams. In particular, the control of execution is extended such that running actions can be disabled. In UML, the control is limited to the determination of the moment when actions start. In SysML a behavior may not stop itself. Instead it can run until it is terminated externally. For this purpose SysML introduces *control operators*, i.e., behaviors which produce an output controlling the execution of other actions.
- **Requirements:** One of the major improvements SysML brings to UML is the support for representing requirements and relating them to the models of a system, the actual design and the test procedures. UML does not address how to trace the requirements of a system from informal specifications down to the individual design elements and test cases. Requirements are often only traced to UML use cases but not to the design. Adding design rationale information which captures the reasons for design decisions made during the creation of development artifacts, and linking these to the requirements help analyze the consequences of a requirement change. SysML introduces for this purpose the *Requirement Diagram*, and defines several kinds of relationships improving the requirement traceability. The aim is not to replace existing requirements management tools, but to provide a standard way of linking the requirements to the design and the test suite within UML and a unified design environment.
- **Allocations:** The concept of *allocation* in SysML is a more abstract form of deployment than in UML. It is a relationship established during the design phase between model elements. An allocation provides the generalized capability to a source model element to a target model element. For example, it can be used to link requirements and design elements, to map a behavior into the structure implementing it, or to associate a piece of software with the hardware deploying it.

SysML presents clear advantages. It simplifies UML in several aspects, as it actually removes more diagrams than it introduces. Furthermore, SysML can support the application of Systems Engineering approaches to SoC design. This feature is particularly important, since the successful construction of complex SoC systems requires a cross-functional team with system design knowledge combined with experienced SoC design groups from hardware and software domains which are backed by an integrated tool chain. By encouraging a Systems Engineering perspective and by providing a common notation for different disciplines, SysML allows facing the growing complexity of electronic systems and improving communication among the project members.

However, SysML remains a semi-formal language, like UML. Although SysML contributes to the applicability of UML to non-software systems, it remains a semi-formal language since it lacks associated semantics. For instance, SysML blocks allow unifying the representation of the structure of heterogeneous systems but have weak semantics, in particular in terms of behavior. As another example, the specification of timing aspects is considered out of scope of SysML and must be provided by another profile. The consequence is a risk of discrepancies between profiles which have been developed separately. SysML can be customized to model domain specific applications, and in particular support code generation towards SoC languages. First signs of interest in this direction are already visible [21, 33, 49, 59, 64].

SysML allows integrating heterogeneous domains in a unified model at a high abstraction level. In the context of SoC design, the ability to navigate through the system architecture both horizontally (inside the system at a given abstraction level) and vertically (through the abstraction levels) is of major importance. The semantic integrity of the model of a heterogeneous SoC could be ensured if tools supporting SysML take advantage of the allocation concept in SysML and provide facilities to navigate through the different abstraction layers into the underlying structure and functionality of the system. Unfortunately, such tool support is not yet available at the time of this writing.

3.3 UML Profile for MARTE

The development of the UML profile for MARTE (Modeling and Analysis of Real-Time and Embedded systems) was initiated by the ProMARTE partners in 2005. The specification was adopted by OMG in 2007 and has been finalized in 2009 [43]. The general purpose of MARTE is to define foundations for the modeling and analysis of real-time embedded systems (RTES) including hardware aspects. MARTE is meant to replace the UML profile for Schedulability, Performance and Time (SPT) and to be compatible with the QoS and SysML profile, as conceptual overlaps may exist.

MARTE is a complex profile with various packages in the areas of core elements, design, and analysis with a strong focus on generic hardware/software component models and schedulability and performance analysis (Fig. 2.6). The profile is structured around two directions: the modeling of features of real-time and embedded systems, and the annotation of the application models in order to support the analysis of the system properties.

The types introduced to model hardware resources are more relevant for multi-chip board level designs rather than for chip development. The application of MARTE to SystemC models is not investigated, so that MARTE is complimentary to the UML profile for SoC. MARTE is a broad profile and its relationship to the RTES domain is similar to the one between UML and the system and software domain: MARTE paves the way for a family of specification formalisms.

3.4 UML Profile for IP-XACT

IP-XACT was created by the SPIRIT Consortium as an XML-based standard data format for describing and handling intellectual property that enables automated con-

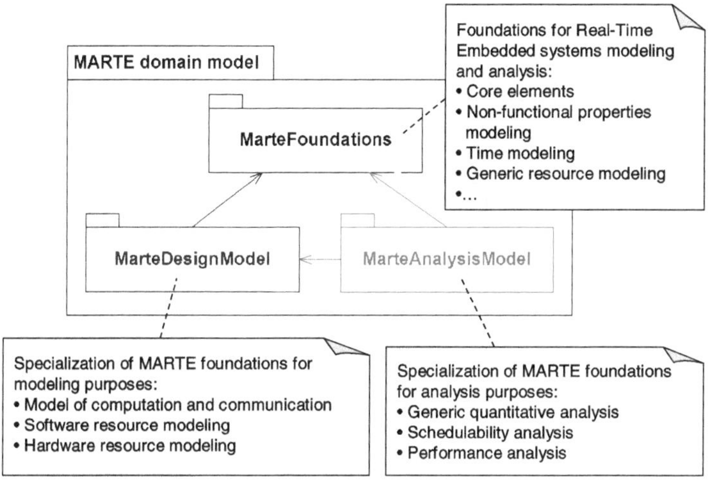

Fig. 2.6 Organization of the MARTE profile

figuration and integration. As such, IP-XACT defines and describes electronic components and their designs [46]. In the context of the SPRINT project an IP-XACT UML profile was developed to enable the consistent application of the UML and IP-XACT so that UML models provide the same information as their corresponding IP-XACT description [54]. For this, all IP-XACT concepts are mapped to corresponding UML concepts as far as possible. The resulting UML-based IP description approach enables the comprehensible visual modeling of IP-XACT components and designs.

4 Automatic SoC Code Generation from UML Models

The relationship between UML models and text code can be considered from a historical perspective as an evolution towards model centric approaches. Originally (Fig. 2.7.a), designers were writing code having in mind their own representation of its structure and behavior. Such approach did not scale with large systems and prevented efficient communication of design intent, and the next step was code visualization through a graphical notation such as UML (Fig. 2.7.b). Round trip capability between the code and the UML model, where UML models and code remain continuously synchronized in a one-to-one relationship (Fig. 2.7.c), is supported today for software languages by several UML tools. Though technically possible [19], less tool support is available for code generation towards SoC languages. The final step in this evolution is a model-centric approach where code generation is possible from the UML model of the system towards several target languages of choice (Fig. 2.7.d) via one-to-many translation rules. Such flexible generation is still in an infancy stage. The need to unify the different semantics of the target languages and hardware/software application domains with UML constitutes here a

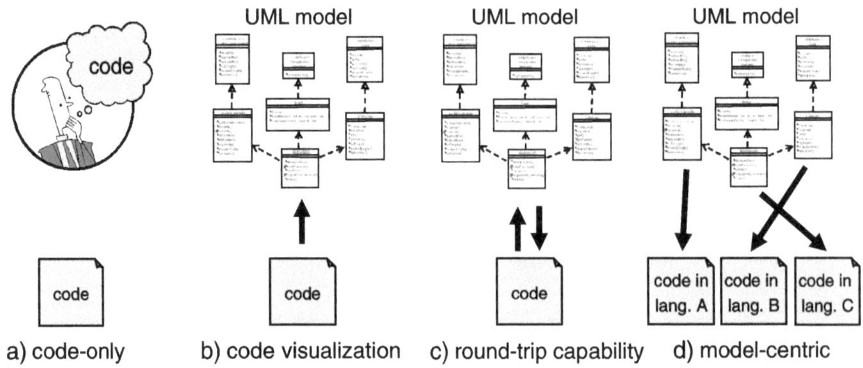

Fig. 2.7 Relationship between UML models and code

major challenge. Furthermore, the models from which code is supposed to be generated must have fully precise semantics, which is not the case with UML. Outside of the UML domain, interestingly, tools such as MATLAB/Simulink now support automatic code generation towards hardware (VHDL/Verilog) *and* software (C/C++) languages from the same model [56]. The quality of the generated code is increasing with the tool maturity, and such achievement proves the technical feasibility of model-centric development. This result has been achieved by narrowing the application domain to signal processing intensive systems, and by starting from models with well defined semantics.

In the following sections, we will investigate various combinations of UML models and SoC languages, and the associated support for code generation.

4.1 One-to-One Code Generation

A language can only be executed if its syntax and semantics are both clearly defined. UML can have its semantics clarified by customizing it towards an unambiguous executable language, i.e., modeling constructs of the target language are defined within UML, which inherits the execution semantics of that language. This procedure is typically done via the extension mechanisms of UML (stereotypes, constraints, tagged values) defined by the user or available in a profile. This one-to-one mapping between code and UML, used here as a notation complementing code, allows for reverse engineering, i.e., generation of UML diagrams from existing code (Fig. 2.7.b), as well as the automatic generation of code frames from a UML model. The developer can add code directly to the UML model or in separate files linked to the output generated from the models. The UML model no longer reflects the code if the generated output is changed by hand. Such disconnect is solved by round-trip capability supported by common UML tools (Fig. 2.7.c). This approach is typically used in the software domain for the generation of C, C++ or Java code. In the SoC context, UML has been associated with register-transfer level (RTL) as well as electronic system level (ESL) languages. The abstraction level which can be reached in the UML models is essentially limited by the capabilities of the target language.

In Sect. 5, we introduce in more detail the application of a one-to-one code generation from the SATURN project [53]. The application is based on the extension and integration of commercial tools for SystemC/C++/Simulink co-modeling, co-simulation, and co-synthesis.

UML and RTL Languages Initial efforts concentrated on generating behavioral VHDL code from a specification expressed with UML models in order to allow early analysis of embedded systems by means of executable models [36]. However, the main focus was always to generate synthesizable VHDL from StateCharts [24] and later from UML State Machines [2, 8, 13, 14]. In the context of UML, the Class and State Machine Diagrams were the main diagrams used due to their importance in the first versions of UML. UML classes can be mapped onto VHDL entities, and associations between classes onto signals. By defining such transformation rules, VHDL code can be generated from UML models, which inherit the semantics from VHDL. Similarly, the association between UML and Verilog has also been explored.

UML and C/C++ Based ESL Languages In the late 90s, several SoC design languages based on C/C++ (e.g., SpecC, Handel-C, ImpulseC, SystemC) were developed in order to reach higher abstraction levels than RTL and bridge the gap between hardware and software design by bringing both domains into the same language base. These system level languages extend C/C++ by introducing a scheduler, which supports concurrent execution of threads and includes a notion of time. Besides these dialects, it is also possible to develop an untimed model in plain C/C++, and let a behavioral synthesis tool introduce hardware related aspects. Mentor Graphics CatapultC, Cadence C-to-Silicon Compiler and NEC CyberWorkBench are examples of such tools starting from C/C++. In all these cases, users develop a model of the system using a language coming actually from the software field. As the roots of UML lie historically in this domain, it is natural to associate UML with C/C++ based ESL languages. Although the first generation of behavioral synthesis tools in the 1990s was not a commercial success, a second generation has appeared in the recent years and is increasingly used by leading semiconductor companies for dataflow driven applications with good quality of results. In addition to the advantage of having a unified notation and a graphical representation complementary to the ESL code, it is now easier to bridge the gap between a high level specification and a concrete implementation. It is indeed possible to express the former as high level UML models, refine these and the understanding of the system until the moment where ESL code can be generated, verify the architecture and the behavior by executing the model, and eventually synthesize it. Such design flow is essentially limited by the capabilities of the chosen behavioral synthesis tool.

In the last decade, SystemC emerged as one of the most prominent ESL languages. Tailoring UML towards SystemC in a 1-to-1 correspondence was first investigated in [5, 20, 47]. Several benefits were reported when UML/SysML is asso-

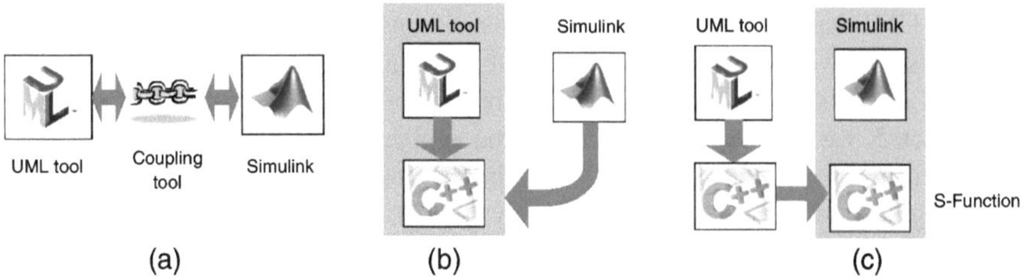

Fig. 2.8 UML and MATLAB/Simulink

ciated with SystemC, including a common and structured environment for the documentation of the system specification, the structure of the SystemC model and the system's behavior [47]. These initial efforts paved the way for many subsequent developments, whereas the introduction of several software-oriented constructs (e.g., Interface Method Calls) in SystemC 2.0 and the availability of UML 2.x contributed to ease the association between UML and SystemC. For example, efforts at Fujitsu [20] have been a driving factor for the development of the UML profile for SoC (Sect. 3.1), and STMicroelectronics developed a proprietary UML/SystemC profile [51]. Additionally, NXP and the UML tool vendor Artisan collaborated in order to extend the C++ code generator of Artisan so that it can generate SystemC code from UML models [19, 48]. This work was the starting point for further investigations which are presented in Sect. 5. It is furthermore possible to rely on a code generator which is independent of the UML tool and takes as input the XML Metadata Interchange (XMI) file format for UML models, which is text based [10, 37, 67]. The aim of all these works is to obtain quickly a SystemC executable model from UML, in order to verify as soon as possible the system's behavior and performance. UML can also be customized to represent SpecC [29, 32] or ImpulseC [65] constructs, which allows a seamless path towards further synthesis of the system. Other efforts to obtain synthesizable SystemC code from UML have also been reported [55, 67].

UML and MATLAB/Simulink Two main approaches allow coupling the execution of UML and MATLAB/Simulink models: co-simulation, and integration based on a common underlying executable language (typically C++) [60].

In the case of co-simulation (Fig. 2.8.a), Simulink and the UML tool communicate with each other via a coupling tool. Ensuring a consistent notion of time is crucial to guarantee proper synchronization between the UML tool and Simulink. Both simulations exchange signals and run concurrently in the case of duplex synchronization, while they run alternatively if they are sequentially synchronized. The former solution increases the simulation speed, whereas the time precision of the exchanged signals is higher in the latter case. As an example, the co-simulation approach is implemented in Exite ACE from Extessy [17], which allows, e.g., coupling a Simulink model with Artisan Studio [57] or IBM Rational Rhapsody [26]. Exite ACE will be further introduced in the application example given in Sect. 5. A similar simulation platform is proposed in [25] for IBM Rational Rose RealTime.

The alternative approach is to resort to a common execution language. In absence of tool support for code generation from UML, the classical solution is to generate C/C++ code from MATLAB/Simulink, using MATLAB Compiler or Real-Time Workshop, and link it to a C++ implementation of the UML model. The integration can be done from within the UML tool (Fig. 2.8.b) or inside the Simulink model (Fig. 2.8.c). This solution was formerly adopted, for instance, in the Constellation framework from Real-Time Innovation, in the GeneralStore integration platform [50], or in IBM's Telelogic Rhapsody and Artisan Software Studio. Constellation and GeneralStore provide a unified representation of the system at model level on top of code level. The Simulink subsystem appeared in Constellation as a component, which can be opened in MATLAB, whereas a UML representation of the Simulink subsystem is available in GeneralStore, based on precise bidirectional transformation rules.

The co-simulation approach requires special attention to the synchronization aspect, but allows better support for the most recent advances in UML 2.0, the UML profile for SoC and SysML, by relying on the latest commercial UML tools. On the other hand, development frameworks which rely on the creation of a C++ executable model from UML and MATLAB/Simulink give faster simulations.

One of the advantages of combining UML with Simulink compared to a classical Simulink/Stateflow solution is that UML offers numerous diagrams which help tie the specification, architecture, design, and verification aspects in a unified perspective. Furthermore, SysML can benefit from Simulink by inheriting its simulation semantics in a SysML/Simulink association. UML tool vendors are working in this direction and it will be possible to plug a block representing a SysML model into Simulink. Requirements traceability and documentation generation constitute other aspects for potential integration between SysML and Simulink, as several UML tool vendors and Simulink share similar features and 3rd party technology.

4.2 One-to-Many Code Generation

Some UML tools, such as Mentor Graphics Bridgepoint [11] or Kennedy Carter iUML [30], support the execution of UML models with the help of a high-level action language whose semantics is defined by OMG, but not its syntax. As a next step, code in a language of choice can be generated from the UML models by a *model compiler* (Fig. 2.7.d). In contrast to the one-to-one relationship described in previous section, there is not necessarily a correspondence between the structure of the model and the structure of the generated code, except that the behavior defined by the model must be preserved. Such an approach, often called *executable* (xUML) or *executable and translatable UML* (xtUML), is based upon a subset of UML which usually consists of Class and State Machine Diagrams. The underlying principle is to reduce the complexity of the UML to a minimum by limiting it to a semantically well-defined subset, which is independent of any implementation language. This solution allows reaching the highest abstraction level and degree of independence

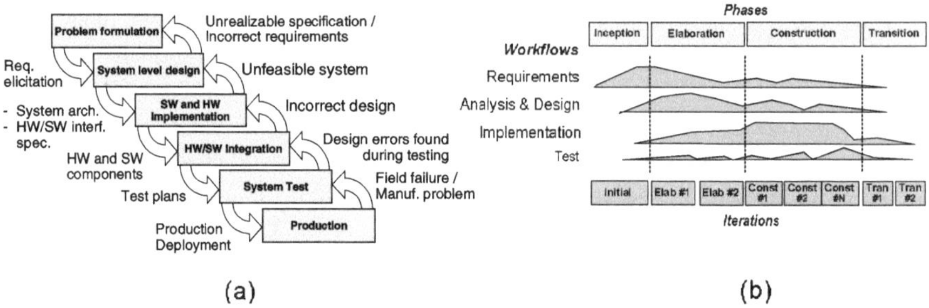

Fig. 2.9 Waterfall vs. iterative development processes (adapted from [31])

with respect to implementation details. However, this advantage comes at the cost of the limited choice of modeling constructs appropriate to SoC design and target languages available at the time of writing (C++, Ada, for example). Still, recent efforts such as [56] confirm that approaches based on a one-to-many mapping may gain maturity in the future and pave the road towards a unified design flow from specification to implementation. In particular, a behavioral synthesis technology from UML models towards both RTL languages and SystemC has become available recently [3]. Provided that synthesis tools taking as input C or C++ based SoC languages gain more popularity, xtUML tools could as well support in theory flexible generation of software and hardware implementations, where the software part of the system is produced by a model compiler optimizing the generated code for an embedded processor, while the hardware part is generated targeting a behavioral synthesis tool.

4.3 *Methodological Impact of Code Generation*

UML is often and wrongly considered as a methodology. UML is essentially a rich and complex notation that can address complex systems and help improve cross-disciplinary communication. Its application should be guided by a development process that stipulates which activities should be performed by which roles during which part of the product development. The absence of a sound methodology and poor understanding of the purposes of using UML lead inevitably to failures and unrealistic expectations [7].

Nevertheless, the possibility of generating code from UML models has a methodological impact, by enabling an iterative design flow instead of a sequential one. Modern development processes for software [31], embedded software [15], and systems engineering [4] follow iterative frameworks such as Boehm's spiral model [9]. In disciplines such as automotive and aerospace software development, however, we can still find processes relying on sequential models like the waterfall [52] and the V-model [18], due to their support of safety standards such as IEC 61508, DIN V VDE 0801, and DO 178-B. A traditional waterfall process (Fig. 2.9.a) assumes a clear separation of concerns between the tasks which are executed sequentially. Such a process is guaranteed to fail when applied to high risk projects that use innovative technology, since developers cannot foresee all upcoming issues and pitfalls. Bad design decisions made far upstream and bugs introduced during requirements elicitation become extremely costly to fix downstream. On the contrary, an iterative process is structured around a number of iterations or microcycles, as illustrated on Fig. 2.9.b with the example of the Rational Unified Process [31]. Each of these involves several disciplines of system development running in parallel, such as requirements elicitation, analysis, implementation, and test. The effort spent in each of these parallel tasks depends on the particular iteration and the risks to be mitigated by that iteration. Large-scale systems are incrementally constructed as a series of smaller deliverables of increasing completeness, which are evaluated in order to produce inputs to the next iteration. The underlying motivation is that the whole system does not need to be built before valuable feedback can be obtained from stakeholders inside (e.g., other team members) or outside (e.g., customers) the project.

Iterative processes are not restricted to the software domain or to UML: as an example, model-centric design flows based on Simulink [56], where models with increasing levels of details are at the center of the specification, design, verification, and implementation tasks, belong to the same family of design flows. The possibility to generate C/C++ and VHDL/Verilog code from Simulink models share similarities with the code generation capability of UML tools. In the context of SoC design, executable models based on UML and ESL languages provide a means to support iterative development process customized towards SoC design, as proposed in [47]. Automatic code generation from UML models enables rapid exploration of design alternatives by reducing the coding effort. Further gain in design time is possible if UML tools support code generation towards both hardware and software implementation languages, and if the generated code can be further synthesized or cross-compiled. Further examples of SoC design flows based on UML can be found in [6, 28, 61, 66].

5 Application Design Example

In the remainder of this chapter, we will present a complete application example illustrating the configuration of a UML editor for the purpose of SystemC-based modeling, automatic one-to-one code generation, simulation and synthesis. The approach has been developed in the ICT project SATURN (FP7-216807) [53] to close the gap between UML based modeling and simulation/synthesis of embedded systems. More precisely, SATURN extends the SysML editor ARTiSAN Studio for the co-modeling of synthesizable SystemC, C/C++, and Matlab/Simulink; the generated code implements a SystemC/C/C++/Simulink co-simulation based on EXITE ACE from EXTESSY.

Before we go into technical details, we first present the SATURN design flow and introduce the different SATURN UML profiles.

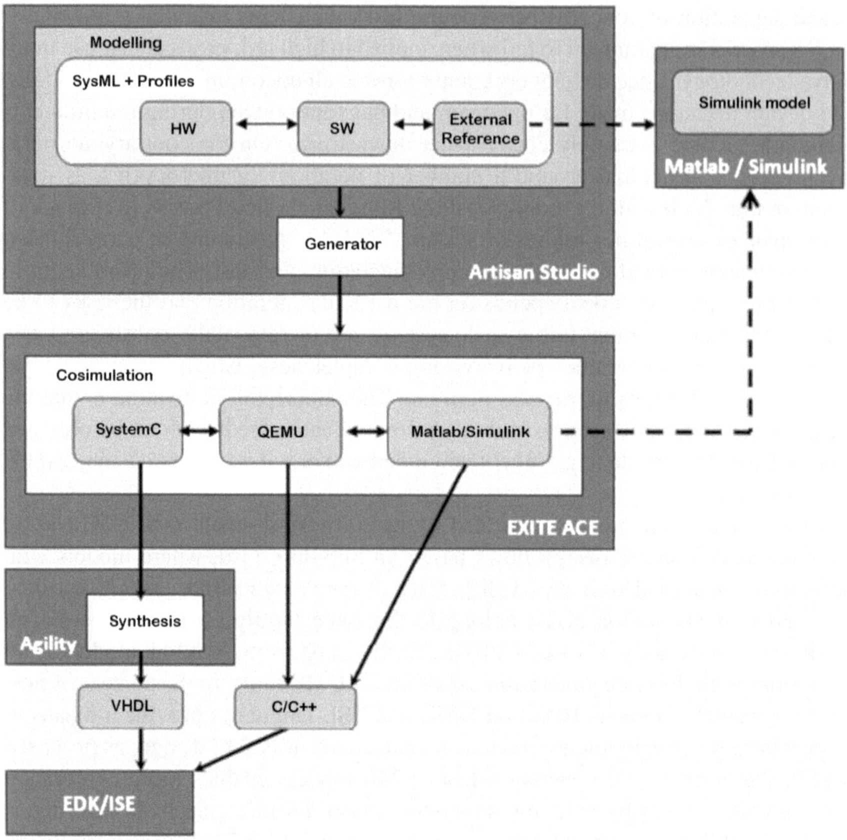

Fig. 2.10 The SATURN design flow

5.1 Methodology

The SATURN design flow, shown in Fig. 2.10, is defined as a front-end flow for industrial designs based on FPGAs with integrated microcontrollers such as the Xilinx Virtex-II-Pro or Virtex-5 FXT, which integrate PowerPC 405 and PowerPC 440 microcontrollers. The flow starts with the SysML editor Artisan Studio, which was customized by additional UML profiles for synthesizable SystemC, C/C++ and MATLAB/Simulink. As such, the developer takes early advantage of UML/SysML to capture the system requirements and proceeds to hardware/software partitioning and performance estimation without changing the UML-based tool environment. To support IP integration, the reference to different external sources is supported, i.e., MATLAB/Simulink models and C/C++ executables running on different CPUs and operating systems. Though the design flow is directed towards the SystemC subset synthesizable by the Agility SystemC compiler [1] extended by the special features of the synthesis tool, the general principles are not limited to synthesizable SystemC and Agility. Other back-ends tools and synthesizable subsets, such as Mentor Graphic's CatapultC and FORTE's Cynthesizer, could be supported as well through additional UML profiles. After creating the model, a one-to-one code generation is carried out by the ACS/TDK code generation framework. The code generator is implemented by the Template Development Kit (TDK). The Automated Code Synchronization (ACS) automatically synchronizes the generated code with the model.

In a first step, code generation is applied for simulation purposes. ACS generates SystemC models for simulation as well as interface software for full system mode co-simulation with the QEMU software emulator. The additionally generated makefiles and scripts implement the design flow automation such as the compilation of the C/C++ files to an executable and the OS image generation for QEMU. This tool flow also covers C code generated from MATLAB/Simulink models, e.g., by Mathworks RealTime Workshop or dSPACE TargetLink which can be compiled for the target architecture and executed by QEMU and co-simulated with SystemC.

QEMU is a software emulator based on binary code translation which is applied in replacement to an Instruction Set Simulator. It supports several instruction set architectures like x86, PPC, ARM, MIPS, and SPARC. There is typically no additional effort to port the native binaries from QEMU to the final platform. The simulation is currently based on the semantics of a TLM 1.0 blocking communication. The integration with QEMU applies shared memory communication with QEMU in a separate process. The co-simulation with other simulators like Simulink is supported by means of the EXITE ACE co-simulation environment, e.g., for test-bench simulation.

After successful simulation, the synthesizable SystemC code can be further passed to Agility for VHDL synthesis. The design flow follows then conventional lines, i.e., the Xilinx EDK/ISE tools takes the VHDL code as input and generates a bitstream file which is finally loaded with the OS image to the FPGA.

The next section will outline more details of the SATURN UML profiles, before we describe a modeling example and provide further details on code generation.

5.2 The SATURN Profiles

The SATURN profile is based on SysML and consists of a set of UML profiles:

- UML profile for synthesizable SystemC
- UML profile for Agility
- UML profile for C/C++ and external models

UML Profile for Synthesizable SystemC The UML Profile for synthesizable SystemC is introduced as a pragmatic approach with a focus on structural SystemC models. Graphical symbols for some stereotypes like interfaces and ports are inherited from the SystemC drawing conventions. The stereotypes of the profile provide a semantics oriented towards SystemC to SysML constructs in SysML Internal Block

Table 2.2 UML profile for synthesizable SystemC

SystemC concept	UML stereotypes	Base class
sc_main	*«sc_main»*	*Class*
sc_module	*«sc_module»*	*Class*
sc_interface	*«sc_interface»*	*Interface*
sc_port	*«sc_port»*	*Port*
sc_in	*«sc_in»*	*Port*
sc_out	*«sc_out»*	*Port*
sc_out	*«sc_out»*	*Port*
sc_signal	*«sc_signal»*	*Property, Connector*
sc_fifo	*«sc_fifo»*	*Property, Connector*
sc_clock	*«sc_clock»*	*Class*
sc_method	*«sc_method»*	*Action*
sc_trace	*«sc_trace»*	*Property*

Diagrams, such as blocks, parts, and flowports. Table 2.2 gives an overview of all stereotypes for synthesizable SystemC.

A stereotype *«sc_main»* defines the top-level module containing the main simulation loop with all of its parameters as attributes. The top level module may be composed of a set of *sc_modules* as the fundamental building blocks of SystemC. For this purpose, the *«sc_module»* stereotype is defined and applied to a SysML block. The debugging of tracing signals and variables is supported through the application of the *«sc_trace»* stereotype.

In order to connect modules, dedicated stereotypes for in, out, and inout ports are provided. Those stereotypes allow refining a SysML flowport as a SystemC primitive port. In SystemC, the *sc_in*, *sc_out*, and *sc_out* ports indicate specialized ports using the interface template like $sc\_signal\_in\_if\langle T\rangle$. The *«sc_port»* stereotype is applied to a SysML standard port through which SystemC modules can access a channel interface.

Ports connect to channels or other ports, optionally via interfaces. Regarding channels, the profile supports signals and complex channels like fifos. The *«sc_clock»* stereotype is applied to declare clocks in the SystemC model. Although clocks are not synthesizable, they are required for simulation purposes.

In order to model the system behavior, SystemC provides *sc_threads*, *sc_cthreads*, and *sc_methods*. The SystemC profile currently only supports *sc_methods* in its first version. As *sc_methods* do neither include wait statements nor explicit events, this limitation makes designs less error-prone and simplifies the task of code generation.

UML Profile for Agility SATURN currently applies the Agility SystemC compiler [1] to transform TLM-based SystemC models to RTL or EDIF netlists, which can be further processed by Xilinx ISE. The tool specific properties of Agility has been defined by a separate UML profile. Alternative synthesis tools like CatapultC

Table 2.3 UML profile for synthesizable SystemC

Agility concept	UML stereotypes	Base class
ag_main	*«ag_main»*	*Class*
ag_global_reset_is	*«ag_global_reset_is»*	*Port*
ag_ram_as_blackbox	*«ag_black_box»*	*Property*
ag_add_ram_port	*«ag_add_ram_port»*	*Property*
ag_constrain_port	*«ag_constrain_port»*	*Port*
ag_constrain_ram	*«ag_constrain_ram»*	*Property*

Table 2.4 UML profile for C/C++ extensions and external models

UML stereotypes	Base class
«cpu»	*Class*
«executable»	*Action*
«external»	*Class*

can be integrated by the definition of alternative UML profiles along the lines of the following approach. In order to allow the understanding of the code by Agility, a few extensions to the SystemC profile have to be defined and are summarized in Table 2.3. The designer is indeed able to insert some statements with *ag_* prefix into the SystemC code, which will be processed by the Agility compiler. These statements are basically pragmas which are transparent for simulation and only activated by the Agility compiler during synthesis. These pragmas provide additional synthesis information for SystemC entities. As a result, Agility stereotypes can only be assigned to object which have already a stereotype from the SystemC profile.

Agility identifies *ag_main* as a top level module for synthesis. An asynchronous global reset of internal registers and signals is defined by *ag_global_reset_is*. Additionally, Agility supports references to native implementations of RAMs through *ag_ram_as_blackbox*. In VHDL, for instance, this generates a component instantiation with appropriate memory control/data ports. The internal behavior could then, for instance, be linked with a netlist of a platform specific RAM. Through *ag_add_ram_port*, an array can be declared as a single or dual port RAM or ROM. *ag_constrain_port* allows assigning manually a specific VHDL type to a port, different to the standard port types which are *std_logic* for single bit ports and *numeric_std.unsigned* otherwise. By default *ag_constrain_ram* declares an array as a single-port RAM with one read-write port. However, most RAMs such as the BlockRAM of the Xilinx Virtex series are also configurable as multi-port memories. Through the corresponding stereotype, ROMs as well as RAMs with true dual port capabilities can be implemented.

UML Profile for C/C++ and External Models Additional basic extensions to the SystemC profile have to be defined for the purpose of hardware/software comodeling. They are listed in Table 2.4.

A basic feature for software integration to TLM-based SystemC models is supported by «*cpu*» which indicates a SysML block as a Central Processing Unit characterized by (i) its architecture (Register, RISC, CISC, etc.) and (ii) the Operation System (OS) running on top of its hardware.

For a CPU the «*executable*» stereotype is used to define an instance of a C/C++ application (process) which is cross-compiled for the specific hardware under the selected operating system. In order to support software reuse «*executable*» simply refers to existing C/C++ source code directories managed by makefiles. Though currently not supported the stereotype can be easily extended for full UML software modeling by means of activity or state machine diagrams. Finally, the «*external*» stereotype is introduced to interface the design with arbitrary native models which are supported by the underlying simulation and synthesis framework. Currently, the additional focus is on MATLAB/Simulink models as their integration is covered by the EXITE ACE co-simulation environment.

5.3 Co-modeling

Modeling starts in ARTiSAN Studio by loading the individual UML profiles and libraries which are hooked on to the integrated SysML profile.

Thereafter, as a first step, the design starts with the specification of a SysML Block Definition Diagram (BDD), which is based on the concepts of structured UML classes. In a BDD, users specify modules and clocks as blocks as well as their attributes and operations. A relationship between different blocks indicates the hierarchical composition. Figure 2.11 shows the BDD of a simple example, consisting of a design with a top level block, a PPC405 CPU, and a SystemC model for an FPGA, which has several subcomponents like clock, PLB bus and some transactors.

For the definition of the architecture of a design expressed in SystemC, C or Simulink, SysML Internal Block Diagrams (IBDs) are applied in a second step. Hereby, IBD blocks and parts are defined as instances of the BDD. Each SystemC block is defined by a simple activity diagram with one action for each method. Each method is defined as plain sequential ASCII code. This approach is motivated by several studies which have shown that it is more efficient to code SystemC at that level as textual code rather than by activity or state machine diagrams. Additional studies have shown that it is not very effective to represent 1-dimensional sequential code through 2-dimensional diagrams. Non-trivial models may easily exceed the size of one A4 pages which is hard to manage as a diagram.

In order to map software executables to processor components, i.e., blocks stereotyped with «*cpu*», the SATURN profile applies SysML allocations. Figure 2.12 shows the principles mapping a SysML block stereotyped with «*executable*» to a processor instance. In IBDs such an association is indicated by the name of the allocated software executable in the *allocatedFrom* compartment. Additionally, the *allocatedTo* compartment of the software block lists the deployment on the processor platform. As it is shown in the properties of a software block, each executable

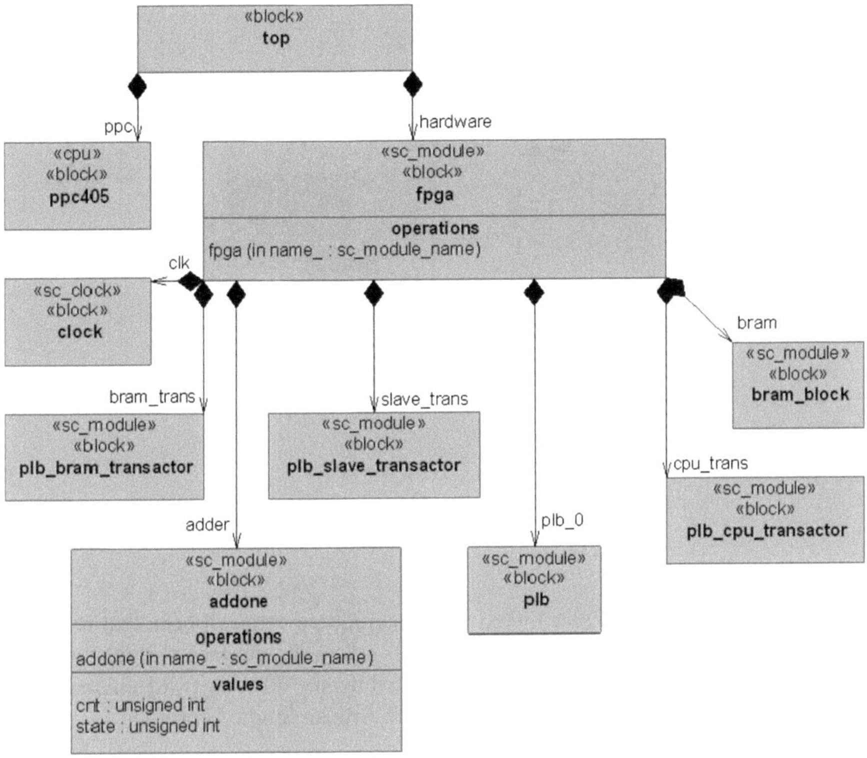

Fig. 2.11 Block definition diagram example

Fig. 2.12 Software allocation example

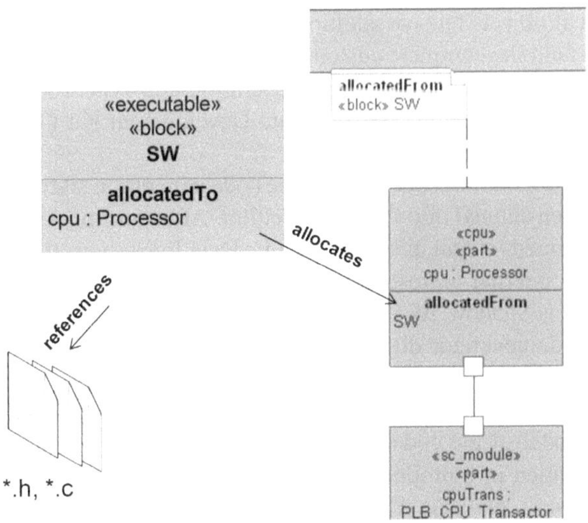

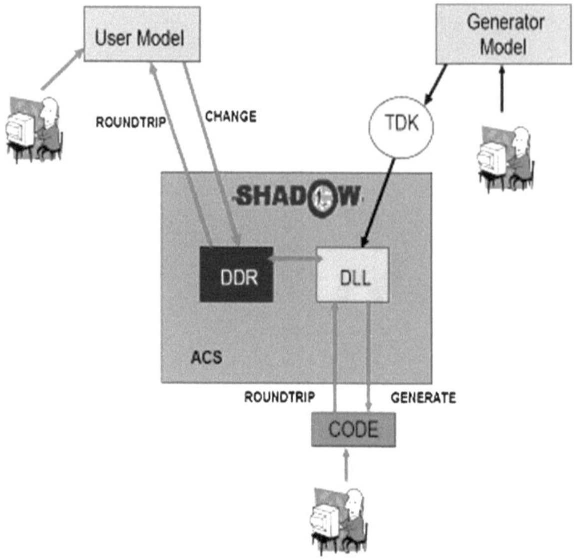

Fig. 2.13 ARTiSAN studio code generation

has the tagged value *directory* linked to the stereotype «*executable*» that refers to the directory of the source code. This provides a flexible interface to integrate arbitrary source code which also could be generated by any other software environment or any UML software component based on the Artisan Studio C profile.

5.4 Code Generation

The previous UML profiles are introduced to give adequate support for SystemC/C++-based code generation. By means of the introduced stereotypes, individual objects receive additional SystemC/C++ specific information. To better understand the complete customization, we briefly outline the concepts of the ARTiSAN Studio's retargetable code generation and synchronization, which is composed of two components: the Template Development Kit (TDK) and the Automated Code Synchronization (ACS).

As outlined in Fig. 2.13, the code generation starts with the user model which has been entered into the SysML editor. After starting ACS, the user model is first transformed into an internal Dynamic Data Repository (DDR), which saves the model in an internal representation of the user model. Each time the user model is modified, Shadow ACS is triggered, the DDR updated, and new code generated by a code generator dll. For reverse-engineering, the ACS can also be triggered by the changes of the generated code finally updating the user model. The code generator itself is defined by a *Generator Model* through TDK. A *Generator Model* is a model of the code generation, which is mainly composed of transformation rules written in a proprietary code generation description language, i.e., the SDL Template Language. This language has various constructs, through which all elements and properties of the user model can be retrieved and processed. The following is a list of main constructs like conditional statements, for loops and an indicator of the current object. Note that all keywords are identified by %.

- **%if ... %then ... {%elseif ... %then ...} [%else ...] %endif**
  Conditional Statement
- **%for (<listexpr>) ... %endfor**
  A loop through all objects in <listexpr>.
- **%current**
  Variable identifying the current object.

The following example shows an SDL excerpt for generating of SystemC code from SystemC stereotyped objects. The specification first goes through all classes of the model and checks them for individual stereotypes for generating different code segments. The example also sketches how to write the code of an sc_module header and the opening and closing brackets into a file.

```
%for "Class"
    %if %hasstereotype "sc_module" %then
        %file %getvar "FileName"
            "class␣" %getlocalvar "ClassName"
                "␣:\n\tpublic␣sc_module\n{"
                ...
            "}"
        %endfile
    %else
        ...
    %endif
%endfor
```

Figure 2.14 gives a more complex example which takes the block name specified in the user model as the class name and generates an sc_module inheritance. All declarations of operations and attributes as well as implementations of constructors are exported to the header .h file of an individual block. All implementations of operations are written to the .cpp source file.

5.5 Co-simulation

The hardware-software co-simulation for the generated code (cf. Fig. 2.15) is implemented by means of EXITE ACE, which is a simulator coupling framework developed by EXTESSY [17]. EXITE ACE provides an execution environment allowing for co-simulation of heterogeneous components, which are defined as the smallest function units of a system. Currently components from several tools (MATLAB, Simulink, TargetLink, ASCET, Dymola, Rhapsody in C, etc.) are supported. Each component is composed of an interface specification and a function implementation. The interface specification, which can be taken from the UML/SysML model,

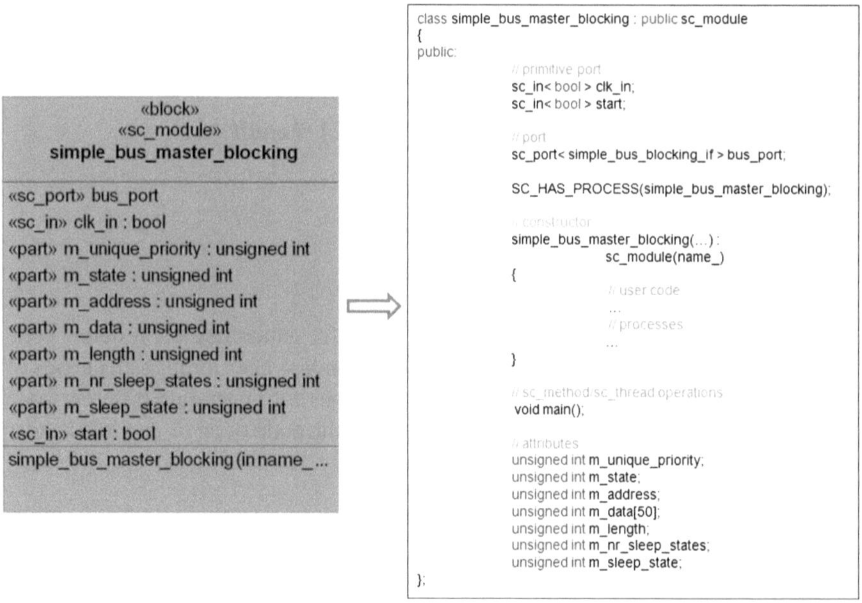

Fig. 2.14 SystemC code generation of a SysML block

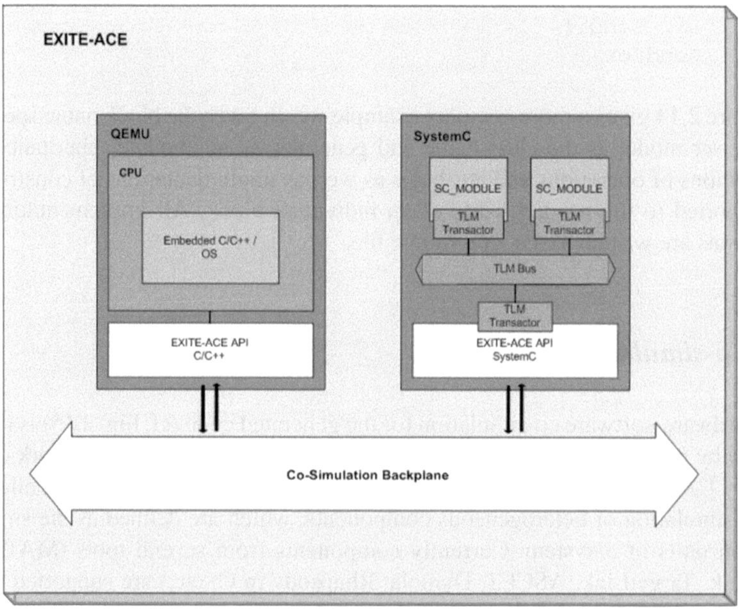

Fig. 2.15 SystemC–C/C++ co-simulation by EXITE ACE

describes the communication interface in form of ports definition. The function implementation usually refers to an executable model (for instance DLL or MDL file) incorporating computation algorithm. In the context of SystemC based verification, EXITE ACE is extended to support SystemC and QEMU components in order to allow for hardware-software co-simulation. Additionally, we extended QEMU for a blocking communication with the backplane and implemented a SystemC transactor to interface with the EXITE ACE. The transactor has to implement the individual communication policy, such as blocking or non-blocking. This architecture also supports the direct communication via shared memory between the SystemC simulator and QEMU in order to avoid the overhead of the simulation backplane.

6 Conclusions

The adoption of UML in an organization provides an opportunity to adopt new design practices and to improve the quality of the final product. Several efforts from the academic and industrial user community as well as UML tool vendors have been carried out in the recent years to investigate how tools could be extended, developed, and associated, in order to ease the use of UML for the design of electronic systems. Although UML still appears as a risky technology in this context, the situation is likely to change with the growing complexity of electronic designs and the need to specify efficiently heterogeneous systems. In addition, the increasing quality of system-level tools from EDA vendors and the expansion of UML tool vendors towards the market of electronic system design give the opportunity to bridge the gaps between the different development phases, and between the application domains. The perspective of having a unified framework for the specification, the design and the verification of heterogeneous electronic systems is gradually becoming reality. The application presented in the last section gave a first impression on the extension and integration of commercial tools into a coherent design flow for SystemC based designs. However, this is just a first step and some issues as traceability and management of synthesized objects through UML front-ends require further investigations and presumably a deeper integration of the tools.

Acknowledgements The work described in this chapter was partly funded by the German Ministry of Education and Research (BMBF) in the context of the ITEA2 project TIMMO (ID 01IS07002), the ICT project SPRINT (IST-2004-027580), and the ICT project SATURN (FP7-216807).

References

1. Agility: http://www.mentor.com
2. Akehurst, D., et al.: Compiling UML state diagrams into VHDL: an experiment in using model driven development. In: Proc. Forum Specification & Design Languages (FDL) (2007)
3. Axilica FalconML: http://www.axilica.com

4. Bahill, A., Gissing, B.: Re-evaluating systems engineering concepts using systems thinking. IEEE Trans. Syst. Man Cybern., Part C, Appl. Rev. **28**, 516–527 (1998)
5. Baresi, L., et al.: SystemC code generation from UML models. In: System Specification and Design Languages. Springer, Berlin (2003). Chap. 13
6. Basu, A.S., et al.: A methodology for bridging the gap between UML & codesign. In: Martin, G., Mueller, W. (eds.) UML for SoC Design. Springer, Berlin (2005). Chap. 6
7. Bell, A.: Death by UML fever. ACM Queue **2**(1) (2004)
8. Björklund, D., Lilius, J.: From UML behavioral descriptions to efficient synthesizable VHDL. In: 20th IEEE NORCHIP Conf. (2002)
9. Boehm, B.: A spiral model of software development and enhancement. Computer **21**(5), 61–72 (1988)
10. Boudour, R., Kimour, M.: From design specification to SystemC. J. Comput. Sci. **2**, 201–204 (2006)
11. Bridgepoint: http://www.mentor.com/products/sm/model_development/bridgepoint
12. CATS XModelink: http://www.zipc.com/english/product/xmodelink/index.html
13. Coyle, F., Thornton, M.: From UML to HDL: a model driven architectural approach to hardware–software co-design. In: Proc. Information Syst.: New Generations Conf. (ISNG) (2005)
14. Damasevicius, R., Stuikys, V.: Application of UML for hardware design based on design process model. In: Proc. Asia and South Pacific Design Automation Conf. (ASP-DAC) (2004)
15. Douglass, B.: Real Time UML. Addison-Wesley, Reading (2004)
16. Electronics Weekly & Celoxica: Survey of System Design Trends. Technical report (2005)
17. Extessy: http://www.extessy.com
18. Forsberg, K., Mooz, H.: Application of the "Vee" to incremental and evolutionary development. In: Proc. 5th Annual Int. Symp. National Council on Systems Engineering (1995)
19. From UML to SystemC—model driven development for SoC. Webinar, http://www.artisansw.com
20. Fujitsu: New SoC design methodology based on UML and C programming languages. Find **20**(4), 3–6 (2002)
21. Goering, R.: System-level design language arrives. EE Times (August 2006)
22. Grell, D.: Wheel on wire. C't **14**, 170 (2003) (in German)
23. Grötker, T., Liao, S., Martin, G., Swan, S.: System Design with SystemC. Springer, Berlin (2002)
24. Harel, D.: Statecharts: a visual formalism for complex systems. Sci. Comput. Program. **8**(3), 231–274 (1987)
25. Hooman, J., et al.: Coupling Simulink and UML models. In: Proc. Symp. FORMS/FORMATS (2004)
26. IBM Rational Rhapsody: http://www.ibm.com/developerworks/rational/products/rhapsody
27. IEEE Std 1666–2005 SystemC Language Reference Manual (2006)
28. Kangas, T., et al.: UML-based multiprocessor SoC design framework. ACM Trans. Embed. Comput. Syst. **5**(2), 281–320 (2006)
29. Katayama, T.: Extraction of transformation rules from UML diagrams to SpecC. IEICE Trans. Inf. Syst. **88**(6), 1126–1133 (2005)
30. Kennedy Carter iUML. http://www.kc.com
31. Kruchten, P.: The Rational Unified Process: An Introduction. Addison-Wesley, Reading (2003)
32. Kumaraswamy, A., Mulvaney, D.: A novel EDA flow for SoC designs based on specification capture. In: Proc. ESC Division Mini-conference (2005)
33. Laemmermann, S., et al.: Automatic generation of verification properties for SoC design from SysML diagrams. In: Proc. 3rd UML-SoC Workshop at 44th DAC Conf. (2006)
34. Martin, G., Mueller, W. (eds.): UML for SoC Design. Springer, Berlin (2005)
35. McGrath, D.: Unified Modeling Language gaining traction for SoC design. EE Times (April 2005)
36. McUmber, W., Cheng, B.: UML-based analysis of embedded systems using a mapping to VHDL. In: Proc. 4th IEEE Int. Symp. High-Assurance Systems Engineering (1999)

37. Nguyen, K., et al.: Model-driven SoC design via executable UML to SystemC (2004)
38. OMG: OMG Systems Modeling Language Specification 1.1
39. OMG: UML 2.0 Testing Profile Specification v2.0 (2004)
40. OMG: UML Profile for Modeling QoS and Fault Tolerance Characteristics and Mechanisms (2004)
41. OMG: UML Profile for Schedulability, Performance, and Time (SPT) Specification, v1.1 (2005)
42. OMG: UML Profile for System on a Chip (SoC) Specification, v1.0.1 (2006)
43. OMG: A UML Profile for MARTE (2009)
44. OMG: UML v2.2 Infrastructure Specification (2009)
45. OMG: UML v2.2 Superstructure Specification (2009)
46. Open SoC Design Platform for Reuse and Integration of IPs (SPRINT) Project. http://www.sprint-project.net
47. Pauwels, M., et al.: A design methodology for the development of a complex System-on-Chip using UML and executable system models. In: System Specification and Design Languages. Springer, Berlin (2003). Chap. 11
48. Ramanan, M.: SoC, UML and MDA—an investigation. In: Proc. 3rd UML-SoC Workshop at 43rd DAC Conf. (2006)
49. Raslan, W., et al.: Mapping SysML to SystemC. In: Proc. Forum Spec. & Design Lang. (FDL) (2007)
50. Reichmann, C., Gebauer, D., Müller-Glaser, K.: Model level coupling of heterogeneous embedded systems. In: Proc. 2nd RTAS Workshop on Model-Driven Embedded Systems (2004)
51. Riccobene, E., Rosti, A., Scandurra, P.: Improving SoC design flow by means of MDA and UML profiles. In: Proc. 3rd Workshop in Software Model Engineering (2004)
52. Royce, W.: Managing the development of large software systems: concepts and techniques. In: Proc. of IEEE WESCON (1970)
53. SATURN Project: http://www.saturn-fp7.eu
54. Schattkowsky, T., Xie, T., Mueller, W.: A UML frontend for IP-XACT-based IP management. In: Proc. Design Automation and Test Conf. in Europe (DATE) (2009)
55. Tan, W., Thiagarajan, P., Wong, W., Zhu, Y.: Synthesizable SystemC code from UML models. In: Proc. 1st UML for SoC workshop at 41st DAC Conf. (2004)
56. The Mathworks: Model-based design for embedded signal processing with Simulink (2007)
57. Thompson, H., et al.: A flexible environment for rapid prototyping and analysis of distributed real-time safety-critical systems. In: Proc. ARTISAN Real-Time Users Conf. (2004)
58. UML-SoC Workshop Website. http://www.c-lab.de/uml-soc
59. Vanderperren, Y.: Keynote talk: SysML and systems engineering applied to UML-based SoC design. In: Proc. 2nd UML-SoC Workshop at 42nd DAC Conf. (2005)
60. Vanderperren, Y., Dehaene, W.: From UML/SysML to Matlab/Simulink: current state and future perspectives. In: Proc. Design Automation and Test in Europe (DATE) Conf. (2006)
61. Vanderperren, Y., Pauwels, M., Dehaene, W., Berna, A., Özdemir, F.: A SystemC based System-on-Chip modelling and design methodology. In: SystemC: Methodologies and Applications, pp. 1–27. Springer, Berlin (2003). Chap. 1
62. Vanderperren, Y., Wolfe, J.: UML-SoC Design Survey 2006. Available at http://www.c-lab.de/uml-soc
63. Vanderperren, Y., Wolfe, J., Douglass, B.P.: UML-SoC Design Survey 2007. Available at http://www.c-lab.de/uml-soc
64. Viehl, A., et al.: Formal performance analysis and simulation of UML/SysML models for ESL design. In: Proc. Design, Automation and Test in Europe (DATE) Conf. (2006)
65. Wu, Y.F., Xu, Y.: Model-driven SoC/SoPC design via UML to impulse C. In: Proc. 4th UML-SoC Design Workshop at 44th DAC Conf. (2007)
66. Zhu, Q., Oishi, R., Hasegawa, T., Nakata, T.: Integrating UML into SoC design process. In: Proc. Design, Automation and Test in Europe (DATE) Conf. (2005)
67. Zhu, Y., et al.: Using UML 2.0 for system level design of real time SoC Platforms for stream processing. In: Proc. IEEE Int. Conf. Embedded Real-Time Comp. Syst. & Appl. (RTCSA) (2005)

Chapter 3
Executable Specifications for Heterogeneous Embedded Systems

An Answer to the Design of Complex Systems

Yves Leduc and Nathalie Messina

1 Introduction

The semiconductor industry is facing formidable issues during the development of complex Systems on Chip (SoC). Efforts to deliver such systems on time drain a significant part of resources while failure can push a company out of the market.

It is easy to recognize the heterogeneity of a system when it is composed of analog and digital modules, or software and hardware, or mechanical, optical and electronic technologies. But heterogeneity is the essence of all embedded systems, a subtle combination of data processing and control management. Specifying data and control requires specific expertise, modeling and tools. Designers must deal with numerous objects or concepts of various natures.

System development relies currently on a set of "human language" based specifications. If specifications are not properly validated and verified, circuits may not fully correspond to customer requirements (*not doing the right product*) or could lead to the late discovery of unexpected bugs (*not doing the product right*). This is a risk we cannot afford in the lengthy development of complex SoCs. As embedded systems intimately combine control and data processing, it is necessary to provide a precise description of both domains, stressing the subtle interactions between data and control flows. This description must be carefully debugged for it to be an accurate representation of the required system.

An executable specification is an obvious solution to the development of complex systems since it is inherently unambiguous and mathematically consistent. However it is important to know how to write it safely and correctly. This is the objective of this chapter. It presents a comprehensive methodology to build an executable specification of a heterogeneous system which is represented by interconnected data and control parts.

Y. Leduc (✉) · N. Messina
Advanced System Technology, Wireless Terminal Business Unit, Texas Instruments, Villeneuve-Loubet, France
e-mail: y-leduc@ti.com

G. Nicolescu et al. (eds.), *Design Technology for Heterogeneous Embedded Systems*,
DOI 10.1007/978-94-007-1125-9_3, © Springer Science+Business Media B.V. 2012

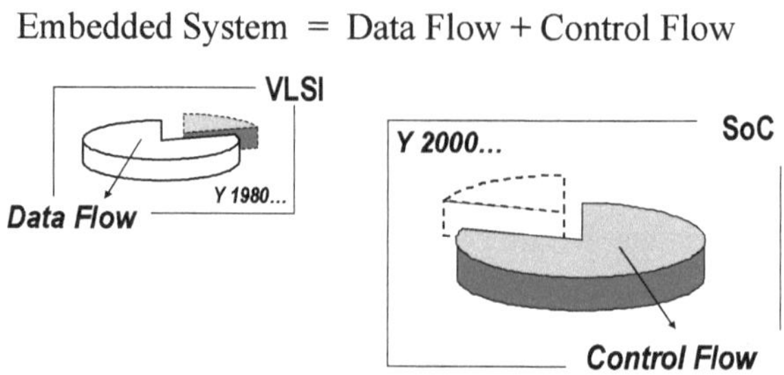

Fig. 3.1 Data and control flows

2 Risks Associated with Complex Critical Systems

2.1 A Few Concepts and Definitions

A *critical system* is a system which is vital to the customer and/or his supplier. Criticality may be understood as a life threatening risk or a business killer.

A *complex system* is a system whose complexity is much larger than the complexity of the modules of which it is comprised. Described in another way, it would be difficult or impossible to expect to catch all system architecture issues during the design phase of its component modules, but instead such issues only become apparent after the integration of the modules together.

Combining both aspects, *critical and complex systems* represent therefore major risks which cannot be underestimated.

Complexity itself adds another risk. Designers are used to perform the *verification of the system implementation* [1]. They must answer the question "are we building the product right?" They are not responsible for checking the quality of the system's specification itself. The *validation of the specification* [1] of a complex system, i.e. "are we building the right product?" becomes a task of prime importance in addressing complex development risks. This task is often underestimated and should be addressed carefully.

2.2 System Development Discontinuity

A system is a combination of data and control flows. Before the SoC era, integrated circuits were specialized in control processing such as micro-controllers, or in data processing such as mixed signal chips and digital signal processors. The first attempts to integrate embedded systems included mostly data flow with a very small control part. The advent of SoCs has dramatically changed the landscape, with IC's now largely dominated by control flow (Fig. 3.1).

Data flow design remains a task for signal processing gurus, for data conversion, digital or analog filters, algorithms, etc. The control flow is the domain of specialists in very large scale integration and large state machines. With the strong increase in

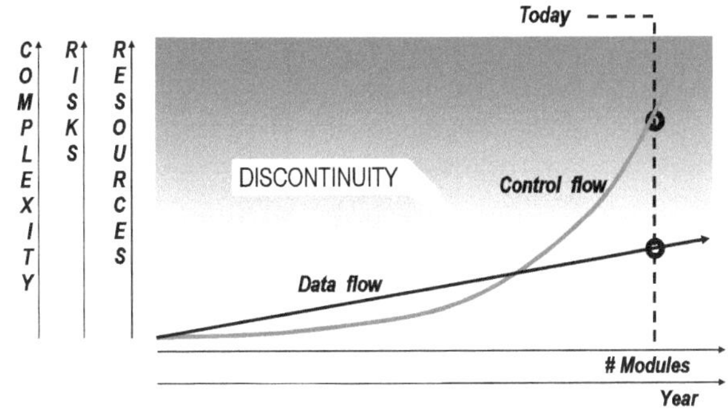

Fig. 3.2 Control flow vs. data flow

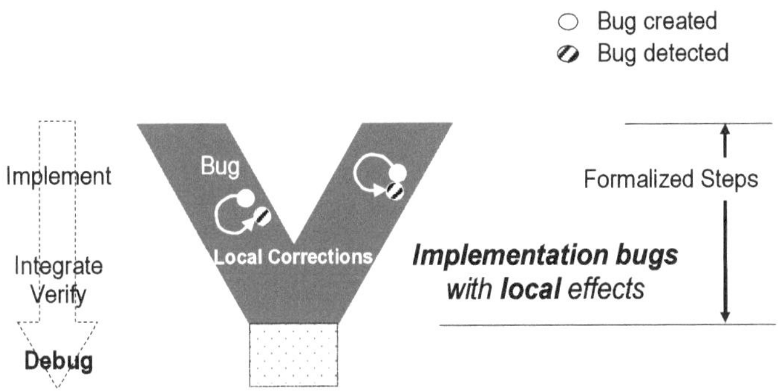

Fig. 3.3 V development methodology with late integration

the number of functions being integrated, state machine design by its combinatorial nature now dominates the development of complex SoCs (Fig. 3.2), such that demand in design resources is increasing exponentially.

It is not surprising that methodology must therefore change. This is the kind of discontinuity which is causing a revolution in the design community and impacts schedules, costs and risks.

2.3 *Risks in Traditional System Development*

Traditional development methodologies were established to organize the design of data-flow centric IC's. It is of course essential that the data processing meets its targets. All efforts of the designers were focused to build bug free modules. Integration of modules in an IC did not produce too much burden as the overall structure was fairly simple. Bugs having a local impact only, there was a good probability that the development of such IC's was under control. Such a methodology is a traditional "V shape" methodology and corresponds to *late integration* (see Fig. 3.3).

When dealing with complex embedded systems, the limitations of V development flow are immediately exposed (see Fig. 3.4). Module design is well under control but many bugs are introduced lately during the integration phase. These bugs

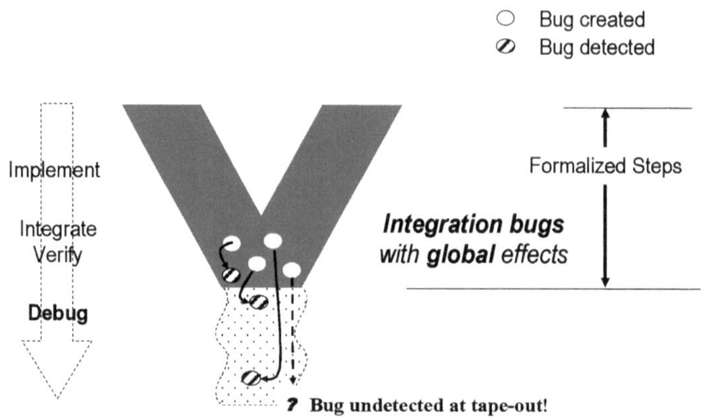

Fig. 3.4 V development methodology applied to complex systems

are introduced by the connection of the modules which create large hidden state machines with an ill-predicted behavior. Designers have little chance to catch these bugs during the development of their modules. Integration bugs are hard to find and may be left undetected until the IC is sold to end-customers. Design managers recognize now that although they were pretty confident during the development of the modules, they lose control of the development of their complex products during this integration phase. Bugs pop up randomly, diagnostics are fuzzy and last but not least, project managers are not able to fix a reliable date for a release to production.

3 Moving Forward

3.1 Replace Late Integration by Early Integration Methodology

Another solution needs to be found to handle the development of complex embedded systems.

Here it is proposed to use a "Y shape" development methodology based on *early integration* and, in addition to improve validation of the specifications, by a preliminary analysis of the customer's requirement (see Fig. 3.5). The entire concept relies on the assumption that we are able to describe and simulate the behavior of the system BEFORE it is designed. This is the basis of a true *executable specification* [2].

This proposal, which we name AUGIAS ("**A** **U**nified **G**uidance for the **I**ntegration of **A**dvanced **S**ystems"), defines a methodology flow from needs identification to IC implementation (see Fig. 3.6).

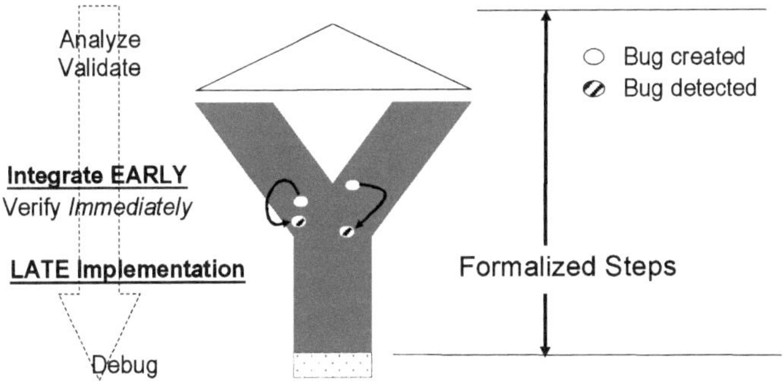

Fig. 3.5 Y development methodology based on early integration

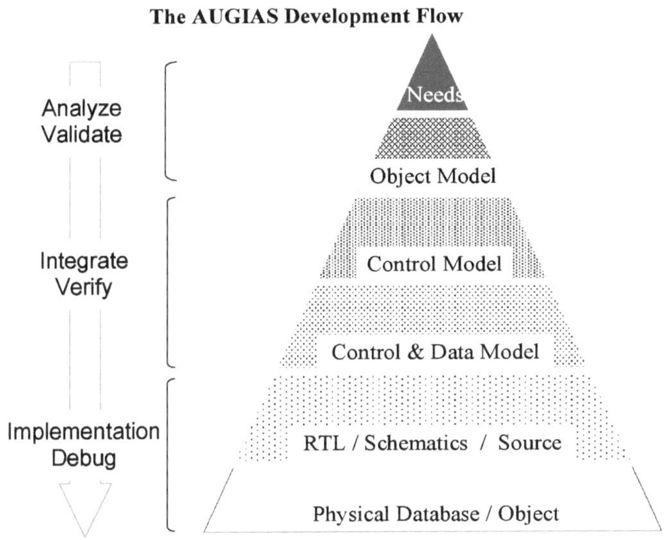

Fig. 3.6 Proposed development flow

3.2 *Expected Benefits*

The proposed methodology secures the system development phase, naturally encourages strong dialog among a large team, and formalizes the specification from the beginning of the project. The specification cannot be complete at the moment the project starts. It is the work of each team member to build the complete specification by adding his own expertise at each step of the project development. The specification will evolve throughout the development. At each level of the development flow, the upstream specification requirements must be met, and an augmented specification must be provided downstream to the next lower level of the project. It

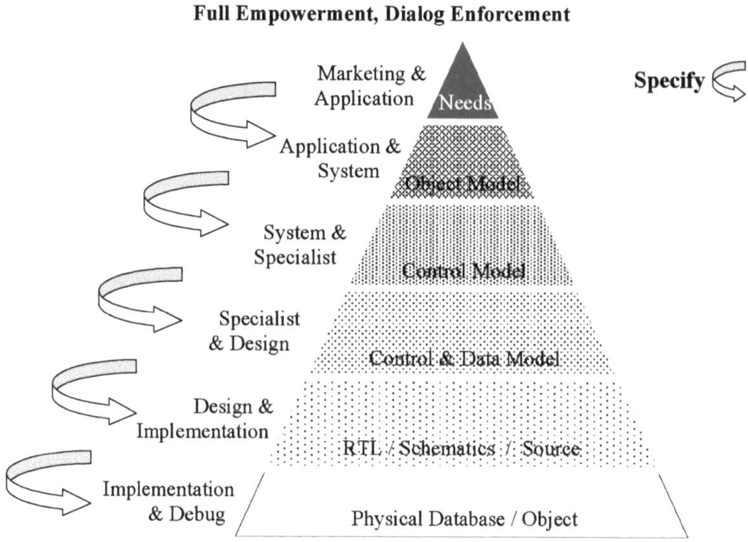

Fig. 3.7 Empowerment and dialog enforcement

is everyone's responsibility to build (i.e. *deploy*) the specification at each step of the project.

Early integration will immediately catch structural bugs in the state machines created by the connections of the modules. At each level of the proposed development flow, responsibilities are clearly identified and the deliverable model is checked with appropriate tools (see Fig. 3.7).

By overlapping responsibilities between adjacent levels, this methodology insures for instance that the application engineer understands and accurately captures customer requirements in a formalized way. The system engineer can now work with a solid basis to partition and create the internal architecture. As the resulting model is executable, it provides to all a non-ambiguous reference specification of the SoC being developed. This model is an ideal support to validate, verify and simulate the behavior or the expected performances of the system to be designed.

Following this methodology, the hardware or software design engineer can feel confident that they have implemented and validated a precise specification, while the system engineer is confident to know that the system validated will exactly match his specifications. By using appropriate hierarchical modeling, module reuse is possible and encouraged, and can take the form of reconfigurable, target, independent, or frozen modules.

And finally, customers will appreciate the solidity of a design flow under control and will be comforted early-on that their system requirements are being carefully respected.

3.3 *Methodology Streamline Based on a Strict Data & Control Separation*

Using high-level behavioral simulations and formal verification, an early validation of the complete system secures the design and speeds up the release of the product to the market thanks to a true executable specification.

The flow is based on a widely accepted approach to create and validate software. We have adapted and completed this methodology to our core competency: the co-development of systems including heterogeneous technologies, hardware and software, analog and digital, data and control. This methodology starts with an UML description (Unified Modeling Language [3]) for capturing customer requirements and for object-oriented partitioning analysis (Refer to *Needs Model* and *Object Model* in Fig. 3.7). We describe the system essentially using the UML class, sequence and state machine diagrams. The results of this analysis are used to build an executable model (*Control Model* in Fig. 3.7). Formal verification [4] is not a simple task and it becomes immensely complicated when handling data. We therefore propose a complete abstraction of the data domain into the control domain to create an executable specification built only with control state machines. This abstraction must describe the interactions between data and control signals, replacing the values of the data by their abstracted 'qualities'. In fact we refuse to introduce any data values in the control model. We propose here to *qualify the data* to allow their abstraction to be described as pure control signals. After verification of the model with formal proof and an appropriate set of test vectors, we will carefully introduce the values of the data to complete the description (*Control & Data Model* in Fig. 3.7) without modifying the state machines previously described and fully verified.

We will now illustrate the how these Control and Control & Data Models can be constructed.

4 The Control Model: A Practical Example

4.1 *Control Domain*

In a module, some signals can represent pure control. A control signal is represented simply by its presence or absence. The difficulty comes with the values of the data signals. Many data signals contain information related also to the control domain. We will focus in this paragraph on the abstraction of data in a pure control model description and particularly on data which impacts the behavior of the system.

4.2 *The Concept of Data Qualification*

We will show that it is possible to describe system behavior at the highest level of abstraction with "appropriate" representation of the data signals in the control

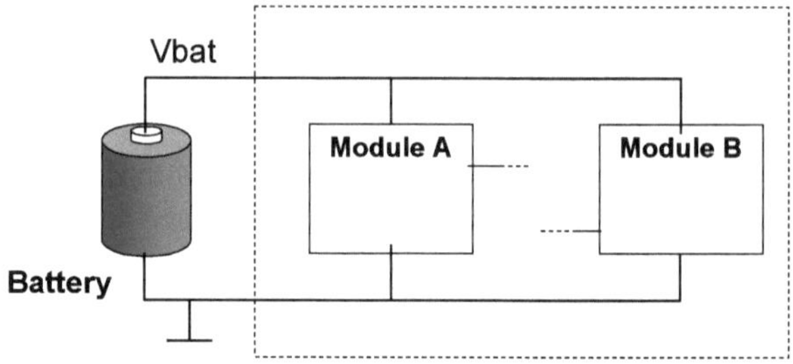

Fig. 3.8 Hardware to abstract

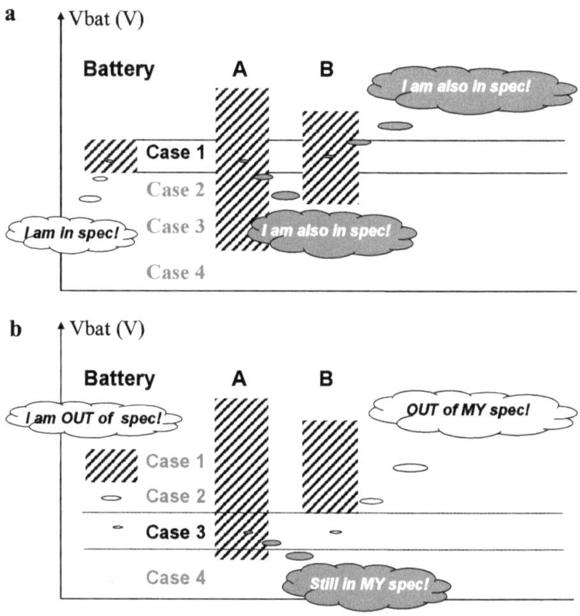

Fig. 3.9 Data qualification: (**a**) case '1'; (**b**) case '3'

domain. What is important at this point is not the value of the data but how the data signal interacts with the system. We will therefore aggregate data in data domains where data values may fluctuate without interfering with the global system behavior. So for now we will decide to ignore data values themselves.

At each data domain we will associate a *signal qualifier* which behaves as a control signal. A description of the system via a pure control model therefore becomes possible. We will illustrate the concept of data qualification via the interaction between a simple system made from a battery and two modules connected to this battery (Fig. 3.8).

Data qualification is not often straightforward as shown in Figs. 3.9a, b. In this example, the battery and the modules A and B are described with their respective specifications. It is expected that the battery and the modules themselves are operating within their individual specifications (see Fig. 3.9a). This is the description of a typical operation's condition.

However there is no reason that a module which is dependent on the battery voltage has exactly the specification as the battery specification. It is therefore possible that module A or B could still be observed as operating within specification while the supply voltage is outside of the battery specification (see Fig. 3.9b). The qualification "in spec" or "out of spec" cannot be considered as absolute but is instead relative to the characteristics of each module. As we are specifying a system which is not yet designed, we must describe all possibilities that the assembly of the modules could create. We will leave the choice of the correct combination of characteristics to the formal verification phase of the specification.

Some combinations can be rejected immediately: we must reject a module which does not match the battery characteristics. This combination of values will therefore be described in the Control Model as an invalid choice.

4.3 Local Requalification of the Data

It is a formidable task to describe systematically all the possible states of a system. We propose a "divide & conquer" method to avoid any incomplete description. Each module should be described as independent building blocks. As it is the assembly of the modules which will create complex control structures, we have to rely on the assembly of these modules in the Control Model to produce an exhaustive state machine, which should be complete by construction. The method is illustrated in Fig. 3.10. In this simple example, we describe the qualification of the battery voltage and the operating voltage of the modules A and B. The control variable describing the qualification of the signals will take respectively a "true"/"false" value corresponding respectively to "inspec"/"outspec" voltage. Still, as illustrated in this figure, there is no reason that the battery and the modules are inside or outside their own specifications in the same domains of voltage. We have not enough information at this point to fix the voltage ranges and we will describe here all the possibilities, leaving the decision when the system or the subsystem will be assembled. It will then be possible to decide and to prove the exactness of the choices on a rational basis.

In the example of the Fig. 3.10, if the battery is inside its specification (*case 1*), it is mandatory that the modules A and B accept the battery voltage value as valid: the module A and B should obviously be designed to work with this battery. When the battery voltage is outside the specification of the battery, modules A and B may or may not operate correctly. Module A and B could operate correctly together (*case 2*), only one of them could operate correctly, we have illustrated one of the possibility (*case 3*), or none of them are operating (*case 4*).

On the right side of Fig. 3.10, we introduce an auxiliary control signal for each of the modules powered by the battery. This is the *requalification signal*. Depending on the value "true" or "false" of this auxiliary signal, we can express simply whether or not the module can remain "inspec" when the main incoming signal is qualified as "outspec" (Fig. 3.11). With this auxiliary signal **Requalify**, we are now able to

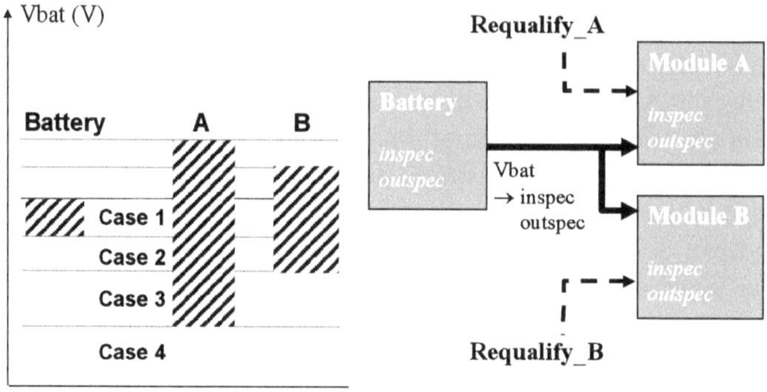

Fig. 3.10 Covering all state space with local requalification signals

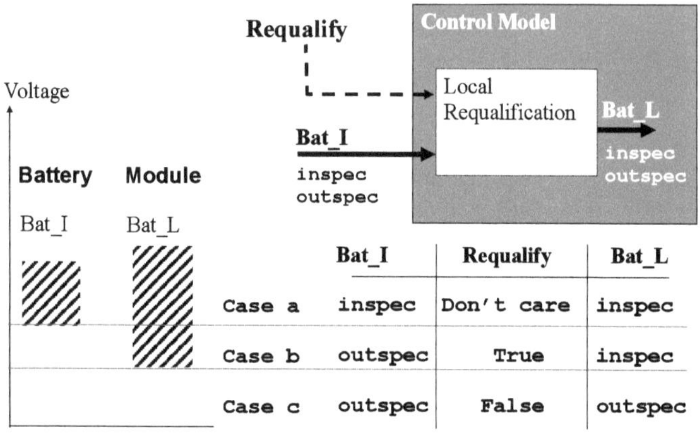

	Bat_I	Requalify	Bat_L
Case a	inspec	Don't care	inspec
Case b	outspec	True	inspec
Case c	outspec	False	outspec

Fig. 3.11 Auxiliary requalification signal

explore all possible cases of the state space of the system as shown in the figure. Plugging a new module to the small system of the example above introduces by construction, a new set of possibilities. The assembly of the blocks automatically constructs all possible states.

There is a great side effect: these requalification signals represent the choices still open at this stage of the design flow. It will be the responsibility of the designer assembling the system to analyze these possibilities, and to reject the combinations which cause an error at the system level by defining each requalification signal value. At the completion of the Control Model all of these auxiliary signals will have been assigned a value "true" or "false".

In some occasions, other qualifications will be added. An example is a module that could be powered off. In this case, in addition to "inspec" and "outspec", a third quality should be added, for example the qualification "zero".

We will illustrate this concept by creating the Control Model of a bandgap voltage reference. This module receives the battery voltage and produces a voltage reference (Fig. 3.12).

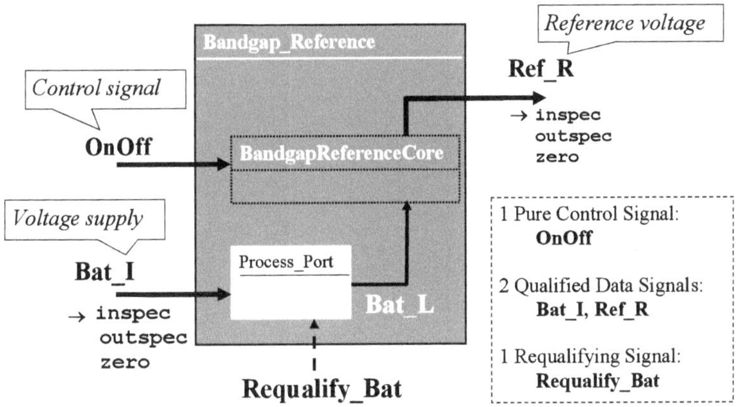

Fig. 3.12 A bandgap voltage reference

We have to model the fact that the battery could be disconnected: so in addition to "inspec" and "outspec" qualities, we add a third state "zero". In addition to the supply, the module receives a pure control signal **OnOff**. We expect three qualities to the output of the module, again "inspec" and "outspec" but also a "zero" state when the module is powered off by the control signal **OnOff** or if the battery voltage is "zero". **Bat_I** being a data signal, we add an auxiliary signal **Requalify_Bat** for signal requalification. The module is therefore an assembly of two sub modules: the "Process_Port" responsible for handling the local data requalification and the "BandgapReferenceCore" responsible for the description of the voltage reference behavior.

We have chosen to use an Esterel [5] description of the state machine as this language is perfectly suited to this job and has access to formal proof engine [6, 7]. As we will not use any of the advanced features of Esterel here, the description is self-explanatory. In Fig. 3.13, the Esterel code specifies that the requalified signal **Bat_L** is "inspec" if the input battery voltage is "inspec", but is also "inspec" if the input battery voltage **Bat_I** is "outspec" while the auxiliary input signal **Requalify_Bat** is "true". It specifies also that the **Bat_L** is "zero" if the voltage **Bat_I** is "zero" and that **Bat_L** is "outspec" in all other cases.

It is important to note that the true input of the core submodule defining the behavior of the bandgap voltage reference is now this internally requalified signal **Bat_L**. After requalification of its inputs, the modules intrinsic behavior is no longer complicated by the multiple combinations of input states. It is now straightforward to describe its behavior (Fig. 3.14) as a control block. The voltage reference will be "zero" if the module is not powered on or if the input battery voltage is "zero". When powered on, the voltage reference will be "inspec", "outspec" or "zero" following the quality of the requalified battery voltage.

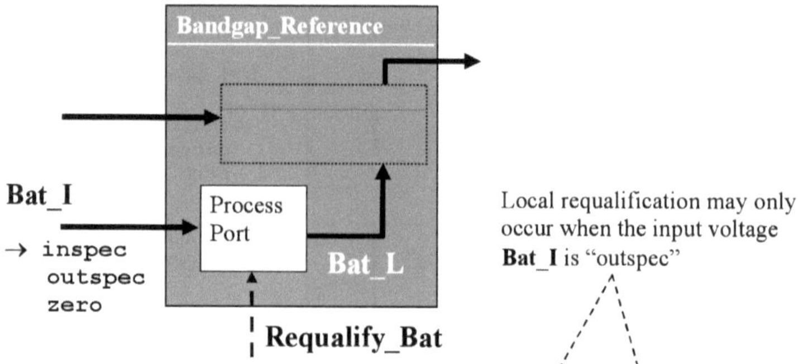

Process Port — *Esterel*

```
sustain {
  Bat_L.zero
   if Bat_I.zero,
  Bat_L.outspec
   if (Bat_I.outspec and not Requalify_Bat),
  Bat_L.inspec
   if (Bat_I.inspec or(Bat_I.outspec and Requalify_Bat))
       }
```

Fig. 3.13 Data requalification of the battery voltage input

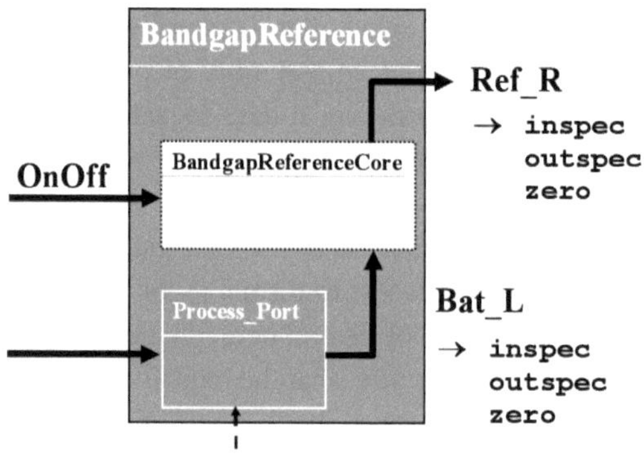

BandgapReferenceCore — *Esterel*

```
sustain {
  Ref_R.zero     if (not OnOff or Bat_L.zero),
  Ref_R.outspec  if (Bat_L.outspec and OnOff),
  Ref_R.inspec   if (Bat_L.inspec and OnOff)
        }
```

Fig. 3.14 Specification of the BandgapReference core

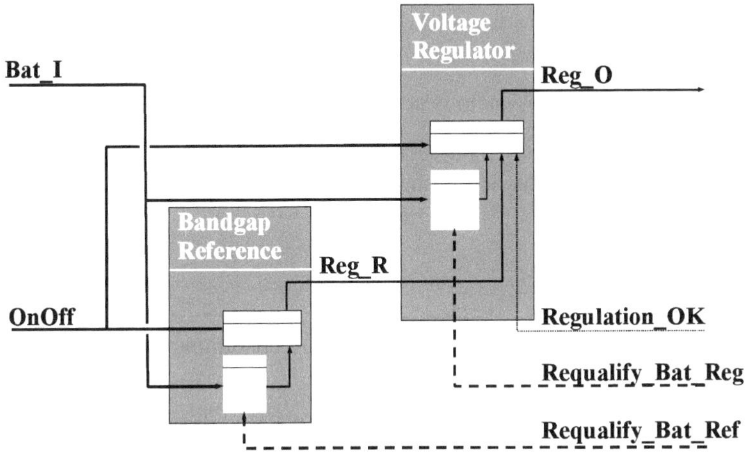

Fig. 3.15 Assembly of the voltage supply subsystem

4.4 Hierarchical Description

Figure 3.15 illustrates how such a module is inserted as a building block in a subsystem. Here we add a second module representing a voltage regulator. The regulator module is a little more complex. The designer responsible for this specification indicates that a voltage regulator needs some electrical and timing requirements to guarantee correct regulation by specifying another auxiliary signal **Regulation_OK**, The regulator output will be within specification if, in addition to the other input conditions, this module correctly regulates its output voltage.

An attentive reader will already have noticed in Fig. 3.15 that there is no requalification signal at the voltage reference input **Reg_R** of the regulator. Typically, a reference cannot be requalified as "inspec" by any module if it is declared as "outspec".

By assembling blocks together, the assembled control descriptions eventually combine to become complex state machines. These state machines will describe the very precise behavior of the system. By running appropriate test vectors, designers or customers will have a precise and non-ambiguous answer to any system or specification related question.

It is good practice to run verification at each level of hierarchy. It allows some requalification signals to be fixed as soon as possible, keeping complexity at its minimum. Many auxiliary requalification signals may be defined very early-on during bottom-up subsystem design.

4.5 Assertion and Verification

This simple model is already quite rich and we may naively think there is no room for introducing bugs. However if we run a formal verification tool on this simple model we will soon have a few surprises. For example let's verify the following property:

- "The voltage supply subsystem is never outside the specification"

In other words, we authorize the regulator to either output a zero voltage, or to be inside the specification. This is translated here by a simple assertion to be verified in all situations:

If ***Reg_O*** *is "outspec" then signal an error.*

The formal verification tool produces two counterexamples:

If the signal ***OnOff*** *is "true",* ***Bat_I*** *is not "zero" and the* ***Regulation_OK*** *is "false" then the output of the subsystem* ***Reg_O*** *is "outspec".*

If the signal ***OnOff*** *is "true",* ***Bat_I*** *is "outspec",* ***Requalify_Bat_Ref*** *is "false" and* ***Requalify_Bat_Reg*** *is "true" then the output of the subsystem* ***Reg_O*** *is "outspec".*

The first affirmation is not surprising as it clearly states that a faulty regulation cannot produce a correct result. The second one is more subtle, it indicates that the regulator may carefully follow an out of specification reference voltage. The model, as assembled, propagates the "outspec" quality of the voltage reference to the output. This is not a mistake in the design of a module but a bug introduced by the assembly of these modules.

This illustrates the responsibility of the designer in charge of the Control Model. Some combinations of values of these auxiliary signals will produce incorrect behavior of the voltage regulator subsystem. Such combinations must be avoided by appropriate choice of the auxiliary signals. In this particular situation, the designer will add two specifications to the model. These specifications will be used by the designer of the Data & Control Model when data will be added to the model as will be described in the next paragraph. Here is the Boolean rule against the second counterexample:

Requalify_Bat_Reg $\Rightarrow$ ***Requalify_Bat_Ref***

It specifies that the voltage regulator should be designed in such a way that it is *less* robust than the bandgap voltage reference. In other words, when the regulator is operating within its own specification, it is now guaranteed to be receiving a good voltage reference or a zero. The regulator will be designed to switch off when it is operating outside its own specification. This is a local problem implying a local solution. The possibility to issue an out of specification signal no longer exists. When the Data & Control Model is being written, its designer will have to strictly follow this explicit rule by selecting the appropriate electrical specifications of the bandgap voltage reference and of the voltage regulator. The behavior of the subsystem will now be correct in any situation. We avoid a classical but dangerous mistake here: the electrical designer could be tempted to make a better-than-specified voltage regulator. Such a regulator, being more robust than the voltage reference, will introduce a bug. The explicit specification associated to the relation between requalification signals ***Requalify_Bat_Reg*** and ***Requalify_Bat_Ref*** guarantees that this situation is under control in this design and in all variants we may generate in the future.

In other situations, the designer will have to add postulates to prevent some particular configurations. These are typically conditions of use (or abuse) of the end product. Such explicit postulates will specifically fix one or several auxiliary signals and provide useful documentation for the designers and/or the end users of the system.

This is of course an oversimplified example, but complex models may be derived by the assembly of such small control blocks. Many of these blocks will be generic and could advantageously be constituent blocks of a design library.

The Control Model is completed when it is fully verified against specifications issued from the next higher-level in the modeling flow, the Object Model. Relationships between some auxiliary signals have to be added to remove incorrect combinations. The designer of the next lower-level in the modeling flow, the Data & Control Model, will have to strictly follow this exhaustive set of specifications when he introduces the data values.

4.6 Façade Model, Equivalence Model Checking

It is important to verify in a bottom-up manner to limit the complexity of the verification. Some subsystems are often complex in their realization but simple in their behavior. A good example is an algorithmic analog to digital converter. Such a converter may be quite a sophisticated design but its behavior may be simple: after a determined number of clock cycles, a digital signal is output if the converter is "on", if it is correctly powered and if a few other conditions are valid. It is therefore recommended to build such a model as a façade model instead of a detailed model. If details do not bring any value to the verification of the complete system but instead make the system more complex, then this is a good reason to describe the system at a higher level of abstraction. This façade model will become eventually the reference structure for a second more detailed model describing the subsystem which can be developed later as necessary. Verification this second description must be done via a careful equivalence model check against the façade model.

4.7 Benefits of the Control Model

The Control Model is extremely important as it describes the behavior of the system which *will* be designed in minute detail and in a non-ambiguous form. Being a true executable specification, it is an immensely valuable tool to prove the correctness of the concept and the validity of the specifications to match the customer's requirements. The completion of the Control Model is an important milestone in the design of the system. As we have seen, auxiliary signals pop up naturally during the design phase as indications of degrees of freedom or due to the discovery of issues which are normally well hidden.

Critics used to say that complex systems are too complex to describe. But it is hard to believe that it is possible to successfully design something which no one can describe. The Control Model, as proposed here, forces naturally simple descriptions using well-known "divide & conquer" techniques. The result is a clean and safe description of a system under control. Anything too complex has been broken down to simple pieces and strange and weird algorithms are replaced by smarter solutions.

5 The Data and Control Model

5.1 Coexistence of Models

The designers in charge of the Data & Control Model will refer strictly to the Control Model as their specifications. Their responsibility is to plug appropriate data values into the Control Model without interference: any modification to the state machine of an already proven Control Model will invalidate it. This is a challenge we have to address.

Hierarchy of the design has been decided during the Control Model construction. The Data & Control Model will be an external wrapper respecting the hierarchy and also the interfaces of the Control Model. In addition to control values, such as "inspec" or "outspec", the signals exchanged between modules will transport data values (such as voltage, current, digital values, etc.). To facilitate team work sharing and reuse, a Data & Control Model should be capable of exchanging signals from other Data & Control Models which are already finalized, but equally should be capable of exchanging signals from a Control Model whose Data & Control Model has not yet been built.

A Data & Control Model will therefore be designed to be capable of receiving signals from a Control Model or from a Data & Control Model. This means that a Data & Control module which expects to process data, will instead receive only control signals from a Control Model. It will be the responsibility of the designer of the Data & Control module to create typical default values representing the signal for each data qualification on-the-fly. For example, a typical operating voltage could be chosen automatically on-the-fly when this voltage has been qualified as "inspec" while a low voltage such as zero in our example above could be chosen to represent an "outspec" qualification. With such a scheme, the designer will be able to run initial simulations and verifications of the system before getting access to a complete set of Data & Control modules (Fig. 3.16).

It is also possible that a Data & Control Model could be used to drive a Control-only Model as the qualification of the signals is given together with the value of the

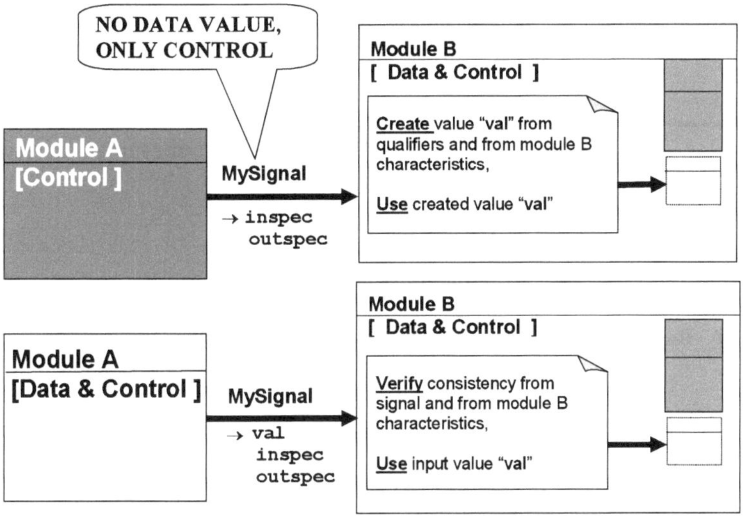

Fig. 3.16 Coexistence of Control and Data & Control Model

data itself. Figure 3.17 shows how a missing data is created from two parameters. To be fully consistent with its Control Model counterpart, the Data & Control Model description still accepts the auxiliary control signal for the requalification of the input even though it is not internally connected.

To replace the auxiliary control signal, a new signal is internally generated, here **Requalify_BatL** which is dependent on the value of the input signal. It is the responsibility of the designer of the Data & Control Model both to choose the data values and to specify the ranges of operation. This local creation of missing data solves the problem of compatibility in a system under construction where some of the modules already have a Data & Control Model while for others one does not yet exist.

5.2 The Core of the Data & Control Model

To build a Data & Control module we propose to *instantiate* the Control module inside it. This Control module has been fully proven during the verification of the system at the control level. By the instantiation we are certain that we do not in any way modify the already proven behavior. The Control module has been built such that its makes decisions about its output depending on the qualifications of its input signals. The Data & Control Model *respects* the decisions of its instantiated Control module. The responsibility of the Data & Control part is therefore easy to understand now: it has to qualify the signal by analyzing the data values and transfer these qualifications to the internal Control module which will decide the behavior of the module. We may compare this cooperation as a team of two persons,

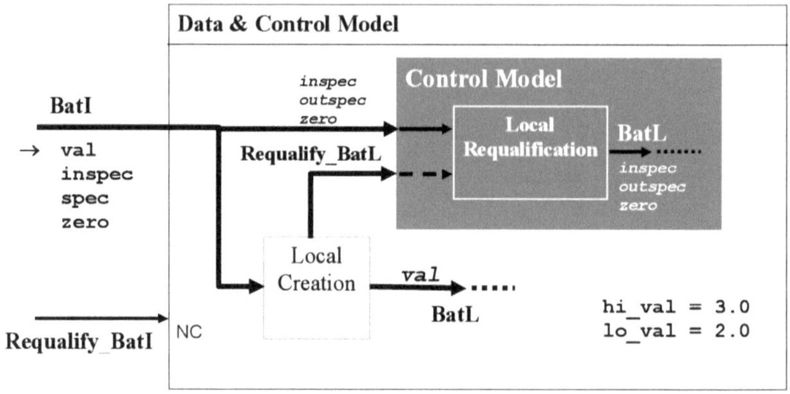

Esterel

```
sustain {
 ?BatL.val= ?BatI.val if BatI.val,
 ?BatL.val= (lo_val + hi_val)/2.0 if (BatI.inspec and not BatI.val),
 ?BatL.val= 0.0 if (BatI.zero and not BatI.val),
 ?BatL.val= lo_val/2.0 if (BatI.outspec and not BatI.val),
 Requalify_BatL if BatI.val andthen (?BatI.val>=lo_val and ?BatI.val<=hi_val)
        }
```

Fig. 3.17 Local creation of the data in a Data & Control Model

a specialist capable of interpreting the data (e.g. this value is OK, but this next one is at a dangerous level), and his manager who depends on the know-how of his specialists to qualify the situation and is the *only* person authorized to take decisions. As the Control module decision can impact the value of a data signal at the output, the responsibility is then transferred to the Data & Control part for processing.

Figure 3.18 shows an example of a simple Data & Control Model of a voltage regulator. The Control Model is instantiated as the behavior manager. Two input modules are responsible for creating the values of data if necessary and to qualify the values for further processing by the Control module. The core of the Data & Control Model is its handling of the data.

The description of the Data & Control Model is the domain of the design specialist and can be as accurate as is required. Statistical analysis or sophisticated mathematical formula may be plugged at this level if there is a need to estimate the yield of the future system.

It is strictly forbidden to introduce any new state machines at this stage since the core of the Data & Control Model is not allowed to modify the behavior of the module. Figure 3.19 shows an example of a simple piece-wise linear model describing the electrical behavior of the voltage regulator under specification. In this model a simple equation determines the control signal **RegulationL_OK** from the values of the data. This in turn determines an auxiliary input signal which is one of the inputs in the Control module instantiated in this Data & Control Model. This

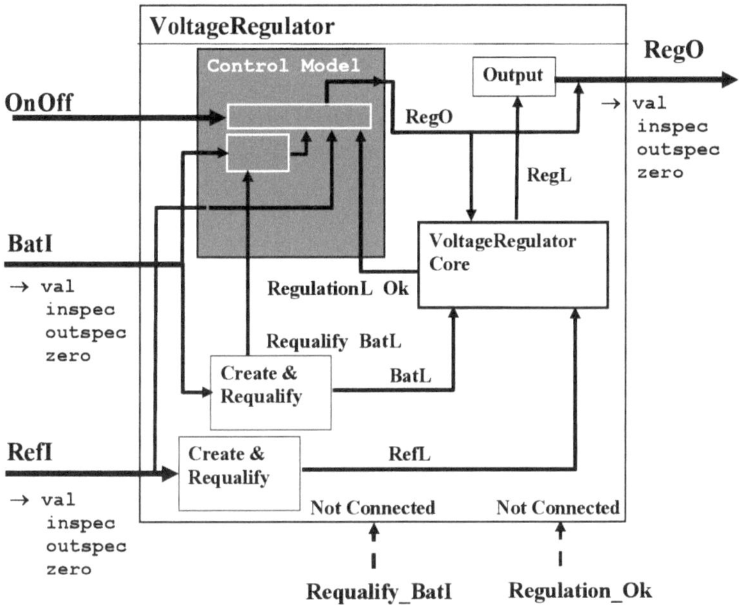

Fig. 3.18 The Data & Control Model of the voltage regulator

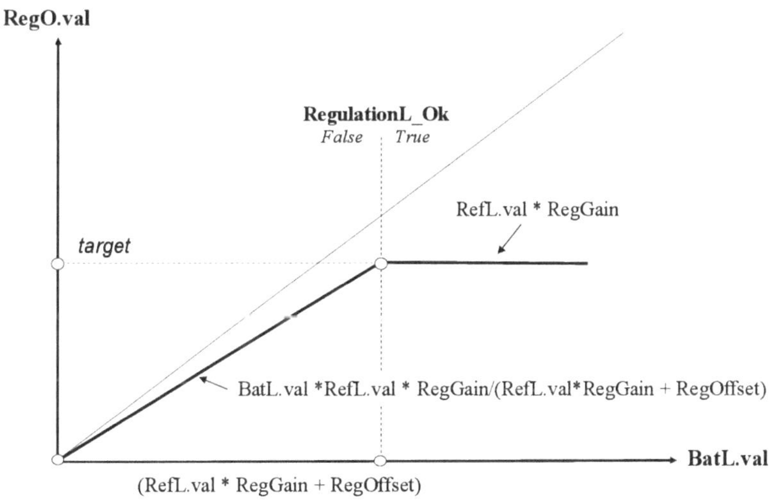

Fig. 3.19 Simplified high level data model of the voltage regulator core

simple model does not interfere with the behavior already described in the Control Model but completes the description.

The design specialist will work with the system engineer to build the Data & Control Model. Since he will be the person who will effectively design the regulator as soon as the Data & Control Model of the system is completed, validated and verified, we may safely consider that the behavior of this model is realistic and realizable.

All submodules of a system must have a Data & Control Model description. These submodules have already been connected in a hierarchical description during the construction of the Control Model. The Data & Control Model makes direct benefit of this Control Model structure and is not allowed to make any change to its hierarchy.

5.3 Benefits of the Data & Control Separation

The Data & Control Model benefits from the complete verification of the Control Model. There is a clean and clear separation of responsibility. The separation of data and control makes the description more robust and more reusable. Designers will rapidly discover that the major system decisions have already been made and validated at the Control Model level. Of course, it is possible that the design specialist might discover that the Control Model describes a system impossible to build. Since he does not have responsibility to change the Control Model himself, he must refer to the person in charge of this model and to suggest appropriate modifications.

6 Conclusion

We have used here several pieces of code written in Esterel. This language is a good candidate for describing the Control and Data & Control models as it excels in the description and verification of state machines. Being a rigorous language, Esterel descriptions also have access to formal proof engines.

This methodology does not depend on technology. We have applied this separation of data and control to various systems such as a sophisticated serial digital bus, and on several SoC power management modules.

We do not claim to have made the description and verification of complex systems an easy task. But we have proposed a practical methodology to effectively separate data and the control in a divide & conquer manner making it possible to raise abstraction advantageously to a higher level.

We have been surprised by the remarkable power of the Control Model and how comfortable we were when this model has been solidly verified. The Control Model of a subsystem represents a very valuable IP and an extraordinary reusable piece of work. It also forces a deep understanding of the behavior of a system and automatically creates complex state machines representing the behavior of the future IC by the assembly of simple subsystems. We have detected dangerous behavior we did not expect to discover in supposedly simple modules. No individual module was incorrect, but the simple connection of a few modules together generated crucial integration bugs which we were able to correct in the Control model before starting the design phase.

What we propose is a modest revolution in design habits, but as with all revolutions, it will take time to be fully accepted. It is particularly important to note that we distribute responsibilities in the design team. Those in charge of system specifications will receive more responsibility while everyone's work will be strongly identified. A designer's life may become less comfortable and therefore this methodology could encounter some resistance to change!

References

1. Capability Maturity Model Integration Product Team: CMMI® for Development, Version 1.02. Carnegie Mellon Software Engineering Institute, Carnegie Mellon University, August 2006
2. Schubert, P.J., Vitkin, L., Winters, F.: Executable specs: what makes one, and how are they used. In: SAE® 2006 Conference, Society of Automotive Engineers, Inc., 2006-01-1357 (2006)
3. Object Management Group™: Unified Modeling Language (UML), version 2.2, February 2009
4. Labbani, O., Dekeyser, J.-L., Rutten, É.: Separating control and data flow: methodology and automotive system case study. INRIA report 5832, February 2006
5. Berry, G., Gonthier, G.: The Esterel synchronous programming language: design, semantics, implementation. Sci. Comput. Program. **19**(2), 87–152 (1992)
6. Hales, T.: Formal proof. Not. Am. Math. Soc. **55**(11), 1370 (2008)
7. Harrison, J.: Formal proof—theory and practice. Not. Am. Math. Soc. **55**(11), 1395 (2008)

Chapter 4
Towards Autonomous Scalable Integrated Systems

Pascal Benoit, Gilles Sassatelli, Philippe Maurine, Lionel Torres, Nadine Azemard, Michel Robert, Fabien Clermidy, Marc Belleville, Diego Puschini, Bettina Rebaud, Olivier Brousse, and Gabriel Marchesan Almeida

1 Entering the Nano-Tera Era: Technology Devices Get Smaller (NANO), Gizmos Become Numerous (TERA) and Get Pervasive

Throughout the past four decades, silicon semiconductor technology has advanced at exponential rates in performance, density and integration. This progress has paved the way for application areas ranging from personal computers to mobile systems. As scaling and therefore complexity remains the main driver, scalability in the broad sense appears to be the main limiting factor that challenges complex system design methodologies. Therefore, not only technology (*fabricability*), but also structure (*designability*) and function (*usability*) are increasingly questioned on scalability aspects, and research is required on novel approaches to the design, use, management and programming of terascale systems.

1.1 The Function: Scalability in Ambient Intelligence Systems

Pervasive computing is a novel application area that has been gaining attention due to the emergence of a number of ubiquitous applications where context awareness is important. Examples of such applications range from ad-hoc networks of mobile terminals such mobile phones to sensor network systems aimed at monitoring

P. Benoit (✉) · G. Sassatelli · P. Maurine · L. Torres · N. Azemard · M. Robert · D. Puschini · B. Rebaud · O. Brousse · G.M. Almeida
LIRMM, UMR 5506, CNRS–Université Montpellier 2, 161 rue Ada, 34095 Montpellier Cedex 5, France
e-mail: Pascal.Benoit@lirmm.fr

F. Clermidy · M. Belleville · D. Puschini · B. Rebaud
CEA Leti, MINATEC, Grenoble, France

G. Nicolescu et al. (eds.), *Design Technology for Heterogeneous Embedded Systems*,
DOI 10.1007/978-94-007-1125-9_4, © Springer Science+Business Media B.V. 2012

geographical or seismic activity. This new approach to computing considerably enhances knowledge necessary for devising solutions capable of meeting application requirements. This is due to the emergence of uncertainty in such systems where environmental interactions and real-time conditions may change rapidly. Further, even though the problem may remain tractable for the small-scale systems used, solutions are not adapted, do not scale well and therefore face the curse of dimensionality.

A number of scientific contributions aimed at facilitating specification of applications [1] and formalizing the problem have emerged over the past decade, such as agent orientation, which promotes a social view of computing in which agents exchange messages, exhibit behaviors such as commitment, etc. The underlying challenge to the efficient design of such systems concerns the concepts behind autonomous systems able to monitor, analyze and make decisions. Machine learning/artificial intelligence techniques and bio-inspiration are among possible solutions that have been investigated for tackling such problems.

1.2 The Technology: Scalability in Semiconductor Technologies

Similarly, with the continued downscaling of CMOS feature size approaching the nanometer scale, the recurrent methods and paradigms that have been used for decades are increasingly questioned. The assumption of the intrinsic reliability of technology no longer holds [2] with increasing electric and lithographic dispersions, failure rates and parametric drifts. Beyond technological solutions, there is growing interest in the definition of self-adaptive autonomous tiles capable of monitoring circuit operation (delays, leakage current, etc.) and taking preventive decisions with respect to parameters such as voltage and frequency.

1.3 The Structure: Scalability in On-chip Architectures

Even though the abstraction of such technological issues may prove tractable, efficiently utilizing the ever-increasing number of transistors proves difficult. To this end, one popular design style relies on devising multicore/multiprocessor architectures [3]. Although such solutions have penetrated several market segments such as desktop computers and mobile terminals, traditional architectural design styles are challenged in terms of scalability, notably because of shared-memory oriented design, centralized control, etc. In this area again, there is growing interest for systems endowed with decisional capabilities. It is often believed that autonomous systems are a viable alternative that could provide adaptability at the chip level for coping with various run-time issues such as communication bottlenecks, fault tolerance, load balancing, etc.

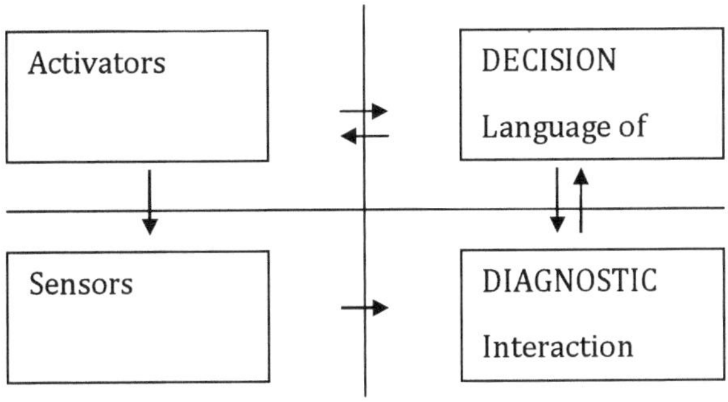

Fig. 4.1 Autonomous system infrastructure

1.4 Towards Multi-scale Autonomous Systems: A General Scheme

Autonomy is the faculty attributed to an entity that can be self-sufficient and act within its environment to optimize its functions. Autonomy also describes a system that can manage itself using its own rules. Autonomy is practiced by living organisms, people, and institutions but not yet by machines. However, the role of the mind in the architecture of autonomic systems is questioned.

In order to apply this concept to the reality of technological systems, this study will start with the abstract view of a system architecture while applying the notion of autonomy.

Figure 4.1 gives a synthetic view of autonomy: the activator creates the physical state of the system and the diagnosis motivates it. In microelectronics, and therefore for an SoC (System On a Chip), autonomy is represented by the fact that a calculation is distributed. In robotics, autonomy is the way to differ, deport or perform sequences of actions without risking damaging to the machine. In both cases, the command language must be able to schedule actions in parallel or in sequence. Autonomy can be used to lower energy consumption in microelectronics. In robotics, the challenge is to increase performance in different environments. In the artificial intelligence domain, autonomy is the consequence of a life cycle where the sensors observe, the diagnosis gives direction, the language of command orders and activators act.

Our objective is to design a fully scalable system and apply autonomy principles to MPSoC (Multiprocessor System-on-Chip). In this chapter, we will first discuss our vision of the infrastructure required for scalable heterogeneous integrated systems. Then we will provide a general model for self-adaptability, exemplified with respect to variability compensation, dynamic voltage and frequency scaling and task migration. Finally, an example of an autonomous distributed system that has been developed in the Perplexus European project will be provided.

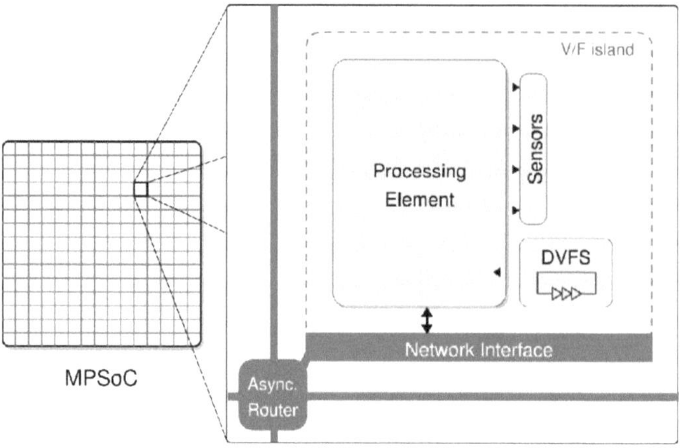

Fig. 4.2 Generic MPSOC architecture

2 Distributed MPSoC Systems

In this section, we discuss our vision of a generic MPSoC architecture supported by two examples. We analyze and suggest the features of a possible scalable and self-adaptive model suitable for future autonomous systems.

2.1 Generic MPSoC Architecture

This section describes a generic MPSoC by only introducing the key elements allowing formulating valid hypotheses on the architecture. The considered MPSoC is composed of several Processing Elements (PE) linked by an interconnection structure as described in Fig. 4.2.

2.1.1 Processing Elements

PEs of an MPSoC depend on the application context and requirements. There are two architecture families. The first includes heterogeneous MPSoCs composed of different PEs (processors, memories, accelerators and peripherals). These platforms were pioneered by the C-5 Network Processor [4], Nexperia [5] and OMAP [6]. The second family represents homogeneous MPSoCs, e.g. as proposed by the Lucent Daytona architecture [3], where the same tile is instantiated several times. This work targets both topologies. Thus, Fig. 4.2 represents a homogeneous or heterogeneous design.

2.1.2 Interconnection

The PEs previously described are interconnected by a Network-on-Chip (NoC) [7–10]. A NoC is composed of Network Interfaces (NI), routing nodes and links. NI implements the interface between the interconnection environment and the PE domain. It decouples computation from communication functions. Routing Nodes are in charge of routing the data between the source and destination PEs through links. Several network topologies have been studied [11, 12]. Figure 4.2 represents a 2D mesh interconnect. We consider that the offered communication throughput is enough for the targeted application set. NoC fulfills the "Globally Asynchronous Locally Synchronous" (GALS) concept by implementing asynchronous nodes and asynchronous-synchronous interfaces in NIs [13, 14]. As in [15], GALS properties allow MPSoC partitioning into several Voltage Frequency Islands (VFI). Each VFI contains a PE clocked at a given frequency and voltage. This approach allows real fine-grain power management.

2.1.3 Power Management

Dividing the circuit into different power domains using GALS has facilitated the emergence of more efficient designs that take advantage of fine-grain power management [16]. As in [17, 18], the considered MPSoC incorporates distributed Dynamic Voltage and Frequency Scaling (DVFS): each PE represents a VFI and includes a DVFS device. It consists of adapting the voltage and frequency of each PE in order to manage power consumption and performance. A set of sensors integrated within each PE provides information about consumption, temperature, performance or any other metric needed to manage the DVFS.

2.2 Examples: Heterogeneous and Homogeneous MPSoC

Nowadays, there are several industrial and experimental MPSoC designs targeting different application domains that fulfill part or all of the characteristics enumerated in the previous section. We briefly describe two examples: ALPIN from CEA-LETI and HS-Scale from LIRMM.

2.2.1 Alpin

Asynchronous Low Power Innovative Network-on-Chip (ALPIN) is a heterogeneous demonstrator [17, 18] developed by CEA-LETI. The ALPIN circuit is a GALS NoC system implementing adaptive design techniques to control both dynamic and static power consumption in CMOS 65 nm technology. It integrates 6 IP (Intellectual Property) units: a TRX-OFDM unit, 2 FHT units, a MEMORY unit,

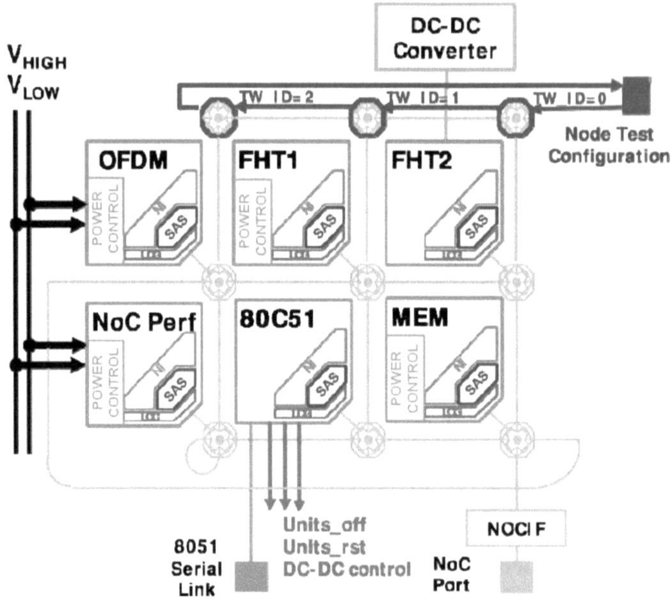

Fig. 4.3 ALPIN architecture

a NoC performance analysis unit and a 80c51 for power mode programming, as shown in Fig. 4.3. The interconnection is provided by 9 NoC asynchronous nodes and one NOC synchronous external interface. The asynchronous Network-On-chip provides 17 GBit/s throughput and automatically reduces its power consumption by activity detection.

Both dynamic and static power are reduced using adaptive design techniques. ALPIN IP units handle 5 distinct power modes. Using VDD-Hopping, dynamic power consumption can be reduced by 8-fold. By using Ultra-Cut-Off, static power consumption can be reduced by 20-fold.

2.2.2 HS-Scale: A Homogeneous MPSoC from LIRMM

Hardware-Software Scalable (HS-Scale) is a regular array of building blocks (Fig. 4.4) [19, 20]. Each tile is able to process data and to forward information to other tiles. It is named NPU (Network Processing Unit) and is characterized by its compactness and simplicity.

The NPU architecture is represented in Fig. 4.4. This architecture contains: a processor, labeled PE in Fig. 4.4; memory to store an Operating System (OS), a given application and data; a routing engine which transfers messages from one port to another without interrupting processor execution; a network interface between the router and the processor based on two hardware FIFOs; an UART that allows uploading of the operating system and applications; an interrupt controller to manage interrupt levels; a timer to control the sequence of an event; and a decoder to address

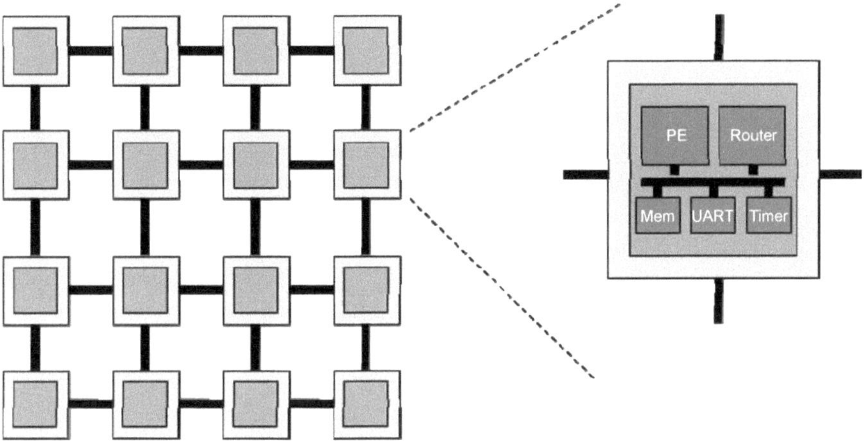

Fig. 4.4 HS-Scale architecture (from [19])

these different hardware entities. An asynchronous wrapper interfaces the processor with the routing engine, allowing several frequency domains and guaranteeing GALS behavior. The system is controlled by a distributed OS specifically designed for this platform, which provides self-adaptive features. It ensures load balancing by implementing task migration techniques.

2.3 Conclusion

Most industrial approaches for embedded systems are heterogeneous. Performance, power efficiency and design methods have been the biggest drivers of such technologies, but the lack of scalability is becoming a major issue with respect to tackling the inherent complexity [21]. Regular designs based on homogeneous architectures such as HS-Scale provide potential scalability benefits in terms of design, verification, program, manufacturing, debug and test. Compared to heterogeneous architectures, the major drawback could be the performance and power efficiency, but our goal is homogeneity in terms of regularity: each processing element could be "heterogeneous", i.e. composed of several processing engines (general purpose processor, DSP, reconfigurable logic, etc.) and instantiated many times in a regular design: we talk about globally homogeneous and locally heterogeneous architectures.

With a homogeneous system, each task of a given application can potentially be handled by any processing element of the system. Assuming that we can design a self-adaptive system with many possible usages, as illustrated in Fig. 4.5: task migration to balance workload and reduce hot spots, task remapping after a processing element failure, frequency and voltage scaling to reduce the power consumption, etc. But to benefit such a potential, we need to define a complete infrastructure, as outlined in the next section.

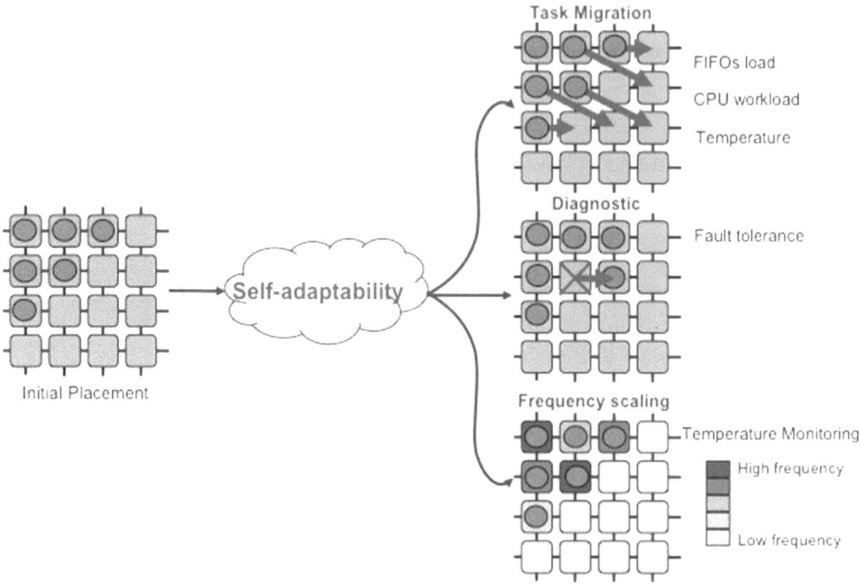

Fig. 4.5 Potential usage of a self-adaptive homogeneous system

3 Self-adaptive and Scalable Systems

In our approach, the infrastructure of a self-adaptive system should enable monitoring of the system, diagnosis, and the optimization process for making the decisions to modify a set of parameters or actuators. Applied to a homogeneous MPSOC system, this infrastructure is presented in Fig. 4.6 and should be completely embedded into the system itself.

In the following sections, we present three contributions to this infrastructure at three levels:

- the design of sensors for process monitoring allowing PVT (Process Voltage Temperature) compensation
- the implementation of a distributed and dynamic optimization inspired by Game-Theory for power optimization
- the implementation of task migration based on software monitors to balance the workload

3.1 Dynamic and Distributed Monitoring for Variability Compensation

To move from fixed integrated circuits to self-adaptive systems, designers must develop reliable integrated structures providing, at runtime, the system (or any PVT

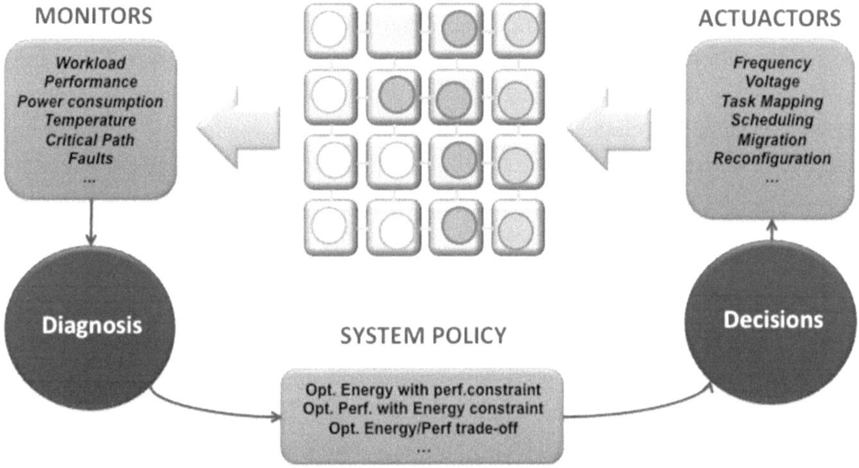

Fig. 4.6 Infrastructure of a self-adaptive system

hardware manager) with trustable and valuable information about the state of the hardware.

Monitoring Clocked System on Chips made of a billion transistors at a reasonable hardware and performance cost is an extremely difficult task for many reasons. Among them one may found, the increasing random nature of some process parameters, the spatial dependence of process (including aging), voltage and temperature variations, but also the broad range of time constants characterizing variations in these physical quantities.

Two different approaches to the monitoring problem can be found in the literature. The first one consists of integrating specific structures or sensors to monitor, at runtime, the physical and electrical parameters required to dynamically adapt the operating frequency and/or the supply voltage and/or the substrate biasing. Several PVT sensors commonly used for post fabrication binning have been proposed in the literature [22–26] for global variability compensation. However, there are some limitations to the use of such PVT sensors.

First, their area and power consumption may be high, so their number has to be limited. Second, their use requires: (a) integration of complex control functions in LUT, and (b) intensive characterization of the chip behavior w.r.t. the considered PVT variables.

Finally, another limitation of this approach concerns the use of Ring Oscillator (RO) structures [22–24] to monitor the circuit speed since RO may be sensitive to PVT variables which are quite different from those of data paths. However, this second limitation can be overcome by adopting a replica path approach, as proposed in [25]. It involves monitoring the speed of some critical paths which are duplicated in the sensors to replace the traditional RO.

The second approach, to compensate for PVT variations and aging effects, is to directly monitor sampling elements of the chip (Latches or D-type Flip Flop) to detect delay faults. This can be achieved by inserting specific structures or using ad-hoc sampling elements [26, 27] to detect a timing violation by performing a delayed comparison or by detecting a signal transition within a given time window.

This approach has several advantages, with the main one being its ability to detect the effects of local and dynamic variations (such local hot spots, localized and brief voltage drops) on timings. A second and significant advantage is the interpretation of the data provided by the sensors, which is simple and binary.

However, this second approach has some disadvantages. One of them is that a high number of sensors might be required to obtain full coverage of the circuit. Therefore, these structures must be as small as possible and consume a small amount of energy when the circuit operates correctly. A second and main disadvantage of such kind of sensors [26, 27] is that error detection requires full replay of the processor instruction at a lower speed. However, this replay is not necessarily possible if the 'wrong' data has been broadcasted to the rest of the chip.

In this setting, a solution is to monitor the timing slack pattern with PVT variations of the critical sampling elements of circuits rather than detecting errors. We thus developed a new monitoring structure, in line with [26–29] concepts, aimed at anticipating timing violations over a wide range of operating conditions. This timing slack monitor, which is compact and has little impact on the overall power consumption, may allow application of dynamic voltage and/or frequency scaling as well as body bias strategies.

Figure 4.7 shows the proposed monitoring system and its two blocks detailed in [30]: the sensor and the specific programmable Clock-tree Cell (CC). The sensor, acting as a stability checker, is intended to be inserted close to the D-type Flip-Flops (DFF) located at the endpoints of the critical timing paths of the design while the CC are inserted at the associated clock leaves. Note that critical data paths to be monitored can be chosen by different means such through the selection of some critical paths provided either by a usual STA (Static Timing Analysis) or a SSTA (Statistical STA).

To validate the monitoring system and its associated design flow, the monitoring system has been integrated, according to [31], in an arithmetic and reconfigurable block of a 45 nm telecom SoC. This block contains about 13400 flip-flops, which leads to a $600 \times 550\ \mu m^2$ core floorplan implementation. Intensive simulations of certain *part* of the arithmetic block demonstrated the efficiency of the monitoring system which allows anticipating timing violations (a) over the full range of process and temperature conditions considered to validate actual designs, and (b) for supply voltage values ranging from 1.2 V to 0.8 V thanks to the programmable CC.

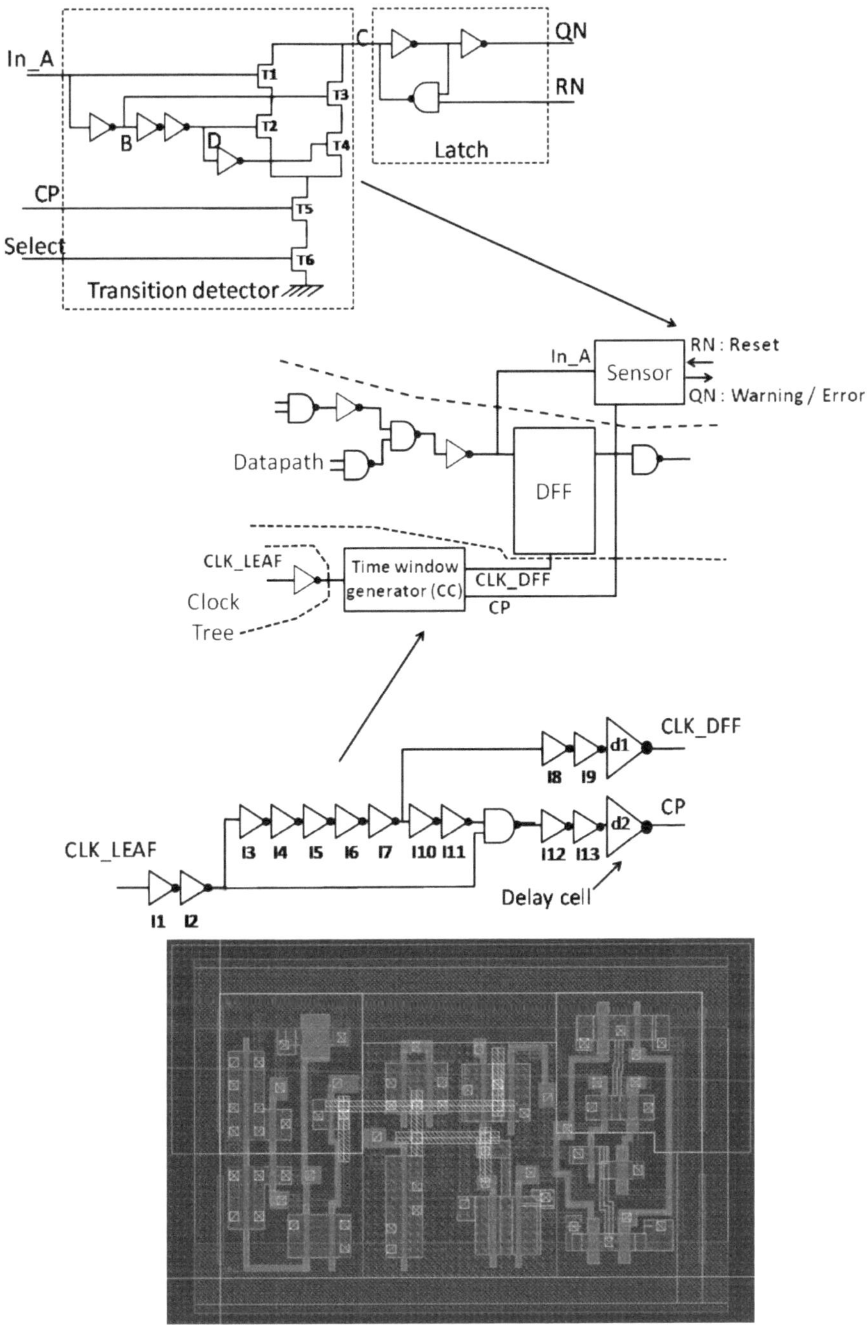

Fig. 4.7 Monitoring system implemented on a single path and the sensor layout in 45 nm technology

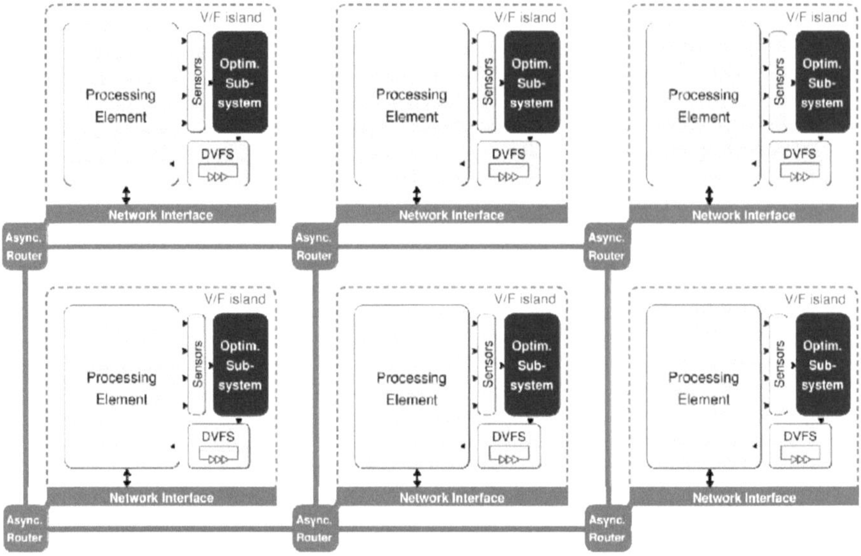

Fig. 4.8 Distributed dynamic optimization of MPSoC

3.2 Dynamic and Distributed Optimization Inspired by Game Theory

3.2.1 Distributed and Dynamic Optimization

Existing methods [32–40], even if they operate at run time, are not based on distributed models. An alternative solution to centralized approaches is to consider distributed algorithms. Our proposal is to design an architecture, as illustrated in Fig. 4.8, where each processing element of an MPSoC embeds an optimization subsystem based on a distributed algorithm. This subsystem manages the local actuators (DVFS in Fig. 4.8) that take the operating conditions into account. In other words, our goal is to design a distributed and dynamic optimization algorithm.

3.2.2 Game Theory as a Model for MPSoC Optimization

Game theory involves a set of mathematical tools that describe interaction among rational agents. The basic hypothesis is that agents pursue well-defined objectives and take their knowledge and behaviors of other agents in the system into account to make their choices. In other words, it describes interactions of players in competitive games. Players are said to be rational since they always try to improve their score or advance in the game by making the best move or action. Game theory is based on a distributed model: players are considered as individual decision makers. For these reasons, game theory provides a promising set of tools to model distributed optimization on MPSOC and, moreover, this is an original approach in this context.

(a) Players analyse the scenario in order to choose the best action (b) Players choose and play their actions

Fig. 4.9 A non-cooperative simultaneous game

As illustrated in [41], a non-cooperative strategic game Γ is composed of a set N of n players, a set of actions per player S_i and the outcomes u_i, $\forall i \in N$. In such a game, players N interact and play through their set of actions S_i in a non-cooperative way, in order to maximize u_i. Consider the non-cooperative game of Fig. 4.9 consisting of 4 players. In Fig. 4.9(a), players analyze the scenario. Each one incorporates all possible information by communicating or estimating it. The information serves to build a picture of the game scenario, to analyze the impact of each possible action on the final personal outcome. Finally, each player chooses the best action that maximizes his/her own outcome. Then, as shown in Fig. 4.9(b), players play their chosen actions and recalculate the outcome. Note that due to a set of interactions and choices of other players, the results are not always the estimated or desired ones. If this sequence is repeated, players have a second chance to improve their outcomes, but then they know the last movements of the others. Thus, players improve their chances of increasing their outcomes when the game is repeated several times. In other words, they play a repetitive game. After a given number of repetitions, players find a Nash equilibrium solution if it exists. At this time, players no longer change their chosen action between two cycles, indicating that they can no longer improve their outcomes.

Consider now that the game objective is to set the frequency/voltage couple of each processing element of the system represented in Fig. 4.10 through the distributed fine-grain DVFS. The figures represent a MPSoC integrating four processing elements interconnected by an NoC. The aim of the frequency selection is to optimize some given metrics, e.g. power consumption and system performance. These two metrics usually depend not only on the local configuration but also on the whole system due to the applicative and physical interactions.

In such scenarios, each processing element is modeled as a player in a game like the one in Fig. 4.9. In this case, the set of players N consists of n tiles of the system ($n = 4$ in the figure). The set of actions S_i is defined by each possible frequency set by the actuator (DVFS). Note that now communications between players are made through the interconnection system. In Fig. 4.10(a), tiles analyze the scenario like in Fig. 4.9(a). They estimate the outcome of each possible action depending on the global scenario in terms of the optimization metrics (energy consumption and performance). The estimation is coded in the utility function u_i. Then, in Fig. 4.10(b), processing elements choose the actions that maximize the outcome. Finally, they execute them, like in Fig. 4.9(b).

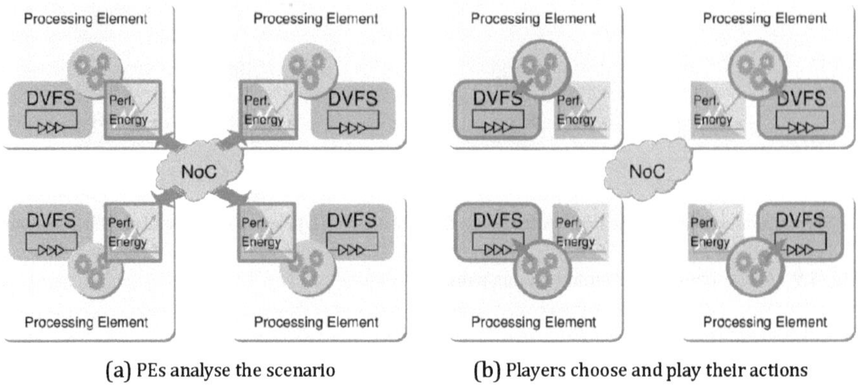

Fig. 4.10 MPSoC modeled as a non-cooperative simultaneous game

MPSoC are distributed architectures. In addition, the presence of distributed actuators such as fine-grain DVFS, justifies the use of a non-cooperative models. These models are based on the principle that decisions are made by each individual in the system. This hypothesis matches the described MPSoC scenario. In MPSoCs, tiles cannot be aware of the state of the whole system and decisions of others, but they have partial information. This is the case of incomplete and imperfect information games. If players do not have a correct picture of the whole scenario, the NE can be hardly reached in the first turn of the game. An iterative algorithm providing several chances to improve the choices will also provide more chances to reach the NE. The distributed nature of MPSoCs also make it hard to synchronize the decision time of all players. In other words, no playing order is set in order to avoid increasing the system complexity. Players are allowed to play simultaneously. For these reasons, our proposal is based on a non-cooperative simultaneous repetitive game.

3.2.3 Scalability Results

The evaluation scenario proposed in [41] illustrates the effectiveness of such techniques. The objective of this proof of concept is to provide a first approach and to characterize its advantages and problems. The metric models used in this formulation are very simple, offering a highly abstracted view of the problem. However, they provide a strong basis for presenting our approach. The statistical study proved the scalability of our method (Fig. 4.11). An implementation based on a well-known microcontroller has highlighted its low complexity. This conclusion comes from an abstracted analysis. In addition, the statistical study showed some deficiencies in terms of convergence percentage, leading to the development of a refined version of the algorithm.

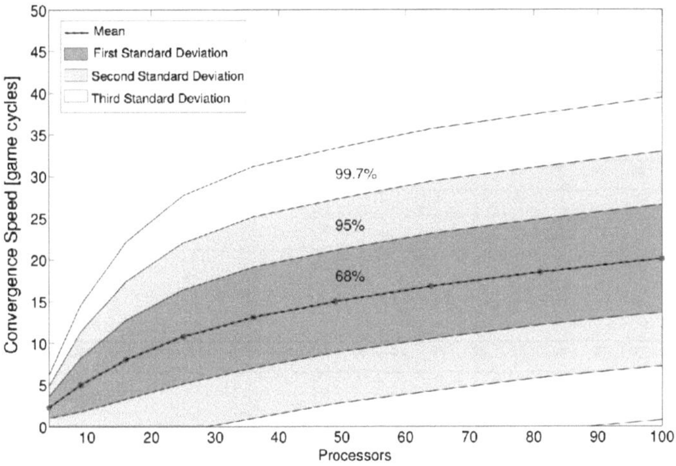

Fig. 4.11 Convergence speed from 4 to 100 PEs

3.2.4 Energy and Latency Optimization Results

In [42] and [43], a new algorithm has been proposed. The number of comparisons per iteration cycle has been markedly reduced, thus simplifying the implementation. The new procedure was examined using four TX 4G telecommunication applications. The results (Fig. 4.12) show that the system adapts the performances when the application changes during execution time. The proposed procedure adapts, in a few cycles, the frequency of each PE. Moreover, when the external constraints (energy and latency bounds) change, the system also reacts by adapting the frequencies. For the tested applications, we have observed improvements of up to 38% in energy consumption and 20% in calculation latency. Compared to an exhaustive optimal search, our solution is less than 5% of the Pareto optimal solution.

3.3 Workload Balancing with Self-adaptive Task Migration

As the key motivations of HS-Scale [19, 20] are scalability and self-adaptability, the system is built around a distributed memory/message passing system that provides efficient support for task migration. The decision-making policy that controls migration processes is also fully distributed for scalability reasons. This system therefore aims at achieving continuous, transparent and decentralized run-time task placement on an array of processors for optimizing application mapping according to various potentially time-changing criteria.

Each NPU has multitasking capabilities, which enable time-sliced execution of multiple tasks. This is implemented thanks to a tiny preemptive multitasking Operating System, which runs on each NPU. Structural (a) and functional (b) views of the NPU are depicted in Fig. 4.13. The NPU is built around two main layers, the

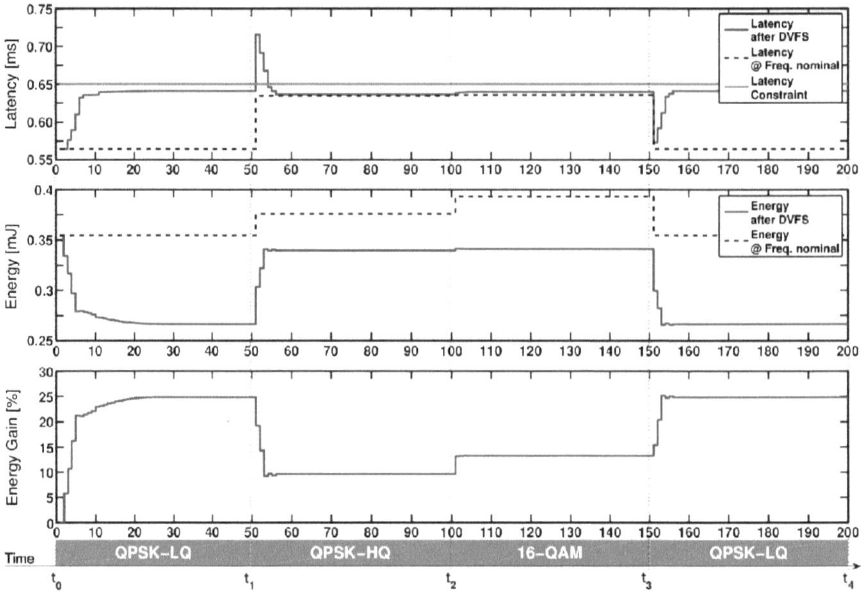

Fig. 4.12 Energy consumption minimization under latency constraints

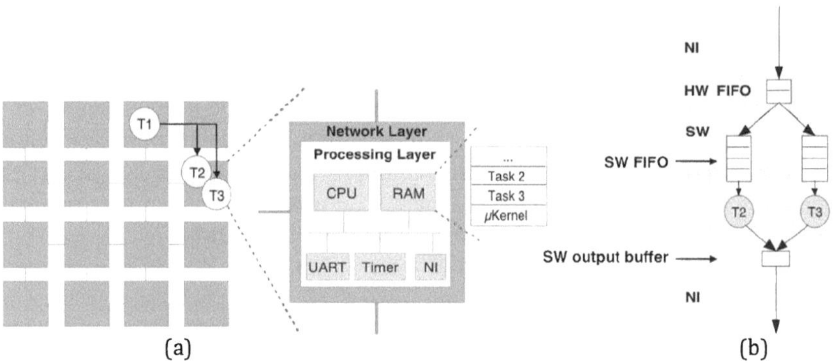

Fig. 4.13 HS-scale principles

network layer and the processing layer. The Network layer is essentially a compact routing engine (XY routing). Packets are read from incoming physical ports, and then forwarded to either outgoing ports or the processing layer. Whenever a packet header specifies the current NPU address, the packet is forwarded to the network interface (NI). The NI buffers incoming data in a small hardware FIFO (HW FIFO) and simultaneously triggers an interrupt to the processing layer. The interrupt then activates data de-multiplexing from the single hardware FIFO to the appropriate software FIFO (SW FIFO), as illustrated. The processing layer is based on a simple and compact RISC microprocessor, its static memory, and a few peripherals (one

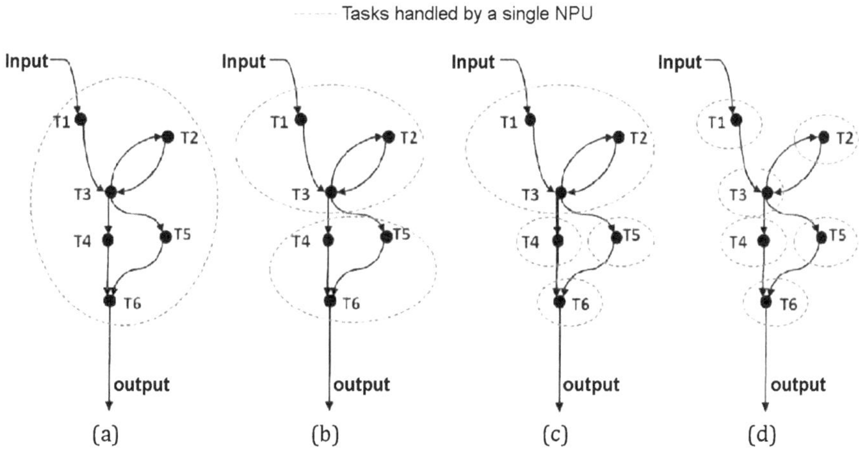

Fig. 4.14 Dynamic task-graph mapping

timer, one interrupt controller, one UART). A multitasking microkernel (μKernel) implements the support for time-multiplexed execution of multiple tasks.

The platform is entitled to make decisions that relate to application implementation through task placement. These decisions are taken in a fully decentralized fashion as each NPU is endowed with equivalent decisional capabilities. Each NPU monitors a number of metrics that drive an application-specific mapping policy. Based on this information, an NPU may decide to push or attract tasks, which results in respectively parallelizing or serializing the corresponding task executions, as several tasks running on the same NPU are executed in a time-sliced manner.

Figure 4.14 shows an abstract example showing that upon application loading the entire task graph runs on a single NPU, subsequent remapping decisions then tend to parallelize application implementation as the final step exhibits one task per NPU. Similarly, whenever a set of tasks become subcritical the remapping could revert to situation (c), where T1, T2 and T3 are hosted on a single NPU while the other supposedly more demanding tasks do not share NPU processing resources with other tasks. These mechanisms help in achieving continuous load-balancing in the architecture but can, depending on the chosen mapping policy, help in refining placement for lowering contentions, latency or power consumption.

3.3.1 Task Migration Policies

Mapping decisions are specified on an application-specific basis in a dedicated operating system service. Although the policy may be focused on a single metric, composite policies are possible. Three metrics are available to the remapping policy for making mapping decisions:

- **NPU load:** The NPU operating system has the capability of evaluating the processing workload resulting from task execution.

- **FIFO queue filling level:** As depicted in Fig. 4.13, every task has software input FIFO queues. Similarly to NPU load, the operating system can monitor the filling of each FIFO.
- **Task distance:** The distance that separates tasks is also a factor that impacts performance, contentions in the network and power consumption. Each NPU microkernel knows the placement of other tasks of the platform and can calculate the Manhattan distance with the other tasks it communicates with.

The code below shows an implementation of the microkernel service responsible for triggering task migration. The presented policy simply triggers task migration in case one of the FIFO queues of a task is used over 80%.

```
void improvement_service_routine(){
   int i, j;
   //Cycles through all NPU tasks
   for(i=0; i < MAX_TASK; i++){
   //Deactivates policy for dead/newly instantiated tasks
      if(tcb[i].status != NEW && tcb[i].status != DEAD){
         //Cycles through all FIFOs
         for(j=0; j < tcb[i].nb_socket; j++){
            //Verifies if FIFO usage > MAX_THRESHOLD
            if(tcb[i].fifo_in[j].average > MAX_THRESHOLD){
               //Triggers migration procedure if task
               //is not already alone on the NPU
               if(num_task > 1)
                  request_task_migration(tcb[i].task_ID);
            }
         }
      }
   }
}
```

The request task migration() call then sequentially emits requests to NPUs in proximity order. The migration function will migrate the task to the first NPU which has accepted the request, the migration process is started according to the protocol described previously in Sect. 4.2. This function can naturally be tuned on an application/task specific basis and select the target NPU while taking not only the distance but also other parameters such as available memory, current load, etc., into account.

We also implemented a migration policy based on the CPU load. The idea is very similar to the first one and it consists of triggering a migration of a given task when the CPU load is lower or greater than a given threshold. This approach may be subdivided in two subsets:

(1) Whenever the tasks time $\geq$ MAX THRESHOLD, this means that tasks are consuming more than or equal to the maximum acceptable usage of the CPU time;
(2) Whenever the tasks time $<$ MIN THRESHOLD, this means the tasks are consuming less than the minimum acceptable usage of the CPU time.

For both subsets, the number of tasks inside one NPU must be verified. For the first subset, it is necessary to have at least two tasks running in the same NPU. For

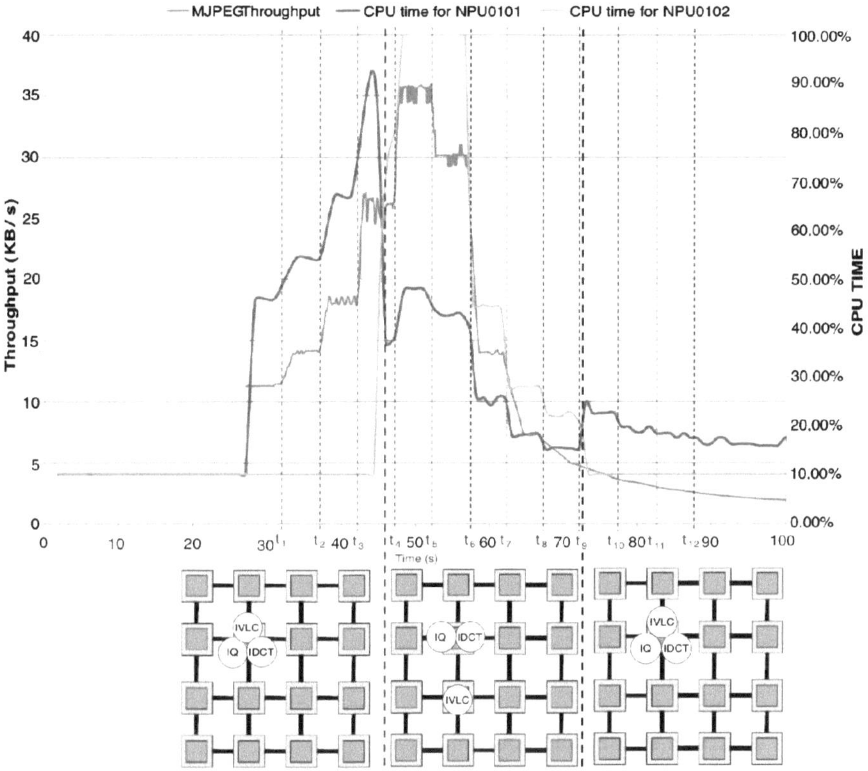

Fig. 4.15 MJPEG throughput with the diagnosis and decision based on CPU workload

the second subset, the migration process may occur whenever there are one or more tasks in the same NPU.

In the same way, the migration process occurs whenever the CPU load is less than MIN_ THRESHOLD (20%). When this occurs, the migration function must look for an NPU that is being used at given CPU usage threshold, i.e. 60% usage in this case. To keep tasks with less than MIN_THRESHOLD from migrating every time, we inserted a delay to reduce the number of migrations.

3.3.2 Results: Task Migration Based on CPU Workload

The example on Fig. 4.15 shows the results of applying this migration policy based on the CPU workload. The experimental protocol used for these results involves varying the input data rate to observe how the system adapts.

At the beginning, all tasks (IVLC, IQ and IDCT) are running on the same NPU(1,1) but the input throughput on the MJPEG application is lower, so the CPU time consumed is around 47%. The input throughput is increased at each step (t_1, t_2 and t_3) so we can see an increase in the CPU time consumed step by step. When the

CPU time used exceeds the threshold (i.e. 80%), the operating system detects that the NPU(1,1) is overloaded (at 45 s), so it decides to migrate the task which uses the most CPU time on a neighboring NPU. In this example, IVLC tasks migrate on NPU(1,2), which decreases the CPU time used by NPU(1,1) by around 35% and increases the CPU used by NPU(1,2) by around 80%. At t_4, the input throughput increases more, which leads to an MJPEG throughput increase of around 35 KB/s and overloads the NPU(1,2) at 100% but no migration is triggered because just one task is computed. From t_5 to t_{12}, the input throughput of the MJPEG application is decreased step by step and, when the CPU time of NPU(1,2) is less than 20% (at 72 s), the operating system decides to move task on the same NPU (the NPU(1,1)). After this migration, we can see a decrease in CPU time used by the NPU(1,2) and an increase in CPU time used by NPU(1,1) but without saturating it. We can observe that the MJPEG application performance is lower than in the static mode because the operating system uses more CPU time (around 10%) to monitor CPU time.

4 Towards Autonomous Systems

The growing interest in pervasive systems that seamlessly interact with their environment motivates research in the area of self-adaptability. Bio-inspiration is often regarded as an attractive alternative to the usual optimization techniques since it provides capability to handle scenarios beyond the initial set of specifications. Such a feature is crucial in multiple domains such as pervasive sensor networks where nodes are distributed across a broad geographical area, thus making on-site intervention difficult. In such highly distributed systems, the various nodes are loosely coupled and can only communicate by means of messages. Further, their architecture may differ significantly as they may be assigned tasks of different natures. One interesting opportunity is to use agent-orientation combined with bio-inspiration to explore the resulting adaptive characteristics.

4.1 Bio-inspiration & Agent-Orientation: at the Crossroads

Programming distributed/pervasive applications is often regarded as a challenging task that requires a proper programming model capable of adequately capturing the specifications. Agent-oriented programming (AOP) derives from the initial theory of agent orientation, which was first proposed by Yoav Shoham [44]. Agent-orientation was initially defined for promoting a social view of computing and finds natural applications in areas such as artificial intelligence or social behavior modeling. An AOP computation consists of making agents interact with each other through typed messages of different natures: agents may be informing, requesting, offering, accepting, and rejecting requests, services or any other type of information. AOP also sets constraints on the parameters defining the state of the agent (beliefs, commitments and choices).

For exploring online adaptability, bio-inspiration appears to be an attractive alternative that has been used for decades in many areas. Optimization techniques such as genetic programming, artificial neural networks are prominent examples of such algorithms. There are several theories that relate to life, its origins and all of its associated characteristics. It is, however, usually considered that life relies on three essential mechanisms, i.e. phylogenesis, ontogenesis and epigenesis [45] (referred to as P, O and E, respectively, throughout this chapter):

- Phylogenesis is the origin and evolution of a set of species. Evolution gears species towards a better adaptation of individuals to their environment; genetic algorithms are inspired from this principle of life.
- Ontogenesis describes the origin and the development of an organism from the fertilized egg to its mature form. Biological processes like healing and fault tolerance are ontogenetic processes.
- Epigenesis refers to features that are not related to the underlying DNA sequence of an organism. Learning as performed by Artificial Neural Networks (ANN) is a process whose scope is limited to an individual lifetime and therefore is epigenetic.

4.2 The Perplexus European Project

The PERPLEXUS European project aims at developing a platform of ubiquitous computing elements that communicate wirelessly and rely on the three above-mentioned principles of life. Intended objectives range from the simulation of complex phenomena such as culture dissemination to the exploration of bio-inspiration driven system adaptation in ubiquitous platforms.

Each ubiquitous computing module (named Ubidules for Ubiquitous Modules) is made of an XScale microprocessor that runs a Linux operating system and a bio-inspired reconfigurable device that essentially runs Artificial Neural Networks (ANN). The resulting platform is schematically described in Fig. 4.16, which shows the network of mobile nodes (MANET) that utilize moving vehicles, and the Ubidules that control them.

4.3 Bio-mimetic Agent Framework

The proposed framework is based on the JADE (Java Agent DEvelopment kit) open-source project. The lower-level mechanisms such as the MANET dynamic routing engine are not detailed here, refer to [46] for a complete description. This section focuses on two fundamental aspects of the proposed BAF: on one hand a description of the BAF and overview of the provided functionality, on the other a description of POE specific agents. Further information on BAF can be found in [47].

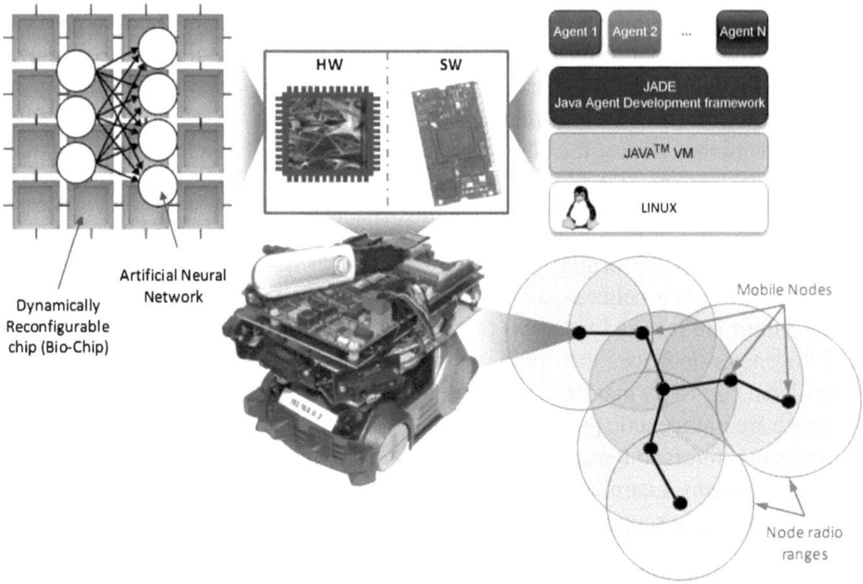

Fig. 4.16 Overview of the Perplexus platform

As bio-inspiration and the three fundamentals of life are at the core of the project, the proposed framework extends JADE default agents by defining agents whose purpose is related to both interfacing and bio-inspired (POE) mechanism support as well as pervasive computing platform management agents.

The BAF specifies 7 agents belonging to 2 families:

- Application agents: Phylogenetic agent(s), Ontogenetic agent(s) and Epigenetic agent(s).
- Infrastructure agents: UbiCom agent(s), Interface agent(s), Network agent(s) and Spy agent(s).

Figure 4.17 shows both the infrastructure and application agents and their interactions (for clarity, JADE-specific agents are omitted):

- P agent: The Phylogenetic agent is responsible for execution of the distributed Genetic Algorithms: it calculates the local fitness of the individual (the actual Ubidule) and synchronizes this information with all other Ubidules. It is responsible for triggering the death (end of a generation) and birth of the embodied individual hosted on the Ubidule.
- O agent: The Ontogenetic agent is tightly coupled to the P agent: it takes orders from this agent and has the capability of creating other software agents (in case of full software implementation).
- E agent: The Epigenetic agent embodies the individual and its behavior: it is a software or hardware neural network.

Fig. 4.17 BAF agents at the Ubidule-level

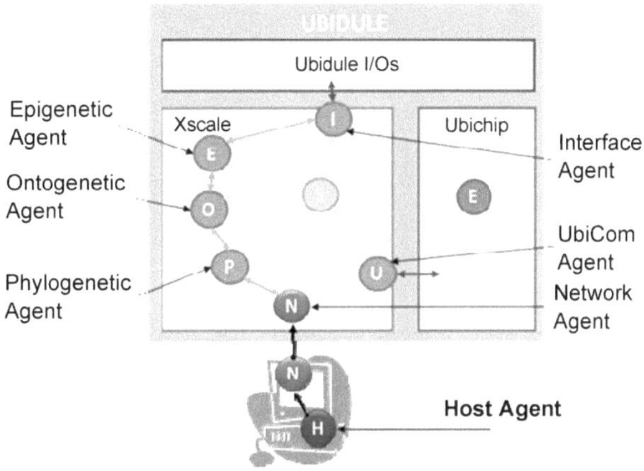

Next to the three POE agents, there are four additional agents for interfacing and networking purposes:

- I agent: The Interface agent provides a set of methods for issuing commands to the actuators or retrieving data from the Ubidule sensors.
- U agent: The UbiCom agent provides software API-like access to the Ubichip and manages hardware communications with the chip.
- S agent: The Spy agent provides information on the platform state (agent status/results, activity traces, bug notification).
- N agent: The Network agent provides a collection of methods for network-related aspects: time-synchronization of data among Ubidules, setting/getting clusters of Ubidules, obtaining a list of neighbors, etc. As it requires access to low-level network-topology information, it also implements MANET functionalities.

Finally, a Host agent (H agent) instantiated on a workstation allows remote control of the PERPLEXUS platform (Start/Stop/Schedule actions).

4.4 Application Results: Online Collaborative Learning

Figure 4.18 schematically depicts the robots used, their sensors and actuators, as well as the framework agents presented previously. Robots use online learning (Epigenesis) to improve their performance. Robots are enclosed in an arena scattered with obstacles (collision avoidance is the main objective here). As this application only targets learning, the P and O agents are not used here.

Besides the three front sensors that return the distance to the nearest obstacle, a bumper switch is added to inform the robot whenever a collision with an object occurs; it is located on the front side of the robot. These robots move by sending speed commands on each of the two motors. As depicted in Fig. 4.18 an Artificial Neural Network (ANN) controls the robot movement: the E agent is a multi-layer Perceptron ANN that uses a standard back-propagation learning algorithm.

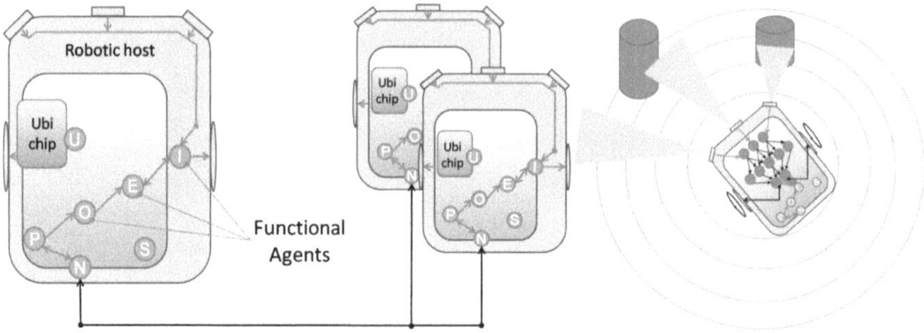

Fig. 4.18 Mapping agents onto the robots and overview of the obstacle avoidance application

Inputs of the ANN are the three values measured by the sensors, five areas have been defined for each sensor, area 0 means that an obstacle is present within a distance of less than 200 mm, subsequent areas are 200 mm deep, therefore enabling detection of objects at up to 800 mm distance. The ANN outputs are speed values sent to the two motors, with each being set as an integer value from -7 to $+7$, -7 being the maximum negative speed of a wheel (i.e. fast backward motion), and $+7$ being with the maximum positive speed of a wheel (i.e. fast forward motion). The robot can turn by applying two different speeds on the motors.

Robots are moving in an unknown environment. Each time they collide into an obstacle, a random modification of the relevant learning pattern is applied and an ANN learning phase is triggered online. The robot then notifies all its peers that this pattern shall be modified, and the modification is registered by all robots, therefore collectively speeding up convergence toward a satisfactory solution.

Our experiments show that this technique exhibits a speedup (versus a single robot) that is almost linear with the number of robots used. Furthermore, it has been observed that a convergence threshold is reached after a number of iterations, which is a function of the complexity of the environment. Once this threshold is reached, adding more obstacles in the arena retriggers learning until a new threshold is reached, thus demonstrating the adaptability potential of the proposed solution. Further experiments presented in [48] utilizing evolutionary techniques also show promising results (demonstration videos are available at http://www.lirmm.fr/~brousse/Ubibots).

5 Conclusion

Not only technology but also the rapidly widening spectrum of application domains raises a number of questions that challenge design techniques and programming methods that have been used for decades. Particularly, design-time decisions prove inadequate in a number of scenarios because of the unpredictable dimension of the environment, technology and applicative requirements that often give rise to major scalability issues.

Techniques that rely on assessing the system state and adapting at run-time appear attractive as they relieve designers of the burden of devising tradeoffs that perform reasonably well in a chosen set of likely scenarios.

This chapter stresses two important guidelines that are believed to be the cornerstone to the design of systems for the decade to come: self-adaptability and distributiveness. To this end, the presented work stressed the associated benefits of systems that comply with these two rules.

In some cases in which the scope of monitored parameters is limited, the results are remarkable as they permit to achieve significant improvements with limited overhead.

For the most ambitious techniques that rely on completely distributed decision-making based on heuristics, experiments show promising results but also highlight some limitations such as suboptimality and uncertainty. Such techniques will nevertheless be unavoidable for very-large scale systems that currently only exist in telecommunication networks.

We believe that there is no single technique that will answer every requirement but it is rather important to promote a panel of tools that will be made available to designers for devising systems tailored for a particular application area.

References

1. Complex systems and agent-oriented software engineering. In: Engineering Environment-Mediated Multi-Agent Systems. Lecture Notes in Computer Science, vol. 5049, pp. 3–16. Springer, Berlin (2008)
2. Borkar, S.: Thousand core chips: a technology perspective. In: Annual ACM IEEE Design Automation Conference, pp. 746–749 (2007)
3. Wolf, W., Jerraya, A., Martin, G.: Multiprocessor System-on-Chip (MPSoC) technology. IEEE Trans. Comput.-Aided Des. Integr. Circuits Syst. **27**(10), 1701–1713 (2008)
4. Freescale Semiconductor, Inc.: C-5 Network Processor Architecture Guide, 2001. Ref. manual C5NPD0-AG. http://www.freescale.com
5. Dutta, S., Jensen, R., Rieckmann, A.: Viper: A multiprocessor SOC for advanced set-top box and digital TV systems. IEEE Des. Test Comput. **18**(5), 21–31 (2001)
6. Texas Instruments Inc.: OMAP5912 Multimedia Processor Device Overview and Architecture Reference Guide, 2006. Tech. article SPRU748C. http://www.ti.com
7. Guerrier, P., Greiner, A.: A generic architecture for on-chip packet-switched interconnections. In: DATE '00: Proceedings of the 2000 Design, Automation and Test in Europe Conference and Exhibition, pp. 250–256 (2000)
8. Dally, W.J., Towles, B.: Route packets, not wires: on-chip interconnection networks. In: DAC '01: Proceedings of the 38th Conference on Design Automation, pp. 684–689. ACM, New York (2001)
9. Benini, L., De Micheli, G.: Networks on chips: a new SoC paradigm. Computer **35**(1), 70–78 (2002)
10. Bjerregaard, T., Mahadevan, S.: A survey of research and practices of Network-on-Chip. ACM Comput. Surv. **38**(1), 1 (2006)
11. Pande, P.P., Grecu, C., Jones, M., Ivanov, A., Saleh, R.: Performance evaluation and design trade-offs for network-on-chip interconnect architectures. IEEE Trans. Comput. **54**(8), 1025–1040 (2005)
12. Bertozzi, D., Benini, L.: Xpipes: a network-on-chip architecture for gigascale systems-on-chip. IEEE Circuits Syst. Mag. **4**(2), 18–31 (2004)

13. Beigne, E., Clermidy, F., Vivet, P., Clouard, A., Renaudin, M.: An asynchronous NOC architecture providing low latency service and its multi-level design framework. In: ASYNC '05: Proceedings of the 11th IEEE International Symposium on Asynchronous Circuits and Systems, pp. 54–63. IEEE Comput. Soc., Washington (2005)
14. Pontes, J., Moreira, M., Soares, R., Calazans, N.: Hermes-GLP: A GALS network on chip router with power control techniques. In: IEEE Computer Society Annual Symposium on VLSI, ISVLSI'08, April 2008, pp. 347–352 (2008)
15. Ogras, U.Y., Marculescu, R., Choudhary, P., Marculescu, D.: Voltage-frequency island partitioning for GALS-based Networks-on-Chip. In: DAC '07: Proceedings of the 44th Annual Conference on Design Automation, pp. 110–115. ACM, New York (2007)
16. Donald, J., Martonosi, M.: Techniques for multicore thermal management: Classification and new exploration. In: ISCA '06: Proceedings of the 33rd International Symposium on Computer Architecture, pp. 78–88 (2006)
17. Beigne, E., Clermidy, F., Miermont, S., Vivet, P.: Dynamic voltage and frequency scaling architecture for units integration within a GALS NoC. In: NOCS, pp. 129–138 (2008)
18. Beigne, E., Clermidy, F., Miermont, S., Valentian, A., Vivet, P., Barasinski, S., Blisson, F., Kohli, N., Kumar, S.: A fully integrated power supply unit for fine grain DVFS and leakage control validated on low-voltage SRAMs. In: ESSCIRC'08: Proceedings of the 34th European Solid-State Circuits Conference, Edinburgh, UK, Sept. 2008
19. Saint-Jean, N., Benoit, P., Sassatelli, G., Torres, L., Robert, M.: Application case studies on HS-scale, a mp-soc for embedded systems. In: SAMOS'07: Proceedings of the International Conference on Embedded Computer Systems: Architectures, Modeling, and Simulation, Samos, Greece, July 2007, pp. 88–95 (2007)
20. Saint-Jean, N., Sassatelli, G., Benoit, P., Torres, L., Robert, M.: HS-scale: a hardware-software scalable mp-soc architecture for embedded systems. In: ISVLSI '07: Proceedings of the IEEE Computer Society Annual Symposium on VLSI, pp. 21–28. IEEE Comput. Soc., Washington (2007)
21. ITRS Report/Design 2009 Edition, http://www.itrs.net/Links/2009ITRS/2009Chapters_2009Tables/2009_Design.pdf
22. Nourani, M., Radhakrishnan, A.: Testing on-die process variation in nanometer VLSI. IEEE Des. Test Comput. **23**(6), 438–451 (2006)
23. Samaan, S.B.: Parameter variation probing technique. US Patent 6535013, 2003
24. Persun, M.: Method and apparatus for measuring relative, within-die leakage current and/or providing a temperature variation profile using a leakage inverter and ring oscillators. US Patent 7193427, 2007
25. Lee, H.-J.: Semiconductor device with speed binning test circuit and test method thereof. US Patent 7260754
26. Abuhamdeh, Z., Hannagan, B., Remmers, J., Crouch, A.L.: A production IR-drop screen on a chip. IEEE Des. Test Comput. **24**(3), 216–224 (2007)
27. Drake, A., et al.: A distributed critical path timing monitor for a 65 nm high performance microprocessor. In: ISSCC 2007, pp. 398–399 (2007)
28. Das, S., et al.: A self-tuning DVS processor using delay-error detection and correction. IEEE J. Solid-State Circuits **41**(4), 792–804 (2006)
29. Blaauw, D., et al.: Razor II: In situ error detection and correction for PVT and SER tolerance. In: ISSCC 2008, pp. 400–401 (2008)
30. Rebaud, B., Belleville, M., Beigne, E., Robert, M., Maurine, P., Azemard, N.: An innovative timing slack monitor for variation tolerant circuits. In: ICICDT'09: International Conference on IC Design & Technology (2009)
31. Rebaud, B., Belleville, M., Beigne, E., Robert, M., Maurine, P., Azemard, N.: On-chip timing slack monitoring. In: IFIP/IEEE VLSI-SoC—International Conference on Very Large Scale Integration, Florianopolis, Brazil, 12–14 October 2009, paper 56
32. Niyogi, K., Marculescu, D.: Speed and voltage selection for GALS systems based on voltage/frequency islands. In: ASP-DAC '05: Proceedings of the 2005 Conference on Asia South Pacific Design Automation, pp. 292–297. ACM, New York (2005)

33. Deniz, Z.T., Leblebici, Y., Vittoz, E.: Configurable on-line global energy optimization in multi-core embedded systems using principles of analog computation. In: IFIP 2006: International Conference on Very Large Scale Integration, Oct. 2006, pp. 379–384 (2006)
34. Deniz, Z.T., Leblebici, Y., Vittoz, E.: On-Line global energy optimization in multi-core systems using principles of analog computation. In: ESSCIRC 2006: Proceedings of the 32nd European Solid-State Circuits Conference, Sept. 2006, pp. 219–222 (2006)
35. Murali, S., Mutapcic, A., Atienza, D., Gupta, R.J., Boyd, S., De Micheli, G.: Temperature-aware processor frequency assignment for MPSoCs using convex optimization. In: CODES+ISSS '07: Proceedings of the 5th IEEE/ACM International Conference on Hardware/Software Codesign and System Synthesis, pp. 111–116. ACM, New York (2007)
36. Murali, S., Mutapcic, A., Atienza, D., Gupta, R.J., Boyd, S., Benini, L., De Micheli, G.: Temperature control of high-performance multi-core platforms using convex optimization. In: DATE'08: Design, Automation and Test in Europe, Munich, Germany, pp. 110–115. IEEE Comput. Soc., Los Alamitos (2008)
37. Coskun, A.K., Simunic Rosing, T.J., Whisnant, K.: Temperature aware task scheduling in MPSoCs. In: DATE '07: Proceedings of the Conference on Design, Automation and Test in Europe, pp. 1659–1664. EDA Consortium, San Jose (2007)
38. Coskun, A.K., Simunic Rosing, T.J., Whisnant, K.A., Gross, K.C.: Temperature-aware MP-SoC scheduling for reducing hot spots and gradients. In: ASP-DAC '08: Proceedings of the 2008 Conference on Asia and South Pacific Design Automation, pp. 49–54. IEEE Comput. Soc., Los Alamitos (2008)
39. Ykman-Couvreur, Ch., Brockmeyer, E., Nollet, V., Marescaux, Th., Catthoor, Fr., Corporaal, H.: Design-time application exploration for MP-SoC customized run-time management. In: SOC'05: Proceedings of the International Symposium on System-on-Chip, Tampere, Finland, November 2005, pp. 66–73 (2005)
40. Ykman-Couvreur, Ch., Nollet, V., Catthoor, Fr., Corporaal, H.: Fast multi-dimension multi-choice knapsack heuristic for MP-SoC run-time management. In: SOC'06: Proceedings of the International Symposium on System-on-Chip, Tampere, Finland, November 2006, pp. 195–198 (2006)
41. Puschini, D., Clermidy, F., Benoit, P., Sassatelli, G., Torres, L.: A game-theoretic approach for run-time distributed optimization on MP-SoC. International Journal of Reconfigurable. Computing, ID(403086), 11 (2008)
42. Puschini, D., Clermidy, F., Benoit, P.: Procédé d'optimisation du fonctionnement d'un circuit intégré multiprocesseurs, et circuit intégré correspondant. Report No. PCT/FR2009/050581 32, France (2009)
43. Puschini, D., Clermidy, F., Benoit, P., Sassatelli, G., Torres, L.: Dynamic and distributed frequency assignment for energy and latency constrained MP-SoC. In: DATE'09: Design Automation and Test in Europe (2009)
44. Shoham, Y.: Agent oriented programming. Artif. Intell. **60**, 51–92 (1996)
45. Sanchez, E., Mange, D., Sipper, M., Tomassini, M., Perez-Uribe, A., Stauffer, A.: Phylogeny, ontogeny, and epigenesis: three sources of biological inspiration for softening hardware. In: Higuchi, T., Iwata, M., Liu, W. (eds.) Evolvable Systems: From Biology to Hardware. LNCS, vol. 1259, pp. 33–54. Springer, Berlin (1997)
46. Bellifemine, F.L., Caire, G., Greenwood, D.: Developing Multi-Agent Systems with JADE. Wiley, New York (2007)
47. Brousse, O., Sassatelli, G., Gil, T., Guillemenet, Y., Robert, M., Torres, L., Grize, F.: Baf: A bio-inspired agent framework for distributed pervasive applications. In: GEM'08, Las Vegas, July 2008
48. Sassatelli, G.: Bio-inspired systems: self-adaptability from chips to sensor-network architectures. In: ERSA'09, Las Vegas, July 2009

Chapter 5
On Software Simulation for MPSoC

A Modeling Approach for Functional Validation and Performance Estimation

Frédéric Pétrot, Patrice Gerin, and Mian Muhammad Hamayun

1 Introduction

The recent advances in Very Large Scale Integration (VLSI) technology allow to integrate close to a billion transistors on a single chip. In this context, the development of dedicated hardware is close to impossible, and IP reuse is the rule. Still, the simplest solution for the hardware designer is to put many programmable IPs (processors) on a chip. The famous quote of Chris Rowen: "*The processor is the nand gate of the future*" is indeed being put into practice, as more than 1000 processors are expected to be integrated on a chip by 2020 according to the ITRS roadmap [15], (see Fig. 5.1).

In the integrated systems field, power and yield issues require the use of architectures that must be power-efficient, and thus usually include different types of processors (heterogeneity) and specialized hardware. These circuits, application specific multiprocessors or Multi-Processors System on Chip (MPSoC), are used in many industrial sectors, such as telecommunications, audio/video, aerospace, automotive, military, and so on, as they provide a good power vs computation trade-off. A typical MPSoC circuit contains several CPUs, possibly of different types, that are programmed in an *ad hoc* manner for optimization purposes. The typical architecture is a set of CPU sub-systems, possibly Symmetric Multiple Processor (SMP), organized around a shared interconnect, as illustrated in Fig. 5.2.

The design of new application-specific integrated systems using the ASIC design flow leads to unacceptable cost and delays, because the software part is ignored and

F. Pétrot (✉) · P. Gerin · M.M. Hamayun
TIMA Laboratory, CNRS/Grenoble-INP/UJF, Grenoble, France
e-mail: Frederic.Petrot@imag.fr

P. Gerin
e-mail: Patrice.Gerin@imag.fr

M.M. Hamayun
e-mail: Mian-Muhammad.Hamayun@imag.fr

G. Nicolescu et al. (eds.), *Design Technology for Heterogeneous Embedded Systems*,
DOI 10.1007/978-94-007-1125-9_5, © Springer Science+Business Media B.V. 2012

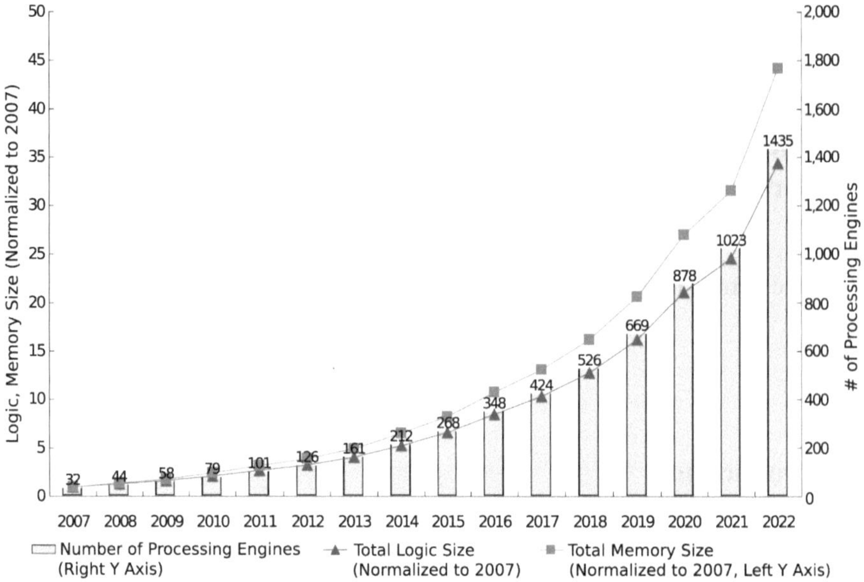

Fig. 5.1 SOC consumer portable design complexity trends

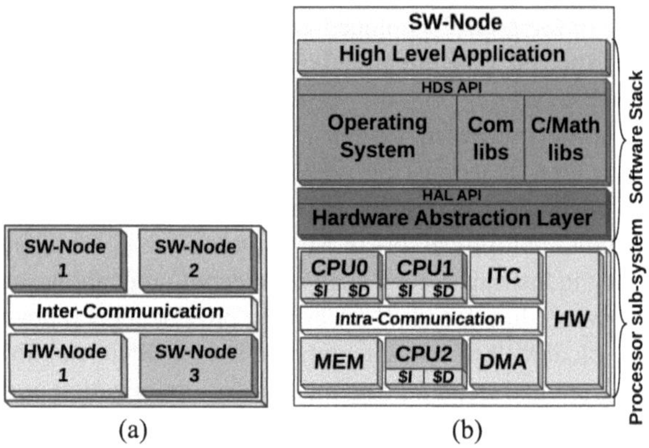

Fig. 5.2 (**a**) MPSoC architecture and (**b**) Software Node sub-system

implemented once the circuits and boards are available. Therefore, new methods that rely on the heavy use of simulation are being developed and transferred to the industry [10, 12]. Such methods are already in use for products today. Simulation relies on models and it can be understood that if the models are close to reality, the simulation will be slow, and if they are abstract, the accuracy of the results may be disputable. However, if a hardware model is used for early software development, it can be functionally accurate without timing and be very useful.

Timing estimation is useful for system dimensioning and applications for which the timing is part of the functionality. The timing performance of a software application depends on multiple static and dynamic factors. Static sources of timing are

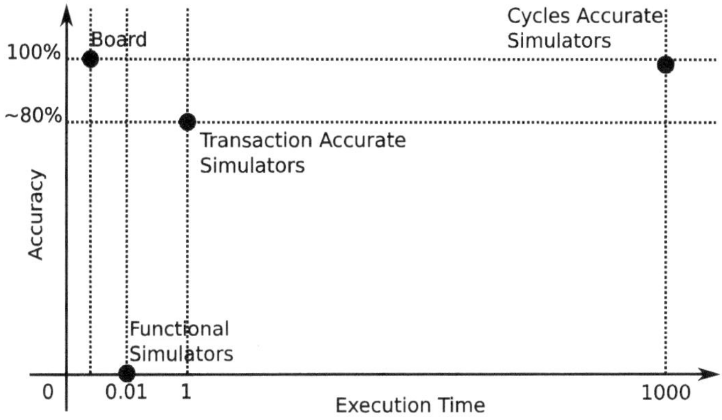

Fig. 5.3 Time vs accuracy comparison of simulation models

mainly given by the instruction set architecture and can be analyzed at compilation time, without needing to execute the program. Uniprocessor Worst Case Execution Time (WCET) techniques are useful in this context but their timing estimates are usually far above (usually at least a factor of 2) the actual execution time, which is not acceptable for consumer products for which cost is a major issue. The dynamic aspect depends on the execution state of the system: contention on the interconnect, pipeline state, cache contents and so on, and can be measured only at runtime.

The rest of the chapter details models for system simulation in which the software is untimed, and then an automatic way of annotating software, to build functional and timing accurate hardware/software system models.

2 System Simulation Models

Software Simulation Models are used to model embedded systems before they are developed and deployed in reality. Broadly speaking these software simulation models can be divided into three types. Each one of these has its own advantages and shortcomings. Figure 5.3 shows a coarse *Time vs Accuracy* comparison of these models.

2.1 Cycle Accurate Simulators

A Cycle-Accurate Bit-Accurate (CABA) simulator is similar in goal to an RTL simulator, but achieves its accuracy using more abstract, Finite State Machine (FSM) based, models. The advancement of simulation time is not due to events occurring on the inputs of gates, but to the edges of a clock shared by the model. This allows for much faster simulation speed, as it is, in general, possible to statically schedule the model execution order and thus avoid events propagation [14]. CABA simulators may make use of Instruction Set Simulator (ISS) to execute software. An Instruction Set Simulator (ISS) is a simulation model which reads the instructions for a specific processor and maintains internal variables which represent the processor's registers. These models may provide an exact execution time estimate (or a quite good one), and are thus considered the best way of proving the timing behavior of a given embedded system. The only bottleneck limiting the usage of these simulators is the speed of simulation, which is inherently very slow because the simulation model takes into account the very low level details of the system. Speed can be a bit higher by loosing some precision, but it is hardly fast. It takes days of processing to complete the simulation of a single complex design solution, which effectively limits their usage for the design space exploration of many possible combinations.

2.2 Functional/Behavioral Simulators

The Functional/Behavioral simulation models lie on the other extreme and provide means to verify the functional properties of a given software system. Simulations using these models are very fast as most of the code executes in *zero-time* and timing information is not part of the simulation framework. This allows to cover the interleaving of tasks execution that would not be easy to produce by other means, and thus features a nondeterminism, due to the possible randomness of the tasks scheduling in the simulator, that is useful for functional validation. However, timing properties cannot be verified using this level of abstraction.

2.3 Transaction Accurate Simulators

Transaction Accurate (TA) models, for which the hardware and communication protocol are not modeled at the bit-level but instead abstractly and that communicate using transactions instead of wire-driven FSMs, are currently in use for early software validation [4]. Although platforms based on TA models are much faster than the ones based on CABA models, they still use ISSs, leading to lengthy simulations when numerous processors are simulated.

To interpret instructions abstractly, three other possibilities have been investigated:

1. The first one is to rely on dynamic binary translation techniques such as the one developed for virtualization technologies. A well documented and open-source example of this approach is available in [2], and its usage in MPSoC simulation is detailed in [13]. Indeed, the binary translation approach can also be used with CABA models for the rest of the system (including cache controllers), but the speed gain is not significant because many synchronizations are necessary at this level.

2. The second one, a flavor of the so called *native* simulation approach [1, 3, 6, 9, 17], consist of wrapping the application code into a SystemC hardware module that represents a processor. This technique is not generalized, as the software (possibly multi-threaded) is tied to a processor and thus it is not possible to mimic the behavior of SMP systems. Furthermore, wrapping the application software requires that some sort of abstract operating system exists and is implemented using the hardware simulator primitives.
3. The third solution is also a *native* simulation approach, but it does not wrap the application software into SystemC hardware modules. Instead, it relies on the fact that SystemC simulation modules can provide very low level primitives, identical to the concept of Hardware Abstraction Layer (HAL) API of an operating system [22]. So, by compiling all the software layers except the HAL implementation for the host computer and have this software execute on-top of the TA models of the hardware through HAL function calls, it is possible to have a fast software simulation.

Using simulations at higher levels of abstraction, it is possible to get a considerable speed-up as compared to the Cycle-Accurate simulations and it also provides a relatively straightforward integration in system simulation environments making it suitable for early design space exploration and validation. However, it is necessary to compromise on the accuracy of the results achieved using these simulations for getting the improvement in terms of simulation performance.

In the rest of this chapter, we focus on the third, more innovative, solution.

3 Principle of Native Execution of Software on TLM-Based Platforms

The CPU Subsystem node models, which are the focus of TA Level, provide the execution environment for the MPSoC application software tasks. Based on the principle of hardware abstraction layer as used in operating systems (OS), to keep the core of OS independent of a specific hardware platform, the node models provide a similar abstraction in addition to the hardware TLM components.

As SystemC has become the preferred development language for MPSoC modeling, we base our explanations throughout this chapter on SystemC TLM.

The TA abstraction can be done only if a well defined layered software is used with predefined application programming interface (API). Starting from this assumption, the lowest abstraction level which can be implemented for native execution has to provide the well known Hardware Abstraction Layer API as illustrated in Fig. 5.4. This API is a set of functions that allow the software to interact with the hardware devices. The HAL is especially important for designing portable OSes for different hardware platforms. This portability remains valid for the Transaction Accurate model because there is no way to access the hardware except through the HAL.

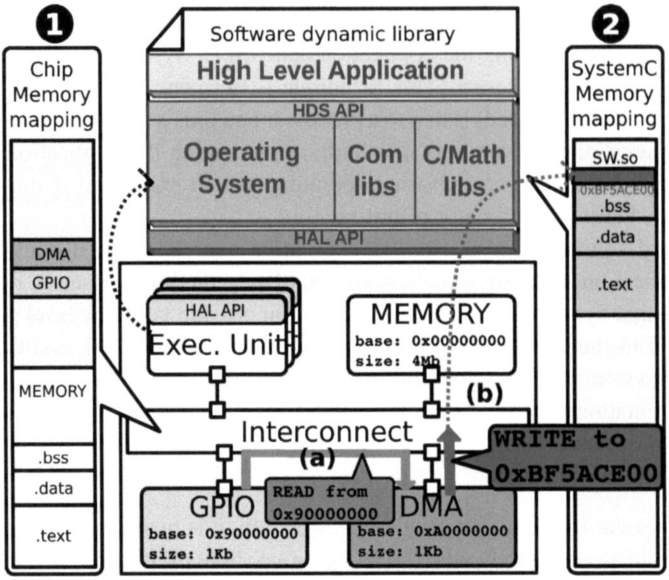

Fig. 5.4 Native execution on TLM-based platforms principle

3.1 Software Execution

SystemC is a hardware oriented language and doesn't provide any construct or method to implement and simulate sequential execution of software. We advocate the use of SystemC only for hardware modeling as opposed to most of the OS modeling works that wrap software threads into SystemC threads. In this case, the software execution is supported by an Execution Unit (EU) (represented in Fig. 5.4 and detailed in Fig. 5.5) which acts as a processor. Therefore it is effectively a hardware model that executes software.

The EU is the only component where software implementation is allowed (Fig. 5.5(a)). This software corresponds to the EU specific HAL API implementation and provides, among others, context switching, interrupt management and I/O access functions. The hardware part of the EU (Fig. 5.5(b)) is composed of a single SystemC thread (`SC_THREAD`) which calls the software entry point. Here the software is an executive containing all software layers, *i.e.* application, libraries and middleware and OS linked together. The start routine is the usual processor and platform bootstrap, as opposed to the usual HW/SW co-simulation [7, 11] approaches where software tasks are wrapped in SystemC modules and OS services implemented in SystemC. From this point on, the software executes natively until it calls a HAL API function, which translates into a read/write access to a port or signal of the modeled hardware.

The key interest of EU based approach lies in the fact that it allows to model Symmetric Multiple Processor like architectures, on which the migration of tasks can occur between identical processors under the control of OS. This cannot be

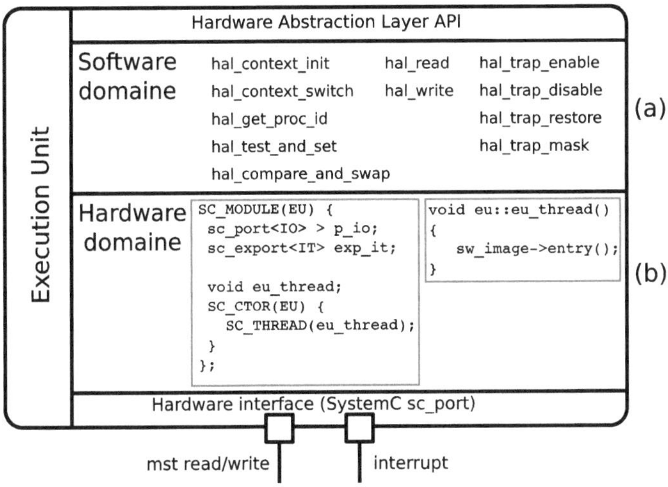

Fig. 5.5 Execution Unit architecture

done by the wrapping techniques that in many cases solely allow a single thread per processor. The above sketched solution allows to model any kind of multiprocessor architecture (heterogeneous and SMP), as it mimics a real architecture and thus the support for multiprocessor depends only on the OS and the hardware capabilities to handle low level synchronization techniques. Since multiple EUs can *boot* the same application, SMP or derived architectures are naturally supported as well as interrupts.

However, solely relying on the HAL API abstraction does not provide a working environment, as the native compilation of the software leads to memory representations that are different in the native (software) and simulated (hardware) parts. A scheme to handle this discrepancy is detailed below.

3.2 Unified Memory Representation

By default, two memory mappings have to be considered in native simulation as depicted in Fig. 5.4:

Platform memory mapping: It is defined by the hardware designer and used by the platform interconnect address decoder at simulation time.

SystemC memory mapping: It is shared by the SystemC process resulting from the compilation of the hardware models and the natively compiled application.

Mixing both memory mappings is the solution that has been chosen in many MP-SoC simulation environments [7, 20]. However this technique is not applicable when software and hardware have to interact closely, for example as it is the case when Direct Memory Access (DMA) devices are used, *i.e.* practically speaking in *all* platforms.

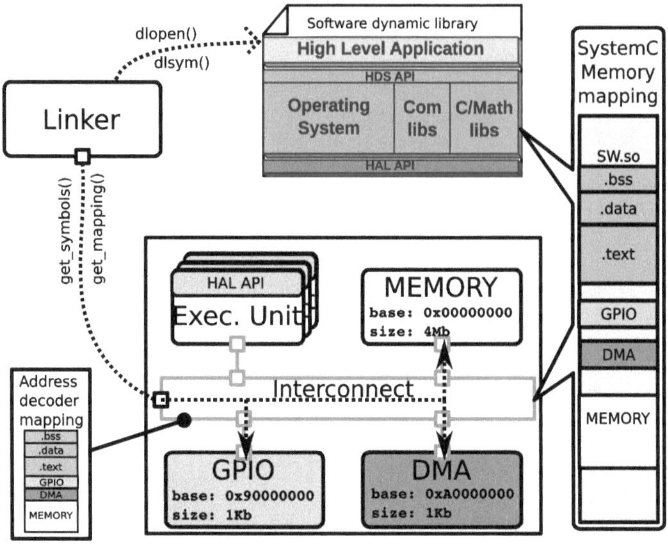

Fig. 5.6 TLM-based platforms with a unified memory representation

Let's consider a DMA transfer from a General Purpose Input/Output device (GPIO) to a memory region allocated by the native software application (Fig. 5.4). The DMA is configured with a source address of `0x90000000` (as it is the GPIO device input register address) and a destination address of `0xBF5ACE00` (which is a natively allocated buffer).

1. Since the GPIO addresses are valid in the platform memory space, the DMA can access source data through the interconnect.
2. The destination address is not valid (even though it may be valid if the hardware platform and native host address spaces collide) in the platform interconnect address decoder. Thus the DMA must not access the destination address as it is defined.

Usually remapping techniques are employed in such cases but they cannot deal with the overlapping between platform and software memory. To solve this problem, native approaches should rely on a consistent and unique memory mapping which is shared from the point-of-view of hardware and software.

The idea is to use the SystemC memory mapping as the unique memory space, as it is shared with the software application, as shown in Fig. 5.6, because the component models themselves are allocated by the SystemC process (Unix-wise). Instead of using a user defined memory mapping for each of the hardware components of the platform, it is the address of the field of the C++ class that models the component that will be used in the interconnect address decoder. Thus, the previous DMA example is automatically supported.

Figure 5.7 clarifies the application memory mapping (❷) within SystemC memory space. The application, along with the OS and all required libraries, is compiled

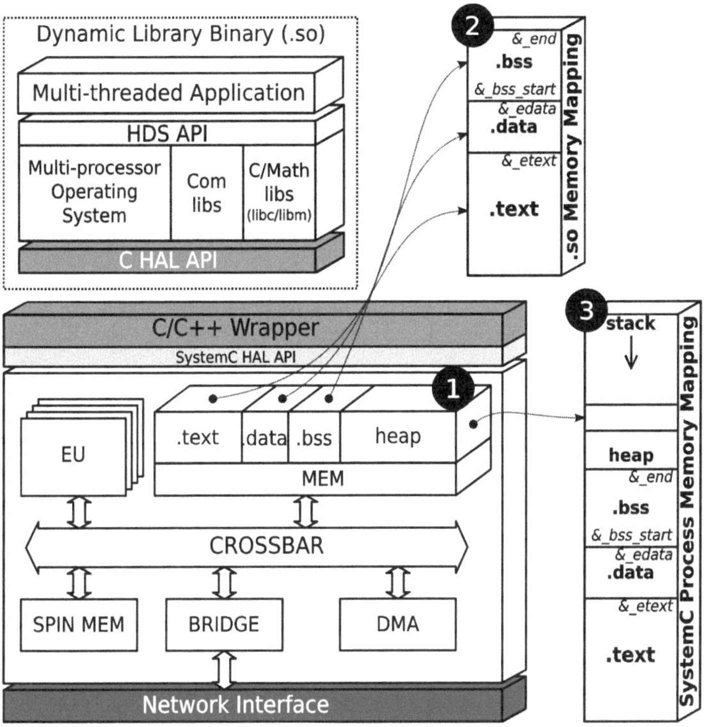

Fig. 5.7 Application and SystemC memory mapping

into a single dynamic library (`.so`) which interacts with the underlying hardware platform model using the SystemC HAL API Layer. The different memory segments of the dynamic library, (the most typical ones being *.bss*, *.data* and *.text*), are attached to the simulated memory component model when the simulator is loaded into memory, as shown in ❶. Each of these segments is identified by a start and an end address which is retrieved by an introspection tool that updates the decoding tables of the interconnect in order to construct the unified memory mapping as shown in part ❸ of the figure. This strategy substitutes the original absolute base addresses of the simulated systems components behind the scene, and thus allows the access of addresses that targets a program segment through the interconnect.

This solution enables to accurately model realistic memory hierarchies, but requires some specific implementation strategies.

3.3 Dynamic Software Linkage

The key issue in using SystemC process memory mapping concerns the base addresses of the memory sections and device registers which are known only when the simulation is ready to start, so precisely speaking at the end of elaboration phase. In existing platforms, these addresses are known by the low-level programmer and are commonly hard-coded as shown in Program 1, line 1.

Program 1: Hard coded access to memory

```
1 *(volatile uint32_t*)0x90000000 |= 0x000000010;
```

A simple solution is to delay the address resolution of hardware devices and memory sections till the initialization of simulation. This can be done using the `extern` declaration of program 2, line 1, and by demanding the driver or OS programmer to use the specific functions (lines 4 and 6) to access the devices.

Program 2: Access to memory based on link-time defined address

```
1 extern volatile uint32_t *GPIO_BASE;
2 ...
3 uint32 status;
4 status = HAL_READ(UINT32, GPIO_BASE+0x04);
5 status |= 0x000000010;
6 HAL_WRITE(UINT32,GPIO_BASE, status);
```

Instead of linking the application according to the hardware memory mapping at the end of the native compilation, a dynamic linker (represented in Fig. 5.6) builds the memory mapping of the simulated platform during the elaboration phase and then resolves the unknown addresses in the software before simulation starts. This technique requires specific support from the hardware component models.

The SystemC slave components must provide the following two methods:

- `get_mapping()`: that returns a list of memory regions occupied by the slave component and defined in terms of base address, size and name.
- `get_symbols()`: that returns a list of symbols defined by the (name, value) pair, which will be resolved in the software application.

At the end of the platform elaboration, the dynamic linker calls the bus or NoC component's `get_mapping()` and `get_symbols()` methods which dispatch this call to all the connected slave components. Using this information, the interconnect builds the platform memory map for address decoding. The linker finally obtains the complete list of memory regions accessible through the interconnect model and the list of all the symbols that must be resolved in the application.

The implementation of such a hardware component in SystemC is exemplified in Program 3. In the constructor function (lines 8 to 18) the memory segment allocated by the GPIO component is initialized and added to the segments list (lines 11 and 12).

Program 3: Slave component implementation example

```
1 #include "systemc.h"
2 SC_MODULE(GPIO)
3 {
4   public:
5     // IO interfaces methods
6     void read(uint32_t *addr, uint32_t *data);
7     void write(uint32_t *addr, uint32_t data);
8     GPIO(sc_module_name name) : sc_module(name)
9     {
10      segment_t *gpio_segment;
11      gpio_segment = new segment_t("GPIO", _registers, 1024);
12      _segments.push_back(gpio_segment);
13
14      symbol<uint32_t*> * gpio_symbol;
15      gpio_symbol = new symbol<uint32_t*> ("GPIO_BASE");
16      gpio_symbol.value = _registers;
17      _symbols.push_back(gpio_symbol);
18    }
19    ~GPIO();
20    std::vector< segment_t*> *get_mapping() {return _segments;}
21    std::vector< symbol_t*> *get_symbols() {return _symbols;}
22  private:
23    std::vector<segment_t*> _segments;
24    std::vector<symbol_t*>  _symbols;
25    uint32_t                _registers[256];
26 };
```

Similarly, the symbols that are to be resolved in the software application are declared and added to the symbols list (lines 14 and 17). In this example, the value of this symbol is supposed to be the base address of the GPIO component registers in memory. Thus the address of the modeled registers allocated in the SystemC process mapping (line 25) is directly assigned to the symbol value (line 16).

Finally, the linker has to set this value to the actual `GPIO_BASE` symbol in the software. To do this, we use the standard POSIX `dlsym` function. This function returns the address at which a symbol is loaded into memory. Program 4 presents a simplified implementation of the linking phase.

The SystemC kernel calls the `start_of_simulation` method of the linker component at the end of platform elaboration phase. The `dlopen` function is used to load the application which gives a handle named `_sw_image`. This application handle is then used in `dlsym` function to obtain the addresses of all unresolved symbols as shown in (line 9). The real implementation has to handle different types of symbols (only `uint32_t` type is handled in this example) and multiple software images.

The interest of this solution is its relative implementation simplicity and its capability to model realistic hardware/software interaction. Hierarchical system interconnect are also supported which allows to model complex architectures with

Program 4: Sketch of the dynamic linking phase

```
void linker::start_of_simulation()
{
  uint32_t i;
  std::vector< symbol_t* > *symbols;
  uint32_t                 *sym_ptr;

  symbols = p_linker->get_symbols();
  for(i= 0 ; i< symbols->size(); i++) {
    sym_ptr = (uint32_t*)dlsym(_sw_image,(*symbols)[i]->_name);
    *sym_ptr = (*symbols)[i]->_value;
  }
}
```

multiple software nodes, each one containing multiple processors, as depicted in the introduction Fig. 5.2. Furthermore, the slave component support for dynamic linking is not intrusive and can be easily added to already existing components.

3.4 Limitations

The approach has some limitations that are worth mentioning, as it opens interesting research perspectives.

- First of all, legacy code containing hard coded addresses, assembly code, and implicit access to memory (bypassing of the HAL layer) cannot be used in this approach.
- Secondly, in some circumstances it is useful to be able to run self modifying code (for example when the dynamic linker resolves a dynamic reference to a function). Even though this applies only to sophisticated OSes, they are now available for embedded devices as well. Thus a solution using an abstract API or some other technique needs to be defined.
- Thirdly, the ability to estimate multiprocessor software performance is useful to take early dimensioning decisions. Even though the native execution approach was primarily targeting functional verification, it is possible to obtain fairly accurate performance estimates. This is the topic of the rest of this chapter.

4 Software Performance Estimation

Estimating the software performance on the native code seems at first infeasible, as the host and target machine may have quite different architectures. The host may include a floating point unit whereas the target may not, the pipeline structure may be very different (simple 5 stage pipeline for the target vs out-of-order superscalar pipeline with several issue slots for the host machine). The overall compilation framework, even if it is the same, will make different optimization choices based on the hardware resources and thus generate a binary code that will have a very different structure.

This being said, it seems clear that performance estimation of the software itself can be done only on the code generated for the target machine. The overall execution delay is computed using the software estimated delays for the target machine and the interactions that take place with the hardware when loads and stores are issued.

Low Level Virtual Machine (LLVM) compilation infrastructure has been used and modified for the implementation of the technique described in this section.

4.1 Performance Estimation Principle for Native Simulation

The binary code generation is based on an intermediate representation called a Control Flow Graph (CFG), a graph whose nodes are sequences of instructions such that there is exactly one entry point at the start of the sequence and a branch at the end (called basic blocks) and whose arcs represent execution dependencies.

The software delay, as opposed to the memory access delays, is obtained by annotating the basic blocks generated by the compiler and gathering the time (or any other physical quantity) at execution. The way to have a meaningful graph of basic blocks, *i.e.* a graph whose traversal behaves at run time on the host as it would do on the target, is to use a cross-intermediate representation approach that constructs and maintains an equivalent CFG between the native and target objects. In this scheme the annotation process is separated from the performance estimation problem and the main idea is to directly annotate the compiler Intermediate Representation (IR) format. Here it should be clear that this scheme is different from Dynamic Binary Translation (DBT) schemes as it inserts the annotation function calls to the IR format and it does not do any kind-of binary translation as the resultant binary is *already* generated for the native machine.

Similar to a classical compiler infrastructure, the language specific front ends for C/C++ are used to transform the source files into the IR. Then the IR is provided as input to the target specific back-ends, which generate the binary code with the help of code emitters. It is possible to replace the final code emitter pass by an annotation pass. However it is necessary to maintain the Cross-IR equivalent, to the target specific IR, throughout the back-end in-order to keep track of the target specific CFG transformations done during optimization. When the annotation pass is included with the compilation passes, it annotates the cross-IR, which is then passed to the native back-end as input for generating native object. This native object can then be simulated on the native simulation platform. Figure 5.8 shows how this scheme works.

It is possible to summarize the timing estimation problem with the following questions:

- How many cycles are needed to execute the current target basic block instructions without external dependencies?

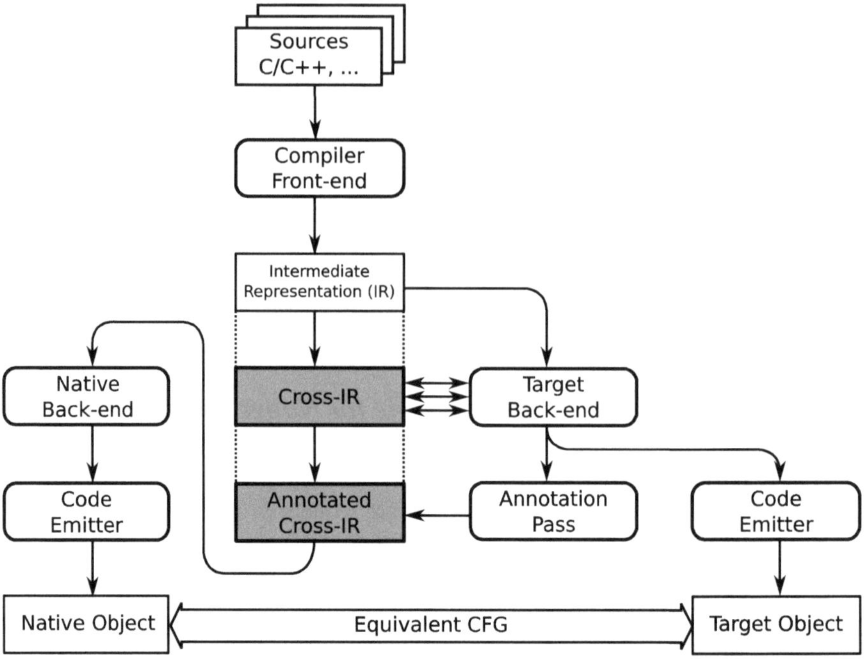

Fig. 5.8 Compiler infra-structure with extended IR for native annotation

- How the cost of execution of a basic block is affected given that it may have multiple predecessor basic blocks and some of them may or may not affect it?
- How many instructions have been fetched from memory to execute this basic block and their addresses?
- How many data memory locations have been accessed for this basic block alongwith their types (read/write) and addresses?

The first two requirements are concerned with the static analysis of each basic block. Each basic block is analyzed independently and the number of cycles required to execute each instruction on the target processor are estimated. This gives an approximate cost of instructions for each basic block. We know that dependencies can exist between different instructions and they can delay the execution of instructions. For example in Fig. 5.9 two basic blocks are shown. The first basic block shows instruction dependencies that should be considered during the estimation process. The first type-of dependency exists when a loaded register is immediately used in the next instruction. Similarly when multiple registers are loaded in an instruction and the last loaded register is used in the next instruction, then extra cycles should be added to the estimation measure. The second basic block shows the dependencies that do exist but do not affect the estimation measure (in case of simple RISC processors). Such dependencies exist when the loaded registers are not used in the immediately next instruction. Similarly, the dependencies do not affect the estima-

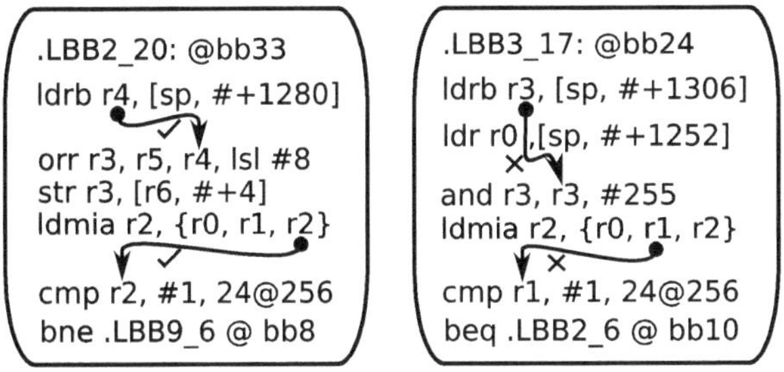

Fig. 5.9 Instruction dependency analysis

tion measure when multiple registers are loaded using an instruction and any of the loaded registers, except the last one, is used in the immediately next instruction.

The last two requirements are more concerned with the dynamic analysis of the performance estimation problem and will be useful for the higher level simulation framework. Firstly the number of instructions that will be fetched from memory for the execution of basic block has to be calculated, in order to see their effect on the performance estimate. Also the locality of basic block instructions in memory may be useful for cache modeling. Lastly, analyzing the number of data memory accesses for each basic blocks will be useful, in case of multiple memory banks and multiple processors.

4.2 Annotation with Static Analysis in Native Simulation

For the sake of clarity, we have based our description on the LLVM compiler infrastructure that is well suited for performing this kind of analysis. The annotation part acts as driver for the target specific implementation of instruction analysis. It produces data-bases that are subsequently used for annotation purposes. Here are the details of the process.

1. **Analyze Basic Block:** Each target specific basic block corresponding to the target independent basic block is analyzed to extract the relevant information for the performance estimation. This includes the number of instructions, number of cycles required to execute these instructions, extra cycles incurred due to dependencies and branch penalty analysis (for relatively simple RISC processors).
2. **Store Database:** A data structure for the database entry is created in the target specific LLVM module and the analysis results are stored in this database. At the end of the analysis, a pointer is returned to the target independent LLVM module for accessing this database entry.
3. **Annotate Basic Block:** In the target independent LLVM module each native basic block is annotated by inserting a call to the `annotate` function before the first instruction of each basic block. The unique argument to this function, which is implemented in the hardware model supporting the native software execution, is the address where the target basic block database is stored in memory.

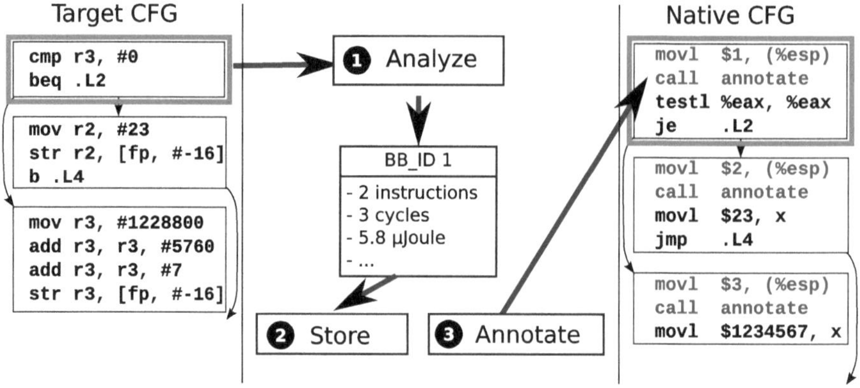

Fig. 5.10 The annotation process

Finally the annotated cross-IR is used as input for the host processor back-end in order to obtain the native object file. See Fig. 5.10 for a visual representation of the annotation process.

Figure 5.11 shows how this annotation information is used at runtime. Each of the basic block has a call to the `annotate` function which it executes once this basic block is entered. Although all basic blocks have been annotated with estimation information, a given execution flow will only execute a certain subset of these basic blocks. The estimation databases of only this subset is taken into account for a particular execution flow. It is also evident that basic blocks can execute more than once and in such cases their corresponding databases will be added-up multiple times to the estimation measure. In the given example only db1, db2, db5 and db6 will be considered for estimation purpose. The total number of instructions and cycles for this execution flow will be $12+5+14+4=35$ instructions and $17+5+18+7=47$ cycles.

This approach ensures a perfect match between the target and native software in terms of CFG execution, which is certainly not a guarantee of timing accuracy but assures similar execution flows that reflect the dynamic aspects of software. For example the data-dependent software execution where the input data determines the control flow of the software at runtime. As already mentioned, the number of processor clock cycles required to execute a given piece of software depends on two independent parts:

1. The internal architecture of the processor on which the software is executed.
2. The external architecture of the hardware platform.

4.3 Internal Architecture Annotation

The internal architecture annotation assumes an ideal architecture, where data and instruction memory accesses take *zero time*. In such an architecture the number of

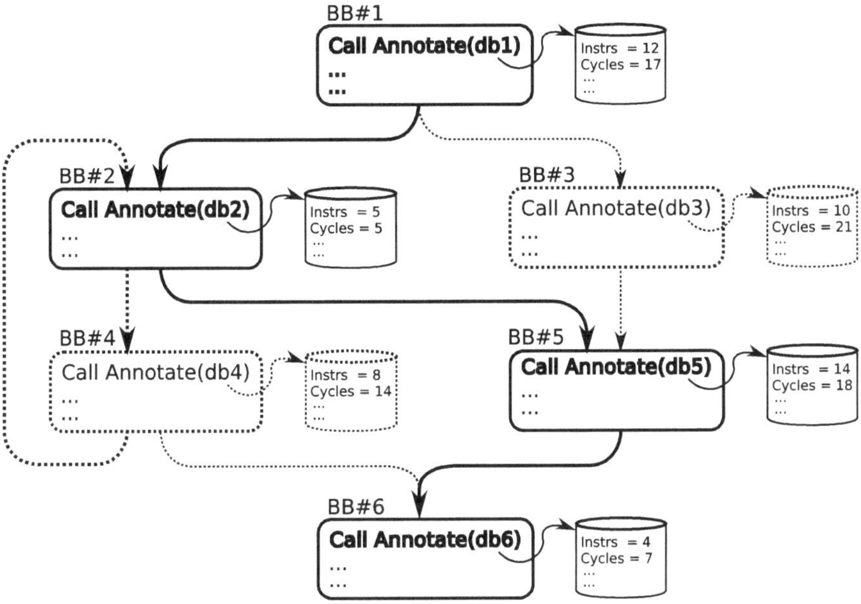

Fig. 5.11 Estimation information usage at simulation time

clock cycles needed to execute a basic block instruction sequence constitute a constant and a variable part. The constant part depends only on the instructions and can be easily determined by static analysis and using the given processor data-sheet. Whereas the accuracy of the variable part of this estimation depends on the processors internal architecture complexity, for example the re-ordering of instructions, branch prediction etc. In RISC processors, there are three major factors which have influence on software performance estimation:

- Instruction Operands
- Inter-instruction Dependencies
- Branch Misprediction Penalties

Using more complex processors would lead to a much greater list of factors. To illustrate the analysis, we take the example of the ARM-9 ISA.

Instruction Operands Every instruction requires some specific number of cycles to execute on a given processor. These cycles not only depend on the instruction class but also on the number, type and value of the operands used in the instruction. We can analyze the number and types of operands at static analysis time, but the values of these operands arc usually unknown so the native source code annotation does not support their analysis. In such cases we consider the worst-case execution cycles for instruction timing estimate. (Here we must not confuse the use of term "worst-case" with Analytical WCET Techniques). Multiply instruction is one such example in the ARM-9 processor where its execution time depends on the value of the operands supplied to it (see Table 5.1 and instruction #4 in Fig. 5.12).

Table 5.1 ARM-9 multiplier early termination

Syntax	$MUL\langle cond\rangle\{S\}Rd, Rm, Rs$
Semantics	$Rd = Rm \times Rs$
Instruction cycles	$2 + M$
Where	$M = 1$ for $-2^8 \leq Rs < 2^8$
	$M = 2$ for $-2^{16} \leq Rs < 2^{16}$
	$M = 3$ for $-2^{24} \leq Rs < 2^{24}$
	$M = 4$ for $-2^{32} \leq Rs < 2^{32}$

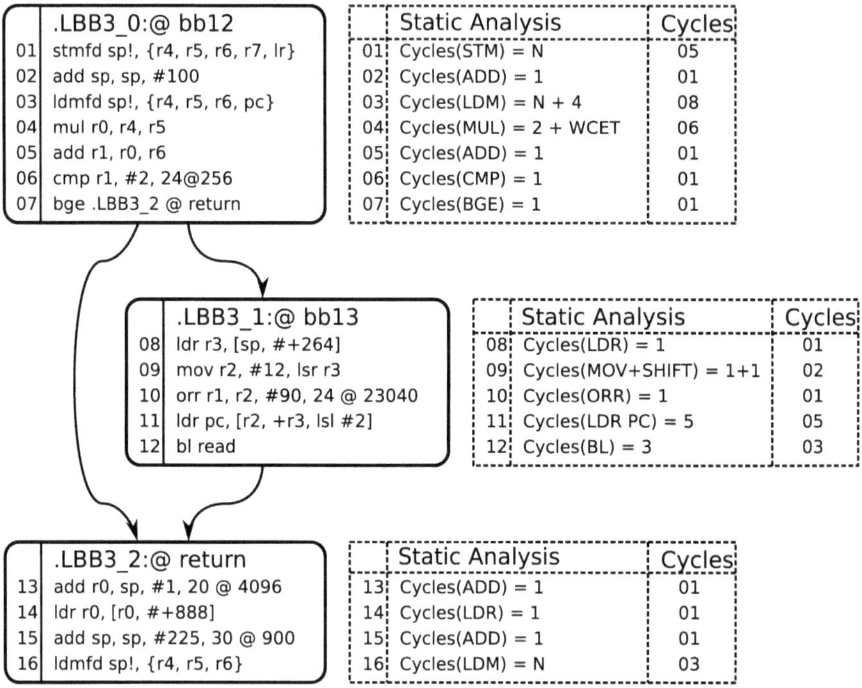

Fig. 5.12 Internal architecture instruction annotation

Similarly the number of operands and their types can also affect the instructions timing estimation. For example in Fig. 5.12 the instructions #1, #3 and #16 are such cases. And further special cases arise when the program counter is also modified in such instructions which is equivalent to a jump instruction and requires 4 extra cycles for execution.

Similarly we have to count one extra cycle if any instruction uses a register specified shift operation in addition to the primary operation. Like the instruction #9 uses a similar shift operation *i.e.* Logical Shift Right (LSR). Similar cases exist for arithmetic and load/store instructions as well, and the number of extra cycles incurred depend on the instruction class.

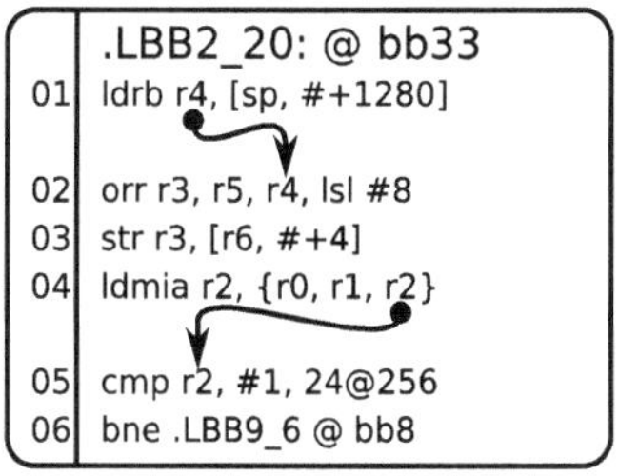

	Basic Cycles	Dep. Cycles	Total
01	Cycles(LDR) = 1	0	01
02	Cycles(ORR+SHIFT) = 2	2	04
03	Cycles(STR) = 1	0	01
04	Cycles(LDM) = 3	0	03
05	Cycles(CMP) = 1	1	02
06	Cycles(B) = 1	0	01

Fig. 5.13 Instruction dependencies and their effect on estimation

Inter-instruction Dependencies In order to achieve higher performance, processors use pipelines and multiple functional units on which several instructions can be executed simultaneously. In this case the execution of an instruction can start only if all of its input data sources have been evaluated by prior instructions. In classical embedded processors, the instruction latency time due to these interlocks between two instructions can be easily determined statically and corresponds to a constant number of extra cycles.

In Fig. 5.13 we can see a dependency relationship between instructions #1, #2 and instruction #4, #5. The first dependency is due to the usage of register $r4$ in the immediately next instruction. Similarly the next dependency is due to the usage of last loaded register $r2$ in the next instruction. The analysis table in the same figure shows the number of instruction-class based cycles and extra cycles incurred due to these dependencies. These extra dependency cycles reflect the stalling of instruction pipeline in the given processor.

In modern processors, instruction execution may start even if the input data is not yet available. Reservation techniques are also used and allow instructions to reserve hardware resources for future use. Using techniques that target the computation of the WCET of basic blocks in super-scalar processors with dynamic instruction scheduling, such as the one described in [21] is a way to address this issue.

Branch Misprediction Penalties During the software execution, branch instructions lead to a non-negligible number of additional clock cycles which can be determined at runtime only. These additional clock cycles depend on whether the branch is *taken* or *not taken*.

The branch prediction policy varies from one processor to another and can be defined using fixed, static or dynamic branch prediction schemes. The ARM-9 processor does not implement any type of branch prediction at all, so we can consider this situation equivalent to *not-taken* branch prediction policy. In this case the processor performs the fetch and decode stages on instructions that follow the branch instruction and discards them if the branch is actually *taken* resulting in a branch misprediction penalty (in the equivalent sense).

Figure 5.14(a) represents a simple graph of basic blocks with two branch instructions. Depending on the branch prediction, the arcs of the target processor CFG have to be annotated by the insertion of additional basic blocks on the corresponding

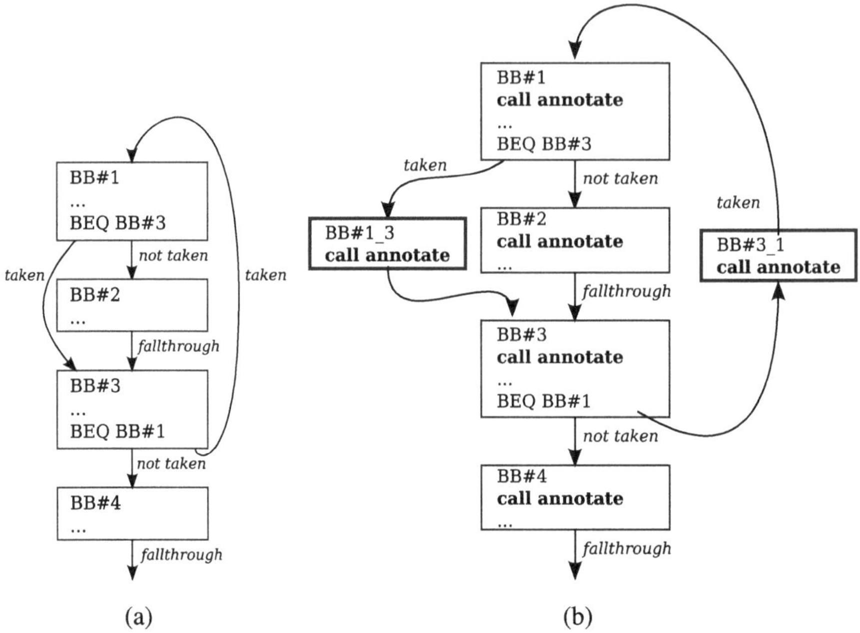

Fig. 5.14 (**a**) Target processor basic block CFG and (**b**) Host processor equivalent CFG with branch penalty annotation

arcs of the host machine CFG. This solution does not affect the software behavior and even-if target and host basic block CFGs are not *isomorphic* anymore, they are still *equivalent* when we consider the instructions executed by the execution path because the newly inserted basic blocks contain the annotation calls only. In Fig. 5.14(b) the host machine CFG is annotated according to the *not taken* branch prediction policy. Thus the misprediction penalties have been added to the *taken* arcs.

From the implementation point of view, we analyze every successor of each target basic block for the possibility of acting as a branch target. In cases where any successor target basic block satisfies this condition, we create a database entry for this path and pass this information to the target independent LLVM module. This module then creates an extra basic block and adds it to the path between the two LLVM target independent basic blocks. This new basic block will be executed each time this branch is *taken* and annotation information will be added in the performance estimation measure.

This type of simple branch prediction is typically used in classical RISC processors, which are the type of processors commonly used in SoCs. More complex prediction techniques like bimodal, local or global may require annotation on both *taken* and *not taken* paths of the CFG and if necessary some support from the underlying hardware model can be added.

4.4 External Architecture Annotation

The processor external architecture is also responsible for a non-negligible number of additional clock cycles required for the instructions execution. These additional clock cycles are spent during the instruction fetches and data memory accesses and lastly I/O operations add considerable number of extra cycles as well.

As the annotation process is independent from the platform on which the software is executed, timing information cannot be collected according to external architecture of the platform using the target processor basic blocks. However, static analysis can extract relevant information from the target processor basic blocks which can assist the underlying hardware model in the estimation process at simulation runtime.

This section briefly describes the approach that we are advocating for our future work and enlists the key ideas that we would be taking into account for the implementation.

Instruction Memory Accesses Another important contributor to the software execution cost is the number of cycles spent for accessing the program instructions from the memory. In most of the current abstract software execution time estimation techniques [5, 8, 16, 18, 19], this sort of timing information is not taken into account. In the approach described here, it is possible to bring these instruction fetch related costs into the performance estimation calculation as well.

As the basic blocks contain only consecutive instructions, the base address of the basic block and its size in the target architecture represents enough information for the underlying hardware platform to model memory accesses. Since the hardware platform model uses the host machine memory mapping, the native base address of the basic block is needed in order to model the accesses to the correct corresponding memory in the hardware platform.

Data Memory Accesses Evaluating data memory accesses latency is mandatory for estimating the performance of software. In comparison with program instructions the data locality is more segregated in nature and knowing the location of one data item does not provide any useful information about the locality of other data items. Furthermore, the locality of data in the memory hierarchy is an important contributor to the data access timings as well. Due to this diversity of data accesses, using a precise annotation scheme would drastically slowdown the simulation speed as each of the memory access (Read/Write) will have to be annotated by an annotation function call. Realistically speaking, for many of the data accesses, the memory addresses needed by the EU cannot be known at the static analysis time. For example, when data memory locations are reached using pointers, their target addresses cannot be known in advance.

A way to handle this issue stems from the consideration that all the memory accesses of the target cross-compiled code also exist in a way in the compiler cross-IR representation that the difference of architecture between target and host processor will result in a different number of memory accesses for the same variable. This is largely due the architectural differences between the two machine types and typically the native execution of a program on an x86 processor will generate more memory accesses. The main reason for this phenomena is the reduced number of registers in comparison to the RISC processors like ARM, Mips and Sparc, which are commonly used in embedded systems. The static analysis extracts the number of data memory accesses (Reads/Writes) in the basic block cross-IR and applies heuristics depending on the target processors to estimate the number of equivalent memory accesses on the target processor platform. The addresses of these accesses are known in the native software simulation and are still valid in the TA memory model.

5 Summary

In this chapter we have reviewed the most commonly accepted techniques for high level simulation of hardware and software in multiprocessor systems. In particular we have focused on a *native* simulation approach in which the software is not cross-compiled and executed on an ISS, instead it is compiled on the host machine and dynamically linked to an event driven simulator executable. This is possible only because the access to the hardware resources is done through the HAL layer that provides a well defined and mandatory API. It is then possible to build hardware simulation models for execution units that implement this API and thus provide the necessary support for OS, middleware and application execution.

Thanks to the higher level of simulation speed in this approach, we can perform early functional validation of the given software system. However, currently it is not possible to determine runtime estimates and it needs further work. The performance analysis on native code can be done by guaranteeing that the execution path of the native program is equivalent to the one that would take place in the target program. However to ensure this, we require access to the internals of a retargettable compiler. Using the intermediate representation of a compiler, it is possible to perform basic block level code annotation, in order estimate instructions count, processor cycles or any other required information. The hardware model can be enhanced with pipeline and cache models to provide a complete framework for simulating an MPSoC from a global perspective.

References

1. Bacivarov, M., Yoo, S., Jerraya, A.A.: Timed HW-SW cosimulation using native execution of OS and application SW. In: HLDVT '02: Proceedings of the Seventh IEEE International High-Level Design Validation and Test Workshop, p. 51. IEEE Comput. Soc., Washington (2002)
2. Bellard, F.: Qemu, a fast and portable dynamic translator. In: USENIX 2005 Annual Technical Conference, FREENIX Track, pp. 41–46 (2005)
3. Benini, L., Bertozzi, D., Bruni, D., Drago, N., Fummi, F., Poncino, M.: SystemC cosimulation and emulation of multiprocessor SoC designs. Computer **36**(4), 53–59 (2003)

4. Cai, L., Gajski, D., Kritzinger, P., Olivares, M.: Top-down system level design methodology using SpecC, VCC and SystemC. In: DATE '02: Proceedings of the Conference on Design, Automation and Test in Europe, p. 1137. IEEE Comput. Soc., Washington (2002)
5. Cai, L., Gerstlauer, A., Gajski, D.: Retargetable profiling for rapid, early system-level design space exploration. In: DAC '04: Proceedings of the 41st Annual Conference on Design Automation, pp. 281–286. ACM, New York (2004)
6. Calvez, J.P., Heller, D., Pasquier, O.: Uninterpreted co-simulation for performance evaluation of HW/SW systems. In: CODES '96: Proceedings of the 4th International Workshop on Hardware/Software Co-Design, pp. 132–139. IEEE Comput. Soc., Washington (1996)
7. Cheung, E., Hsieh, H., Balarin, F.: Fast and accurate performance simulation of embedded software for MPSoC. In: ASP-DAC '09: Proceedings of the 2009 Conference on Asia and South Pacific Design Automation, Yokohama, Japan, pp. 552–557 (2009)
8. Cheung, E., Hsieh, H., Balarin, F.: Fast and accurate performance simulation of embedded software for MPSoC. In: ASP-DAC '09: Proceedings of the 2009 Conference on Asia and South Pacific Design Automation, pp. 552–557. IEEE Press, Piscataway (2009)
9. Chevalier, J., Benny, O., Rondonneau, M., Bois, G., Aboulhamid, E.M., Boyer, F.-R.: SPACE: a hardware/software SystemC modeling platform including an RTOS. In: Forum on Specification and Design Languages, Lille, France, pp. 91–104. Kluwer Academic, Dordrecht (2004)
10. Cornet, J., Maraninchi, F., Maillet-Contoz, L.: A method for the efficient development of timed and untimed transaction-level models of Systems-on-Chip. In: DATE '08: Proceedings of the Conference on Design, Automation and Test in Europe, pp. 9–14. ACM, New York (2008)
11. Ecker, W., Heinen, S., Velten, M.: Using a dataflow abstracted virtual prototype for HDS-design. In: ASP-DAC09: Proceedings of the 2009 Conference on Asia and South Pacific Design Automation, pp. 293–300. IEEE Press, Piscataway (2009)
12. Ghenassia, F. (ed.): Transaction-Level Modeling with SystemC: Tlm Concepts and Applications for Embedded Systems. Springer, New York (2006)
13. Gligor, M., Fournel, N., Pétrot, F.: Using binary translation in event driven simulation for fast and flexible MPSoC simulation. In: CODES+ISSS'09: Proceedings of the 7th IEEE/ACM/IFIP International Conference on Hardware/Software Codesign and System Synthesis, Grenoble, France (2009)
14. Hommais, D., Pétrot, F.: Efficient combinational loops handling for cycle precise simulation of system on a chip. In: Proc. of the 24th Euromicro Conf., Vesteras, Sweden, pp. 51–54 (1998)
15. International technology roadmap for semiconductors. In: System Drivers, p. 7 (2007)
16. Kempf, T., Karuri, K., Wallentowitz, S., Ascheid, G., Leupers, R., Meyr, H.: A SW performance estimation framework for early system-level-design using fine-grained instrumentation. In: DATE '06: Proceedings of the Conference on Design, Automation and Test in Europe, pp. 468–473. European Design and Automation Association, Leuven (2006)
17. Lajolo, M., Lazarescu, M., Sangiovanni-Vincentelli, A.: A compilation-based software estimation scheme for hardware/software co-simulation. In: CODES '99: Proceedings of the Seventh International Workshop on Hardware/Software Codesign, pp. 85–89. ACM, New York (1999)
18. Lee, J.-Y., Park, I.-C.: Timed compiled-code simulation of embedded software for performance analysis of SoC design. In: DAC '02: Proceedings of the 39th Conference on Design Automation, pp. 293–298. ACM, New York (2002)
19. Pieper, J.J., Mellan, A.P., JoAnn, M., Thomas, D.E., Karim, F.: High level cache simulation for heterogeneous multiprocessors. In: DAC '04: Proceedings of the 41st Annual Conference on Design Automation, pp. 287–292. ACM, New York (2004)
20. Popovici, K., Guerin, X., Rousseau, F., Paolucci, P.S., Jerraya, A.A.: Platform-based software design flow for heterogeneous MPSoC. ACM Trans. Embed. Comput. Syst. **7**(4), 1–23 (2008)
21. Rochange, C., Sainrat, P.: A context-parameterized model for static analysis of execution times. Trans. High-Perform. Embed. Archit. Compil. **2**(3), 109–128 (2007)
22. Yoo, S., Bacivarov, I., Bouchhima, A., Yanick, P., Jerraya, A.A.: Building fast and accurate SW simulation models based on hardware abstraction layer and simulation environment abstraction layers. In: DATE '03: Proceedings of the Conference on Design, Automation and Test in Europe, pp. 10550–10555 (2003)

Chapter 6
Models for Co-design of Heterogeneous Dynamically Reconfigurable SoCs

Jean-Luc Dekeyser, Abdoulaye Gamatié, Samy Meftali, and Imran Rafiq Quadri

1 Introduction

Since the early 2000s, Systems-on-Chip (or SoCs) have emerged as a new paradigm for embedded systems design. In a SoC, the computing units: programmable processors; memories, I/O devices, etc., are all integrated into a single chip. Moreover, multiple processors can be integrated into a SoC (Multiprocessor System-on-Chip, MPSoC) in which the communication can be achieved through Networks-on-Chips (NoCs). Some examples of domains where SoCs are used are: multimedia, automotive, defense and medical applications.

1.1 SoC Complexity and Need of Reconfiguration

As the computational power increases for SoCs, more functionalities are expected to be integrated in these systems. As a result, more complex software applications and hardware architectures are integrated, leading to a *system complexity* issue which is one of the main hurdles faced by designers. The fallout of this complexity is that the system design, particularly software design, does not evolve at the same pace as

J.-L. Dekeyser · A. Gamatié · S. Meftali · I.R. Quadri (✉)
INRIA Lille Nord Europe–LIFL–USTL–CNRS, Parc Scientifique de la Haute Borne, Park Plaza–Batiment A, 40 avenue Halley, 59650 Villeneuve d'Ascq, France
e-mail: imran.quadri@lifl.fr

J.-L. Dekeyser
e-mail: jean-luc.dekeyser@lifl.fr

A. Gamatié
e-mail: abdoulaye.gamatie@lifl.fr

S. Meftali
e-mail: samy.meftali@lifl.fr

G. Nicolescu et al. (eds.), *Design Technology for Heterogeneous Embedded Systems*,
DOI 10.1007/978-94-007-1125-9_6,

that of hardware. This has become an extremely significant issue and has finally led to the *productivity gap*.

Reconfigurability is also a critical issue for SoCs which must be able to cope with end user environment and requirements. For instance, mode-based control plays an important role in multimedia embedded systems by allowing to describe Quality-of-Service (QoS) choices: (1) changes in executing functionalities, e.g., color or black and white picture modes for modern digital cameras; (2) changes due to resource constraints of targeted platforms, for instance switching from a high memory consumption mode to a smaller one; or (3) changes due to other environmental and platform criteria such as communication quality and energy consumption. A suitable control model must be generic enough to be applied to both software and hardware design aspects.

The reduction in complexity of SoCs, while integrating mechanisms of system reconfiguration in order to benefit from QoS criteria, offers an interesting challenge. Several solutions are presented below.

1.2 Component Based Design

An effective solution to SoC co-design problem consists in raising the design abstraction levels. This solution can be seen through a *top-down* approach. The important requirement is to find efficient design methodologies that raise the design abstraction levels to reduce overall SoC complexity. They should also be able adequately express the control in order to integrate reconfigurability features in modern embedded systems.

Component based design is also a promising alternative. This approach increases productivity of software developers by reducing the amount of efforts needed to develop and maintain complex systems [10]. It offers two main benefits. First, it offers an incremental or *bottom-up* system design approach permitting to create complex systems, while making system verification and maintenance more tractable. Secondly, this approach allows reuse of development efforts as component can be re-utilized across different software products.

Controlling system reconfiguration in SoCs can be expressed via different component models. Automata based control is seen as promising as it incorporates aspects of modularity that is present in component based approaches. Once a suitable control model is chosen, implementation of these reconfigurable SoC systems can be carried out via Field Programmable Gate Arrays (or FPGAs). FPGAs are inherently reconfigurable in nature. State of the art FPGAs can change their functionality at *runtime*, known as Partial Dynamic Reconfiguration (PDR) [28]. These FPGAs also support internal self dynamic reconfiguration, in which an internal controller (a *hardcore/softcore* embedded processor) manages the reconfiguration aspects.

Finally the usage of *high level component based design approach* in development of real-time embedded systems is also increasing to address the compatibility issues related to SoC co-design. High abstraction level SoC co-modeling design approaches have been developed in this context, such as Model-Driven Engineering (MDE) [33] that specify the system using the UML graphical language. MDE enables high level system modeling (of both software and hardware), with the possibility of integrating heterogeneous components into the system. *Model transformations* [30] can be carried out to generate executable models from high level models. MDE is supported by several standards and tools.

Our contribution relates to the proposal of a high level component based SoC co-design framework which has been integrated with suitable control models for expressing reconfigurability. The control models are first explored at different system design levels along with a brief comparison. Afterwards, the control model is explored at another design abstraction level that permits to link the system components with respective implementations. This control model proves more beneficial as it allows to exploit reconfigurability features in SoC by means of partial dynamic reconfiguration in FPGAs. Finally a case study is illustrated which validates our design methodology.

The plan of the chapter is as follows. Section 2 gives an overview of some related works while Sect. 3 defines the notions associated with component based approaches. Section 4 introduces our SoC co-design framework while Sect. 5 illustrates a reactive control model. Section 6 compares control models at different levels in our framework. Section 7 provides a more beneficial control model in our framework, illustrated with a case study. Finally Sect. 8 gives the conclusion.

2 Related Works

There are several works that use component based high abstraction level methodologies for defining embedded systems. MoPCoM [25] is a project that targets modeling and code generation of embedded systems using the block diagrams present in SysML which can be viewed as components. In [7], a SynDEx based design flow is presented to manage SoC reconfigurability via implementation in FPGAs, with the application and architecture parts modeled as components. Similarly in [20], a component based UML profile is described along with a tool set for modeling, verification and simulation of real-time embedded systems. Reconfiguration in SoC can be related to available system resources such as available memory, computation capacity and power consumption. An example of a component based approach with adaptation mechanisms is provided in [42]; e.g. for switching between different resources [11].

In [27, 39], the authors concentrate on verification of real-time embedded systems in which the control is specified at a high abstraction level via UML state machines and collaborations; by using model checking. However, control methodologies vary in nature as they can be expressed via different forms such as Petri Nets [31], or other formalisms such as mode automata [29].

Mode automata extend synchronous dataflow languages with an imperative style, but without many modifications of language style and structure [29]. They are mainly composed of *modes* and *transitions*. In an automaton, each mode has the same interface. Equations are specified in modes. Transitions are associated with conditions, which serve to act as triggers. Mode automata can be composed together in either in parallel or hierarchical manner. They enable formal validation by using the synchronous technology. Among existing UML based approaches allowing for design verification are the Omega project [20] and Diplodocus [1]. These approaches essentially utilize model checking and theorem proving.

In the domain of dynamically reconfigurable FPGA based SoCs, Xilinx initially proposed two design flows, which were not very effective leading to new alternatives. An effective modular approach for 2-D shaped reconfigurable modules was presented in [41]. [5] implemented modular reconfiguration using a horizontal slice based bus macro in order to connect the static and partial regions. They then placed arbitrary 2-dimensional rectangular shaped modules using routing primitives [22]. This approach has been further refined in [40]. In 2006, Xilinx introduced the *Early Access Partial Reconfiguration Design Flow* [44] that integrated concepts of [41] and [5]. Works such as [4, 34] focus on implementing softcore internal configuration ports on Xilinx FPGAs such as Spartan-3, that do not have the hardware Internal Configuration Access Port (ICAP) reconfigurable core, for implementing PDR. Contributions such as introduced in [12] and [13], illustrate usage of customized ICAPs. Finally in [24], the ICAP reconfigurable core is connected with Networks-on-chip (NoC) implemented on dynamically reconfigurable FPGAs.

In comparison to the above related works, our proposed contributions take into account the following domains: SoC co-design, control/data flow, MDE, UML MARTE profile, SoC reconfigurability and PDR for FPGAs; which is the novelty of our design framework.

3 Components

Components are widely used in the domain of component based software development or component based software engineering. The key concept is to visualize the system as a collection of *components* [10]. A widely accepted definition of components in software domain is given by Szyperski in [43]:

> *A component is a unit of composition with contractually specified interfaces and fully explicit context dependencies that can be deployed independently, and is subject to third-party composition.*

In the software engineering discipline, a component is viewed as a representation of a self-contained part or subsystem; and serves as a building block for designing a complex global system. A component can provide or require *services* to its environment via well-specified *interfaces* [10]. These interfaces can be related to *ports* of the component. Development of these components must be separated from the development of the system containing these modules. Thus components can be used in different contexts, facilitating their reuse.

The definition given by Szyperski permits to separate the component *behavior* and the component *interface*. Component *behavior* defines the functionality or the executable realization of a component. This can be viewed as associating the component with an *implementation* such as compilable code, binary form, etc.; depending upon the component model. This notion enables to link the component to user defined or third party implementations or *intellectual properties* (IPs). A component *interface* represents the properties of the component that are externally visible to other parts of the system.

Two basic prerequisites permit integration and execution of components. A *component model* defines the semantics that components must follow for their proper evolution [10]. A *component infrastructure* is the design-time and run-time infrastructure that allows interaction between components and manages their assembly and resources. Obviously, there is a correspondence between a component model and the supporting mechanisms and services of a component framework.

Typically, in languages such as *Architecture Definition Languages* (ADLs), description of system architectures is carried out via compositions of hardware and software modules. These components follow a component model; and the interaction between components is managed by a component infrastructure [3].

For describing hardware components in embedded systems, several critical properties, such as timing, performance and energy consumption, depend on characteristics of the underlying hardware platform. These extra functional properties such as performance cannot be specified for a *software* component but are critical for defining a hardware platform.

3.1 Component Models

A component model determines the behavior of components within a component framework. It states what it means for a component to implement a given interface, it also imposes constraints such as defining communication protocols between components etc. [10]. We have already briefly described the use of components in software engineering. There exit many component models such as COM (Component Object Model), CORBA, EJB and .NET. Each of these models have distinct semantics which may render them incompatible with other component models. As these models prove more and more useful for the design, development and verification of complex software systems, more and more research is being carried out by hardware designers in order to utilize the existing concepts present in software engineering for facilitating the development of complex hardware platforms.

Already hardware and system description languages such as VHDL and SystemC which support incremental modular structural concepts can be used to model embedded systems and SoCs in a modular way.

3.2 Component Infrastructure

A component infrastructure provides a wide variety of services to enforce and support component models. Using an simple analogy, components are to infrastructures what processes are to an operating system. A component infrastructure manages the resources shared by the different components [10]. It also provides the underlying mechanisms that allow component interactions and final assembly. Components can be either *homogeneous*: having the same functionality model but not the same behavior; or *heterogeneous*. Examples of homogeneous components can be found in systems such as grids and cubes of computing units. In systems such at TILE64 [6], homogeneous instances of processing units are connected together by communication media. These types of systems are partially homogeneous concerning the computation units but heterogeneous in terms of their interconnections. Nowadays, modern embedded systems are mainly composed of heterogeneous components. Correct assembly of these components must be ensured to obtain the desired interactions. A lot of research has been carried out to ensure the correctness of interface composition in heterogeneous component models. Enriching the interface properties of a same component enables in addressing different aspects, such as timing and power consumption [15]. The semantics related to component assembly can be selected by designers according to their system requirements. The assembly can be either static or dynamic in nature.

3.3 Towards SoC Co-design

It is obvious that in the context of embedded systems, information related to hardware platforms must be added to component infrastructures. Properties such as timing constraints and resource utilization are some of the integral aspects. However, as different design platforms use different component models for describing their customized components, there is a lack of consensus on the development of components for real-time embedded systems. Similarly interaction and interfacing of the components is another key concept.

Dynamic Reconfiguration Dynamic reconfiguration of component structure depends on the context required by designer and can be determined by different Quality-of-Service (QoS) criteria. The dynamic aspects may require the integration of a *controller* component for managing the overall reconfiguration. The semantics related to component infrastructure must take into consideration several key issues: instantiation and termination of these components, deletion in case of user requirement etc. Similarly communication mechanisms such as message passing, operation calls can be chosen for inter and intra communication (in case of composition hierarchy) of components.

In case of embedded systems, a suitable example can be of FPGAs. These reconfigurable architectures are mainly composed of heterogeneous components, such as processors, memories, peripherals, I/O devices, clocks and communication media such as buses and Network-on-Chips. For carrying out internal dynamic reconfiguration, a controller component: in the form of a hard/soft core processor, can be integrated into the system for managing the overall reconfiguration process.

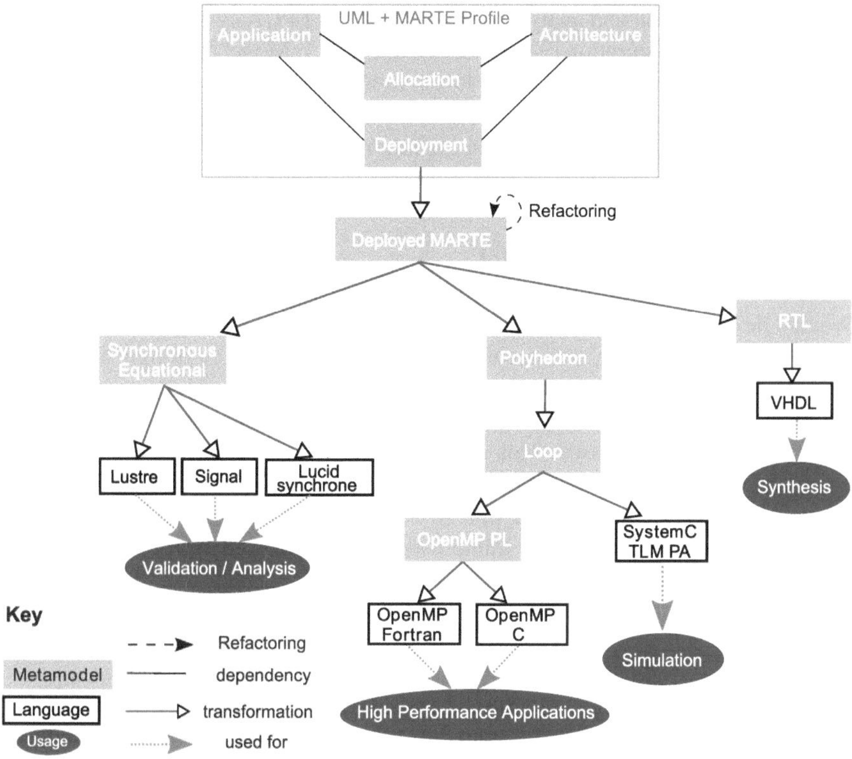

Fig. 6.1 A global view of the Gaspard2 framework

4 Gaspard2: A SoC Co-design Framework

Gaspard2 [14, 17] is a SoC co-design framework dedicated to parallel hardware and software and is based on the classical Y-chart [16]. One of the most important features of Gaspard2 is its ability for system co-modeling at a high abstraction level. Gaspard2 uses the Model-Driven Engineering methodology to model real-time embedded systems using the UML MARTE profile [32]; and UML graphical tools and technologies such as Papyrus and Eclipse Modeling Framework.

Figure 6.1 shows a global view of the Gaspard2 framework. Gaspard2 enables to model *software applications*, *hardware architectures* and their *allocations* in a concurrent manner. Once models of software applications and hardware architectures are defined, the functional parts (such as application tasks and data) can be mapped onto hardware resources (such as processors and memories) via *allocation(s)*. Gaspard2 also introduces a *deployment* level that allows to link hardware and software components with intellectual properties (IPs). This level is elaborated later in Sect. 7.

For the purpose of automatic code generation from high level models, Gaspard2 adopts MDE model transformations (*model to model* and *model to text* transformations) towards different execution platforms, such as targeted towards synchronous

domain for validation and analysis purposes [19]; or FPGA synthesis related to partial dynamic reconfiguration [38], as shown in Fig. 6.1. Model transformation chains allow moving from high abstraction levels to low enriched levels. Usually, the initial high level models contain only domain-specific concepts, while technological concepts are introduced seamlessly in the intermediate levels.

5 A Reactive Control Model

We first describe the generic control semantics which can be integrated into the different levels (application, architecture and allocation) in SoC co-design. Several basic control concepts, such as `Mode Switch Component` and `State Graphs` are presented first. Then a basic composition of these concepts, which builds the mode automata, is discussed. This modeling derives from mode concepts in mode automata. The notion of exclusion among modes helps to separate different computations. As a result, programs are well structured and fault risk is reduced. We then use the Gaspard2 SoC co-design framework for utilization of these concepts.

5.1 Modes

A mode is a distinct method of operation that produces different results depending upon the user inputs. A mode switch component in Gaspard2 contains at least more than one mode; and offers a switch functionality that chooses execution of one mode, among several alternative present modes [26]. The mode switch component in Fig. 6.2 illustrates such a component having a *window* with multiple tabs and interfaces. For instance, it has an m (mode value input) port as well as several data input and output ports, i.e., i_d and o_d respectively. The switch between the different modes is carried out according to the mode value received through m.

The modes, $M_1, \ldots, M_n$, in the mode switch component are identified by the mode values: $m_1, \ldots, m_n$. Each mode can be hierarchical, repetitive or elementary in nature; and transforms the input data i_d into the output data o_d. All modes have the same interface (i.e. i_d and o_d ports). The activation of a mode relies on the reception of mode value m_k by the mode switch component through m. For any received mode value m_k, the mode runs exclusively. It should be noted that only mode value ports, i.e., m; are compulsory for creation of a mode switch component, as shown in Fig. 6.2. Thus other type of ports are represented with dashed lines.

5.2 State Graphs

A state graph in Gaspard2 is similar to state charts [21], which are used to model the system behavior using a state-based approach. It can be expressed as a graphical

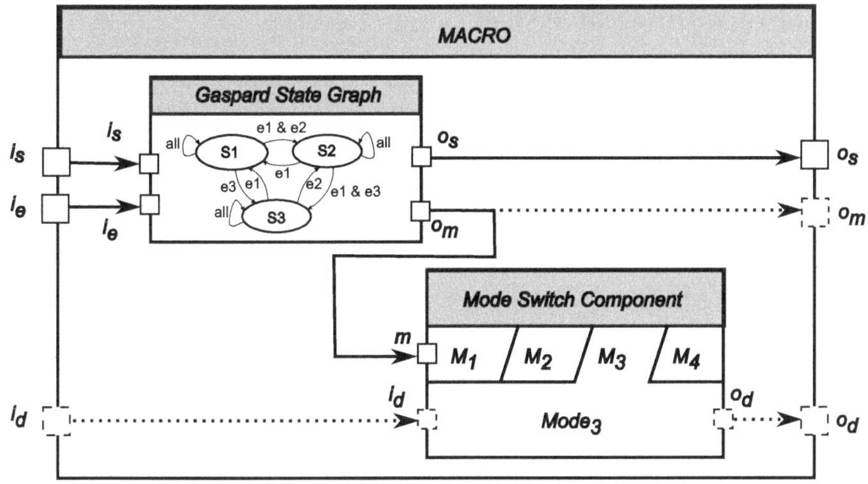

Fig. 6.2 An example of a macro structure

representation of transition functions as discussed in [18]. A state graph is composed of a set of vertices, which are called *states*. A state connects with other states through directed edges. These edges are called *transitions*. Transitions can be conditioned by some events or Boolean expressions. A special label *all*, on a transition outgoing from state s, indicates any other events that do not satisfy the conditions on other outgoing transitions from s. Each state is associated with some mode value specifications that provide mode values for the state. A state graph in Gaspard2 is associated with a `Gaspard State Graph` as shown in Fig. 6.2.

5.3 Combining Modes and State Graphs

Once mode switch components and state graphs are introduced, a `MACRO` component can be used to compose them together. The MACRO in Fig. 6.2 illustrates one possible composition. In this component, the Gaspard state graph produces a mode value (or a set of mode values) and sends it (them) to the mode switch component. The latter switches the modes accordingly. Some data dependencies (or connections) between these components are not always necessary, for example, the data dependency between I_d and i_d. They are drawn with dashed lines in Fig. 6.2. The illustrated figure is used as a basic composition, however, other compositions are also possible, for instance, one Gaspard state graph can also be used to control several mode switch components [37].

6 Control at Different System Design Levels

The previously mentioned control mechanisms can be integrated in different levels in a SoC co-design environment. We first analyze the control integration at the application, architecture and allocation levels in the particular case of the Gaspard2 framework, followed by a comparison of the three approaches.

6.1 Generic Modeling Concepts

We first present some concepts which are used in the modeling of mode automata. Gaspard2 uses the Repetitive Structure Modeling (RSM) package in the MARTE UML profile to model intensive data-parallel processing applications. RSM is based on Array-OL [9] that describes the *potential parallelism* in a system; and is dedicated to data intensive multidimensional signal processing. In Gaspard2, data are manipulated in the form of multidimensional arrays. For an application functionality, both data parallelism and task parallelism can be expressed easily via RSM. A `repetitive component` expresses the data parallelism in an application: in the form of sets of input and output `patterns` consumed and produced by the repetitions of the interior `part`. It represents a regular scalable component infrastructure. A `hierarchical` component contains several `parts`. It allows to define complex functionalities in a modular way and provides a structural aspect of the application. Specifically, task parallelism can be described using a hierarchical component in our framework.

The basic concepts of Gaspard2 control have been presented in Section V, but its complete semantics have not been provided. Hence, we propose to integrate mode automata semantics in the control. This choice is made to remove design ambiguity, enable desired properties and to enhance correctness and verifiability in the design. In addition to previously mentioned control concepts, three additional constructs as present in the RSM package in MARTE, namely the `Interrepetition` dependency (IRD), the `tiler` connector and `defaultLink` concepts are utilized to build mode automata.

A `tiler` connector describes the tiling of produced and consumed arrays and thus defines the shape of a data pattern. The `Interrepetition` dependency is used to specify an acyclic dependency among the repetitions of the same component, compared to a tiler, which describes the dependency between the repeated component and its owner component. The interrepetition dependency specification leads to the sequential execution of repetitions. A `defaultLink` provides a default value for repetitions linked with an interrepetition dependency, with the condition that the source of dependency is absent.

The introduction of an interrepetition dependency serializes the repetitions and data can be conveyed between these repetitions. Hence, it is possible to establish mode automata from Gaspard2 control model, which requires two subsequent steps. First, the internal structure of `Gaspard Mode Automata` is presented by the

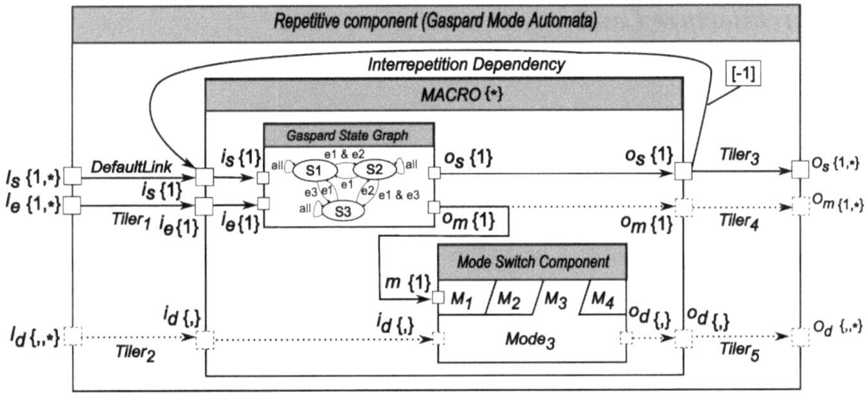

Fig. 6.3 The macro structure in a repetitive component

MACRO component illustrated in Fig. 6.2. The Gaspard state graph in the macro acts as a state-based controller and the mode switch component achieves the mode switch function. Secondly, interrepetition dependency specifications should be specified for the macro when it is placed in a repetitive context. The reasons are as follows. The macro structure represents only a single transition between states. In order to execute continuous transitions as present in automata, the macro should be repeated to have multiple transitions. An interrepetition dependency forces the continuous sequential execution. This allows the construction of mode automata which can be then executed.

6.2 Application Level

With previous presented constructs, the modeling of Gaspard mode automata, which can be eventually translated into synchronous mode automata [29], is illustrated with an example in Fig. 6.3, where the assembly of these constructs is presented. An interrepetition dependency connects the repetitions of MACRO and conveys the current state. It thus sends the target state of one repetition as the source state for the next repetition of the macro component as indicated by the value of -1. The states and transitions of the automata are encapsulated in the Gaspard state graph. The data computations inside a mode are set in the mode switch component. The detailed formal semantics related to Gaspard mode automata can be found in [18]. It should be noted that parallel and hierarchical mode automata can also be constructed using the control semantics.

The proposed control model enables the specification of system reconfigurability at the application level [45]. Each mode in the switch can have different effects with regards to environmental or platform requirements. A mode represents a distinct algorithm to implement the same functionality as others. Each mode can have a different demand of memory, CPU load, etc. Environmental changes/platform requirements are captured as events; and taken as inputs of the control.

6.3 Architecture Level

Gaspard2 uses the *Hardware Resource Modeling* (or HRM) package in the MARTE profile in combination with the RSM package to model large regular hardware architectures (such as multiprocessor architectures) in a compact manner. Complex interconnection topologies can also be modeled via Gaspard2 [35].

Control semantics can also be applied on to the architectural level in Gaspard2. As compared to the integration of control in other modeling levels (such as application and allocation), the control in architecture is more flexible and can be implemented in several forms. A controller can modify the structure of the architecture in question such as modifying the communication interconnections. The structure can be either modified globally or partially. In case of a global modification, the reconfiguration is viewed as static and the controller is present exterior to the targeted architecture. If the controller is present inside the architecture, then the reconfiguration is partial and could result in partial dynamic reconfiguration. However, the controller can be related to both the structural and behavioral aspects of the architecture. An example can be of a controller unit present inside a processing unit in the architecture for managing *Dynamic frequency scaling* [46] or *Dynamic voltage scaling* [23]. These techniques allow power conservation by reducing the frequency or the voltage of an executing processor.

6.4 Allocation Level

Gaspard2 uses the Allocation Modeling package (Alloc) to allocate SoC applications on to the targeted hardware architectures. Allocation in MARTE can be either *spatial* or *temporal* in nature [32].

Control at the allocation level can be used to decrease the number of active executing computing units to reduce the overall power consumption levels. Tasks of an application that are executing parallely on processing units may produce the desired computation at an optimal processing speed, but might consume more power, depending upon the inter-communication between the system. Modification of the allocation of the application on to the architecture can produce different combinations and different end results. A task may be switched to another processing unit that consumes less power, similarly, all tasks can be associated on to a single processing unit resulting in a temporal allocation as compared to a spatial one. This strategy may reduce the power consumption levels along with decrease in the processing speed. Thus allocation level allows to incorporate *Design Space Exploration* (DSE) aspects which in turn can be manipulated by the designers depending upon their chosen QoS criteria.

	Impact on other models	Conditions	Consequences
Application	Local	—	Change in application functionality/structure occurs, the functionality can be related to QoS criteria as well
Architecture	Local	QoS variation only	Variability in observed performance
	Global	Modification in structure	Allocation model needs to be modified as well
Allocation	Local	Other models are fixed	Variability in observed performance
	Global	Other models modifiable	Change in functionality, variation in QoS possible

Fig. 6.4 Overview of control on the first three levels of a SoC framework

6.5 Comparison of Control at the Three Levels

Integrating control at different aspects of system (application, architecture and allocation) has its advantages and disadvantages as briefly shown in the Fig. 6.4. With respect to control integration, we are mainly concerned with several aspects such as the range of impact on other modeling levels. We define the impact range as either *local* or *global*, with the former only affecting the concerned modeling level while the later having consequences on other modeling levels. These consequences may vary and cause changes in either *functional* or *non-functional aspects* of the system. The modification in application may arise due to QoS criteria such as switching from a high resolution mode to a lower one in a video processing functionality. However, the control model may have consequences, as change in an application functionality or its structure may not have the intended end results.

Control integration in an architecture can have several possibilities. The control can be mainly concerned with modification of the hardware parameters such as voltage and frequency for manipulating power consumption levels. This type of control is local and mainly used for QoS, while the second type of control can be used to modify the system structure either globally or partially. This in turn can influence other modeling levels such as the allocation. Thus allocation model needs to be modified every single time when even there is a slight modification in the structure of the execution platform.

Control at the allocation is local only when both the application and architecture models have been pre-defined to be static in nature which is rarely the actual scenario. If either the application or the architecture is changed, the allocation must be adapted accordingly.

It is also possible to form a merged control by combining the control models at different aspects of the system to form a mixed-level control approach. However, detailed analysis is needed to ensure that any combination of control levels does not cause any unwanted consequences. This is also a tedious task. During analysis, several aspects have to be monitored, such as ensuring that no conflicts arise due to a merged approach. Similarly, redundancy should be avoided: if an application control and architecture control produce the same result separately; then suppression of control from one of these levels is warranted. However, this may also lead to an instability in the system. It may be also possible to create a global controller that is responsible for synchronizing various local control mechanisms. However, clear semantics must be defined for the composition of the global controller which could lead to an overall increase in design complexity.

The global impact of any control model is undesirable as the modeling approach becomes more complex and several high abstraction levels need to be managed. A local approach is more desirable as it does not affect any other modeling level. However, in each of the above mentioned control models, strict conditions must be fulfilled for their construction. These conditions may not be met depending upon the designer environment. Thus an ideal control model is one that has only a local impact range and does not have any strict construction conditions.

7 Control at Deployment Level

In this section we explain control integration at another abstraction level in SoC co-design. This level deals with linking the modeled application and architecture components to their respective IPs. We explain the component model of this *deployment* level in the particular case of the Gaspard2 framework within the context of dynamic reconfiguration.

For dynamic reconfiguration in modern SoCs, an embedded controller is essential for managing a dynamically reconfigurable region. This component is usually associated with some control semantics such as state machines, Petri nets etc. The controller normally has two functionalities: one responsible for communicating with the FPGA *Internal Configuration Access Port* hardware reconfigurable core or ICAP [8] that handles the actual FPGA switching; and a state machine part for switching between the available configurations. The first functionality is written manually due to some low level technological details which cannot be expressed via a high level modeling approach.

The control at the deployment level is utilized to generate the second functionality automatically via model transformations. Finally the two parts can be used to implement partial dynamic reconfiguration in an FPGA that can be divided into several static/reconfigurable regions. A reconfigurable region can have several implementations, with each having the same interface, and can be viewed as a mode switch component with different modes. In our design flow, this dynamic region is generated from the high abstraction levels, i.e., a complex Gaspard2 application specified using the MARTE profile. Using the control aspects in the subsequently explained Gaspard2 deployment level, it is possible to create different configurations of the modeled application. Afterwards, using model transformations, the application can be transformed into a hardware functionality, i.e., a dynamically reconfigurable hardware accelerator, with the modeled application configurations serving as different implementations related to the hardware accelerator.

We now present integration of the control model at the deployment level. We first explain the deployment level in Gaspard and our proposed extensions followed by the control model.

7.1 Deployment in Gaspard2

The Gaspard2 deployment level enables one to precise a specific IP for each elementary component of application or architecture, among several possibilities [2]. The reason is that in SoC design, a functionality can be implemented in different ways. For example, an application functionality can either be optimized for a processor, thus written in C/C++, or implemented as a hardware accelerator using *Hardware Description Languages* (HDLs). Hence the deployment level differentiates between the hardware and software functionalities; and allows moving from platform-independent high level models to platform-dependent models for eventual implementation. We now present a brief overview of the deployment concepts, as viewed in Fig. 6.5.

A `VirtualIP` expresses the functionality of an elementary component, independently from the compilation target. For an elementary component *K*, it associates *K* with all its possible IPs. The desired IP(s) is (are) then selected by the SoC designer by linking it (them) to *K* via an `implements` dependency. Finally, the `CodeFile` (not illustrated in the chapter) determines the physical path related to the source/binary code of an IP, along with required compilation options.

7.2 Multi-configuration Approach

Currently in deployment level, an elementary component can be associated with only one IP among the different available choices (if any). Thus the result of the application/architecture (or the mapping of the two forming the overall system) is a static one. This collective composition is termed as a *Configuration*.

Integrating control in deployment allows to create several configurations related to the modeled application for the final realization in an FPGA. Each configuration is viewed as a collection of different IPs, with each IP associated with its respective elementary component. The end result being that one application model is transformed by means of model transformations and intermediate metamodels into a dynamically reconfigurable hardware accelerator, having different implementations equivalent to the modeled application configurations.

A `Configuration` has the following attributes. The name attribute helps to clarify the configuration name given by a SoC designer. The ConfigurationID attribute permits to assign unique values to each of the modeled `Configuration`, which in turn are used by the control aspects presented earlier. Theses values are used by a Gaspard state graph to produce the mode values associated with its corresponding Gaspard state graph component. These mode values are then sent to a mode switch component which matches the values with the names of its related collaborations as explained in [37]. If there is a match, the mode switch component switches to the required configuration. The InitialConfiguration attribute sets a Boolean value to a configuration to indicate if it is the initial configuration to be

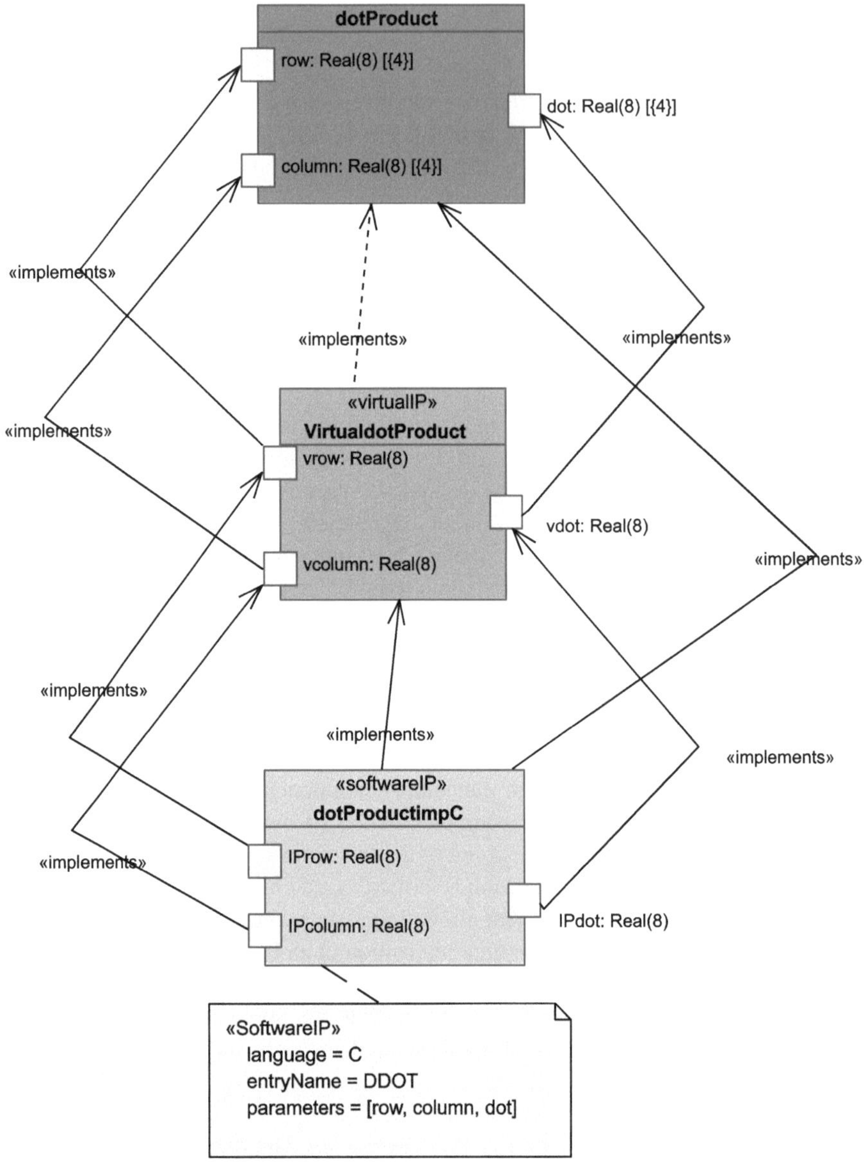

Fig. 6.5 Deployment of an elementary `dotProduct` component in Gaspard2

loaded onto the target FPGA. This attribute also helps to determine the initial state of the Gaspard state graph.

An elementary component can also be associated with the same IP in different configurations. This point is very relevant to the semantics of partial bitstreams, e.g., FPGA configuration files for partial dynamic reconfiguration, supporting *glitchless dynamic reconfiguration*: if a configuration bit holds the same value before and after

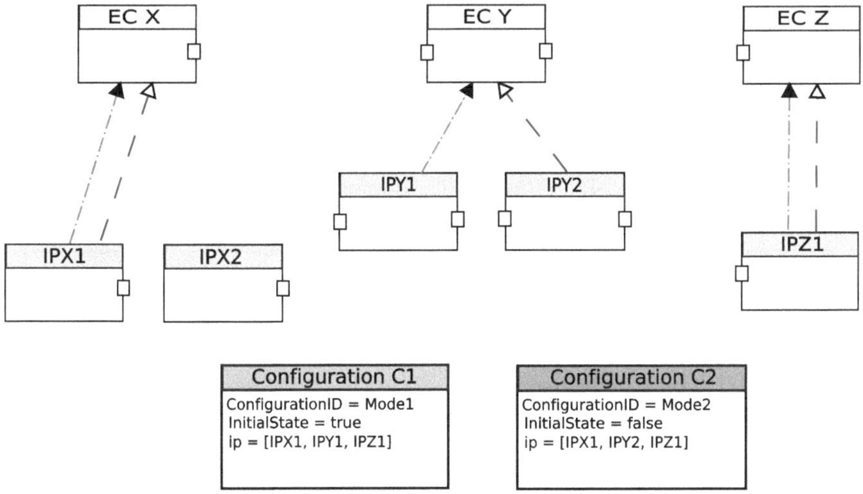

Fig. 6.6 Abstract overview of configurations in deployment

reconfiguration, the resource controlled by that bit does not experience any discontinuity in operation. If the same IP for an elementary component is present in several configurations, that IP is not changed during reconfiguration. It is thus possible to link several IPs with a corresponding elementary component; and each link relates to a unique configuration. We apply a condition that for any n number of configurations with each having m elementary components, each elementary component of a configuration must have *at least* one IP. This allows successful creation of a complete configuration for eventual final FPGA synthesis.

Figure 6.6 represents an abstract overview of the configuration mechanism introduced at the deployment level. We consider a hypothetical Gaspard2 application having three elementary components *EC X*, *EC Y* and *EC Z*, having available implementations *IPX1*, *IPX2*, *IPY1*, *IPY2* and *IPZ1* respectively. For the sake of clarity, this abstract representation omits several modeling concepts such as `VirtualIP` and `Implements`. However, this representation is very close to UML modeling as presented earlier in the chapter. A change in associated implementation of any of these elementary components may produces a different end result related to the overall functionality, and different Quality of Service (QoS) criteria such as effectively consumed FPGA resources.

Here two configurations *Configuration C1* and *Configuration C2* are illustrated in the figure. *Configuration C1* is selected as the initial configuration and has associated IPs: *IPX1*, *IPY1* and *IPZ1*. Similarly *Configuration C2* also has its associated IPs. This figure illustrates all the possibilities: an IP can be globally or partially shared between different configurations: such as *IPX1*; or may not be included at all in a configuration, e.g., case of *IPX2*.

Once the different implementations are created by means of model transformations, each implementation is treated as a source for a partial bitstream. A bitstream contains packets of FPGA configuration control information as well as the configu-

	FPGA resources			Dynamic Power (mW)	Mean Reconfiguration Time (secs)
	Slices	Slice Flip Flops	LUTs		
Configuration 1 : Multiplication IP written with DSP functionality	1272/13696 (9.287%)	2084/27392 (7.608%)	1584/27392 (5.782%)	221.42	1.45
Configuration 2 : Multiplication IP written with if-else functionality	1186/13696 (8.659%)	1944/27392 (7.096%)	1836/27392 (6.702%)	214.56	1.41

Fig. 6.7 An overview of the obtained results

ration data. Each partial bitstream signifies a unique implementation, related to the reconfigurable hardware accelerator which is connected to an embedded controller. While this extension allows to create different configurations, the state machine part of the controller is created manually. For automatic generation of this functionality, the deployment extensions are not sufficient. We then make use of the earlier control semantics at the deployment level.

7.3 Implementation

Once control has been integrated at deployment level, it helps to switch between the different modeled configurations [38]. The configurations relate to a Gaspard2 application modeled at the high abstraction levels. This application is transformed into a hardware functionality, i.e., a hardware accelerator, by means of the model transformations, as stated earlier.

The application targeted for the validation of our methodology is a delay estimation correlation module integrated in an anti-collision radar detection system. Our radar uses a PRBS (Pseudorandom binary sequence) of length of 127 chips. In order to produce a computation result, the algorithm requires 127 multiplications between the 127 elements of the reference code that is generated via MATLAB and the last 127 received samples. The result of this multiplication produces 64 data elements. The sum of these 64 data elements produces the final result. This result can be sent as input to other parts of our radar detection system [36] in order to detect the nearest object. The different configurations related to our application change the IPs related to the elementary components, which in turn allow us to manipulate different QoS criteria such as consumed FPGA resources and overall energy consumption levels. The partially reconfigurable system has been implemented on a Xilinx XC2VP30 Virtex-II Pro FPGA with a hardcore PowerPC 405 processor as a reconfiguration controller with a frequency of 100 MHz. We implemented two configurations on the targeted architecture, two with different IPs related to an multiplication elementary component in the application and a blank configuration. The results are shown in Fig. 6.7.

7.4 Advantages of Control Deployment Level

The advantage of using control at deployment is that the impact level remains local and there is no influence on other modeling levels. Another advantage is that the application, architecture and allocation models can be reused again and only the necessary IPs are modified. As we validate our methodology by implementing partial dynamic reconfigurable FPGAs, we need to clarify about the option of choosing mode-automata.

Many different approaches exist for expressing control semantics, mode automata were selected as they clearly separate control/data flow. They also adapt a state based approach facilitating seamless integration in our framework; and can be expressed at the MARTE specification levels. The same control semantics are then used throughout our framework to provide a single homogeneous approach. With regards to partial dynamic reconfiguration, different implementations of a reconfigurable region must have the same external interface for integration with the static region at run-time. Mode automata control semantics can express the different implementations collectively via the concept of a mode switch, which can be expressed graphically at high abstraction levels using the concept of a mode switch component. Similarly a state graph component expresses the controller responsible for the context switch between the different configurations.

8 Conclusion

This chapter presents a high abstraction level component based approach integrated in Gaspard2, a SoC co-design framework compliant with the MARTE standard. The control model is based on mode automata, and takes task and data parallelism into account. The control semantics can be integrated into various levels in Gaspard2. We compare the different approaches with respect to different criteria such as impact on other modeling levels. Control integration in application level allows dynamic context switching. In addition, safety of the control can be checked by tools associated with synchronous languages when the high-level model is transformed into synchronous code. Control at the architectural level can be concerned with QoS criteria as well as structural aspects. Similarly, control at the allocation level offers advantages of *Design Space Exploration*. Finally we present control semantics in the deployment level which offer reuse of application, architecture and allocation models. This control model makes it possible to support partial dynamic reconfiguration in reconfigurable FPGAs. A case study has also been briefly presented to validate our design methodology. Currently we have only focused on isolating controls at different levels in Gaspard2. An ideal perspective could be a combination of the different control models to form a merged approach.

References

1. Apvrille, L., Muhammad, W., Ameur-Boulifa, R., Coudert, S., Pacalet, R.: A UML-based environment for system design space exploration. In: 13th IEEE International Conference on Electronics, Circuits and Systems, ICECS '06, Dec. 2006, pp. 1272–1275 (2006)
2. Atitallah, R.B., Piel, E., Niar, S., Marquet, P., Dekeyser, J.-L.: Multilevel MPSoC simulation using an MDE approach. In: SoCC 2007 (2007)
3. Bass, L., Clements, P., Kazman, R.: Software Architecture in Practice. Addison-Wesley, Reading (1998)
4. Bayar, S., Yurdakul, A.: Dynamic partial self-reconfiguration on Spartan-III FPGAs via a Parallel Configuration Access Port (PCAP). In: 2nd HiPEAC Workshop on Reconfigurable Computing, HiPEAC 08 (2008)
5. Becker, J., Huebner, M., Ullmann, M.: Real-time dynamically run-time reconfigurations for power/cost-optimized virtex FPGA realizations. In: VLSI'03 (2003)
6. Bell, S., et al.: TILE64-processor: a 64-core SoC with mesh interconnect. In: IEEE International Digest of Technical Papers on Solid-State Circuits Conference (ISSCC 2008), pp. 88–598 (2008)
7. Berthelot, F., Nouvel, F., Houzet, D.: A flexible system level design methodology targeting run-time reconfigurable FPGAs. EURASIP J. Embed. Syst. **8**(3), 1–18 (2008)
8. Blodget, B., McMillan, S., Lysaght, P.: A lightweight approach for embedded reconfiguration of FPGAs. In: Design, Automation & Test in Europe, DATE'03 (2003)
9. Boulet, P.: Array-OL revisited, multidimensional intensive signal processing specification. Research Report RR-6113, INRIA (February 2007). http://hal.inria.fr/inria-00128840/en/
10. Brinksma, E., Coulson, G., Crnkovic, I., Evans, A., Gérard, S., Graf, S., Hermanns, H., Jézéquel, J., Jonsson, B., Ravn, A., Schnoebelen, P., Terrier, F., Votintseva, A.: Component-based design and integration platforms: a roadmap. In: The Artist Consortium (2003)
11. Buisson, J., André, F., Pazat, J.-L.: A framework for dynamic adaptation of parallel components. In: ParCo 2005 (2005)
12. Claus, C., Muller, F.H., Zeppenfeld, J., Stechele, W.: A new framework to accelerate Virtex-II Pro dynamic partial self-reconfiguration. In: IPDPS 2007, pp. 1–7 (2007)
13. Cuoccio, A., Grassi, P.R., Rana, V., Santambrogio, M.D., Sciuto, D.: A generation flow for self-reconfiguration controllers customization. In: Forth IEEE International Symposium on Electronic Design, Test and Applications, DELTA 2008, pp. 279–284 (2008)
14. DaRT team: GASPARD SoC Framework, 2009. http://www.gaspard2.org/
15. Doyen, L., Henzinger, T., Jobstmann, B., Petrov, T.: Interface theories with component reuse. In: EMSOFT'08: Proceedings of the 8th ACM International Conference on Embedded Software, pp. 79–88. ACM, New York (2008)
16. Gajski, D.D., Khun, R.: New VLSI tools. Computer **16**, 11–14 (1983)
17. Gamatié, A., Le Beux, S., Piel, E., Etien, A., Atitallah, R.B., Marquet, P., Dekeyser, J.-L.: A model driven design framework for high performance embedded systems. Research Report RR-6614, INRIA (2008). http://hal.inria.fr/inria-00311115/en
18. Gamatié, A., Rutten, É., Yu, H.: A model for the mixed-design of data-intensive and control-oriented embedded systems. Research Report RR-6589, INRIA (July 2008). http://hal.inria.fr/inria-00293909/fr
19. Gamatié, A., Rutten, É., Yu, H., Boulet, P., Dekeyser, J.-L.: Synchronous modeling and analysis of data intensive applications. EURASIP J. Embedded Syst. (2008, to appear). Also available as INRIA Research Report: http://hal.inria.fr/inria-00001216/en/
20. Graf, S.: Omega—correct development of real time embedded systems. Softw. Syst. Model. **7**(2), 127–130 (2008)
21. Harel, D.: Statecharts: a visual formalism for complex systems. Sci. Comput. Program. **8**(3), 231–274 (1987). Article note found
22. Huebner, M., Schuck, C., Kiihnle, M., Becker, J.: New 2-dimensional partial dynamic reconfiguration techniques for real-time adaptive microelectronic circuits. In: ISVLSI'06 (2006)

23. Im, C., Kim, H., Ha, S.: Dynamic voltage scheduling technique for low-power multimedia applications using buffers (2001). http://citeseerx.ist.psu.edu/viewdoc/summary?doi=10.1.1.59.1133
24. Koch, R., Pionteck, T., Albrecht, C., Maehle, E.: An adaptive system-on-chip for network applications. In: IPDPS 2006 (2006)
25. Koudri, A., et al.: Using MARTE in the MOPCOM SoC/SoPC co-methodology. In: MARTE Workshop at DATE'08 (2008)
26. Labbani, O., Dekeyser, J.-L., Boulet, P., Rutten, É.: Introducing control in the Gaspard2 data-parallel metamodel: synchronous approach. In: Proceedings of the International Workshop MARTES: Modeling and Analysis of Real-Time and Embedded Systems (2005)
27. Latella, D., Majzik, I., Massink, M.: Automatic verification of a behavioral subset of UML statechart diagrams using the SPIN model-checker. In: Formal Aspects Computing, vol. 11, pp. 637–664 (1999)
28. Lysaght, P., Blodget, B., Mason, J.: Invited paper: enhanced architectures, design methodologies and CAD tools for dynamic reconfiguration of Xilinx FPGAs. In: FPL'06 (2006)
29. Maraninchi, F., Rémond, Y.: Mode-automata: about modes and states for reactive systems. In: European Symposium on Programming, Lisbon (Portugal), March 1998. Springer, Berlin (1998)
30. Mens, T., Van Gorp, P.: A taxonomy of model transformation. In: Proceedings of the International Workshop on Graph and Model Transformation, GraMoT 2005, pp. 125–142 (2006)
31. Nascimento, B., et al.: A partial reconfigurable architecture for controllers based on Petri nets. In: SBCCI '04: Proceedings of the 17th Symposium on Integrated Circuits and System Design, pp. 16–21. ACM, New York (2004)
32. OMG. Modeling and analysis of real-time and embedded systems (MARTE). http://www.omgmarte.org/
33. OMG. Portal of the Model Driven Engineering Community, 2007. http://www.planetmde.org
34. Paulsson, K., Hubner, M., Auer, G., Dreschmann, M., Chen, L., Becker, J.: Implementation of a virtual internal configuration access port (JCAP) for enabling partial self-reconfiguration on Xilinx Spartan III FPGA. In: International Conference on Field Programmable Logic and Applications, FPL 2007, pp. 351–356 (2007)
35. Quadri, I.-R., Boulet, P., Meftali, S., Dekeyser, J.-L.: Using an MDE approach for modeling of interconnection networks. In: The International Symposium on Parallel Architectures, Algorithms and Networks Conference (ISPAN 08) (2008)
36. Quadri, I.R., Elhillali, Y., Meftali, S., Dekeyser, J.-L.: Model based design flow for implementing an anti-collision radar system. In: 9th International IEEE Conference on ITS Telecommunications (ITS-T 2009), (2009)
37. Quadri, I.R., Meftali, S., Dekeyser, J.-L.: Integrating mode automata control models in SoC co-design for dynamically reconfigurable FPGAs. In: International Conference on Design and Architectures for Signal and Image Processing (DASIP 09) (2009)
38. Quadri, I.R., Muller, A., Meftali, S., Dekeyser, J.-L.: MARTE based design flow for partially reconfigurable systems-on-chips. In: 17th IFIP/IEEE International Conference on Very Large Scale Integration (VLSI-SoC 09) (2009)
39. Schäfer, T., Knapp, A., Merz, S.: Model checking UML state machines and collaborations. In CAV Workshop on Software Model Checking, ENTCS 55(3) (2001)
40. Schuck, C., Kuhnle, M., Hubner, M., Becker, J.: A framework for dynamic 2D placement on FPGAs. In: IPDPS 2008 (2008)
41. Sedcole, P., Blodget, B., Anderson, J., Lysaght, P., Becker, T.: Modular partial reconfiguration in virtex FPGAs. In: International Conference on Field Programmable Logic and Applications, FPL'05, pp. 211–216 (2005)
42. Segarra, M.T., André, F.: A framework for dynamic adaptation in wireless environments. In: Proceedings of 33rd International Conference on Technology of Object-Oriented Languages (TOOLS 33), pp. 336–347 (2000)
43. Szyperski, C.: ACM/Addison-Wesley, New York (1998)

44. Xilinx. Early access partial reconfigurable flow (2006). http://www.xilinx.com/support/prealounge/protected/index.htm
45. Yu, H.: A MARTE based reactive model for data-parallel intensive processing: Transformation toward the synchronous model. PhD thesis, USTL (2008)
46. Yung-Hsiang, L., Benini, L., De Micheli, G.: Dynamic frequency scaling with buffer insertion for mixed workloads. IEEE Trans. Comput.-Aided Des. Integr. Circuits Syst. **21**(11), 1284–1305 (2002)

Chapter 7
Wireless Design Platform Combining Simulation and Testbed Environments

Alain Fourmigue, Bruno Girodias, Luiza Gheorghe, Gabriela Nicolescu, and El Mostapha Aboulhamid

1 Introduction

Wireless is ubiquitous and new applications are multiplying every day. Remote medical assistance, just-in-time logistic systems, mobile live video streaming conferencing will all take advantage of this technology [1].

Slowly but surely, 3G systems are becoming integrated into our daily lives. Meanwhile, 4G systems in preparation promise to integrate various wireless technologies such as WiFi, WiMAX or GSM/WCDMA [2]. Given the demand, these technologies are becoming increasingly complex. The Physical (PHY) layer needs to be reconfigurable and the Media Access Control (MAC) layer needs to support security and quality of service (QoS).

In the wireless domain, there still is a gap between low-level simulators with sophisticated PHY modeling and high-level simulators with poor support for it. Simulation tools such as Matlab/Simulink [5] are frequently used to model PHY layers and IF/RF interfaces. However, these tools are forced to have a very high abstraction of the upper layers and may compromise important details. Moreover, the emergence of cross-layer designs necessitates a tightly coupled simulation/test of all the protocols layers. For a complete design flow, model-based design simulation requires seamless integration to a testbed platform.

In order to ensure realistic experiments, three requirements have to be taken into consideration:

A. Fourmigue (✉) · B. Girodias · L. Gheorghe · G. Nicolescu
Department of Computer Engineering, Ecole Polytechnique Montreal,
2500 Chemin de Polytechnique Montreal, Montreal, Canada H3T 1J4
e-mail: Alain.fourmigue@polymtl.ca

E.M. Aboulhamid
Department of Computer Science and Operations Research, University of Montreal,
2920 Chemin de la Tour Montreal, Montreal, Canada H3T 1J4

G. Nicolescu et al. (eds.), *Design Technology for Heterogeneous Embedded Systems*,
DOI 10.1007/978-94-007-1125-9_7, © Springer Science+Business Media B.V. 2012

1. MAC and PHY layers under development must interact with the upper layers using the same mechanisms as those that can be found in a final implementation of protocols.
2. The performance of the simulated PHY layer has to be comparable to that of a hardware device. These two requirements ensure that the simulation of the PHY layer is as transparent as possible from the point of view of the upper layers.
3. Each entity in the wireless network (i.e. subscriber station, access point) has to be emulated on a separate physical machine running an instance of the platform. This last requirement ensures that the testbed environment exploits all the computational resources available on the machine efficiently.

This chapter presents a platform for designing wireless protocols. The proposed platform combines a simulation environment based on Matlab/Simulink with a testbed environment based on the GNU/Linux system. Matlab/Simulink provides model-based design, while Linux offers a flexible environment with real-world applications as well as a complete implementation of network protocols. In the proposed platform, the PHY layer is simulated through the simulation environment and the MAC layer is integrated into the testbed environment. The platform provides all the interfaces required to evaluate the MAC and PHY layers working together with the Linux TCP/IP stack, processing real data produced by any of the various network applications that run under Linux.

The chapter is organized as follows. In Sect. 2 we present the existing works on simulation and testbed environments, and in Sect. 3 we give the basic concepts used in our work. Section 4 overviews the architecture of the proposed platform, the Linux Networking Stack. Section 5 shows a wireless protocol implementation with the proposed platform. Section 6 presents the applications and the configuration of the experiment and Sect. 8 gives the results. Section 8 summarizes and concludes the chapter.

2 Related Work: Current Simulation and Testbed Environments in the Networking Domain

Currently, a large number of network simulators and testbeds can be found on the market [6, 7, 9–12]. Network simulators are programs which model the behavior of networks using mathematical formulae. Testbeds are real systems built in a controllable environment.

Two of the most popular network simulators are NS-2 [6] and OPNET [7]; both support simulation of TCP/IP protocols and classic MAC protocols. They are flexible, autonomous applications executed in a self-confined context (i.e. they do not need to interact with any other component to simulate a network).

The TrueTime tool consists of a Simulink block library used to simulate networked control systems [9]. It is based on a kernel block to simulate a real-time environment. However, compared to NS-2 and OPNET, its layers models are not as accurate [8].

Although network simulators provide a good degree of flexibility and reusability, they are often criticized for their lack of accuracy [10, 12]. Therefore, network simulators may not be able to truthfully capture a real wireless environment, since they lack accurate protocol implementations.

Testbed platforms allow for the evaluation of protocols with representative traffic and popular real-world applications. Many testbeds in the literature are based on the Linux platform [10–12]. The possibility to integrate specific hardware such as antennas or FPGAs through a device driver to a real networking implementation (i.e. the Linux TCP/IP stack) makes it a highly attractive environment for researchers and designers. Any application built on TCP/IP can be used as an application for testing purposes.

However, producing testbeds is time consuming and requires significant effort, while pure simulation has the advantage of being flexible and reusable. Therefore, a growing number of platforms for designing wireless protocols attempt to merge simulation and testbed environments [10] to extract the best of both worlds.

Hydra [12] is based on a flexible Linux-based testbed that can be used to validate wireless protocols. Hydra uses software-based MAC and PHY layers. To facilitate the experiments, Hydra uses the Click framework [13] for its MAC implementation. Click allows users to write modular packet processing modules [12].

The platform proposed in this chapter differs from the existing testbeds because it allows more flexible designs by combining a testbed environment with a simulation environment. Compared with Hydra, the simulation environment used in our platform allows model-based design, while the testbed environment enables the reuse of an existing wireless MAC implementation. Moreover, the proposed platform allows more accurate testing of the designed protocols by combining a simulation environment with a testbed environment.

3 Basic Concepts

3.1 Wireless Technologies

This section introduces the basic concepts required to understand the key features of wireless technologies.

3.1.1 Key Wireless Technologies

Nowadays, a large number of technologies such as GSM/UMTS, WiMAX, WiFi and BlueTooth are available to transfer data over a wireless network. To understand the differences and the similarities between them, we examine the networks they target. Depending on their size, four types of wireless networks can be observed: wide, metropolitan, local and personal area networks. GSM/UMTS is used in cellular networks which are wide area networks (WAN) that can cover areas as broad as countries. WiMAX is a recent wireless technology designed to provide mobile, high-quality and high data-rate service in metropolitan networks (MAN), ranging from campus networks to entire cities. WiFi is the most commonly used wireless technology in local area networks (LAN), which are usually limited to a building. Bluetooth is used in personal networks (PAN) that are very small networks, mostly used to connect electronic devices to a computer in a master-slave relationship.

Contrary to wired technologies, such as Ethernet or ADSL, which use a protected medium, wireless technologies use an unprotected medium (mostly radio waves) which is subject to noise or interference. Therefore, wireless technologies include more sophisticated communication mechanisms than wired technologies. Designing wireless protocols is a complex task which requires a good knowledge of all the aspects involved in wireless technologies.

3.1.2 Key Aspects of Wireless Technologies

To understand the key aspects of wireless technologies, we can use the Open System Interconnect (OSI) model to analyze the structure of wireless technologies. The OSI model is a theoretical model describing network activities through a layered architecture. Each layer includes several protocols and provides services to the upper layer. The two layers that define a network technology (and are tightly coupled) are the physical (PHY) layer and the Data Link Layer.

The PHY Layer The PHY Layer is responsible for the physical transmission of the data regardless of the content; it deals only with signal processing. Wireless PHY layers use advanced coding and modulation techniques to strengthen the robustness of the communications in a wireless environment. A single wireless technology often uses several PHY layers. For instance, the IEEE 802.16 standard (WiMAX) defines several PHY layers to be used with different frequency ranges and applications.

The Data Link Layer The Data Link Layer is responsible for making the transmission medium a trustable link for the upper layers and it includes various protocols to control the data flow and to detect transmission errors. In MAN and LAN, the Data Link layer is further divided into a Medium Access Control (MAC) layer and a Logical Link Control (LLC) layer. While the main purpose of the MAC layer is to provide the access method and arbitrate the access to the medium, the LLC layer is mainly used to interface the MAC layer with the various network protocols such as IP or IPX used in the upper layer. The LLC layer is usually implemented within the operating system's networking stack; therefore, the designers of wireless technologies are more concerned with the protocols of the MAC layer. The rest of this chapter focuses on the MAC layer rather than on the LLC layer.

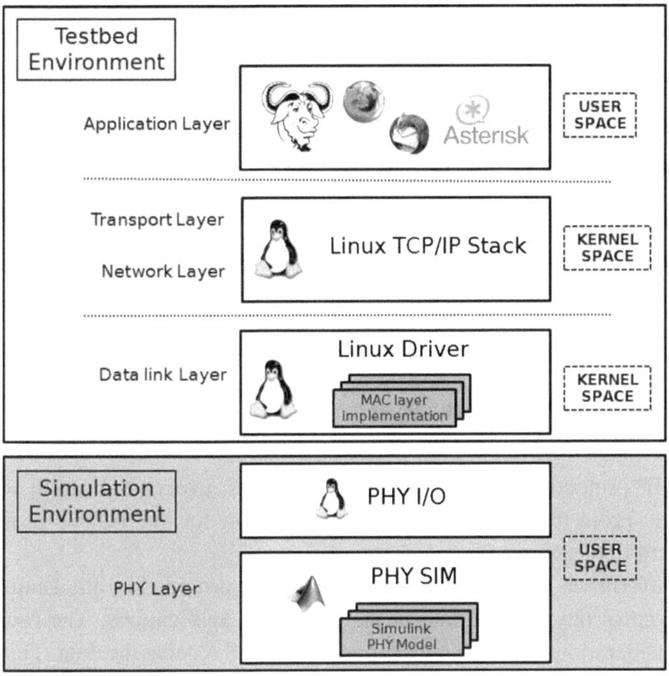

Fig. 7.1 Platform overview

4 Wireless Design Platform

4.1 Platform Overview

The platform presented in this chapter offers a comprehensive suite of development tools for modeling and exploring MAC protocols, as well as PHY layers. A simulation environment based on Matlab/Simulink is integrated seamlessly to a testbed environment based on the GNU/Linux system. The platform allows accurate testing of protocols with real-world applications communicating over a network. Each entity in the wireless network (i.e. subscriber station, access point) is emulated on a separate physical machine running an instance of the platform. Figure 7.1 presents an overview of the platform. It shows the different layers and their respective running address space in Linux. Like most modern operating systems, Linux uses different address spaces. Users' applications execute in user-space where they run in an unprivileged mode. They cannot interact directly with the hardware, whereas the kernel uses a protective memory space called kernel space and has complete access to the hardware.

4.1.1 Application Layer

Deploying Voice over IP (VoIP) and Internet Protocol for Television (IPTV) over wireless media has raised unprecedented quality and performance issues. To face theses challenges, tomorrow's protocols have to be validated with real-world applications. Skype [14] and Asterisk [15] are two popular applications using VoIP. Since these applications are available under Linux, the proposed platform allows for the use of either in order to evaluate protocol designs. More traditional applications like HTTP or FTP clients are also available.

4.1.2 Transport and Network Layer

The TCP/IP protocol suite is at the core of almost every application written for networking. From the early years of Linux, support for the TCP/IP protocols has been present in the kernel.

The platform we propose uses the routing table provided by the Linux network stack to control the flow of data produced by the applications. The routing table allows consideration of which network interface will receive the data. This is a convenient way to choose which MAC and PHY layers will process the data.

4.1.3 MAC Layer

The MAC layer is integrated to the testbed environment and is implemented as part of a Linux driver. A Linux driver can be compiled as a kernel module and loaded on demand. Therefore, the MAC implementation is a piece of code that can be added to the kernel at runtime without rebooting the machine.

The proposed platform provides a generic Linux driver that can embed a custom MAC implementation through a well-defined interface. The MAC implementation can even be written in an object oriented style, although this is not the approach followed by the most common drivers. Designers can plug their own MAC implementation into the platform as long as they respect the basic rules of the interface provided by the platform.

Usually, the drivers are written to access hardware modules that are eliminated in the platform presented here; there is however a simulation environment based on Matlab/Simulink that is executed in the user space. Therefore, the main role of the driver is to forward the data to the user space so that it can be processed by the simulation environment. The driver also declares a network interface which is the entry point to the kernel network stack and the embedded MAC layer can communicate with the kernel stack. The privileged role of the driver makes the simulation of the PHY layer as transparent as possible to the upper layers.

4.1.4 PHY Layer

The PHY layer is integrated into the simulation environment and is composed of two main elements: PHY I/O and PHY SIM.

The PHY I/O is a user process that serves three purposes:

1. The collection of data produced by the testbed environment via the interface with the testbed.
2. The accessibility of the collected data for the PHY SIM component via a co-simulation interface.
3. The transfer of data to other instances of the platform via an interface called the inter-node interface.

The PHY SIM is at the heart of the simulation environment. It uses Matlab/Simulink to simulate the functionality of the PHY layer. Matlab/Simulink favors model-based design; hence the model of the system is at the center of the design process, from specification capture to final tests.

The proposed platform allows for the abstraction of the PHY layer. Thus, designers can concentrate on MAC layers and quickly test the concepts without an accurate PHY model.

4.2 Platform Interfaces

This section presents the three different interfaces required by the proposed platform. These interfaces allow the different components in the platform to work together.

4.2.1 Interface I: Between the Testbed and the Simulation Environments

Interface I is the interface between the testbed environment and the simulation environment and is in charge of data transfer between the MAC layer and PHY I/O. In modern architectures, the MAC layer usually interacts with a hardware device via a DMA controller. In the proposed platform, the hardware device is simulated through an environment which runs in user space. The MAC layer, integrated within the testbed environment and implemented in a driver, is executed in kernel space. Therefore, an interface is required to enable effective data transfers between the kernel address space and the user address space. To keep the design as realistic as possible, the proposed platform models DMA-like transactions using a memory mapping between the MAC layer embedded in the driver and the PHY I/O. The memory mapping is a mechanism provided by the Linux kernel to enable effective data transfers that allows the mapping of a device memory directly into a user process address space. To set up a memory mapping, the proposed driver implements the *mmap* method. The PHY I/O is then able to read/write directly to the driver's memory using the *mmap* system call.

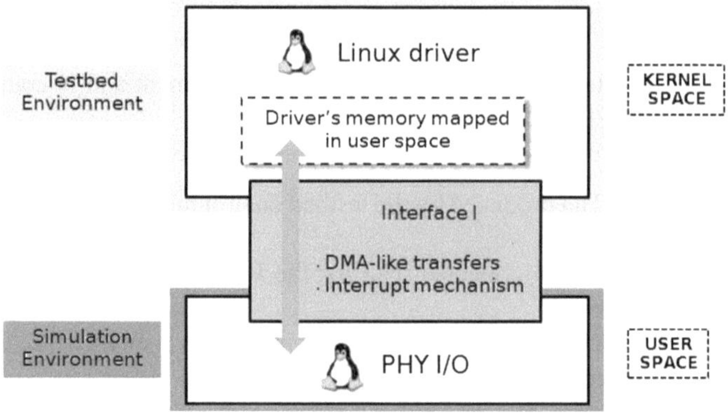

Fig. 7.2 Memory mapping between the MAC layer and the PHY I/O component

Figure 7.2 represents the MAC/PHY interactions allowed by the interface. When a package is ready to be sent, the driver notifies the PHY I/O which starts copying the data into the user space and simulates the interruption the hardware would normally trigger.

4.2.2 Interface II: A Co-simulation Interface Inside the Simulation Environment

Interface II is implemented inside the simulation environment and is in charge of the synchronization of the data transfers between the PHY I/O and the PHY SIM.

In our previous work, we proposed a generic co-simulation interface able to communicate and synchronize data and events from/to Simulink to/from a Linux user process. The behavior of this interface was formally defined and verified in [16] and implemented in [17]. This approach is applied for the definition of the interface II.

The interface II is implemented as a Simulink S-function block programmed in C++. The communication and synchronization is ensured using the triggered subsystem component from the Simulink library and the Inter-Process Communication mechanisms (i.e. shared memories, semaphores) available in Linux.

4.2.3 Interface III: Between Different Instances of the Platform

Interface III is implemented inside the simulation environment and is in charge of data transfer between different instances of the platform running on separate physical machines.

The proposed platform is designed to test wireless protocols over a network. Each entity of a wireless network (e.g. subscriber station, access point) is emulated by a dedicated machine running an instance of the platform. Therefore, the data need to be physically carried from one machine to another.

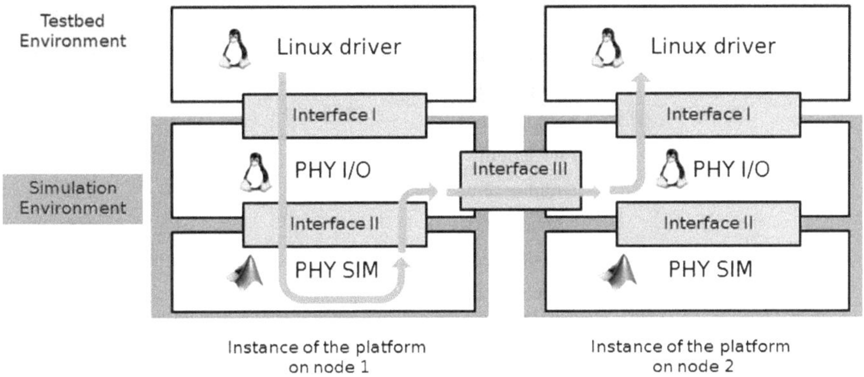

Fig. 7.3 Data flow between two instances of the platform

Each machine runs an instance of the simulation environment which simulates three phenomena: frame transmission by the sender, frame propagation through free space and frames reception by the receiver. Figure 7.3 shows the data flow when node 1 sends data to node 2. The frame transmission (TX), the frame propagation and the frame reception (RX) are modeled in the PHY SIM on node 1. The data are then transferred directly from the PHY I/O on node 1 to the PHY I/O on node 2.

5 Implementation of the IEEE 802.16 Protocol

To platform was used to implement the IEEE 802.16 protocol [18]. It is a very complex protocol; hence, we focus only on mandatory features. The simulation environment executes a Simulink model of the 802.16 PHY, while the testbed environment integrates a MAC implementation of the 802.16 protocol.

5.1 Using the Simulation Environment

The simulation environment executes a Simulink model of the 802.16e PHY layer, available through Mathworks File Exchange [19, 20].

The model respects the IEEE 802.16-2004 standard [18] for the PHY layer but does not support MAC operations. It consists of three main components: transmitter, channel and receiver. The transmitter and the receiver consist of channel coding and modulation sub-components, whereas the channel is modeled by Simulink's AWGN channel block [20]. The AWGN channel block adds white noise to the input signal using the Signal Processing Blockset Random Source block. The model supports all the basic blocks: randomization, Reed-Solomon codec, convolutional codec and interleaving, and includes mandatory modulation: BPSK, QPSK, 16-QAM and 64-QAM. It simulates OFDM (orthogonal frequency division multiplexing) transmission with 200 subcarriers, 8 pilots and 256-point FFTs (see Table 7.1).

Table 7.1 Characteristics of the simulated 802.16 OFDM PHY

PHY	OFDM
Modulation	QPSK-3/4
Channel bandwidth (MHz)	10
Sampling factor	57/50
FFT length	256
Number of used data subcarriers	192
Cyclic prefix	1/8
OFDM symbol duration (μs)	25
Frame duration (ms)	5
Number of OFDM symbols per frame	200
Date rate (Mbps)	11.52
Duplexing mode	TDD
DL/UL ratio	2:1
SNR (dB)	30

5.2 Using the Testbed Environment

This subsection presents an implementation of the 802.16 MAC layer and its integration to the testbed environment as part of a Linux driver. To demonstrate the possibility to re-use existing MAC implementations, we have chosen to re-use, with a few minor modifications, the design of an existing WiMAX driver to implement the 802.16 MAC layer. Our implementation is based on the design of the driver for Intel's WiMAX Connection 2400 baseband chips.

Recently, Intel announced integrated WiMAX network adaptors in its new Centrino 2 platform [21]. These chipsets (codenamed "Echo Peak" and "Baxter Peak") are certified by the WiMAX Forum [22] and will soon be embedded in Intel's new Centrino 2 platform. To support the imminent arrival of WiMAX devices, the Linux kernel since version 2.6.29 includes a WiMAX stack and provides drivers for devices based on the Intel WiMAX Connection 2400 baseband chip.

Intel's WiMAX driver is split into two major components: the module i2400m.ko which acts as a glue with the networking stack and the module i2400m-usb.ko as a USB-specific implementation. As the WiMAX PHY model is not designed to process Intel's device-specific commands, we ignore all these control messages. We use Intel's design to transfer data to the simulation environment to model realistic transactions.

6 Application

This section presents the applications in their context, the configurations required to test our platform and gives details on the QoS settings and the implemented scheduling algorithm.

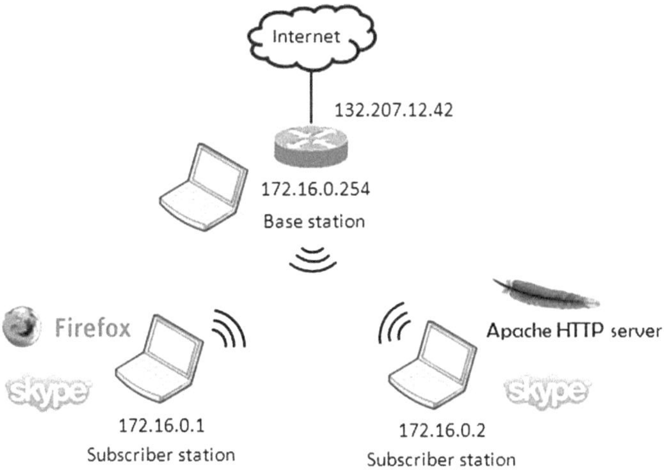

Fig. 7.4 Simulated WiMax network

Table 7.2 Technical specifications

	Base station	Subscriber station
Processor	4 Xeon 3.4 GHz	Intel Core 2 duo 2.0 GHz
L2 Cache (MB)	1	1
RAM (MB)	2048	2048

6.1 WiMax Configuration

This subsection presents the configurations necessary to ensure the realism of the network used to carry out the tests. These configurations are specific to WiMAX networks and deal with address resolution, Maximum Transmission Unit (MTU) and subnetting.

To setup a realistic network, we reproduce the typical WiMAX network architecture which is based on a point-to-multi-point topology. The 802.16 standard defines two logical entities, the base station (BS) and the subscriber station (SS). The subscriber station (SS) is an end-user equipment that provides connectivity to the IEEE 802.16 networks. It can be either fixed/nomadic or mobile equipment. The base station (BS) represents generic equipment that provides connectivity, management, and control between the subscriber stations and the IEEE 802.16 network.

Figure 7.4 shows the WiMAX network used for the demonstration. It is composed of a BS and two fixed SSs. We emulate each SS on a laptop and the BS on a desktop computer. Each SS is connected to the BS by a simulated 802.16 point-to-point link. The BS acts as a gateway to provide internet access to the SSs. There is no direct connection between the SSs. The Linux machine which emulates the BS is configured as a router.

Table 7.2 lists the hardware specifications.

Table 7.3 Traffic classification on the first SS

Application	Use case	Service class
Skype	video call (24 s)	UGS
VLC media player	streaming a media file	rtPS
FTP client	uploading a 1 MB file	nrtPS
Firefox	web surfing	BE

The IEEE 802.16 standard provides two solutions for the transmission of IPv4 packets: they can be carried directly over IEEE 802.16 links, or they can be encapsulated in Ethernet frames carried over 802.16 links. We have chosen to implement WiMAX devices as pure IP devices. IPv4 packets are carried directly over the simulated 802.16 point-to-point links. IPv4 packets are not encapsulated in Ethernet frames.

Since we use point-to-point links, address resolution is not needed, thus eliminating the need to carry ARP packets. However, the point-to-point link model raises some issues. DHCP messages can be carried over 802.16 frames but common DHCP implementations only understand the Ethernet frame format. To circumvent this issue, we assign static IP address to the stations within the range 172.16.0.{0–16}. Since we use a point-to-point link model, each SS resides on a different IP subnet.

The Maximum Transmission Unit (MTU) is another source of concern, since IPv4 packets define the maximum size of IP payloads in 802.16 MAC PDUs. This parameter, which is configurable, has a significant impact on the generated traffic since all the IP packets larger than the MTU will be fragmented. The Internet-Draft [23] strongly recommends the use of a default MTU of 1500 bytes for IPv4 packets over an IEEE 802.16 link. However, the WiMAX forum has already defined a network architecture where the transmission of IPV4 packets over IEEE 802.16 links uses a MTU of 1400 bytes. To increase the realism of the experiment, we have chosen to use this latter figure for MTU.

6.2 Traffic Load

In order to evaluate the 802.16 protocol with representative traffic we used typical real-world applications with different QoS requirements. The IEEE 802.16d standard defines four traffic classes (UGS, rtPS, nrtPS and BE) which provide different levels of QoS. For each traffic class, we use a typical application which should benefit from the QoS provided by the service (Table 7.3).

Unsolicited Grant Service (UGS) connections have a fixed data-rate and are recommended for transferring data at a constant rate. VoIP applications which do not use silence suppression should benefit from UGS connections. We use Skype, a VoIP application to make a video call between the two SSs.

Real-time Polling Service (rtPS) supports delay-sensitive applications that generate variable-sized packets. rtPS connections are a good choice to carry bursty traffic

Table 7.4 Network stream

Traffic	UGS	rtPS	nrtPS	BE
Minimum reserved traffic rate (kbps)	512	512	256	0
Maximum sustained traffic rate (kbps)	512	1024	512	128
Maximum latency (ms)	10	2000	∞	∞

such as streaming videos. We run VLC media player as a server on one SS to stream a media file and as a client on the other SS to receive the streaming video. Table 7.4 presents the characteristics of the media file chosen for the simulation.

Non real-time Polling Service(nrtPS) is designed for non-delay-tolerant data streams for which a minimum data-rate is required. FTP traffic is a typical application that requires the use of the nrtPS class. We use the classical FTP client available under Linux to upload a 1MB file from the SS to the BS.

Best Effort (BE) service has no minimum reserved traffic rate and is simply a best effort service. A BE connection can be used to carry HTTP traffic which has no strong QoS requirement. We use the popular web browser Firefox to simulate a SS downloading a 71 KB HTTP page from an Apache HTTP server running on the other SS.

6.3 QoS Settings

The IEEE 802.16 standard does not specify a scheduling algorithm. The implementation of QoS scheduling algorithms is left to the vendors, for product differentiation. This subsection presents the QoS settings in the simulation and explains the scheduling decisions made by the BS and the SSs.

The IEEE 802.16 standard uses the concepts of minimum reserved traffic rate and maximum sustained traffic rate for a service. The minimum reserved traffic rate is guaranteed to a service over time, while the maximum sustained traffic rate is the rate the service would expect to transmit at, in the absence of bandwidth demand from the other services. UGS has a minimum reserved traffic rate that is always equal to its maximum sustained traffic, while rtPS, nrtPS and BE services have variable reserved traffic rates.

Table 7.5 presents the QoS parameters of the four connections to support the applications chosen earlier. Based on these settings, the SS requests the bandwidth from the BS and makes its own scheduling decisions to use the bandwidth granted.

Our implementation guarantees that all the services receive their minimum reserved traffic rate, thus eliminating the risk of starvation. Excess bandwidth is then distributed proportionally to the services' priority. Our implementation monitors the queues to ensure that the maximum latency for the UGS and rtPS connections is not exceeded. If UGS or rtPS queues are too long to provide tolerable latency, the SS is configured to request more bandwidth.

Table 7.6 shows the scheduling decisions made by the SS to use the 1728 kbps bandwidth granted by the BS (on the uplink). Our simulated network involves only

Table 7.5 QoS settings at the SS

Video stream		Audio stream	
Codec	DIVX	Codec	mpga
Resolution	512 × 336	Sample rate (KHz)	44
Frame Rate (fps)	23,976	Bitrate (kbps)	192

Table 7.6 QoS scheduling decisions made by the SS

Total bandwidth requested by the SS (kbps)				1728
Total bandwidth granted by the BS (kbps)				1728
Total minimum reserved traffic rates (kbps)				1280
Excess bandwidth (kpbs)				448
Traffic class	UGS	rtPS	nrtPS	BE
Minimum reserved traffic rate (kbps)	512	512	256	0
Distribution of excess bandwidth (kbps)	0	256	128	64
Traffic rate (kbps)	512	768	384	64

three stations, therefore the SS can request and obtain more bandwidth. To study queue buffering, we configure the SS to underestimate the bandwidth required for the rtPS connection deliberately. The SS allocates a bandwidth of 768 kbps for the rtPS connection, with a minimum reserved traffic rate of 512 kbps and a maximum sustained traffic rate of 1024 kbps. The VLC media player obviously requires more than 768 kbps to stream the chosen media file. We expect the rtPS queue to quickly saturate, the latency of the rtPS connection to increase exponentially and the VLC media player to quit. Our objective is to evaluate if the queue monitoring allows the SS to request more bandwidth before the VLC media player application suffers too much from the poor QoS of the connection.

7 Experimental Results

This section presents the experimental results. The first subsection provides a low-level analysis of package transmission at the MAC layer level. The second subsection demonstrates the importance of PHY layer modeling through the simulation of noise on the wireless channel.

7.1 Results Related to the MAC Layer

The objective of the experiment is to analyze the behavior of our MAC implementation under real traffic conditions. This subsection provides a low-level evaluation of packet transmission at the driver level.

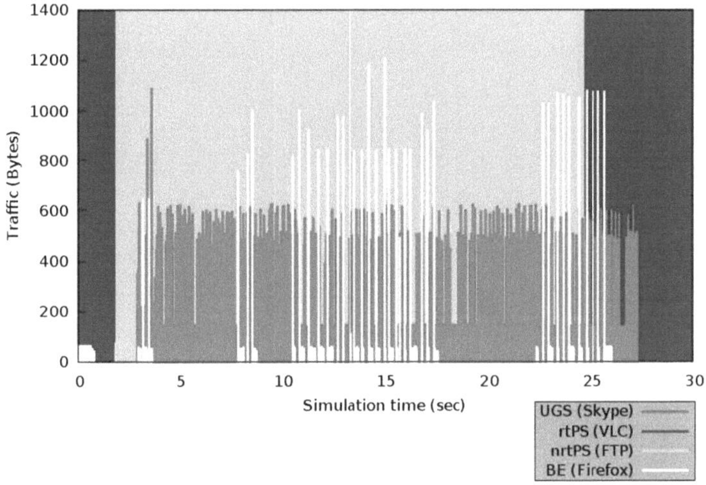

Fig. 7.5 SS's outgoing traffic with QoS support

Figure 7.5 gives an overview of the traffic generated by all four applications chosen for the simulation. Skype traffic (UGS traffic) is composed mainly of two types of packet: video packets interleaved with voice packets. The video packets have an average size of 500 bytes while the voice packets have an average size of 130 bytes. rtPS traffic shows that the VLC media player always uses the maximum size allowed for IP packets (1400 bytes) to stream the media file. The same conclusion applies to FTP traffic (nrtPS traffic). Firefox's HTTP requests can be clearly identified in BE traffic. The requests correspond to the large packets (800 bytes and more), while the short packets (around 50 bytes) acknowledge the receipt of the data.

Figure 7.6 gives accurate timing information for each packet sent by the SS. Low level timing analysis allows the measurement of the traffic bursts generated by the applications. The maximum traffic burst is the largest burst that can be expected at the incoming port of the services and depend on the applications' behavior.

Figure 7.7 and Table 7.7 show the variation in queue lengths during the simulation, and provide information that may help vendors to design their products. Queues can grow very quickly if the transmission rate is not appropriate to the level of traffic. Because we deliberately set a low transmission rate, the rtPS connection quickly reached a maximum of 135 packets queued.

As queue monitoring is enabled, the SS requests for more bandwidth and manages to contain the flow of packets produced by VLC media player. After the Skype video call terminates, the SS allocates the bandwidth reserved for the UGS connection to the rtPS connection. The length of the rtPS queue quickly decreases as the rtPS connection can transmit at its maximum sustained rate. The nrtPS queue increases linearly with time which means that the rate chosen to support FTP traffic is also underestimated. It takes about 20 s for the FTP client to send the 1MB file, then it stops generating packets which allows the scheduler to drain the nrtPS queue. Figure 7.8 shows the time required to transfer a package to the PHY layer once it has

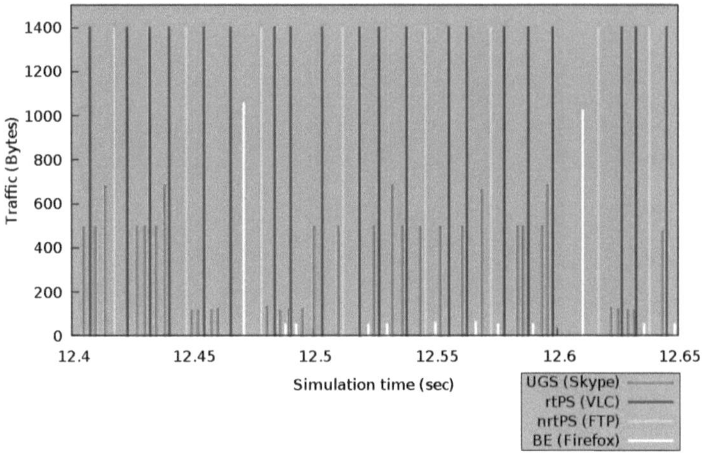

Fig. 7.6 Accurate analysis of SS's outgoing traffic

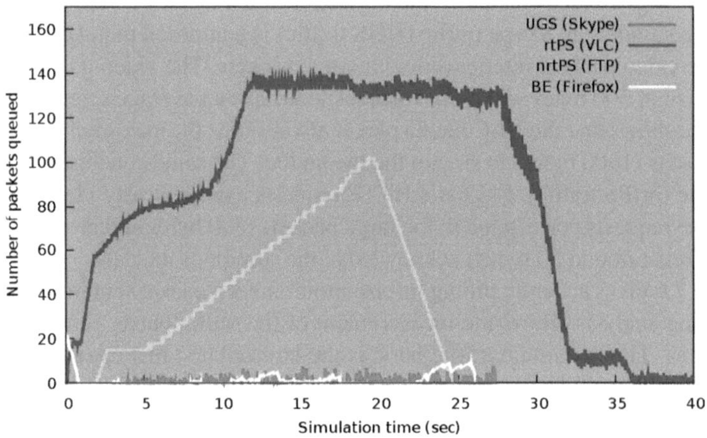

Fig. 7.7 Evolution of queue length during the simulation

Table 7.7 Number of packets in the various queues

Simulation time (s)	5	12	20	24	30	35
UGS (Skype)	0	2	4	1	0	0
rtPS (VLC)	77	135	134	130	85	15
nrtPS (FTP)	15	54	96	9	0	0
BE (Firefox)	0	3	0	8	0	0

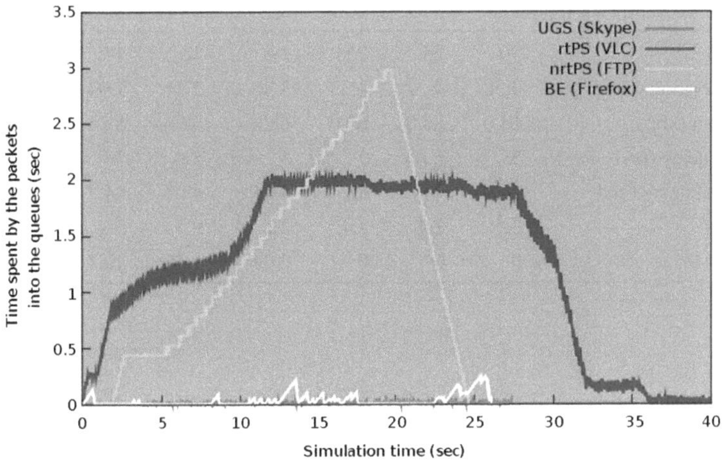

Fig. 7.8 Time spent by the packets in the queues before being transmitted

been queued. The latency incurred by packet queuing depends on the traffic class. As expected, the UGS traffic benefits from the best QoS with the smallest latency. The BE connection shows peaks of latency as soon as there are a few packets to transmit, which was expected too, since this connection only provides a best effort service. As expected, the latency of the rtPS connection quickly increases. The maximum latency of this connection (set to be 2000 ms) is reached after 11 s. The SS uses this information to estimate the rate of VLC media player's traffic. The SS requests for appropriate bandwidth and maintains the latency within acceptable limits. By contrast, the latency of the nrtPS connection stops increasing only because the FTP client has finished to send the file. Maximum latency is not a QoS parameter for nrtPS connections, therefore the queue would keep increasing if the file was larger. Our implementation of QoS provides appropriate scheduling of the traffic classes.

Throughout this experiment, our objective was to evaluate the kind of information the platform can provide. The analysis of packet transmission and queue buffering gave accurate information on the various stages of packet transmission with the 802.16 protocol. This experiment demonstrates that the proposed platform can be used to conduct simulations where end-user applications provide a Simulink model of the PHY layer with real-time traffic.

7.2 Results Related to the PHY Layer

To demonstrate the importance of modeling the PHY layer, we propose to simulate different qualities of the wireless signal using the simulation environment. The 802.16 PHY model used for the demonstration allows the configuration of the Signal-to-Noise ratio (SNR) of the channel. The model also computes the transmission errors and modifies the frames delivered by the network stack. Afterwards,

Table 7.8 IxChariot's settings

SNR (dB)	30	25	20	18	16	14	12	10
MOS estimation (out of 5)	4.38	4.27	4.23	3.36	2.24	1.94	1.27	1.0
Throughput (kbps)	64.0	64.0	64.0	63.8	57.6	53.4	55.1	53.2
One-way delay (ms)	5	5	7	5	5	5	5	6
End-to-end delay (ms)	44.7	44.6	45	44.8	44.7	44.7	44.7	45.2
Jitter (ms)	2.5	2.4	3.4	2.1	9.5	3.2	4.6	2.7
Lost data (%)	0	0	0	0.096	1.45	10.8	34.8	47.0

the frames are returned to the network stack. Therefore, we expect the transmission errors calculated by the Simulink PHY model to be handled in a realistic manner. Throughout this experiment, our objective is to investigate the effect of poor wireless channel conditions in WiMAX networks.

To measure the network performance with varying levels of noise, we use a dedicated software called IxChariot [24]. IxChariot is a popular test tool which allows the assessment of network performance under realistic load conditions. It can emulate real application flows to measure various parameters such as response time, network throughput or lost data and is often used to determine whether a network can support specific applications such as VoIP applications or streaming multimedia applications.

We simulate poor channel conditions through the Simulink model and we use IxChariot for testing the resulting environment. IxChariot consists of a console and a set of endpoints; the console instructs endpoints to run a test and return the results. To test the simulated 802.16 link between the SS and the BS, a first endpoint is installed on the first SS and a second endpoint is installed on the BS. The IxChariot console is installed on another computer on the same network.

We configure the Simulink model to simulate a TX signal of 10 mW. We start our experiments with a SNR of 30 dB. Then, we decrease step by step the SNR. We use IxChariot to determine whether the simulated 802.16 link can still support VoIP applications using the G.711u codec. Table 7.8 shows IxChariot's settings.

Table 7.9 shows various parameters measured by IxChariot to assess the network performance. IxChariot estimates the Mean Opinion Score (MOS) which gives a numerical indication of the perceived voice quality in the different tests. As expected, the MOS estimated by IxChariot decreases while the data lost increases. When the percentage of lost data becomes too high, the link cannot even support the 64 kbps throughput which is required by the G.711u codec. This experiment demonstrates that PHY modeling facilitates the simulation of changes in environmental conditions. Coupled to a real network implementation which can process the data in a realistic manner, a model-based approach for the PHY layer enables simple and accurate testing of PHY concepts.

Table 7.9 IxChariot's results

Test duration (s)	60
Type of application	VoIP
Network protocol	RTP
Codec	G.711u
Packet size (Bytes)	200

8 Conclusion and Future Work

This chapter presents a platform for designing and testing wireless protocols. Inspired by existing projects, we propose a novel approach combining a simulation environment with a testbed environment. The simulation environment uses Matlab/Simulink, and in order to enable rapid modeling of the PHY layers using model-based design, the testbed environment is based on the Linux platform and is as close as possible to final implementation. The platform provides all the necessary interfaces to combine the simulation and the testbed environments.

From the illustrative example of the IEEE 802.16 protocol, we were able to define the issues in the building of wireless protocols and demonstrate the concepts put forward in this chapter. The experimental results show that it is possible to conduct realistic experiments through accurate PHY modeling, MAC layer support and real upper layer implementation. Despite a few compromises, the platform achieves its objective: to fill the gap between low level simulators with poor support of upper layers and high level simulators with poor support for PHY modeling.

Future work will focus on refining the interface between layers to ease the exploration of cross layer designs. Work will also be done to make the platform able to support other network topologies such as the mesh topology.

References

1. Karlson, B., et al.: Wireless Foresight: Scenarios of the Mobile World in 2015. Wiley, Chichester (2003)
2. Qaddour, J., et al.: Evolution to 4G wireless: problems, solutions, and challenges. In: ACS/IEEE International Conference on Computer Systems and Applications, p. 78 (2005)
3. Keating, M., Bricaud, P.: Reuse Methodology Manual for System-on-a-Chip Designs. Kluwer Academic, Boston (2002)
4. Nicolescu, G., Jerraya, A.A.: Global Specification and Validation of Embedded Systems. Springer, Berlin (2007)
5. Matlab/Simulink's web site. http://www.mathworks.com (2009)
6. NS-2's web site. http://www.isi.edu/nsnam/ns/ (2009)
7. OPNET's web site. http://www.opnet.com (2009)
8. Lucio, G.F., et al.: OPNET modeler and Ns-2-Comparing the accuracy of network simulators for packet-level analysis using a network testbed. In: 3rd WEAS International Conference on Simulation, Modelling and Optimization (ICOSMO), pp. 700–707 (2003)
9. Cervin, A., et al.: Simulation of networked control system using True-Time. In: Proc. 3rd International Workshop on Networked Control Systems: Tolerant to Faults (2007)
10. Jansang, A., et al.: Framework architecture for WLAN testbed. In: AMOC 2004 (2004)

11. Armstrong, D.A., Pearson, M.W.: A rapid prototyping platform for wireless medium access control protocols. In: IEEE International Conference on Application-Specific Systems, Architectures and Processors, pp. 403–408 (2007)
12. Mandke, K., et al.: Early results on Hydra: a flexible MAC/PHY multihop testbed. In: Proc. of the IEEE Vehic. Tech. Conference, Dublin, Ireland, April 23–25 (2007)
13. Kohler, E., et al.: The Click modular router. ACM Trans. Comput. Syst. **18**(3), 263–297 (2000)
14. Skype's web site. http://www.skype.com
15. Asterisk's web site. http://www.asterisk.org
16. Gheorghe, L., et al.: Semantics for model-based validation of continuous/discrete systems. In: DATE, pp. 498–503 (2008)
17. Bouchhima, F., et al.: Generic discrete-continuous simulation model for accurate validation in heterogeneous systems design. Microelectron. J. **38**(6–7), 805–815 (2007)
18. IEEE 802.16-2004. http://standards.ieee.org/ (2007)
19. MathWorks: File-Exchange. http://www.mathworks.com/matlabcentral/ (2008)
20. Khan, M.N., Gaury, S.: The WiMAX 802.16e physical layer model. In: IET International Conference on Wireless, Mobile and Multimedia Networks, January 2008
21. Intel's white paper, Delivering WiMAX faster. http://download.intel.com/technology/wimax/deliver-wimax-faster.pdf (March 2009)
22. WiMAX forum web's site. http://www.wimaxforum.org (2009)
23. Internet-Draft, 16ng Working Group,Transmission of IPv4 packets over IEEE 802.16's IP Convergence Sublayer (October 2008)
24. IxChariot, Ixia Leader in IP Performance Testing. http://www.ixiacom.com/products/performance_applications

Chapter 8
Property-Based Dynamic Verification and Test

Dominique Borrione, Katell Morin-Allory, and Yann Oddos

1 Introduction

Systems on a chip today consist in possibly dozens of interconnected active components that communicate through sophisticated communication infrastructures. To guarantee the correct functionality of such a system is an ever increasing complex task that demands a rigorous design and verification methodology supported by a large variety of software tools.

The verification problem is made difficult in particular because of two factors: (1) the complexity in terms of state number and verification scenarios increases exponentially with the system size; (2) constructing systems on chip often involves the insertion of design IPs provided by external sources which are given as black-boxes, so that the internal state encoding of such IPs is unknown.

A straightforward approach is the decomposition of the verification problem along the lines of the system structure: modules are first verified independently, possibly using formal verification techniques; then the correctness of their composition focuses on the module interactions, assuming the correctness of each individual module.

Property Based Design is increasingly adopted to support this compositional verification approach [4, 10]. In this context, the properties we refer to are functional properties which express relationships between the values of design objects (signals, variables), either at the same time or at different times. A property is written in a declarative style rather than as an algorithm. Early property specification languages

D. Borrione (✉) · K. Morin-Allory · Y. Oddos
TIMA Laboratory, 46 Avenue Félix Viallet, 38031 Grenoble Cedex, France
e-mail: Dominique.Borrione@imag.fr

K. Morin-Allory
e-mail: Katell.Morin-Allory@imag.fr

Y. Oddos
e-mail: Yann.Oddos@imag.fr

G. Nicolescu et al. (eds.), *Design Technology for Heterogeneous Embedded Systems*,
DOI 10.1007/978-94-007-1125-9_8,

were the so called temporal logics (LTL, CTL, ACTL, ...) [13] in which logic formulas are preceded with temporal modalities (always, eventually!, next and until are the four temporal modalities of these logics).

With complex designs, specifications include several hundred properties, some of which are carried from one design to the next one. More user-friendliness and property reuse are required for the adoption of temporal properties by designers: IBM defined the Sugar language [14] that replaced tens of elementary modalities with the use of one complex temporal operator. Two standard languages have been derived from Sugar: PSL [11] (Property Language Specification, now part of VHDL) and SVA [19] (SystemVerilog Assertion). In this chapter, all properties are written in PSL, but all methodological considerations hold for SVA as well.

Properties may be placed inside or outside a component description.

- For the designer who has the complete control over the design of the component (white box description), assertions placed in the design itself are useful both for simulation and formal verification of his design.
- For the user of an IP provided by an external source (black-box), assertions are reduced to the observation of constraints and protocols on the component interface. The properties are written outside of the box.
- Properties can be used also as a support to the dynamic verification of simulation scenarios, both to constrain the inputs to the meaningful test cases, and to state the expected results and automate the simulation results analysis.

A large variety of software tools have been built. Properties can be simulated together with the design they refer to in most commercially available RTL simulators. Properties may also be translated to observers that can be synthesized and emulated on gate-level accelerators. After re-writing them in terms of elementary operators of CTL or LTL, properties may also be fed to formal verification tools (model-checker or bounded model-checker) [3].

Writing formal properties is error-prone: some tools have been specially designed to visualize compliant timing diagrams as an aid to property debugging [5]. Finally, when sufficiently complete, a set of properties may be synthesized into a compliant hardware module for correct by construction design prototyping [1, 6, 7, 9, 16, 18].

Figure 8.1 shows the processing of properties written during the specification phase of a design. Properties are a formal expression of the traditional text and drawing specifications: they must be checked for consistency and for completeness. During this phase two important questions are in order: have you written enough properties, and have you written the right properties?

Tools such as waveform simulation and absence of contradiction are needed to help answer these questions. In particular it is necessary to show that properties can be satisfiable in a non trivial way. As an example, consider two properties (where $\rightarrow$ means logical implication):

$$\text{property P1 is } A \rightarrow B$$

$$\text{property P2 is } C \rightarrow \text{not } B$$

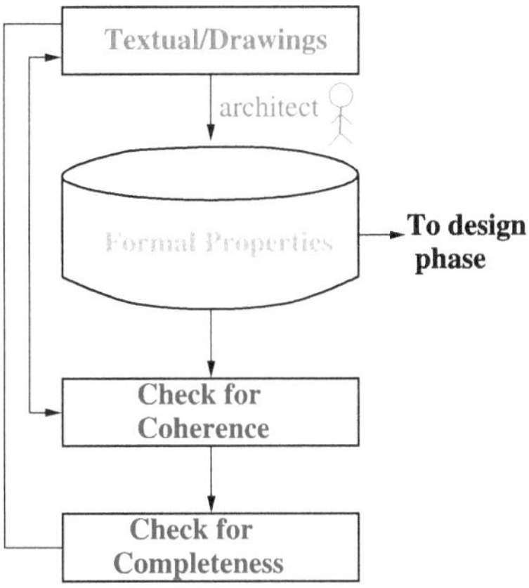

Fig. 8.1 Assertions in the specification phase

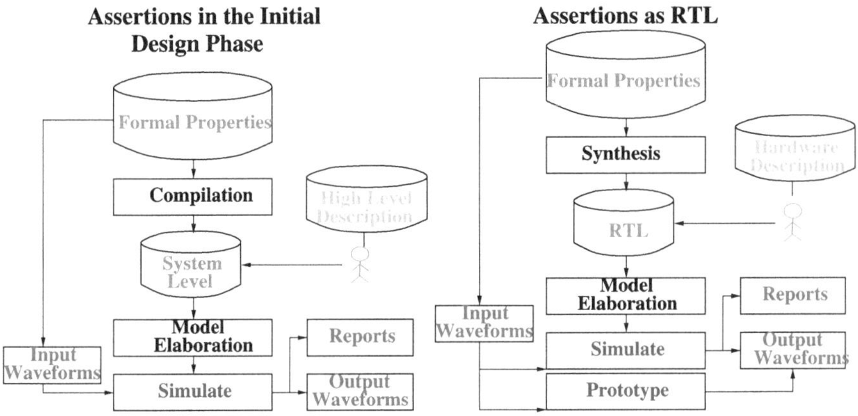

Fig. 8.2 Assertions in design phases

P1 and P2 are vacuously true if A and C are always '0'. Conversely P1 and P2 are contradictory when A and C are both '1'. A specification containing two properties of this form should be complemented with a third property stating for instance that A and C are never simultaneously '1'.

Performing this kind of analysis using formal techniques produces better and more trustable specifications that will be propagated along the successive design steps.

A second step shows the use of the formal properties for checking the initial design written in some high level design language (see left part of Fig. 8.2). After compilation of both the properties and the design, the combined system level model may be simulated with input waveforms that are directly derived from the formal properties about the design environment.

The same formal properties may be used at more detailed design levels such as the well established register transfer level (RTL) as shown on the right part of Fig. 8.2. At this stage, the formal properties are synthesized under the form of ob-

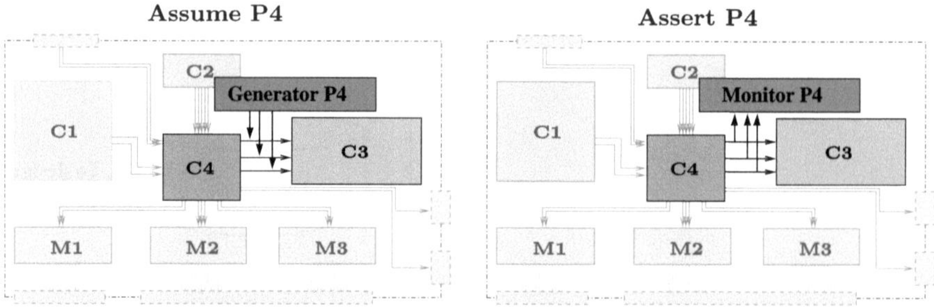

Fig. 8.3 Assume-guarantee paradigm for SoC modules verification

servers that are linked to the RTL design for simulation or prototyping. Once again, the input waveforms may be automatically generated from the subset of the properties that constrain the design inputs.

Figure 8.3 shows the application of this concept to a system in which 3 components C1, C2, C3 are interconnected to 3 memories (M1, M2, M3) via a communication device C4. In the process of checking the interactions between the individual modules, properties may be considered as behavior requirements or restrictions on the environment, depending on the component that is being verified.

As an example, consider C3 and C4 that communicate through signals A, B and C and a property such as:

$$\text{property P4 is always } \{A\} \mid\Rightarrow \{B; C\}$$

P4 has the following meaning: it is always the case that each time A is '1', at the next time B is '1' and C is '1' one time later; here time means a clock cycle, or any other synchronization mechanism that needs to be specified by the context. Property P4 connected to the outputs of C4 is an *assertion* written "assert P4" which specifies an expected behavior for C4. The same property P4 related to the inputs of C3 written "assume P4" is an *assumption* about the environment of C3, i.e. about the signals that C3 should receive.

This distinction between assertions and assumptions is key to the assume-guarantee paradigm. If C4 can be proven to adhere to its assertions, then C3 can be proven to behave as expected when connected to C4, taking the assumptions as an hypothesis. This assume-guarantee paradigm allows the separate checking of each component, replacing the others by their asserted properties while performing formal verification, or by test sequence generators derived thereof in the case of dynamic verification. This greatly simplifies the complexity of the verification model. In addition, it is the only solution in the presence of black-box IPs. This verification paradigm will be best explained on an illustrative example.

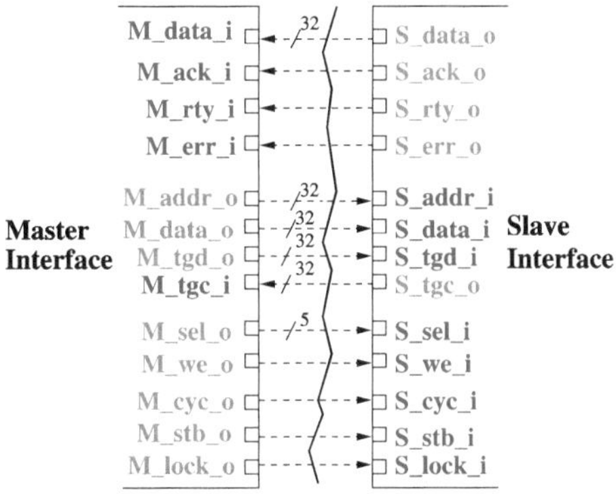

Fig. 8.4 Master and slave Wishbone interfaces

2 The Running Example: CONMAX-IP

The Wishbone Communication Protocol The OpenCores project[1] makes available a set of IPs that may be reused to build a complex system on a chip. In order to ease their interconnection, all IPs comply to the Wishbone standard [12], which defines a generic interface and a communication protocol. In particular, the Wishbone standard distinguishes two component types: masters and slaves, and defines for each type the interface signals dedicated to communications, and their protocol.

Any Wishbone component takes synchronization inputs *Clk* and *reset_i*, 32 bits input $data_i$ and output $data_o$ data ports, and information ports associated to the data ports tgc_i and tgd_o. In addition, a master has an output port $addr_o$ to designate the slave that is requested for the communication, and the register where the slave is to place the data.

The other ports are for the communication protocol. Each output of a master has a corresponding input for the slave, and vice-versa, as shown in Fig. 8.4.

Seen from the master, these communication ports have the following meaning:

- ack_i is '1' if the transfer ended correctly;
- err_i is '1' if the transfer failed;
- rty_i is '1' if the transfer could not start because the slave was not ready;
- cyc_o is '1' if a transfer is being processed between a master and a slave;
- $lock_o$ is '1' if a transfer is not interruptible; the bus is locked as long as this signal or cyc_o is set;
- sel_o indicates that the master has put valid data on $data_o$ in the case of a write, or that the slave must put valid data on $data_i$ in the case of a read.;
- stb_o is '1' to indicate a valid transfer;
- we_o is '1' for a write, 0 for a read;

In the following, for space reasons, we slightly simplified the Wishbone protocol: we do not specify *lock* (only used in critical transfers), and we omit the address and

[1] www.opencores.org.

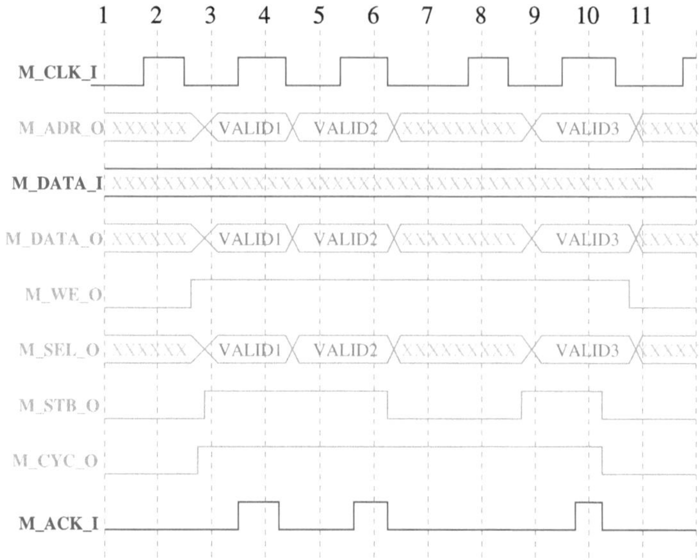

Fig. 8.5 Wishbone protocol: write-burst example

data tags *tga* and *tgc* (they contain information that is useful for the component that receives the data, but do not affect the protocol).

The communication protocol is similar to the AMBA-AXI bus. Figure 8.5 shows a burst write on a slave. At cycle ♯3, signal we_o takes value '1', which selects a write for request stb_o that is set at the same time. The burst write starts at cycle ♯3 and ends at cycle ♯10, according to the value of signal cyc_o. Two successive writes occur between cycles ♯3 and ♯6, acknowledged at cycles ♯4 and ♯6. The third one is acknowledged at cycle ♯10.

The conmax_ip Controller The conmax_ip controller [21] allows the communications between up to 8 masters and up to 16 slaves on a crossbar switch with up to 4 levels of priority. The four most significant bits on *M_addr_o* address the slave. The selection of the master that will own a slave is based on two rules:

- **Priorities:** Each master has a priority that is stored in an internal register `CONF` of the controller. The master *i* priority is given by `CONF[2i..2i-1]`. At each cycle, the master with the greatest priority gets the slave.
- Among masters of equal priority, a **round-robin** policy is applied.

3 The Property Specification Language PSL

PSL has been the first declarative specification language to undergo a standardization process. We provide a quick overview of its main characteristics, and refer the

```
vunit conmax_spec(conmax_ip) {
    default clock is rising_edge(Clk);
    property Reset_Mj is assert always(reset_i →
      ((not M^j_cyc_o and not M^j_stb_o)until not reset_i));
    property No_Sharing_Sk is assume never
      (M^i_cyc_o and M^j_cyc_o and M^i_addr_o = M^j_addr_o);
}
```

Fig. 8.6 Verification Unit example for the conmax_ip component

reader to the IEEE standard for a complete definition. PSL comes with five syntactic flavors to write the basic statements, among which VHDL, Verilog and SystemC. PSL comes with formally defined semantics over *traces*: a *trace* is a sequence of values of the design objects (signals and variables in VHDL, registers and wires in Verilog). Traces may be finite (in simulation or emulation) or infinite (in formal verification).

PSL language is built in **4 layers** [8]:

- **Boolean:** classic Boolean expressions. They are computed using the current values of all the operand objects at the same cycle time. As an example: (not $M^j\_cyc\_o$ and not $M^j\_stb\_o$).
- **Temporal:** expresses temporal relationships between Boolean expressions that are computed at different cycles. This layer is composed of three subsets: FL (Foundation Language), SERE (Sequential Extended Regular Expression) and OBE (Optional Branching Extension). The FL and SEREs subsets are based on LTL and well suited for dynamic verification. They will be described in more details in the next page. In contrast, OBE will not be discussed further, as it is only intended for static verification by model checking.
- **Verification:** Indicates the intended use of a property. It consists of directives that are interpreted by tools to process properties: assert means that the property should be verified, assume means that the property is a constraint, cover indicates that the number of occurrences of the property should be measured.
- **Modeling:** defines the environment of the Design Under Verification (DUV) (Clock synchronization, Design Initialization, etc.).

Assertions, assumptions and the environment model are often grouped into a Verification Unit, (directive vunit) as shown Fig. 8.6. In this example, the verification unit *conmax_spec* is declared to apply to the module *conmax_ip*.

The default clock directive synchronizes all the properties in the verification unit named conmax_ip with the rising edge of signal *clk*. The assertion Reset_Mj verifies that the two signals *M_cyc_o* and *M_stb_o* are always '0' when *reset_i* is active. In this assertion, the symbol "→" is the implication operator between a Boolean expression and a FL property.

The assumption No_Sharing_Sk guarantees that two different masters never ask the same slave at the same cycle. This can be used to simulate the design preventing any collision on the crossbar.

3.1 The Temporal Layer of PSL

The FL Subset It is composed of the following operators: {always, never, eventually!, before, before_, until, until_, next, next[k], next_a[k:l], next_e[k:l], next_event, next_event[k], next_event_a[k:l], next_event_e[k:l]}. All of these operators span the verification over more than one cycle.

For next, the execution ends one cycle after the operator is evaluated. For next[k],next_a[j..k] and next_e[j..k], if the verification begins at cycle t, it completes at cycle $t + k$. For all the other FL operators, the cycle when the evaluation will end depends on the occurrence of an event, not on a number of cycles. These operators are said to be unbounded.

As an example, the waveform of Fig. 8.5 satisfies property *Keep_request_up* that states: if a valid transfer is '1' and not acknowledged, then the same valid transfer remains '1' at the next cycle.

property *Keep_request_up* is assert always (

(not *M_ack_i* and *M_stb_o*) → next *M_stb_o*);

The SERE Subset A SERE is a kind of regular expression built over temporal sequences of values of Boolean objects. Sequences of values are written between curly brackets, separated by semi-colons. Example: $\{M\_stb\_o; M\_ack\_i\}$. This SERE means: *M_stb_o* followed by *M_ack_i* . In Fig. 8.5 it is satisfied at cycles ♯3, ♯5, ♯9. A wide variety of repetition operators are available. Example: $\{M\_stb\_o[*2 : 4]\}$ means $M\_stb\_o =$ '1' during 2 to 4 consecutive cycles. This SERE is satisfied at cycles ♯3, ♯4, ♯5, ♯9.

Two SEREs may be combined using an implication operator: $|\Rightarrow$ means that the occurrence of the left SERE implies the occurrence of the right SERE starting after one cycle. $|\rightarrow$ is the implication where the last cycle of the left SERE and the first cycle of the right SERE overlap. Example: property *Ack_imm* states that a request is acknowledged at the second cycle it is set to '1'. This property is satisfied by the waveform of Fig. 8.5.

property *Ack_imm* is assert always

{not *M_ack_i* and *M_stb_o*; *M_stb_o*} |→ {*M_ack_i*};

In the following, we shall call *temporal sequence* a PSL expression built from SERE and FL operators.

Strong and Weak Operators in PSL In dynamic verification, waveforms have a finite length. For properties that are still under evaluation when the simulation stops, PSL distinguished the final result according to the temporal sequence strength. For a *weak* SERE or temporal operator (written without '!', the default case), a property that is still on-going and has not yet been contradicted is considered to hold. For a *strong* temporal sequence (written with a final '!' character after the SERE or operator name), an on-going property fails when the simulation stops.

Example: Assume the waveform of Fig. 8.5 ended at cycle ♯9. Property *Ack_imm* above is a weak property, it holds at cycle ♯9. Conversely, its strong version *Ack_imm_strong* fails at cycle ♯9, because it is started, but there is no cycle ♯10 to receive the *M_ack_i* signal.

property *Ack_imm_strong* is assert always

$\{$not M_ack_i and $M\_stb\_o; M\_stb\_o\} \mid\rightarrow \{M\_ack\_i\}!;$

Verification of a Temporal PSL Property A property evaluation is started at the initial cycle and may need multiple cycles to be verified.

If a property written with the FL subset does not start with operator always or never, its evaluation is triggered only once. Conversely, if the property starts with always or never, its evaluation is triggered again at each successive cycle, so that the property may be simultaneously under evaluation for multiple starting cycles. Likewise, a SERE with unbounded repetitions may be satisfied for different numbers of repetitions of the same object value. The status of a property is thus not simply *True* or *False*.

The PSL standard defines more precisely the state of a property through four definitions:

- **Holds Strongly:** the property has been evaluated to *True* and will not be triggered again. It is impossible to violate the property in a future cycle because the verification has ended.
- **Holds:** the property has been verified, but it can be triggered again, so that future object values may violate the property.
- **Pending:** the property is being verified, but the verification is not finished.
- **Failed:** the property has been violated at least once.

Example Consider the following two properties *WR_Ok* and *Sub_WR_Ok*, where *Sub_WR_Ok* is a sub-property of *WR_Ok*.

property *Sub_WR_Ok* is

$\{M\_cyc\_o; M\_stb\_o\} \mid\rightarrow$ (next![2](M_ack_i));

property *WR_Ok* is

always $\{M\_cyc\_o; M\_stb\_o\} \mid\rightarrow$ (next![2](M_ack_i));

Property *WR_Ok* states that each cycle t when a master reserves the bus (signal *M_cyc_o* active) and initiates a transfer (Write or Read) at $t + 1$ (signal *M_stb_o* active), then an acknowledgment must be received from the slave 2 cycles later to signal the correct ending of the transfer (reception of *M_ack_i*).

$\{M\_cyc\_o; M\_stb\_o\}$ is a SERE which means "*M_cyc_o* followed by *M_stb_o* in the next cycle". Property *Sub_WR_Ok* holds on a trace if either $\{M\_cyc\_o; M\_stb\_o\}$ does not hold on its first two cycles, or $\{M\_cyc\_o; M\_stb\_o\}$ holds initially and then *M_ack_i* is '1' two cycles later.

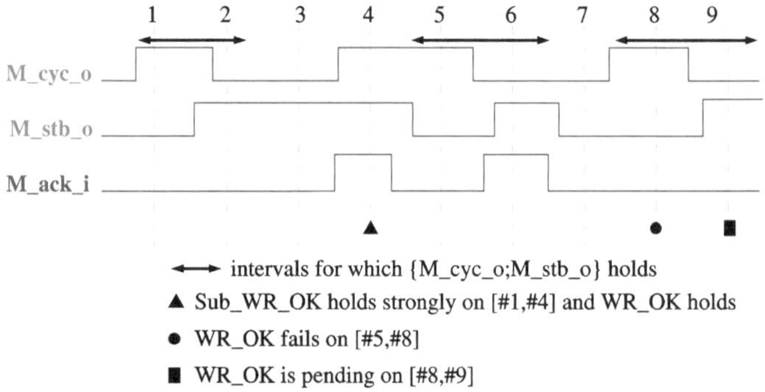

Fig. 8.7 A trace not satisfying *WR_Ok*

Property *WR_Ok* is defined as *Sub_WR_Ok* preceded by the always operator. While *Sub_WR_Ok* is evaluated only with respect to the first cycle of the trace, the evaluation of *WR_Ok* is restarted each cycle: *WR_Ok* holds at cycle T if and only if sub property *Sub_WR_Ok* holds at cycle T and at each subsequent cycle $T' > T$. Figure 8.7 illustrates a trace that does not satisfy *WR_Ok*.

A trace starting at cycle $\sharp i$ and ending at cycle $\sharp j$ will be denoted $[\sharp i, \sharp j]$. On the example of Fig. 8.7:

- *Sub_WR_Ok fails* on $[\sharp 5, \sharp 8]$, because *M_ack_i* is '0' at cycle $\sharp 8$.
- *Sub_WR_Ok holds strongly* on $[\sharp 1, \sharp 9]$: *Sub_WR_Ok* only characterizes the first four cycles of the trace, and evaluates to '1' at cycle $\sharp 4$.
- *WR_Ok holds* on $[\sharp 1, \sharp 5]$ but does not hold strongly: it has been evaluated to '1', but it may fail on an extension of the trace. Indeed, *Sub_WR_Ok* fails on $[\sharp 1, \sharp 8]$ because *WR_Ok* is restarted at cycle $\sharp 5$ and fails at cycle $\sharp 8$.

The PSL Simple Subset The IEEE standard identifies a "simple subset" (PSL_{ss}) of PSL for which properties can be evaluated on the fly, during simulation or execution. In this subset, time advances from left to right through the property. The dynamic verification tools are restricted to formulas written in the subset, which is most widely understood and de facto recommended in the guidelines for writing assertions [10]. The PSL_{ss} subset is obtained by applying the restrictions shown in Table 8.1 to the PSL operators.

4 Synthesis of Temporal Properties

4.1 Turning Assertions into Monitors

A monitor is a synchronous design detecting dynamically all the violations of a given temporal property. We detail here the last release of our approach used to

Table 8.1 Restrictions for the dynamic verification with PSL: PSL_{ss}

PSL operator	Restrictions on operands
Not	Boolean
never	Boolean or sequence
eventually!	Boolean or sequence
Or	at least one Boolean
$\rightarrow$	left hand side Boolean
$\leftrightarrow$	two Boolean operands
until until!	right hand side Boolean
until_ until_!	two Boolean operands
before*	two Boolean operands
next_e	Boolean
next_event_e	right hand side Boolean

Table 8.2 Primitive PSL monitors

Watchers	Connectors
mnt_Signal, iff, eventually!, never	$\rightarrow$, and, or, always, next!,next_a
next_e, next_event_e, before	next_event, next_event_a, until

synthesize properties into hardware monitors. It is based on the principles described in [15].

The monitor synthesis is based on a library of primitive components, and an interconnection scheme directed by the syntax tree of the property. We have defined two types of primitive monitors: connectors and watchers. The first one is used to start the verification of a sub-property. The watcher is used to raise any violation of the property.

The sets of connectors and watchers are given in Table 8.2. The watcher mnt_Signal is used to observe a simple signal.

Primitive monitors have a generic interface depicted in Fig. 8.8.a. It takes as input two synchronization signals *Clk* and *Reset_n*, a *Start* activation signal, and the ports *Expr* and *Cond* for the observed operands. The output ports are: *Trigger* and *Pending* for a connector; *Pending* and *Valid* for a watcher.

The overall monitor is built by post-fixed left to right recursive descent of the property syntax tree. For each node of type connector, its Boolean operand, if any, is connected to input *Cond*. The output *Trigger* is connected to input *Start* of its FL operand. For the watcher type node, its Boolean operands are directly connected to the inputs *Expr* and *Cond* of the current monitor. Its output *Valid* is the *Valid* output of the global monitor.

The couple of signals (*Valid, Pending*) gives the current state of the property at any cycle: failed, holds, holds strongly or pending.

The architecture for the monitor Reset_Mj is depicted in Fig. 8.9.

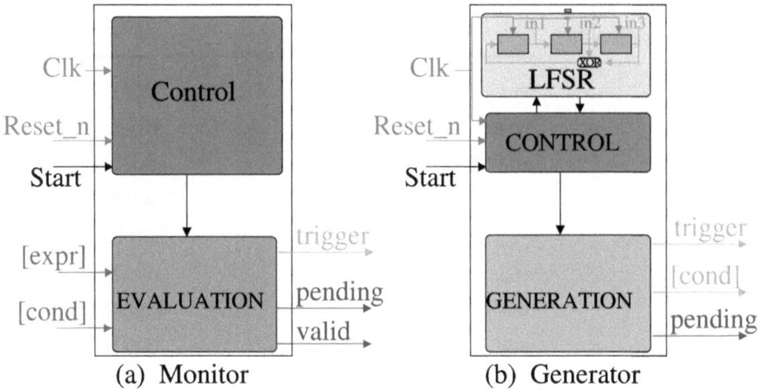

Fig. 8.8 Architectures and interfaces for primitive monitors and generators

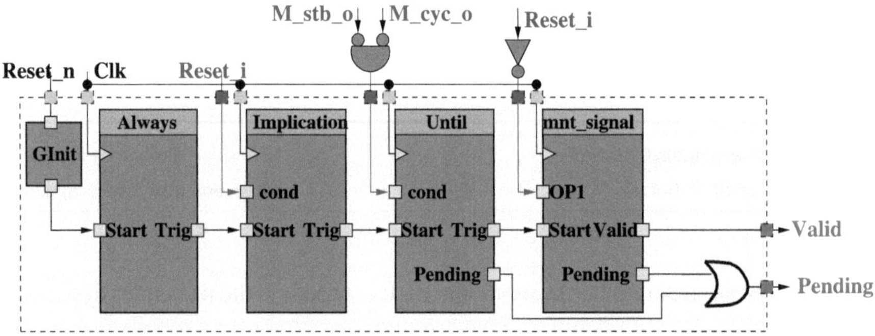

Fig. 8.9 Monitor architecture for Reset_Mj

4.2 Turning Assumptions into Generators

A generator is a synchronous design producing sequences of signals complying with a given temporal property. Their synthesis follows the same global principle as for the monitors: the overall generator is built as an interconnection of primitive generators, based on the syntax tree of the property.

Primitive generators are divided into all the connectors (associated to all PSL operators) and the single type of producer to generate signal values: gnt_Signal.

The interface of primitive generators (Fig. 8.8.b) includes:

- the inputs *Clk*, *Reset_n*, *Start*: same meaning as for monitors.
- the outputs *Trigger* and *Cond* used to launch the left and right operand (for connectors).
- the output *Pending*, to indicate if the current value on *Trigger* and *Cond* are constrained or may be randomly assigned.

Since many sequences of signals can comply with the same property, we need the generators to be able to cover the space of correct traces. To achieve this goal, the gnt_Signal embeds a random number generator (based on a Linear Feedback Shift Register or a Cellular Automaton). By default, the outputs of an inactive complex generator are fixed to '0'. It is possible to produce random values by switching the generic parameter RANDOM to 1. If *Pending* is inactive, the values on *Trigger* and *Cond* are not constrained and produced by the random block.

5 Instrumentation of the conmax_ip

A set of properties has been defined to verify some critical features of the conmax_ip controller.

Initialization Verification: Property Reset_Mj For all masters, signals *M_cyc_o* and *M_stb_o* must be negated as long as *reset_i* is asserted (cf. [12], rule 3.20):

property *Reset_Mj* is assert always
$(reset\_i \rightarrow ((\text{not } M^j\_cyc\_o \text{ and not } M^j\_stb\_o) \text{ until not } reset\_i));$

Connexion Verification: Property LinkMj_Sk It checks the connection between the j-th master and k-th slave by analyzing that each port is well connected.

property *LinkMj_Sk* is assert always
$(M^j\_cyc\_o \text{ and } S^k\_cyc\_i \text{ and } M^j\_addr\_o = S^k\_addr\_i) \rightarrow$
$(M^j\_data\_o = S^k\_data\_i \text{ and } M^j\_data\_i = S^k\_data\_o$
$\text{and } M^j\_sel\_o = S^k\_sel\_i \text{ and } M^j\_stb\_o = S^k\_stb\_i$
$\text{and } M^j\_we\_o = S^k\_we\_i \text{ and } M^j\_ack\_i = S^k\_ack\_o$
$\text{and } M^j\_err\_i = S^k\_err\_o \text{ and } M^j\_rty\_i = S^k\_rty\_o);$

Priorities Verification: Property PrioMj_Mk Assume two masters M^j and M^k have priorities p_j and p_k such that $p_k > p_j$. If M^j and M^k request the same slave simultaneously, M^k will own it first.

property *PrioMj_Mk* is assert always $((M^j\_cyc\_o \text{ and } M^k\_cyc\_o$
$\text{and } CONF[2k..2k-1] > CONF[2j..2j-1]$
$\text{and } M^j\_addr\_o[0..3] = M^k\_addr\_o[0..3])$
$\rightarrow (M^k\_ack\_i \text{ before } M^j\_ack\_i));$

5.1 Modeling Masters and Slaves with Generators

To test the correctness of the conmax_ip controller in isolation, without the overhead of simulating a complete set of masters and slaves, we need to embed the controller in an environment that provides correct test signals. To this aim, we model masters and slaves with generators that must comply with the hand-shake protocol.

5.1.1 Modeling and Launching Masters Actions

Property WriteMj_Sk A write request from the j-th master to the k-th slave is specified by the following property, to which a generator is associated:

property *WriteMj_Sk* is assume
((*M^j_cyc_o* and *M^j_we_o* and *M^j_sel_o* and *M^j_stb_o*
and *M^j_data_o* = VAL_DATA and *M^j_addr_o* = VAL_ADDR_k)
until_ *M^j_ack_i*);

Since we are interested in the communication action, but not in the particular data value being written, the value VAL_DATA that is displayed on the port *M^i_data_o* is a randomly computed constant. The four most significant bits of VAL_ADDR are fixed to select the j-th slave. This property is a simplified model of a master, it does not take into account signals *M^i_rty_i* and *M^i_err_i* (they are not mandatory). These signals would be present in a more realistic model. This property involves the acknowledgment input signal *M^i_ack_i* that stops the constrained generation.

Property GenLaunch The scenario GenLaunch illustrates the request of three masters numbered 0, 1, 2 to the same slave numbered 1. Master 0 first makes a request; then between 16 to 32 cycles later, masters 1 and 2 simultaneously make their request. This scenario is modeled using a property that generates the start signals for three instances of master generators (according to the previously discussed property), and one slave. These different start signals are denoted *start_WriteM0_S1*, *start_WriteM1_S1*, *start_WriteM2_S1*.

property *Gen_Launch*is assume eventually! (*start_WriteM0_S1*
→ next_e[16..32](*start_WriteM1_S2* and *start_WriteM2_S2*));

A large number of scenarios of various complexity have been written, and implemented with generators, in order to have a realistic self-directed test environment. Modeling test scenarios for requests (i.e. read, burst, ...) is also performed with assumed properties, from which generators are produced.

5.1.2 Modeling Slaves Responses

Property Read_Sj For a slave, the most elaborate action is the response to a read request: signal *S_ack_o* is raised and the data is displayed on *S_data_o*. The following property expresses this behavior, at some initial (triggering) time.

property *Read_Sj* is assume
(next_e[1..8](*S^j_ack_o* and *S^j_data_o* = *DATA*));

The generator for property Read_Sj must be triggered each time the slave receives a read request: its start signal is connected to the VHDL expression not *S^j_we_i* and *S^j_cyc_i*.

5.2 Performance Analysis

Monitors can be used to perform measurements on the behavior of the system. To this aim, the Horus platform is instrumented to analyze the monitor outputs, and count the number of times when a monitor has been triggered, and the number of times when a failure has been found.

On the wishbone switch, and assuming that it is embedded in a real environment, it may be useful to test on line the number of times the signal *M_err_i* of a slave is asserted, or how often a slave is requested simultaneously by several masters.

Property CountError The following property is used to count the number of transfers ending with an error:

$$\text{property } CountError \text{ is cover never } (M^0\_err\_i \text{ or } \ldots \text{or } M^7\_err\_i);$$

Property ColliMj_Sk The property allows to know the number of times when more than one master asks for the same slave:

$$\text{property } ColliMj\_Sk \text{ is}$$
$$\text{cover never } (S^j\_cyc\_i \text{ and } S^k\_cyc\_i \text{ and } S^j\_addr\_i = S^k\_addr\_i);$$

5.3 The Horus Flow

The Horus environment helps the user build an instrumented design to ease debugging: it synthesizes monitors and generators, connects them to the DUV and adds a device to snoop the signals of interest. It comes with the VHDL and Verilog flavors. The Horus system has a friendly graphical user interface (GUI) for the generation of the instrumented design in 4 steps. The 4 steps to instrument the conmax_ip are illustrated in Fig. 8.10.

- **Step 1—Design selection**: The DUV (in our case, the conmax_ip), with its hierarchy, is retrieved.
- **Step 2—Generators and Monitors synthesis**: Select properties or property files, define new properties, select target HDL language, synthesize monitors and generators (verification IPs). For the conmax_ip, the properties *Reset_Mj*, *Link_Mj_Sk* and *Prio_Mj_Mk* are synthesized into monitors. *Write_Mj_Sk* and *Read_Mj_Sk* are turned into generators. Finally, the performance properties *CountError* and *Colli_Mj_Sk* are turned into performance-monitors.
- **Step 3—Signal interconnection**: Using the GUI, the user easily connects the monitors and the generators to the DUV. All the signals and variables involved in the DUV are accessible in a hierarchical way. The user needs only select the signals to be connected to each verification IP.

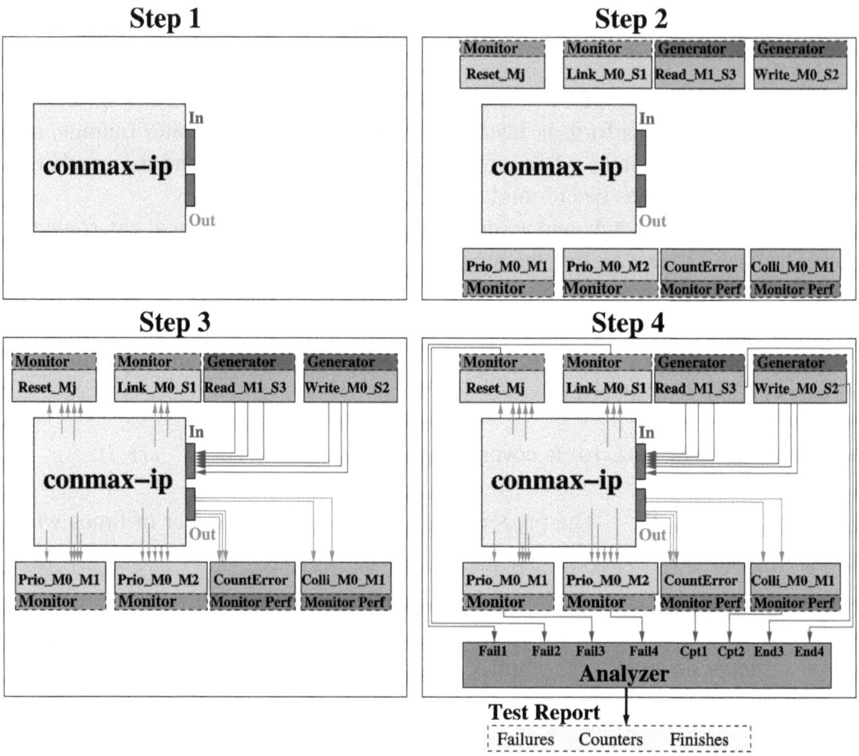

Fig. 8.10 Design instrumentation with Horus

- **Step 4—Generation**: The design instrumented with the verification IPs is generated. When internal signals are monitored, the initial design is slightly modified to make these signals accessible to the monitors.

The outputs of the verification IPs are fed to an instance of a generic Analyzer; this component stores the monitors outputs and sends a global status report on its serial outputs. It also incorporates counters for performance analysis. The instrumented design has a generic interface defined for an Avalon or a Wishbone bus. If the FPGA platform is based on such a bus, the user can directly synthesize and prototype the instrumented design on it.

5.4 Experimental Results

The test platform produced (cf. Fig. 8.10) interconnects the instrumented conmax_ip controller (some internal signals have been made outputs), the monitors, the generators, and the *analyzer* component.

It has been synthesized with QuartusII 6.0 [2] on an Altera DE2 board with a CycloneII EP2C35 FPGA chip set. Tables 8.3, 8.4 and 8.5 show the results in terms

Table 8.3 Synthesis results for performance monitors

Performance monitors	LCs	FFs	Freq.
CountError	5	3	420
CountRetry	5	3	420
ColliMj_Sk	6	3	420
28 props	*168*	*84*	*420*
CountTotal	8	3	420
CountTotalRead	11	3	420
TOTAL	**36**	**15**	**420**
CONMAX	**15084**	**1090**	**130**
IMPACT	**0,2%**	**1,3%**	–

of area and frequency. The LCs (resp. FFs) column gives the number of logical cells (resp. flip-flops) used by each verification IP. Synthesis results are given for each monitor and generator. If several instances of a property are necessary, figures are given for all the instances.

As an example, a property such as *ColliMj_Sk* which involves any two distinct masters and an arbitrary slave is instantiated $M \times N \times (N-1)/2$ times, where N is the number of masters and M the number of slaves.

The last three lines of each table show the total number of monitors or generators, the synthesis results for the controller and the area ratio between the instrumentation and the conmax_ip.

All the components have a high frequency and allow the verification of the design at-speed. Yet a very large number of properties have been implemented in order to build a complete and self-directed testbench for the controller.

As it can be seen in Table 8.3, results for performance analysis are excellent, and the temporal overhead of this method is negligible.

Table 8.4 shows that monitors are very small: the number of flip-flops never exceeds 5 and the number of logical cells never rises above 74. The total instrumentation for monitors is roughly twice the size of the conmax_ip for 564 properties.

In contrast, a similar analysis in Table 8.5 reveals that generators induce a high penalty on the number of registers, which is multiplied by 10: this is due to the LFSRs (Linear Feedback Shift Register) [20] that are implemented in each primitive generator, for producing the outputs at random times, or repeating them a random number of times. The positive counterpart for this penalty is that each generator is fully independent from the others. The generators used to produce scenarios are on the contrary very small and are not included in this analysis.

The reader should not conclude that generators are space inefficient: they indeed replace the masters and slaves modules for testing the conmax_ip. Synthesizing the conmax_ip connected to 8 instances of processors cores and 15 instances of real size memories would have exceeded by one or two orders of magnitude the size of the FPGA platform.

We have compared our results with other tools: FoCs [1] and MBAC [7]. This comparison is done on monitors since no other tool that we are aware of synthesizes

Table 8.4 Synthesis results for monitors

Monitors	LCs	FFs	Freq
Reset_Mj	8	4	420
(4 props)	*32*	*16*	*420*
LinkMj_Sk	74	4	420
(128 props)	*9472*	*512*	*420*
PrioMj_Mk	46	5	329
(432 props)	*19872*	*2160*	*329*
TOTAL(564)	**29379**	**2688**	**329**
CONMAX	**15084**	**1090**	**130**
IMPACT	**194%**	**246%**	–

Table 8.5 Synthesis results for generators

Generators	LCs	FFs	Freq
WriteMj_Sk	81	56	420
(128 props)	*10368*	*7168*	*420*
ReadMj_Sk	57	36	420
(128 props)	*7296*	*4608*	*420*
LaunchGen	224	142	420
TOTAL (256)	**17888**	**11918**	**420**
CONMAX	**15084**	**1090**	**130**
IMPACT	**118%**	**1093%**	–

hardware generators for PSL properties. The results of Horus are equivalent to the results given by MBAC: the FPGA size, in terms of logic cells and registers, are either identical, or within a few percent for complex monitors, the result of MBAC in that case being better optimized.

The positive advantage of the Horus method lies in the fact that the whole construction is formally proven correct. The comparison with FoCs is more difficult since PSL_{ss} is not fully supported by FoCs, and some properties could not be synthesized. When FoCs gives results, Horus is equivalent on simple properties, and significantly better when the property is more complex.

6 Conclusion

A temporal property concisely describes complex scenarios. It is therefore well suited to describe both the expected behaviors and the test input sequences for the design under verification. This is why the property-based design is rapidly developing. The Horus platform eases the test phase by automating test bench creation from temporal properties.

Assertions are turned into monitors. They easily verify the behavior of the DUV, and can perform measurements. Complex test scenarios are efficiently built by synthesizing assumptions into generators.

The verification IPs complexity is linear with the number of temporal operators in the properties, and the time to build monitors and generators is not significant (less than a few seconds, even for complex properties). Moreover, the hardware instrumented design provides a clear test report containing all the interesting informations to ease a successive debug phase, if needed.

While most of the property-based verification tools focus on the RTL level, Horus has a specific module called ISIS which turns assertions into SystemC monitors [17]. The use of assertions can thus start at the initial design phase. Using Horus, property-based design can be applied all along the design flow.

References

1. Abarbanel, Y., Beer, I., Gluhovsky, L., Keidar, S., Wolfsthal, Y.: FoCs—automatic generation of simulation checkers from formal specifications. In: CAV, LNCS, vol. 1855, pp. 538–542. Springer, Berlin (2000). doi:10.1007/10722167_40
2. Altera: Quartus II Handbook v9.1 (Complete Five-Volume Set) (2005). http://www.altera.com/literature/
3. Beer, I., Ben-David, S., Eisner, C., Geist, D., Gluhovsky, L., Heyman, T., Landver, A., Paanah, P., Rodeh, Y., Ronin, G., Wolfsthal, Y.: Rulebase: model checking at IBM. In: Proc. 9th International Conference on Computer Aided Verification (CAV). LNCS, vol. 1254, pp. 480–483. Springer, Berlin (1997)
4. Bergeron, J., Cerny, E., Hunter, A., Nightingale, A.: Verification Methodology Manual for SystemVerilog. Springer, Berlin (2006). ISBN 978-0-387-25556-9
5. Bloem, R., Cavada, R., Eisner, C., Pill, I., Roveri, M., Semprini, S.: Manual for property simulation and assurance tool (deliverable 1.2/4-5). Technical report, PROSYD Project (2004)
6. Borrione, D., Liu, M., Ostier, P., Fesquet, L.: Chapter PSL-based online monitoring of digital systems. In: Applications of Specification and Design Languages for SoCs Selected Papers from FDL 2005, pp. 5–22. Springer, Berlin (2006)
7. Boulé, M., Zilic, Z.: Generating Hardware Assertion Checkers: For Hardware Verification, Emulation, Post-Fabrication Debugging and On-line Monitoring. Springer, Berlin (2008). ISBN 978-1-4020-8585-7
8. Eisner, C., Fisman, D.: A Practical Introduction to PSL (Series on Integrated Circuits and Systems). Springer, New York (2006)
9. Eveking, H., Braun, M., Schickel, M., Schweikert, M., Nimbler, V.: Multi-level assertion-based design. In: 5th ACM & IEEE International Conference on Formal Methods and Models for Co-design MEMOCODE'07, pp. 85–87 (2007)
10. Foster, H., Krolnik, A., Lacey, D.: Assertion-Based Design. Kluwer Academic, Dordrecht (2003)
11. Foster, H., Wolfshal, Y., Marschner, E., IEEE 1850 Work Group: IEEE standard for property specification language PSL (2005). pub-IEEE-STD, pub-IEEE-STD:adr
12. Herveille, R.: WISHBONE system-on-chip (SoC) interconnection architecture for portable IP cores. Technical report (2002). www.OpenCores.org
13. Huth, M.-R.A., Ryan, M.-D.: Logic in Computer Science: Modelling and Reasoning about Systems. Cambridge University Press, Cambridge (1999). ISBN 0521656028
14. Marschner, E., Deadman, B., Martin, G.: IP reuse hardening via embedded sugar assertions. In: International Workshop on IP SoC Design, October 30, 2002. http://www.haifa.il.ibm.com/projects/verification/RB_Homepage/ps/Paper_80.pdf

15. Morin-Allory, K., Borrione, D.: Proven correct monitors from PSL specifications. In: DATE '06: Proceedings of the Conference on Design, Automation and Test in Europe, pp. 1246–1251 (2006)
16. Oddos, Y., Morin-Allory, K., Borrione, D.: On-line test vector generation from temporal constraints written in PSL. In: International Conference on Very Large Scale Integration System on Chip VLSI SoC'06, Nice, France (2006)
17. Pierre, L., Ferro, L.: A tractable and fast method for monitoring SystemC TLM specifications. IEEE Trans. Comput. **57**, 1346–1356 (2008)
18. Schickel, M., Nimbler, V., Braun, M., Eveking, H.: An efficient synthesis method for property-based design in formal verification: on consistency and completeness of property-sets. In: Advances in Design and Specification Languages for Embedded Systems, pp. 179–196. Springer, Berlin (2007). ISBN 978-1-4020-6149-3
19. Srouji, J., Mehta, S., Brophy, D., Pieper, K., Sutherland, S., IEEE 1800 Work Group: IEEE standard for SystemVerilog—unified hardware design, specification, and verification language. Technical report (2005). pub-IEEE-STD:adr
20. Texas Instruments: What's an LFSR? (1996). http://focus.ti.com/general/docs/
21. Usselman, R.: WISHBONE Interconnect Matric IP Core (2002). http://www.opencores.org/projects.cgi/web/wb_conmax/overview

Chapter 9
Trends in Design Methods for Complex Heterogeneous Systems

C. Piguet, J.-L. Nagel, V. Peiris, S. Gyger, D. Séverac, M. Morgan, and J.-M. Masgonty

1 Introduction

With the introduction of very deep submicron technologies as low as 45 and 32 nanometers, or even 22 nanometers, integrated circuit designers have to face two major challenges: first, they have to take into account a dramatic increase in complexity due to the number of components including multi-core processors ("More Moore"), but also due to the significant increase in heterogeneity ("More than Moore"). Secondly, the significant decrease in reliability of the components has to be taken into account, and specifically the behavior of switches which are very sensitive to technology variations, temperature effects and environmental conditions.

This chapter describes the design of SoCs developed at CSEM [1] both for applied research demonstrators and for industrial applications. These chips are clearly heterogeneous by nature, as they contain generally low-power RF blocks such as sub-GHz short-range connectivity radios, high-performance mixed signal blocks such as ADCs and power management units, advanced analog vision sensor circuits, and complex digital blocks such as DSP and control processors as well as embedded memories. In addition, even for relatively conservative CMOS technologies like 180 nm, leakage is an issue for most digital blocks and sleep transistors have to be used to disconnect idle blocks, in particular for portable applications where huge low-activity periods can occur in long-lifetime applications. As these chips are generally operated at low voltage, e.g. 0.9 V, the effects of temperature, supply voltage V_{dd} and technology parameter variations are noticeable and design methodologies have to take them into account even for 180 nm technologies—they become critical beyond. How to handle properly the impact of these low-level effects on high-level design and synthesis still remains an important open question. This chapter will therefore also present CSEM's practical design methodology for SoCs, where a particular focus concerns the description and manipulation of objects

C. Piguet (✉) · J.-L. Nagel · V. Peiris · S. Gyger · D. Séverac · M. Morgan · J.-M. Masgonty
CSEM, Neuchâtel, Switzerland
e-mail: christian.piguet@csem.ch

G. Nicolescu et al. (eds.), *Design Technology for Heterogeneous Embedded Systems*,
DOI 10.1007/978-94-007-1125-9_9, © Springer Science+Business Media B.V. 2012

at increasingly higher levels for complex SoCs and MPSoCs; while on the other hand, an increasing number and significance of low-level effects have to be taken into account for ultra deep-submicron SoCs.

In the last part of the chapter, some examples of complex SoCs designed by our company or jointly with industrial companies will be described. They show the large diversity of applications and technologies used for these portable devices. However, all these SoCs share the fact that they are extremely low-power chips.

2 Context

2.1 Applications

Applications of our customers are very diverse, but the common requirement is extreme low power. The first class of applications is characterized by very low- to medium-speed applications (32 kHz to 10 MHz) for electronic watches, hearing aids, portable medical surveillance devices and wireless sensor networks (WSN). These applications rely on SoCs comprising sensor interfaces, digital processing and wireless communication links. Generally, the duty cycle is very low, down to 1% or even 0.1%, meaning that leakage power is a main issue. A second class of applications is medium- to high- speed applications up to 200 MHz, such as mobile TV circuits, vision sensor-based circuits or arrays of processors for configurable platforms. While low power consumption is still a major preoccupation for the latter applications, another very important requirement is the computation throughput that has to be maximized for a given amount of energy. Therefore, very efficient digital processing in terms of clock cycles for a given algorithm is mandatory.

2.2 Low Power Is a Must

Low-power requirements have to be analyzed through the diverse applications in terms of average power per block. RF wireless links are the most power hungry blocks, reaching peak currents of a few tens to even a few hundreds of mA depending on the required range, carrier frequency and data rate. Low-power short-range connectivity RF transceivers designed by our company for the sub-GHz ISM bands in active mode consume around 2 mA (peak) in reception and 20 mA (peak) in transmission [2]. However, if the RF communication duty cycle is very low, as in the case of wireless sensor networks, the average current consumption of the RF wireless link can be reduced to a few μA. Consequently, a tradeoff has to be found at the system level between the number of transmitted bits and the local computation achieved in the SoC itself. The energy per transmitted bit is roughly 100 nJ per bit, while the energy per operation is roughly 1 nJ/operation in a general purpose processor, 0.25 nJ/operation in a DSP core and 0.001 nJ/operation or less in specific co-processors and random logic blocks [3]. It is therefore mandatory to process and compress the information before sending it through the wireless link. Regarding sensors and actuators, to significantly reduce the total power (sensor, digital processing), it is necessary to move a part of the digital processing to the sensors themselves. This strategy includes low-power data encoding directly within the sensor at the pixel level, hence allowing only the features of interest such as edges and contrasts to be directly extracted. Only the data corresponding to these features of interest (and not the entire bitmap as would be the case with a standard CMOS imager) is provided to the digital processor. Compared with CMOS imager and DSP systems, the digital processing load and power consumption of the vision sensor platform is in this way drastically reduced.

A major system issue is that the power budget of the complete system has to be analyzed to identify the most power-hungry blocks. The challenge consists of reducing the overall power of the principal contributors, as it is useless to save power in a block that consumes only 1% of the total power. Such a global analysis has clearly to include the application issues too, as it is useless to optimize the SoC power consumption when the external system parts may be burning much more power.

2.3 Technologies

The choice of a technology is directly related to the application and to the expected production volumes. This is clearly an issue for small or medium production volume, for which conservative technologies, such as 180, 130 or 90 nanometers, are generally used, while 65 to 32 nm will be used for large production volume. With 180 to 90 nm technologies, leakage problems and technology variations may seem non critical at first sight. Although this is true for nominal supply voltage and high-speed or high-duty cycle applications, the situation changes for very low-voltage chips and very low duty cycles (in the order of 1% or below), as technology variation problems, respectively leakage problems, become a main issue even in such conservative technologies. The situation degrades even further for circuits relying on subthreshold logic [4, 5] and operating at supplies equal to or lower than the MOS threshold voltage. As these SoCs are very low-voltage and low duty cycle chips, design methodologies clearly have to take into account leakage and technology variations problems even in 180 nm.

2.4 Embedded Systems Design

The design complexity of SoC is increasing. Besides the "More Moore" and the "More than Moore" effects, one can say that the relationships and interdependencies between many design levels are largely responsible for this increase of the design complexity. Figure 9.1 shows a list of many problems that appear in the design

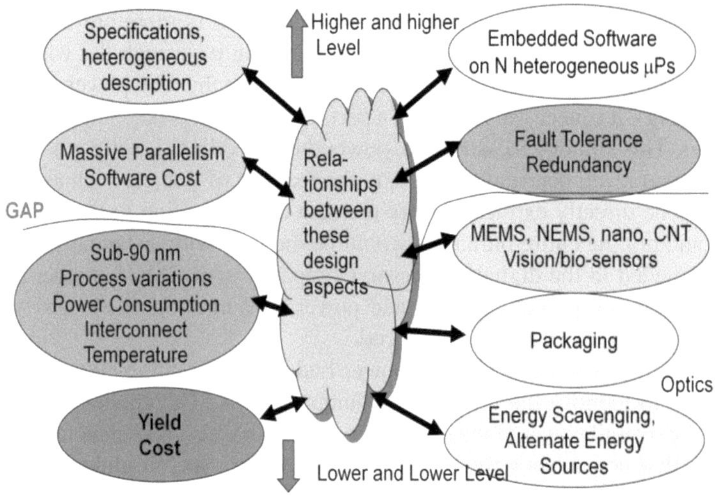

Fig. 9.1 Problems in SoC design

of SoCs. The trend is to describe SoC behavior at increasingly higher levels in order to enable reduced time to market. The gap with low level effects inherent to very deep submicron technologies is widening, as one has to take into account more and more effects like process variations, leakage, temperature, interconnect delays, yield and cost. Furthermore, power management and increased levels of autonomy are more than ever a critical issue, and call for complex management blocks that can accommodate a variety of sources ranging from batteries to energy-scavengers. For these reasons, the relationships between all these design aspects become very complex and clearly need to be addressed using interdisciplinary approaches. This is the essence of heterogeneous SoC design, as pointed out in [6]: "Computer science research has largely ignored embedded systems, using abstractions that actually remove physical constraints from consideration"; "An embedded system is an engineering artifact involving computation that is subject to physical constraints". The same reference adds that: "A lack of adequately trained bicultural engineers causes inefficiencies in industry", referring to people in Control Theory and Computer Engineering. Section 4 will detail further how to take into account, in high-level design, the tough constraints related to the low levels in the context of deep submicron technologies.

3 SoC Design Methodologies

The complexity of the systems designed and their stringent low-power requirements call for a well-defined design methodology. This methodology (Fig. 9.2), used for all the chips described later in Sect. 5, is described below.

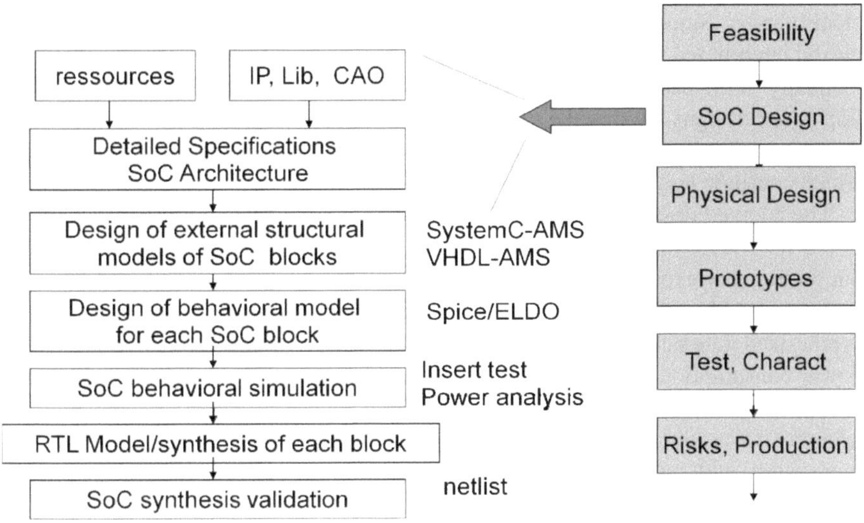

Fig. 9.2 CSEM SoC design methodology

3.1 Top-Down SoC Design Methodology at CSEM

Starting from the detailed system specification, stubs of the main sub-blocks (later replaced by behavioral models, and finally by the detailed design) and a top-level schematic are created. System simulations are performed early in the life-cycle of the design. This is a key issue, since with increase in complexity, the verification stage becomes the design bottleneck.

As illustrated in Sect. 5, some of the designs contain analog blocks with limited digital functionality (e.g. radio SoCs), whereas other systems containing at least one processor and many peripherals are dominated by digital functionality. Depending on this distribution, either a "digital-on-top" or "analog-on-top" design methodology is followed.

In the former, the analog blocks are characterized and black-boxes are instantiated in the top-level netlist, which is either written directly or obtained from a top-level schematic. Behavioral models allow the verification of the design completely in the digital flow. Placement and routing are similarly performed with a digital tool. This flow is especially efficient for timing closure, but special routing of sensitive nets connected to the analog blocks is relatively difficult.

In the analog-on-top methodology, each digital block or sub-block is verified, placed and routed individually. The assembly of the system is performed in a layout editor, with the help of a top-level schematic, where a symbol is created also for digital blocks.

Digital IP blocks (e.g. processor or microcontroller cores, communication peripherals, etc.) are all designed in a latch-based paradigm [7]. The incoming clock cycle with a duty cycle of 50% is internally divided to generate two or four non-overlapping clocks. Data paths then naturally consist of latches enabled by alternate clock phases. Since currently available CAD tools are not able to convert multiplexed latch-based structures to clock-gating structures, clock gating is instantiated manually in the design. The two main advantages of latch-based designs over flip-flop based designs are that the designs are more robust to hold time violations and that the synthesis of a clock tree is less constrained and usually leads to solutions which consume less power [7].

The adopted design flow does not yet rely on high-level exploration tools to derive a RTL level (in VHDL or Verilog) representation based on the specification, but preliminary experiments and evaluation have been carried out with the Orinoco tool proposed by ChipVision. Similarly, power estimation is performed only on the synthesized gate netlist but never at higher abstraction levels. An estimation of the power consumption early in the specification phase thus generally relies solely on the *a priori* knowledge of designers, i.e. based on their past experience. Power management was traditionally introduced manually at the block level of large SoCs, but more rarely with finer granularity, e.g. at the arithmetic unit level.

3.2 Advanced Design Methodology for Power Management

New CAD tools are now available to move power estimation to stages earlier in the design with a tool such as Orinoco. This allows a more in-depth design exploration from a power gating point of view. Orinoco thus fits in CSEM's design flow, between the behavioral block/system description and the actual RTL implementation. The addition of this tool in the flow should allow a more precise estimation of power consumption, an exploration of different architectures and power management possibilities and some improvement in power management, particularly power gating, with a finer grained resolution (currently only performed at the block level, not at the arithmetic level).

Orinoco was evaluated with some dedicated digital arithmetic sub-blocks. The results provided by Orinoco regarding power optimization lived up to expectations. Regarding power gating, it was observed that the energy used by adders in this datapath could be reduced from approximately 2.7 nJ down to 2.0 nJ using this technique. Furthermore, the linear increase in leakage currents when using four adders instead of only two nearly vanishes, because the additional adder instance augments the possibilities for setting components to a sleep state where leakage is reduced. In this sleep state additional leakage currents are very small compared to an additional adder in the active state. This preliminary experiment is thus very encouraging for introducing more power gating in future designs, even at finer design granularities.

In a SoC such as "icycam" (Sect. 5.2), the power consumption of the clock distribution represents more than 50% of the dynamic power consumption in certain operating modes (e.g. stand-by mode). The number of buffer instances inserted during clock tree synthesis is also significant and impacts routing congestion. Therefore, optimization of clock tree topologies, with emphasis on power consumption, is a key issue for upcoming circuits designed at CSEM. The design of low-power processors and IP cores follows a latch-based methodology [7], which yields the following advantages: it decreases constraints on the clock tree generation; it allows efficient gating of the design clocks; and finally, reduced power consumption is possible due to smaller logical depths, hence lower activities. Clock gating is inserted manually, but a tool which may help the designer to find suitable gating conditions in order to regroup several registers gated with close but not completely equal gating conditions would be very interesting. This motivated the evaluation of the LPClock (designed by Politecnico di Torino) and BullDAST flows, even in the context of latch-based designs and of manual clock gating insertion.

The latch-based design flow requires a slightly modified use of the "normal" LPClock flow. The clock activation functions [8] used by LPClock are normally derived automatically by CGCap [9] in a typical flip-flop design. But in latch-based designs it is more difficult to identify activation functions, and as a result a TCL script was developed, which parses the netlist in order to locate the manually instantiated clock gating instances.

The MACGIC DSP was used as a test vehicle for the evaluation of LPClock. This DSP is implemented as a customizable, synthesizable, VHDL software intellectual property (soft IP) core [10]. Experiments were conducted at operating frequencies of 10, 25 and 50 MHz, which is almost the maximum operating frequency for the design using the TSMC 0.18 μm standard cell library. The best results for clock power reduction are obtained when clocks are optimized together (ck1, ck2, ck3 and ck4), because there is more than one optimization and furthermore this is the best way to create a balanced structure for the whole synchronization circuit. As a conclusion, results show that power savings are positive for three of the four clock domains in the MACGIC design and that the best savings reach more than 15% and are obtained when all clocks are optimized.

4 From Low Level Effects to High Level Design

The interdependency between low-level issues mainly originating in very deep submicron technologies, and high-level issues related to SoC design is a major design issue today. It is clear that the gap between low level and high level is increasingly large, with the risk that high-level designers can totally ignore low-level effects and produce nonfunctional SoCs. Leakage power, technology variations, temperature effects, interconnect delay, design for manufacturability, yield, and tomorrow's "beyond CMOS" unknown devices, are the main low-level design aspects that have to be integrated into high-level synthesis. They will impact the high-level design methodologies for instance, by requiring new clocking schemes in processor architectures, by the introducing redundancy and fault-tolerance, by increasing the number of processor cores, by using multiple voltage domains or by using more dedicated techniques to reduce dynamic and static power. An example of the strong impact of the low level on high-level design is in interconnect delays. These increase due to the decreasing section of wires distributing the clock. This is prompting shifts to alternatives such as clockless or asynchronous architectures, moving to multi-cores organized into GALS (Globally Asynchronous and Locally Synchronous) and using Networks-on-Chip.

4.1 Leakage Power Reduction at All Design Levels

There are many techniques [11] at low or at circuit level for reducing leakage, such as using sleep transistors to cut the supply voltage for idle blocks, but other techniques are also available (such as several threshold voltages or bulk biasing). In addition to circuit-level techniques, the total power consumption may also need to be reduced at architectural level. Specific blocks can be operated at optimal supply values (reduced V_{dd} reduces dynamic power), and optimal threshold voltage V_T (a larger V_T reduces static power) for a given speed, in order to find the lowest total power (P_{tot}) depending on the architecture of a given logic block. Therefore, between all the combinations of V_{dd}/V_T guaranteeing the desired speed, only one couple will result in the lowest total power consumption [12–14]. The identification of this optimal working point and its associated total power consumption is tightly related to architectural and technology parameters like activity (a) and logical depth (LD). A reasonable activity is preferred in such a way that dynamic power would not be negligible compared to static power. A small LD is preferred as too many logic gates in series results in gates that do not switch sufficiently. A gate that does not switch is useless as it is only a leaky gate. The ratio between dynamic and static power is thus an interesting figure of merit, and it is linked to the technology $I_{\text{on}}/I_{\text{off}}$ ratio.

4.2 Technology Variations

Technology variations are present from transistor to transistor on the same die, and can be systematic or random due to oxide thickness variations, small differences in W and L transistor dimensions, doping variations, temperature and effects of V_{dd} variations. Many of these variations impact the V_T, which can impact the delay variations by a factor of 1.5 and leakage by a factor of 20. Other effects should not be neglected, such as soft errors. Overall, these effects have a very dramatic impact on yield and consequently on the fabrication cost of the circuits.

In addition to their low-level impacts, the variations described above also affect higher levels. An interesting impact is the fact that multi-core architectures, at the same throughput, are better to mitigate technology variations than single core architectures. With a multi-core architecture, one can work at a lower frequency for the same computation throughput. Consequently, the processor cores (at lower frequencies) are less sensitive to process variations on delay. At very high frequencies, even a very small V_T variation will have a quite large impact on delay variation. It is also better to work at high or nominal V_{dd}, whereas at very low V_{dd} (for instance 0.5 V) any digital block is very sensitive to V_T variation as the speed is inversely proportional to $V_{dd} - V_T$. Logic circuits based on transistors operating in weak inversion (also called the subthreshold regime) therefore offer the minimum possible operating voltage [4], and thereby the minimum dynamic power for a given static power. This technique has been revived recently and applied to complete subsystems operating below 200 mV. It has been demonstrated that minimal energy circuits are those operated in the subthreshold regime with V_{dd} below V_T, resulting in lower frequencies and larger clock period [16, 17]. Therefore, dynamic power and static power are decreased, although the static energy is increased as more time is required to execute the logic function. This means that there is an optimum in energy. As previously indicated, this optimal energy also depends on logic depth LD and activity factor a [15]. The minimal V_{dd} (and minimal energy) is smaller for small logical depth and for large activity factors.

Another approach is to introduce spatial or timing redundancy to implement fault-tolerant architectures. This is a paradigm shift, as any system would not be expected to be composed of completely reliable units, but could still function under the consideration that a number of units could fail, without compromising the functionality of the entire system. One possible architecture is to use massive parallelism while presenting redundant units that could take over the work of faulty units. One can have spatial redundancy (very expensive) or timing redundancy (quite expensive in terms of throughput). However, all redundant architectures face the same problem: the overhead in hardware or in throughput is huge, which is a contradictory effect for an energy-efficient architecture.

4.3 Yield and DFM

For very deep submicron technologies, the smallest dimensions of transistors geometries on the mask set are well below the lithographic light wavelengths. This yields a variety of unwanted effects, such as bad line end extension, missing small geometries, etc. These effects can be corrected by OPC (Optical Proximity Correction) which is a means available to DFM (Design For Manufacturability). However, to facilitate the process of mask correction by OPC, it is recommended to have a regular circuit layout. Regular arrays implementing combinational circuits like PLAs or ROM memories are therefore increasingly attractive. Figure 9.3 shows three examples of a regular layout. A first example from 1988 [18] is shown at the right of Fig. 9.3 in micron-scale technology, and is called the gate-matrix style. It was used to facilitate automatic layout generation. The two other pictures describe a SRAM cell as well as nanowires [19] for which it is mandatory to have very regular structures. This has a huge impact on architectures and systems: SoC architectures should be based on regular arrays and structures, such as PLAs and ROMs for combinational circuits and volatile memories such as SRAM for data storage. Consequently, SoC design should be fully dominated by memories and array structures.

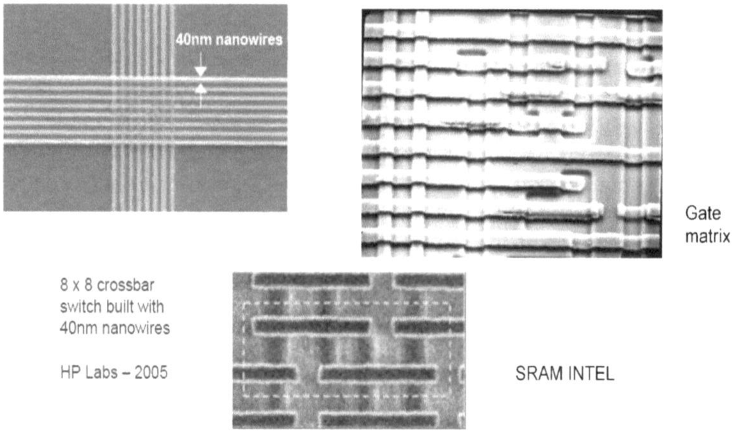

Fig. 9.3 Very regular blocks at layout level for DFM

5 Design Examples

5.1 Wisenet SoC

CSEM has launched a wireless sensor network (WSN) project named WiseNET. A major priority was to achieve extremely low energy consumption both at the circuit level, with the design of the WiseNET SoC [20], and at the system level with the development of the WiseMAC protocol [21].

The WiseNET SoC is a circuit that has been leveraged and industrialized into a home security application for industrial customer. The chip contains an ultra-low-power dual-band radio transceiver (for the 434 MHz and 868 MHz ISM bands), a sensor interface with a signal conditioner and two analog-to-digital converters, a digital control unit based on a CoolRISC microcontroller with SRAM low-leakage memories and a power management block. In terms of power consumption, the most critical block is the RF transceiver. In a 0.18-micrometer standard digital CMOS process, in receive mode, the radio consumes 2.3 mA at 1.0 Volt and 27 mA in transmit mode for 10 dBm emitted power. However, as the duty cycle of any WSN application is very low, using the WiseNET transceiver with the WiseMAC protocol [21], a relay sensor node consumes about 25 microwatts when forwarding 56-byte packets every 100 seconds, enabling several years of autonomy from a single 1.5 V AA alkaline cell. Figure 9.4 shows the integrated WiseNET SoC.

The WiseMAC protocol is a proprietary protocol that is based on the preamble sampling technique. This technique consists of regularly sampling the medium to check for activity. By sampling the medium, is meant listening to the radio channel for the duration required to measure the received power (i.e. a few symbols). All sensor nodes in a network sample the medium with the same constant period. The WiseMAC protocol running on the WiseMAC SoC achieves more than an order of

Fig. 9.4 Wisenet SoC

magnitude better power consumption than standard solutions such as Zigbee, where the protocol is not optimized for extreme low-power.

Within the WiseNET project, it was mandatory to co-specify and co-develop the WiseNET SoC and the WiseMAC protocol, for being able to achieve the lowest possible energy consumption with the WSN project. For instance, the WiseNET SoC is designed to minimize the sleep current consumption and the leakage during the long periods that can separate transmission bursts using the WiseMAC protocol. Also, the RF transceiver, which is the largest contributor of the SoC in terms of peak current consumption, is designed for optimal turn-on, turn-off and Rx-to-Tx (receiver-to-transmitter) turn-around sequences, in order to keep the energy consumption at the lowest possible levels during the "medium sampling" sequences of the protocol, hence limiting the energy waste. Due to the low duty cycle requirement, the sleep current is a key issue which requires the design of dedicated low-leakage SRAM using a bulk-biasing technique, because standard library SRAM cells yield unacceptable static power consumption for WSN applications. Conversely, the protocol is designed to take into account limiting issues related to the RF transceiver circuit. For example, the peak transmit current consumption for achieving 10 dBm output power is much larger than the peak receive current, and therefore the protocol is built on a "periodic listening" scheme for minimizing the transmissions, hence increasing the WSN global life-time. Another example is that the WiseMAC protocol exploits the knowledge of the sampling schedule of its direct neighbors, thanks to a precise crystal-based time reference within the WiseNET SoC. This allows the protocol to use a wake-up preamble of very short length, hence further minimizing the energy wastage. These selected examples show clearly that the complexity of the SoC design extends way beyond the design of the IC by encompassing high-level system issues such as the communication protocol.

5.2 Vision Sensor SoC

Icycam is a circuit combining on the same chip a 32-bit icyflex processor [22] operating at 50 MHz and with a high dynamic range versatile pixel array, integrated in a 0.18 μm optical process. It enables the implementation, on a single chip, of image capture and processing, thus bringing considerable advantages in terms of cost, size and power consumption. Icycam has been developed to address vision tasks in fields such as surveillance, automotive, optical character recognition and industrial control. It can be programmed in assembler or C-code to implement vision algorithms and controlling tasks. It is a very nice example of MPSoC, as there are one processor and one co-processor (icyflex and Graphical Processing Unit (GPU) tailored for vision algorithms), as well as pixel-level data encoding to facilitate further processing (320 × 240 pixels). It is possible to integrate the vision sensor as well as the digital processing functions on the same die. It is also a very representative example of heterogeneous SoC, as the vision sensor is integrated on-chip in an optical 180 nanometers technology. The rest of the circuit (digital, memories, analog) is integrated in the same optical 180 nm technology, so the design methodology is the same as before, with an additional block: the vision sensor.

The heart of the system is the 32-bit icyflex processor clocked at a 50 MHz frequency [22]. It communicates with the pixel array, the on-chip SRAM and peripherals via a 64-bit internal data bus. The pixel array has a resolution of 320 by 240 pixels (QVGA), with a pixel pitch of 14 μm. Its digital-domain pixel-level logarithmic compression makes it a low noise logarithmic sensor with close to 7 decades of intra-scene dynamic range encoded on a 10-bit data word. One can extract on the fly the local contrast magnitude (relative change of illumination between neighbor pixels) and direction when data are transferred from the pixel array to the memory. Thus it offers a data representation facilitating image analysis, without overhead in term of processing time. Data transfer between the pixel array and memory or peripherals is performed by groups of 4 (10 bits per pixel) or 8 (8 bits per pixel) pixels in parallel at the system clock rate. These image data can be processed with the icyflex's Data Processing Unit (DPU) which has been complemented with a Graphical Processing Unit (GPU) tailored for vision algorithms, able to perform simple arithmetical operations on 8- or 16-bit data grouped in a 64-bit word. As the internal SRAM is size consuming, the internal data and program memory space is limited to 128 kBytes. This memory range can be extended with an external SDRAM up to 32 MBytes. The whole memory space is unified which means accessible via the data, program and DMA busses. An internal DMA working on 8/16/32 and 64 bits enables transfers from/to the vision sensor, memories and peripherals with data packing and unpacking features. The chip has been integrated and is pictured in Fig. 9.5.

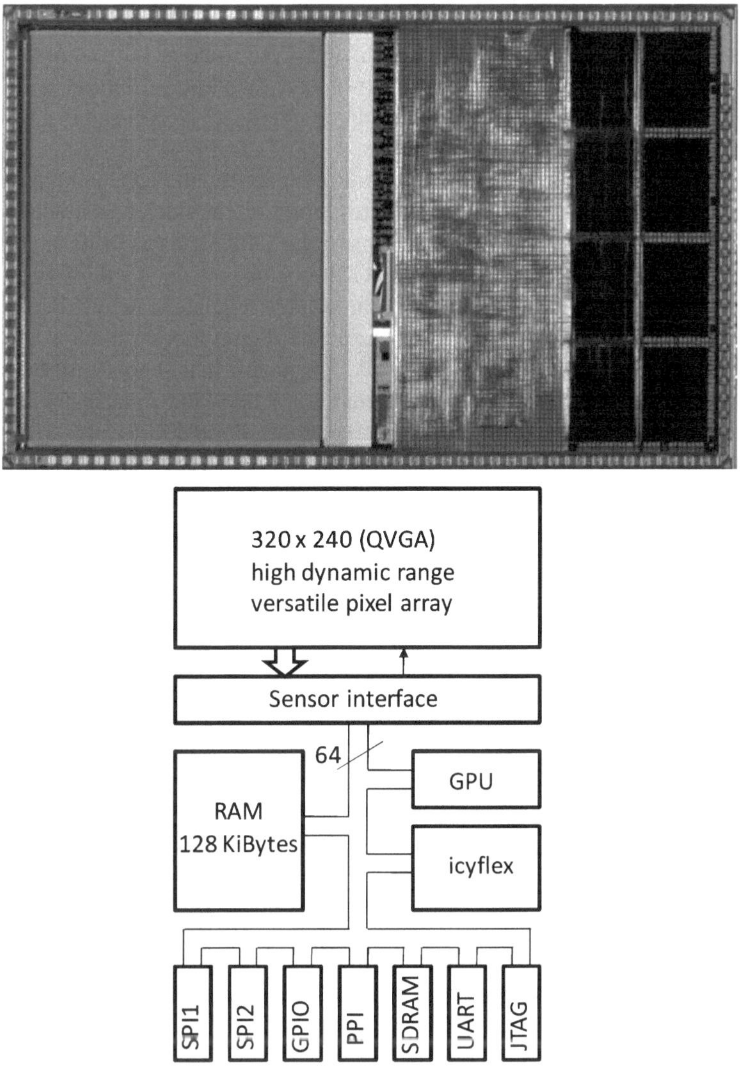

Fig. 9.5 icycam SoC

5.3 DSP and Radio SoC

The icycom SoC chip is also based on the icyflex DSP processor [22], and includes 96 kByte low-leakage SRAM program or data memory. Similarly to Wisenet (Sect. 5.1), the chip contains a RF wireless link for EU&US 863–928 MHz bands. Its data rate is up to 200 kbps with various modulation schemes such as OOK (On Off Keying), FSK (Frequency Shift Keying), MSK (Minimum Shift Keying), GFSK (Gaussian Frequency Shift Keying) and 4-FSK (4-level Frequency Shift Keying). The Rx current is 2.5 mA at 1 V. Many peripheral blocks are available, such as a

10-bit ADC, DMA, IRQ, 4 timers, digital watchdog and real time clock (RTC), as well as standard digital interfaces (I2C, I2S, SPI, UART and 32 GPIO). A set of regulators brings advanced power management features and power modes, and finally a 10-bit ADC is also available. The chip can be interfaced by SPI and/or I2C bus to one or two external non-volatile memories.

Apart from its processing and interconnect capability, the icycom chip also offers some power management functions. Icycom SoC provides power supplies for external blocks by using the digital outputs of the GPIO pads as switchable power supplies. The supply voltage used for the GPIOs is taken from the digital regulator, or one of the three DC-DC converters. Four voltage regulators are on chip: a 0.9 V (switchable) regulator that supplies power to the digital blocks, a 0.8 V regulator that supplies power to the RF VCO, a 0.9 V regulator that supplies to the RF PA (tunable) and a programmable regulator from 1.2 V up to the voltage supply minus 0.1 V. There are three types of clock generators: an internal fast wake-up 16 MHz RC oscillator, a 32–48 MHz Xtal oscillator and a 32 kHz Xtal oscillator.

Icycom SoC offers multiple idle modes for start-up time versus leakage trade-off. In the sleep mode (4 μA), the processor is not clocked but some peripherals may remain clocked. The wake-up is instantaneous with a fast clock (HF Xtal or RC based oscillator). In the frozen mode (2.5 μA), all the digital (including Xtal) oscillators are not clocked (except RTC). The wake-up has to wait for the RC oscillator start-up (typically 0.5 ms). In the hibernation mode, the processor and its peripherals (except RTC) are switched off from V_{dd} to further reduce the leakage down to 1 μA. A reboot is then necessary and the wake-up time depends on the amount of RAM to reload (typically below 500 ms).

At 1 V supply and with a low-power standard cell library in 0.18 μm technology, the maximum frequency is close to 3.4 MHz. The stand-by current, with only RTC running at 32 kHz, is 1 μA. The chip is 5 mm × 5 mm in fully digital 180 nm technology (Fig. 9.6).

5.4 Abilis Mobile TV SoC

CSEM has licensed another DSP core, called MACGIC [10] to Abilis [23], a Swiss company of the Kudelski group. This DSP core has been used in a SoC for broadband communication in a wireless multipath environment using Orthogonal Frequency Division Multiplexing (OFDM). Although the theory of OFDM is well developed, implementation aspects of OFDM systems remain a challenge for supporting many different standards on a single chip, and for reaching ultra low power consumption. The SoC developed by Abilis (Fig. 9.8) is an OFDM digital TV receiver for the European DVB-T/H standards containing a multi-band analog RF tuner, immediately followed by an analog-to-digital-converter (ADC) and a digital front-end implementing time-domain filtering and I/Q channel mismatch correction. Several algorithms are executed on chip, such as mismatch correction, Fast Fourier Transform (FFT), equalizer, symbol de-mapping and de-interleaving, forward error correction (FEC) through Viterbi decoder, de-interleaver and Reed-Solomon decoder.

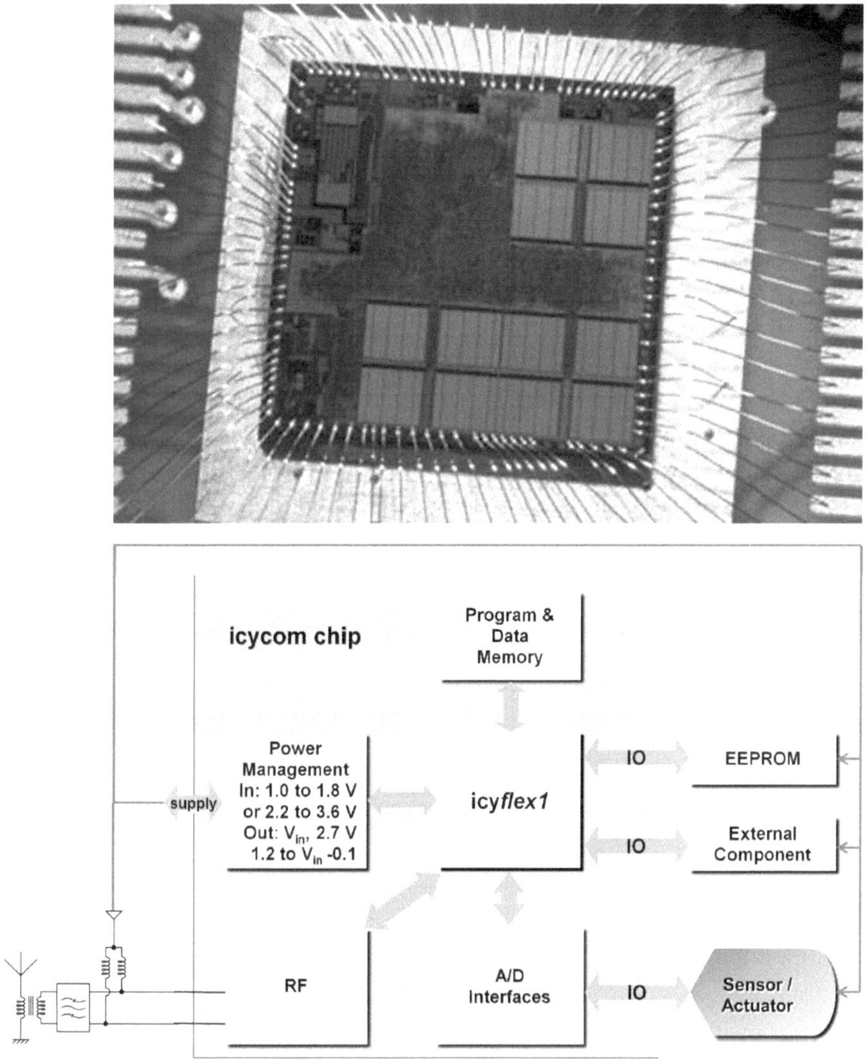

Fig. 9.6 icycom SoC

The main algorithms implemented by the software-programmable OFDM demodulator are the frequency compensation, the FFT and an adaptive channel estimation/equalization. Abilis has designed a 90 nm single-die digital mobile TV receiver platform (Fig. 9.7), from which two different chips, the AS-101 and AS-102 have been developed (for DVB-T/H applications). They both integrate a multi-band RF tuner, an advanced programmable OFDM demodulator, memory and various I/O interfaces. The programmable OFDM demodulator is implemented as a set of 3 MACGIC DSPs customized for OFDM applications. The MPSoC also contains an

Fig. 9.7 Abilis mobile TV SoC with three MACGIC DSP

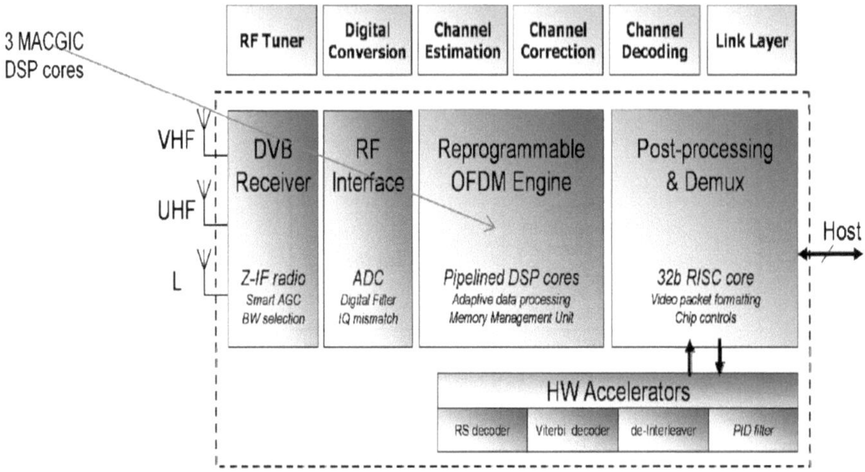

Fig. 9.8 Abilis mobile TV SoC architecture

ARC 32-bit RISC core as well as four hardware accelerators (RS decoder, Viterbi decoder, de-interleaver, PID filter, Fig. 9.8), making this chip a true MPSoC.

6 Disruptive SoC Architectures

6.1 Nanodevices Replacing MOS?

CMOS "scaling" is predicted to reach an end around 11 nanometers, roughly 10 years from now. After 2017, CMOS should move to "Beyond CMOS". However, today, there is no clear alternative route to replace CMOS. The current scientific literature shows that devices such as carbon nanotubes (CNT), nanowires and molecular switches can be fabricated [24] with some fault-tolerance mechanism [25], but it is not yet clear how to interconnect billions of switches and billions of wires in the design of complex architectures and systems. Combining CMOS and nanodevices is also an interesting idea [26] which will further push the heterogeneity of integrated circuits. Nanowires, like very advanced CMOS devices, are non-ideal switches, as they present high subthreshold and gate leakages and are very sensitive to technology variations. It is therefore mandatory to propose design techniques to reduce leakage power drastically and to mitigate the effects of technology variations (Sect. 4.2). Nanowires are similar to gate-all-around (GAA) devices for which it is impossible to apply very well-known technique such as body bias to both reduce leakage and the impact of technology variations. The body bias technique, used both to increase V_T for reduced leakage and to adjust V_T for mitigating technology variations, requires a body terminal that does not exist in GAA devices. Even the very advanced tri-gate CMOS transistors present a much smaller body effect [27]. The nanowire V_T thus cannot be modified by substrate bias [28]. Other techniques have to be proposed, such as the source biasing technique [29] that dynamically modifies the supply voltage by $-\Delta V_p$ and ground voltage by $+\Delta V_n$. Depending on the values $-\Delta V_p$ and $+\Delta V_n$, the leakage and delays of subthreshold CMOS circuits can be adjusted to mitigate the effects of technology variations.

6.2 SoC Dominated by Memories

It is sometimes interesting to revise completely classical ways of thinking and to try to elaborate disruptive heterogeneous and SoC architectures. Disruptive ideas sometimes happen not to be so new, as they may have been proposed a long time ago and forgotten, and may be revived to address tomorrow's challenges. This paragraph focuses on four such ideas.

In the hearing aids market, for example, all competitors have more or less the same hardware needs (size, consumption, digital treatment), but they have all designed their own hardware solutions. Nowadays, all technical requirements and price constraints can be addressed by implementing powerful highly integrated SoCs. As such, hearing aids companies can concentrate on the development of new algorithms on a single chip, which is the same for all new products. The real core business, and generation of added value, has thus become pure software development. Consequently, a first idea could be to design a single universal SoC or MPSoC platform: the motivation is to say that all applications have to rely on the same hardware, and consequently, the design and differentiator between various applications is completely concentrated in embedded software. Such an MPSoC platform would be very expensive to develop (about 100 M€) and one could question whether it remains reasonable for applications sensitive to power consumption or to other specific performance metrics.

A second idea is a SoC or MPSoC dominated by memories. Memories are automatically generated, implying that the hardware part to design is very small and requires a low development effort. This means that one has to maximize the on-chip memory part, with very small processors and peripherals. In this case, the design of a new chip mainly consists of the development of embedded software. It is therefore similar to the first idea, the difference being that a new chip is designed with the required amount of memory, but not more.

A third idea is a SoC or MPSoC with 1000 parallel processors. This is very different from multicore chips with 2 to 32 cores. With 1000 cores, each core is a very small logic block of 50 K gates combined with a lot of memory.

A fourth idea is the design of SoC architectures with nano-elements (Sect. 6.1). The design methodology should be completely different, consisting of a bottom-up design methodology and not of a top-down one. This is due to the fact that the fabrication process will produce many nano-devices, a non-negligible proportion of which will be nonfunctional. As such, the design methodology will consist of checking if the fabricated chip can actually be used for something useful. Hardware will target very regular circuits and layouts. However, the applications are likely to be completely different to existing microprocessors—one can expect to see approaches based on neural nets, biological circuits or learning circuits.

7 Conclusion

The diagnostic is clear: complexity increases, interdisciplinarity too. There are increasingly more interactions between all design levels from application software down to RF-based MPSoC as described with the various design cases developed at CSEM, such as the WiseNET SoC, the Vision Sensor SoC, the icycom SoC and the Mobile TV SoC.

Consequently, engineers have to design towards higher and higher design levels but also down to lower and lower design levels. This widening gap will call for design teams that are increasingly heterogeneous, and with increasingly challenging objectives: to perform focused research for providing outstanding and innovative blocks in a SoC, but also interdisciplinary research which becomes the "key" to successful SoC designs.

Acknowledgements The authors wish to acknowledge the CSEM design teams that contributed to the SoC cases described above: Claude Arm, Flavio Rampogna, Silvio Todeschini, Ricardo Caseiro of the "SoC and Digital Group", Pierre-François Ruedi, Edoardo Franzi, François Kaess, Eric Grenet, Pascal Heim, Pierre Alain Beuchat, of the "Vision Sensor Group", D. Ruffieux, F. Pengg, M. Kucera, A. Vouilloz, J. Chabloz, M. Contaldo, F. Giroud, N. Raemy of the "RF and Analog IC Group" and E. Le Roux, P. Volet of the "Digital Radio Group".

The authors also wish to acknowledge the EU project MAP2 partners (CRAFT-031984), i.e. OFFIS, ChipVision, Politecnico di Torino and BullDAST, for the design methodologies described in Sect. 3. The authors also acknowledge the industrial contributions from Hager and Semtech for the WiseNET SoC, and Abilis for the MACGIC-based SoC for mobile TV.

References

1. www.csem.ch
2. Enz, C., et al.: WiseNET: an ultra-low power wireless sensor network solution. Computer **37**, 62–70 (2004)
3. Rabaey, J.: Managing power dissipation in the generation-after-next wireless systems. In: FTFC'99, Paris, France, June 1999
4. Vittoz, E.: Weak inversion for ultimate low-power logic. In: Piguet, C. (ed.) Low-Power Electronics Design. CRC Press, Boca Raton (2004). Chap. 16
5. Hanson, S., Zhai, B., Blaauw, D., Sylvester, D., Bryant, A., Wang, X.: Energy optimality and variability in subthreshold design. In: Intl. Symp. on Low Power Electronics and Design, pp. 363–365 (2006)
6. Henzinger, T., Sifakis, J.: The discipline of embedded systems design. Computer **40**, 32–40 (2007)
7. Arm, C., Masgonty, J.-M., Piguet, C.: Double-latch clocking scheme for low-power I.P. Cores. In: PATMOS, Goettingen, Germany, September 13–15, 2000
8. Donno, M., Ivaldi, A., Benini, L., Macii, E.: Clock-tree power optimization based on RTL clock-gating. In: Proc. DAC'03, 40th Design Automation Conference (DAC'03), p. 622 (2003)
9. Benini, L., et al.: A refinement methodology for clock gating optimization at layout level in digital circuits. J. Low Power Electron. **6**(1), 44–55 (2010)
10. Arm, C., Masgonty, J.-M., Morgan, M., Piguet, C., Pfister, P.-D., Rampogna, F., Volet, P.: Low-power quad MAC 170 µW/MHz 1.0 V MACGIC DSP core. In: ESSCIRC, Montreux, Switzerland, Sept. 19–22, 2006
11. Roy, K., Mukhopadhyay, S., Mahmoodi-Meimand, H.: Leakage current mechanisms and leakage reduction techniques in deep-submicrometer CMOS circuits. Proc. IEEE **91**(2), 305–327 (2003)
12. Schuster, C., Nagel, J.-L., Piguet, C., Farine, P.-A.: Leakage reduction at the architectural level and its application to 16 bit multiplier architectures. In: PATMOS '04, Santorini Island, Greece, September 15–17, 2004
13. Schuster, C., Piguet, C., Nagel, J.-L., Farine, P.-A.: An architecture design methodology for minimal total power consumption at fixed V_{dd} and V_{th}. J. Low Power Electron. **1**(1), 1–8 (2005)
14. Schuster, C., Nagel, J.-L., Piguet, C., Farine, P.-A.: Architectural and technology influence on the optimal total power consumption. In: DATE 2006, Munich, March 6–10, 2006
15. Zhai, B., Blaauw, D., Sylvester, D., Flautner, K.: Theoretical and practical limits of dynamic voltage scaling. In: DAC 2004, pp. 868–873 (2004)
16. Hanson, S., Zhai, B., Blaauw, D., Sylvester, D., Bryant, A., Wang, X.: Energy optimality and variability in subthreshold design. In: International Symposium on Low Power Electronics and Design, ISLPED 2006, pp. 363–365 (2006)
17. Kwong, J., et al.: A 65 nm Sub-Vt microcontroller with integrated SRAM and switched-capacitor DC-DC converter. In: ISSCC'08, pp. 318–319 (2008)
18. Piguet, C., Berweiler, G., Voirol, C., Dijkstra, E., Rijmenants, J., Zinszner, R., Stauffer, M., Joss, M.: ALADDIN: a CMOS gate-matrix layout system. In: Proc. of ISCAS 88, Espoo, Helsinki, Finland, p. 2427 (1988)
19. Haykel Ben Jamaa, M., Moselund, K.E., Atienza, D., Bouvet, D., Ionescu, A.M., Leblebici, Y., De Micheli, G.: Fault-tolerant multi-level logic decoder for nanoscale crossbar memory arrays. In: Proc. ICCAD'07, pp. 765–772
20. Peiris, V., et al.: A 1 V 433/868 MHz 25 kb/s-FSK 2 kb/s-OOK RF transceiver SoC in standard digital 0.18 µm CMOS. In: Int. Solid-State Circ. Conf. Dig. of Tech. Papers, Feb. 2005, pp. 258–259 (2005)
21. El-Hoiydi, A., Decotignie, J.-D., Enz, C., Le Roux, E.: WiseMAC, an ultra low power MAC protocol for the WiseNET wireless sensor network. In: SenSys'03, Los Angeles, CA, USA, November 5–7, 2003

22. Arm, C., Gyger, S., Masgonty, J.-M., Morgan, M., Nagel, J.-L., Piguet, C., Rampogna, F., Volet, P.: Low-power 32-bit dual-MAC 120 μW/MHz 1.0 V icyflex DSP/MCU core. In: ESSCIRC, Edinburgh, Scotland, UK, Sept. 15–19, 2008
23. http://www.abiliss.com
24. Huang, Yu, et al.: Logic gates and computation from assembled nanowire building blocks. Science **294**, 1313–1316 (2001)
25. Schmid, A., Leblebici, Y.: Array of nanometer-scale devices performing logic operations with fault-tolerant capability. In: Fourth IEEE Conference on Nanotechnology IEEE-NANO (2004)
26. Ecoffey, S., Pott, V., Bouvet, D., Mazza, M., Mahapatra, S., Schmid, A., Leblebici, Y., Declercq, M.J., Ionescu, A.M.: Nano-wires for room temperature operated hybrid CMOS-NANO integrated circuits. In: Solid-State Circuits Conference, ISSCC 2005, 6–10 Feb. 2005, pp. 260–597, vol. 1 (2005)
27. Frei, J., et al.: Body effect in tri- and pi-gate SOI MOSFETS. IEEE Electron Device Lett. **25**(12), 813–815 (2004)
28. Singh, N., et al.: High-performance fully depleted silicon nanowire (diameter < 5 nm) gate-all-around CMOS devices. IEEE Electron Device Lett. **27**(5), 383–386 (2006)
29. Kheradmand Boroujeni, B., et al.: Reverse Vgs (RVGS): a new method for controlling power and delay of logic gates in sub-VT regime. Invited talk at VLSI-SoC, Rhodes Island, Oct. 13–15, 2008

Chapter 10
MpAssign: A Framework for Solving the Many-Core Platform Mapping Problem

Youcef Bouchebaba, Pierre Paulin, and Gabriela Nicolescu

1 Introduction

The current trend for keeping pace with the increasing computation budget requirements of embedded applications consists in integrating large numbers of processing elements in one chip. The efficiency of those platforms, also known as many-core platforms, depends on how efficiently the software application is mapped on the parallel execution resources. In this context, the design of tools that can automate the mapping process is of major importance. The problem of automatic application mapping on many-cores is a non-trivial one because of the number of parameters to be considered for characterizing both the applications and the underlying platform architectures. On the application side, each component composing the application may have specific computation and memory requirements in addition to some real-time constraints. On the platform side, many topological concerns will impact the communication latencies between several processing elements, which may have different computation and memory budgets. Since most of these parameters are orthogonal and do not permit any reductions, the problem of mapping applications on such platforms is known to be *NP-hard* [1, 2].

Recently, several authors [2–5] proposed to use multi-objective evolutionary algorithm to solve this problem within the context of mapping applications on Network-on-Chips (NoC). These proposals are mostly based on the *NSGAII* [6] and *SPEA2* [7] algorithms, which consider only a limited set of application and architecture constraints, and which define only a few objective functions. However, in the case of real life applications many constraints and objective functions need

Y. Bouchebaba (✉) · P. Paulin
STMicroelectronics, 16 Fitzgerald Rd, Ottawa, ON, K2H 8R6, Canada
e-mail: youcef.bouchebaba@st.com

G. Nicolescu
Department of Computer Engineering, Ecole Polytechnique Montreal, 2500 Chemin de Polytechnique Montreal, Montreal, Québec, Canada H3T 1J4

G. Nicolescu et al. (eds.), *Design Technology for Heterogeneous Embedded Systems*, DOI 10.1007/978-94-007-1125-9_10, © Springer Science+Business Media B.V. 2012

to be considered and some of these parameters may be contradictory (e.g. execution speed with memory consumption or load balancing with communication cost) which results in a multi-objective optimization problem [8, 9]. We believe that a good mapping tool should:

- Provide several meta-heuristics. These different meta-heuristics allow exploring different solution spaces.
- Provide several objective functions and architecture (or application) constraints. As shown in Sects. 4.2 and 4.3, several objective functions and architecture constraints are provided by our tool.
- Offer the designer the flexibility to easily add any new objective function and architecture (or application) constraint.
- Offer the designer the flexibility to extend and to adapt the different genetic operators (e.g. mutation, crossover, etc.).

In this chapter, we present a new mapping tool which offers all of the above cited characteristics. Our tool is implemented on top of the *jMetal framework* [10]. *jMetal* offers several benefits. First, it provides an extensible environment. Second, it integrates several meta-heuristics (e.g. *NSGAII* [6], *SPEA2* [7], *MOCELL* [11], *SMPSO* [12], *GDE* [13], *PESA* [14], *FASTPGA* [15], *OMOPSO* [16], etc.), which we can use as a basis for evaluating our proposal against others.

This chapter also presents a parallel implementation of multi-objective evolutionary algorithm, which allows the distribution of several meta-heuristics on different processing islands (to exploit several meta-heuristics at the same time). These islands collaborate during their execution in order to converge to a better solution by leveraging the feedback that they obtain from their neighbors. Indeed, parallel implementation of genetic algorithms is a well-known technique in the literature of combinatorial optimization [17].

In order to evaluate the effectiveness of our tool, we considered the case of an industrial research project aiming at the distribution of parallel streaming applications on a *NoC*-based many-core platform. We present in this chapter the objective and architecture constraint functions that we defined for this case. We also compare the results obtained by several new-meta-heuristics offered by our tool with the results given by the classical meta-heuristics such as *NSGAII* [6] and *SPEA2* [7].

The chapter is organized as follows. Section 2 introduces the application and platform characterization. Section 3 presents an overview of the multi-objective optimization problem and describes one of the meta-heuristics offered by our tool (*MOCELL* [11]). Section 4 presents the implementation of our tool in *jMetal* [10], including the objective and constraint functions. Section 5 evaluates our proposal with comparisons to existing techniques. Section 6 discusses related work. Finally, Sect. 7 concludes the chapter and discusses perspectives for future work.

2 Application and Platform Characterization

There is a plethora of programming models aimed at programming parallel applications. Programming models based on thread-based parallelization (e.g. *POSIX*

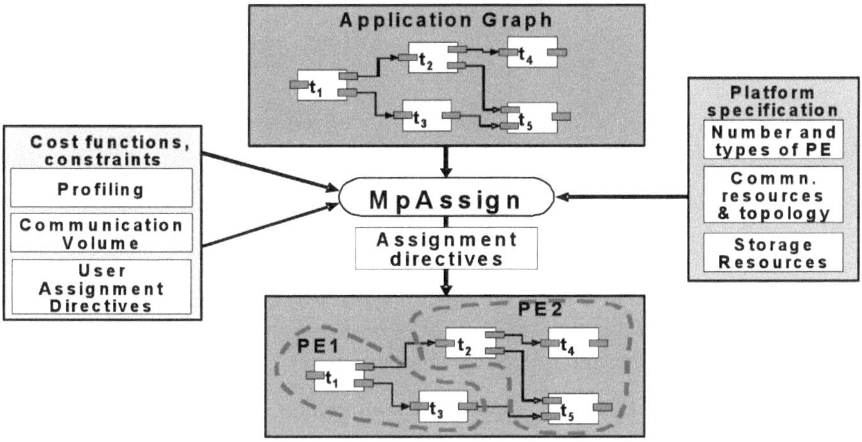

Fig. 10.1 Inputs and outputs of our mapping tool (MpAssign)

threads, *OpenMP* [18]), or message passing (e.g. *MPI* [19]) are widely used. However, the weak encapsulation features of those models becomes a burden for distributing software applications on many-core platform in a systematic way. This motivated recent proposals (e.g. *StreamIt* [20], OpenCL [21]) to identify a more structured way of building applications. Common features of these recent proposals can be summarized as (1) strong encapsulation of program units into well-defined tasks and (2) explicit capture of communication between different tasks. This way of capturing the software architecture is very well aligned with the internal structure of many streaming applications [20]. Moreover, the explicit capture of tasks and their dependencies is key to identifying the tasks that can be executed in parallel and to explore different mapping possibilities on the underlying platform architecture.

We based our work on a stream-based parallel programming framework called *StreamX*. The capture model of *StreamX* can be seen as an extension of the *StreamIt* [20] capture model. It provides support for dynamic dataflow execution and tasks with multiple input/output ports. Since the implementation details of this programming model are not the main concern of this chapter, we present hereafter only the abstractions that are used by the mapping tool. The mapping tool (*MpAssign*) presented in this chapter (Fig. 10.1) receives as input:

- The application capture.
- The high-level platform specification.
- The set of criteria to be optimized (our tool offers several objective functions; the user has the possibility to choose a subset of them or to define new ones).
- The set of architecture constraints (our tool implements several architecture constraints; the user has the possibility to choose a subset of them or to define new ones).

The output of our tool is a set of assignment directives specifying the mapping of the application tasks to the multiple processing elements of the platform.

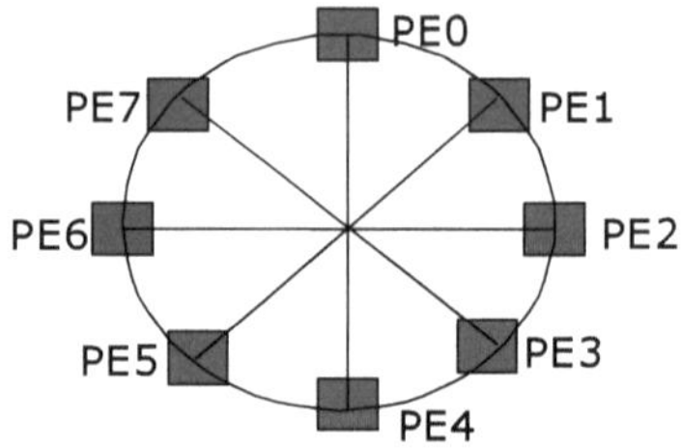

Fig. 10.2 Example of NoC with 8 PEs

An application written in *StreamX* can be captured as a parameterized task graph (T, E) where T is a non-empty set of vertices (tasks) t_i and E is a non-empty set of edges e_i. Each task t_i has the following annotations:

- $Load(t_i)$: the load of the task t_i; corresponding to the amount of clock cycles that are required for executing t_i on a given PE.
- $Memory(t_i)$: the amount of memory used by t_i.
- $PE(t_i)$: the pre-assignment of the task t_i to a processor (i.e. optionally, the user can force the mapping of the task t_i on a given PE).

In addition to the above task annotations, each edge e_i in E is labeled with $volume(e_i)$ representing the amount of data exchanged at each iteration, between the tasks connected by e_i. The platform used to evaluate the mapping tool was developed at *STMicroelectronics* in the context of an industrial research project. It contains a configurable number of symmetric processing elements connected to a *NoC* through a flow controller which can be used to implement hardware *FIFOs*. A host processor is used to load and control the application and can access the computing fabric through a system *DMA*. Finally, each processing node on the *NoC* has a local memory which can be accessed by other cores (*NUMA* architecture). Consequently, the high level platform specification used by the mapping tool contains information on the NoC topology, characteristics of the processing elements and architectural constraints (memory space limit, number of channels in the flow controllers and the *DMA*). The *NoC* we use is called the *STNOC* [22]. It adopts the full Spidergon topology and it connects an even number of routers as a bidirectional ring in both clockwise and counter clockwise directions, with in addition a cross connection for each couple of routers. Figure 10.2 depicts the Spidergon topology graph for an 8 node configuration.

Other types of topologies are also taken into account like a *2D mesh/torus*. In all cases, the latency is approximated based on the topology without taking contention into account.

3 Multi-objective Evolutionary Algorithms

As mentioned previously, our framework offers several new meta-heuristics which are not explored by previous works on the many-core platform mapping problems. Among these new meta-heuristics, we can cite *MOCELL* [11], *SMPSO* [12],

GDE [13], *PESA* [14], *FASTPGA* [15], *OMOPSO* [16], etc. For lack of space, we describe in this section only one of them (*MOCELL* [11]). Prior to this, we will begin with a brief introduction of multi-objective optimization.

3.1 Multi-objective Optimization

The performance of applications executed on many-core platforms depends on a multitude of parameters. As discussed in the previous sections, those parameters may involve the structure of the application (e.g. task decoupling, communication volumes, dependencies, etc.) and the features of the underlying platform (e.g. computation budget, memory budget, communication latencies, NoC topology, etc.). The presence of so many parameters inhibits the definition of one clear optimization objective. Therefore, developers typically aim at optimizing a set of performance criteria among a long list including execution speed, memory consumption, energy consumption, and so forth. Obviously, some of those optimization objectives will often lead to contradictions. For that reason, a good mapping tool must provide solutions for optimizing multiple objectives while providing some support for taking into account some tradeoffs between contradictory objectives. The multi-objective optimization problem [8, 9] can be defined as the problem of finding a vector of decision variables $(x_1, \ldots, x_n)$ which satisfies constraints and optimizes a vector of functions which elements represent m objective functions $f = (f_1(x), \ldots, f_m(x))$.

- The objective functions form a mathematical description of performance criteria (objective functions) that are usually in conflict with each other.
- The constraints define a set of feasible solutions X.

As there is generally, no solution x for which all the functions $f_i(x)$ can be optimized simultaneously, we need to establish certain criteria to determine what would be considered an optimal solution. A way of dealing with the above problem is known as the Pareto optimum [8, 9]. According to it, a solution x dominates another solution y if and only if the two following conditions are true:

- x is not worse than y in any objective, i.e. $f_j(x) \leq f_j(y)$ for $j = 1, \ldots, m$.
- x is strictly better than y in at least one objective, i.e. $f_j(x) < f_j(y)$ for at least one j in $\{1, \ldots, m\}$.

In this context, x is also said to be non-dominated by y and y is dominated by x (Fig. 10.3). Among a set of solutions X, the non-dominated set of solutions P are those that are not dominated by any other member of the set X. When the set X is the entire feasible search space, then the set P is called the global Pareto optimal set. The image $f(x)$ of the Pareto optimal set is called the Pareto front. Each Pareto solution is better than another one for at least one criterion. In the case of our mapping problem, the vector of decision variables will be represented by a vector $x = (x_1, x_2, \ldots, x_n)$, where x_i represents the processor on which will be mapped the task t_i. Each solution (mapping) x can be characterized by a set of objective functions which can be expressed as function of the NoC architecture and the input application task graph. They will be explained in detail in Sect. 4.2.

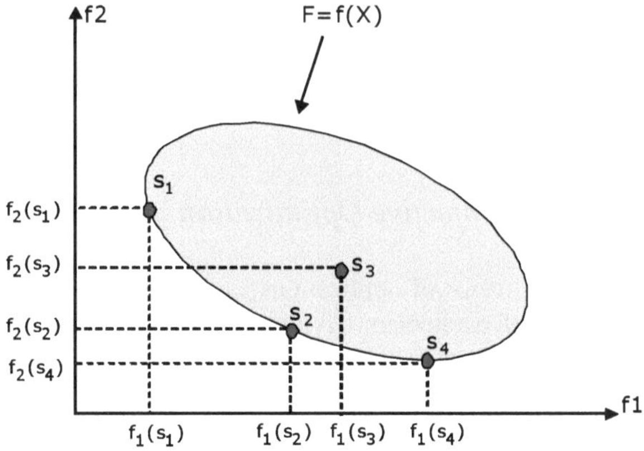

Fig. 10.3 Non-dominated and dominated solution examples: S_1, S_2 and S_4 are non-dominated solutions. S_3 is dominated by S_2

3.2 Cellular Genetic Algorithm

Several evolutionary algorithms are studied in the literature. In this chapter, we introduce the cellular genetic algorithm given by A.J. Nebro et al. [11]. The Cellular Genetic Algorithm (*cGA*) is the canonical basis of the Multi-Objective Cellular Genetic Algorithm (*MOCELL* [11]) presented in the next section. *cGA* is only used for the resolution of single objective optimization problems. Algorithm 1 presents the pseudo-code for the implementation of *cGA*. According to this algorithm, the population is represented by a regular grid of dimension *d* and for each individual belonging to this population a neighborhood function is defined.

Algorithm 1 (Pseudo code for a canonical cGA)

```
1: cGA (Parameter P)
2:    while not Termination_Condition ( ) do
3:       for individual ← 1 to P.popSize do
4:          list ← Get_Neighborhood (individual);
5:          parents ← Selection (list);
6:          offspring ← Recombination (P.pc, parents);
7:          offspring ← Mutation (P.pm, offspring);
8:          Fitness_Evaluation (offspring);
9:          Insert (position(individual), offspring, aux_pop);
10:      end for
11:      pop ← aux_pop;
12:   end while
13: end cGA
```

As depicted in Algorithm 1, the following steps are applied until a termination condition is met (*line 2*).

- For each individual in the grid (*line 3*):

1. Extract the neighborhood list (*line 4*).
2. Select two parents on the neighborhood list (*line 5*).
3. Apply the crossover operator to the parents with probability Pc (*line 6*).
4. Apply the mutation operator to the resulting offspring with probability Pm (*line 7*).
5. Computes the fitness value of the offspring individual (*line 8*).
6. Insert the offspring (or one of them) in new (auxiliary) population (*line 9*). The insertion position is equivalent to the position of the current individual.

- The newly generated auxiliary population becomes the new population for the next generation (*line 11*).

The most often used termination conditions are (1) to perform a maximum number of generations, (2) to have obtained a solution with a better cost than a provided acceptable value or (3) a combination of these.

3.3 Multi-objective Cellular Genetic Algorithm

Algorithm 2 presents the pseudo-code for the implementation of *MOCELL* [11]. This algorithm is very similar to the above single-objective algorithm; the main difference between them is the introduction of a Pareto front to deal with multi-objective optimizations. The Pareto front is an additional population (i.e. external archive) used to contain a number of non-dominated solutions.

Algorithm 2 (Pseudo code of MoCell)

```
 1: MOCELL (Parameter P)
 2:   Pareto_front = Create_Front()
 3:   while not TerminationCondition() do
 4:     for individual ← 1 to P. popSize do
 5:       list ← Get Neighborhood(individual);
 6:       parents ← Selection(list);
 7:       offspring ← Recombination(P.pc, parents);
 8:       offspring ← Mutation(P.pm, offspring);
 9:       Evaluate_Fitness (offspring);
10:       Insert (position(individual), offspring, aux_pop);
11:       Insert_Pareto_Front(individual);
12:     end for
13:     pop ← aux_pop;
14:     pop ← Feedback(ParetoFront);
15:   end while
16: end MOCELL;
```

As depicted in Algorithm 2, an empty Pareto front (*line 2*) is first created and then the following steps are applied until the termination condition is met (*line 3*):

- For each individual in the grid (*line 4*):
  1. Extract the neighborhood list (*line 5*).
  2. Select two parents from its neighborhood list (*line 6*).
  3. Recombine the two selected parents with the probability *pc* in order to obtain an offspring individual (or individuals) (*line 7*).
  4. Mutate the resulting offspring individual (or individuals) with probability *pm* (*line 8*).
  5. Compute the fitness value of the offspring individual (or individuals) (*line 9*).
  6. Insert the offspring individual (or individuals) in both the auxiliary population and the Pareto front if it is not dominated by the current individual (*lines 10, 11*).
- The old population is replaced by the auxiliary one (*line 13*).
- Replace randomly chosen individuals of the population by solutions from the archive (*line 14*).

The major difference between the *MOCELL* [11] and the classical evolutionary algorithms (i.e. NSGAII [6] and *SPEA2* [7]) is that the former intensively exploits the concept of neighborhood. Indeed, in *MOCELL* [11] an individual may only interact with its neighbors in the breeding loop.

4 Implementation

Our approach is implemented on top of the *jMetal* Framework [10] which we have adapted and extended to better suit the many-core platform mapping problem. *jMetal* [10] is an object-oriented Java-based framework that facilitates the development and the experimentation of multi-objective evolutionary algorithms. It includes many meta-heuristics which are not explored by the previous work on many-core platform mapping problems. To integrate our approaches in the *jMetal* [10] framework, we defined:

- The solution coding (chromosome).
- The different objective functions.
- The different architecture and application constraints.
- New mutation and crossover operators.

We also made some adaptations in order to implement parallel versions of multi-objective evolutionary algorithm.

4.1 Solution Coding

jMetal [10] provides many date types (Real, Binary, Integer, etc.) to code the solutions. For our mapping problem, we used an integer coding: each mapping solution

1	2	4	6	2	7	5	3	2	0
xo	x1	x2	x3	x4	x5	x6	x7	x8	x9

Fig. 10.4 Chromosome example where a 10 task graph is mapped on an 8 PE platform

is represented by a tuple $(x_1, x_2, \ldots, x_n)$, where x_i gives the PE on which will be mapped the task t_i. The value of each variable x_i is in the set $\{0, \ldots, P-1\}$ where P is the platform PE number. Since the user has the possibility to force some tasks to be assigned to certain PEs, we modified *jMetal*[10] to support this feature. Figure 10.4 gives a chromosome example where a 10 task graph is mapped on an 8 PEs platform (the PEs are numbered from 0 to 7). For example, $x_5 = 7$ indicates that the task t_5 is mapped on PE $= 7$.

4.2 Objective Functions

We defined a set of objective functions which measure the quality of a given mapping. As mentioned previously, the user has the possibility to define new ones if required. Concretely, we defined the following main objective functions: one objective function destined to better balance the load (increasing the parallelism), one objective function to minimize the communication, one objective function to reduce the energy consumption, and one objective function destined to balance the memory usage. We do not define an execution time objective function in the present chapter since this objective function depends on many parameters. Instead, Sect. 4.2.5 discusses these different parameters and how to exploit them in order to reduce the execution time.

4.2.1 Load Variance

This objective function gives the load variance between the different PEs for a given mapping x. By minimizing the load variance, the amount work will be divided uniformly among the different PEs. In streaming applications where each task is executed several times, this objective function allows to increase the parallelism. The load variance of a given mapping x is defined as follows:

$$\sqrt{\frac{\sum_{i=0}^{P-1}(load(PE_i) - avgload)^2}{P}}$$

where, $load(PE_i)$ represents the weighted sum of all tasks assigned to PE_i, $avgload$ is the average load and P is the number of PEs.

4.2.2 Memory Variance

This objective function gives the memory variance between the different PEs for a given mapping x. By minimizing the memory variance, the amount of required memory will be divided uniformly among the different PEs. The memory variance of a given mapping x, is defined as follow:

$$\sqrt{\frac{\sum_{i=0}^{P-1}(mem(PE_i) - avgmem)^2}{P}}$$

where $mem(PE_i)$ is the memory size needed by all the tasks assigned to PE_i and *avgmem* is the memory size needed by all tasks divided by P (the number of PEs).

4.2.3 Communication Cost

This objective function gives the total amount of communication between all the PEs:

$$\sum_{e_i \in E} Volume(e_i) * Distance[PE(Source(e_i)), PE(Sink(e_i))]$$

where, E is the set of edges in the application task graph, $Volume(e_i)$ is the amount of data exchanged by the tasks connected by the edge e_i. $Source(e_i)$ and $Sink(e_i)$ represent respectively the source and the sink tasks of the edge e_i. $PE(t_i)$ gives the PE on which the task t_i is mapped. $Distance(PE_1, PE_2)$ gives the distance (the hop count number) between PE_1 and PE_2. As one can see, this objective function is in conflict with the load variance since zero communication cost could be achieved by assigning all tasks to the same processor.

4.2.4 Energy Consumption

The energy model developed in our framework is based mainly on the data transfers through the different routers and links of the NoC. The input buffers in the routers of the NoC are implemented using registers, which eliminates the buffering energy consumption. For this type of routers, Ye et al. [23] and Hu et al. [24] proposed a good approximation of the energy consumed when one bit of data is transferred through the router:

$$En_{\text{bit}} = ES_{\text{bit}} + EL_{\text{bit}}$$

where ES_{bit} and EL_{bit} represent respectively the energy consumed on the switch and on the output link of the router. By using the preceding equation, the average energy consumption for sending one bit of data from PE_i to PE_j can be computed as follows:

$$En_{\text{bit}}^{i,j} = (nhops + 1) \cdot ES_{\text{bit}} + nhops \cdot EL_{\text{bit}}$$

where *nhops* is the hop count number from PE_i to PE_j (when $i = j$, $En_{\text{bit}}^{i,j} = 0$). Using the preceding equation, the energy consumption objective function can be modeled as follows:

$$\sum_{e_i \in E} Volumebit(e_i) \cdot En_{\text{bit}}^{PE(Source(e_i)),PE(Sink(e_i))}$$

where E is the set of edges in the application task graph, $Volumebit(e_i)$ is the amount (in bits) of data exchanged by the tasks connected by the edge e_i. $Source(e_i)$ and $Sink(e_i)$ represent respectively the source and the sink tasks of the edge e_i. $PE(t_i)$ gives the PE on which the task t_i is mapped.

Since the target architecture is homogeneous, the energy consumed by the different PEs is not taken into account in our model. The global energy consumed by all PEs in this type of architecture is almost a constant (it does not vary from one mapping to another). This means that it has no impact on the objective function minimization. Of course, other parameters (e.g. NoC contention, scheduling policy, etc.) have a small impact on the global energy consumption. In the current model, these parameters are not considered. However, we believe that the objective function given in this chapter is a good approximation.

4.2.5 Execution Time

The execution time objective function depends on many parameters (e.g. communication cost, load variance, NoC contention, etc.). In data flow streaming applications [20], where each task is executed several times on different data, the two parameters which have the biggest impact on the execution time objective function are the communication cost and the load variance between the different PEs. Unfortunately, these two parameters are in conflict (zero communication cost could be achieved by placing all the tasks on the same PE). In our tool, two approaches can be exploited in order to reduce the execution time:

- Aggregate the load variance and the communication objective functions into only one objective function: $w_1 \cdot load\_variance() + w_2 \cdot communication$, where w_i represents the weight associated with each objective function (these two parameters are given by the user). Unfortunately, this approach has some limitations: (1) the w_i weights are platform and application dependent, (2) the generated solutions for different values of w_i are not Pareto.
- Select the load variance and the communication cost as the objective functions to optimize (of course, the user has the possibility to add other objective functions) in order to generate Pareto set of solutions. Since each Pareto solution is better than another one in one criterion, it is not easy to select a solution which will give the best execution time. By consequence, the user needs to simulate all these solutions (or a sub-set of them) in order to select an appropriate one.

Several other objective functions are defined in our tool in order to help the user to optimize the execution time. However, these objective functions have less impact than the load variance and the communication objective functions:

- *max_loaded_pe*(). Gives the most loaded PE which has an impact on the load variance objective function.
- *max_in_communication*(). Gives the biggest input communication for all the PEs. This objective function has an impact on the system throughput.
- *max_min_load_diff*(). Gives the load difference between the most and least loaded PEs. This is another variant of the load variance objective function.
- *hop_count_number*(). Gives the total hop count number which has an impact on the communication objective function.

4.3 Architecture and Application Constraints

jMetal [10] offers a very efficient mechanism to associate a set of constraints with each solution. These constraints will be evaluated for each solution in several stages of the evolutionary algorithms. Each constraint can be expressed as follows:

$$b - A \cdot Y \geq 0$$

where A is a matrix of constant, b is a real constant and Y is a vector computed as a function of the mapping solution x. To distinguish between the solutions which violate some constraints, extra information is added:

- *violated_number*(). Gives the number of violated constraints for a given solution.
- *overall_violation*(). With each violated constraint C_i, we associate a value $V_i = |b - A \cdot Y|$. The sum of all the values V_i gives *overall_violation*().

These two pieces of information are used during the process selection (*line 6* of Algorithm 2). The solutions with a low number of violated constraints will be favored compared to the solutions with a higher number of violated constraints. The solutions with the same number of violated constraints will be differentiated using the overall violation information.

The constraints considered in this chapter are the channel number, the memory space and the task limit number. Of course, we can add other constraints for the load variance and the commutation cost, but we believe that this is not interesting for the following reason: as these two objective functions are in conflict, adding a constraint for one will prevent finding interesting solutions for the other.

4.3.1 Channel Number Constraint

Each PE in the target platform has access to only *16 DMA* channels which implement FIFOs in hardware. During the task assignment, we need to ensure that the number of incoming and outgoing edges for a given PE does not exceed 16. For each PE_i, we compute the outgoing and incoming edges ($edges(PE_i)$). The constraint associated with PE_i is defined as follows:

$$16 - edges(PE_i) \geq 0$$

4.3.2 Memory Space Constraint

Each PE in the target platform has a limited local memory space (128 KB in this experiment). Therefore, we added a new constraint in order to limit the total memory space used by the tasks assigned to a given PE_i. The constraint associated with PE_i is defined as follow:

$$128 - memory(PE_i) \geq 0$$

where $memory(PE_i)$ is the total memory used by all tasks assigned to PE_i.

4.3.3 Task Number Limit

In some cases, it is necessary to limit the number of tasks assigned to a given PE (for example, in our platform, each PE has 4 hardware contexts). For this, we developed two strategies. In the first strategy (*Genetic_1*), we used the same technique as the preceding constraints, i.e. for each PE_i we compute the number of tasks assigned to it ($tasks(PE_i)$) and we added a new constraint:

$$\max - tasks(PE_i) \geq 0$$

where max is the maximum number of tasks that can be mapped on each PE. Unfortunately, this first strategy is very limited when the constant max takes small values:

- The number of solutions generated in the initial population which violate this constraint is much higher compared to the number of valid solutions.
- Sometimes, the crossover and the mutation operators generate no valid offspring even from valid parents.

To avoid these problems, we introduced a second strategy (*Genetic_2*) in which:

- We modified how the initial population is generated (during this step, we ensure that all the generated solutions satisfy the task limit constraint).
- We modified the crossover and the mutation operators in order to always generate valid solutions.

For example, let's consider a case where we have 10 tasks, 8 PEs (the PEs are numbered from 0 to 7) and the maximum tasks per PE is 3. If, in this example (Fig. 10.5), we want to remap the task t_5 ($x_5 = 7$) on the PE 2 (mutation), this will give an invalid solution, because 4 tasks will be assigned to the PE 2 (t_1, t_4, t_5, t_8). To avoid this problem, we also need to remap one of the tasks (t_1, t_4, t_8) on the PE 7.

We can apply the same procedure to the other constraints, but this is not necessary, because the number of solutions that violate these constraints is not significant compared to the whole search space.

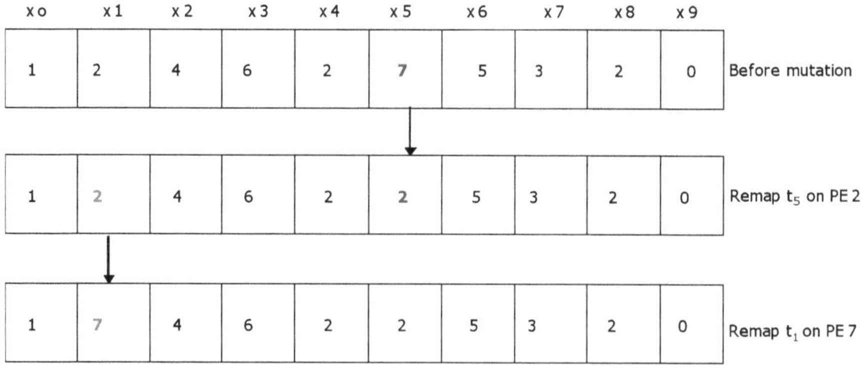

Fig. 10.5 Mutation example repair

4.4 Parallel Multi-objective Evolutionary Algorithm

This section is dedicated to the presentation of a parallel implementation of the multi-objective evolutionary algorithm in order to enlarge the exploration space of possible solutions. Contrary to other works, where only one meta-heuristic is exploited to solve a given problem, our parallel implementation allows exploiting several meta-heuristics at the same time to solve the many-core platform mapping problem. There are three main parallelization models in the literature [17]:

- Global: this model uses parallelism to speed up the sequential genetic algorithm. It uses a global shared population and the fitness evaluation is done on different processors.
- Diffusion: in this model, the population is separated into a large number of very small sub-populations, which are maintained by different processors.
- Island: in this model, the population is divided into a few large independent sub-populations. Each island evolves its own population using an independent serial multi-objective evolutionary algorithm.

In our case, we decided to implement the island model, because (1) several meta-heuristics can be exploited at the same time using this model, and (2) this model seems to be adapted for the problem under study, where the search space is very large and requires a good diversity. According to this model, every processor runs an independent evolutionary algorithm by regularly exchanging migrants (good individuals). This way, a better solution is expected to be found, since more solution space is explored. As depicted in Fig. 10.6, the island model that we implemented is based on a ring topology. In this model, the whole population is divided into multiple sub-populations (i.e. islands). Each processor runs an independent serial multi-objective evolutionary algorithm. Periodically, some of the individuals that are evaluated to be good candidates (non-dominated solutions) are sent to the neighbor island. This operation is called *migration*. The migration of individuals from one island to another is controlled by: (1) The connectivity between the islands (ring

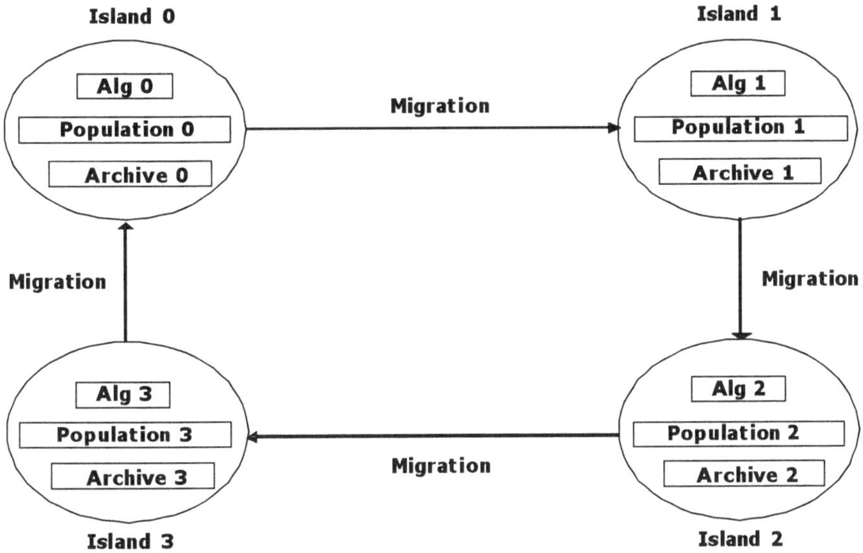

Fig. 10.6 Island parallel model

topology in our case) and (2) The number of individuals to be exchanged (migrated) between the neighbor islands The different islands collaborate by executing on each of them the following steps:

Algorithm 3 (The steps executed by each island)

- *Create an initial population on a random basis.*
- *Evolve its population and archive for a given number of generations.*
- *After each period, send some solutions selected as good candidates from the produced Pareto archive to the neighboring island.*
- *Receive migrating solutions and replace the worst solutions by those immigrants.*

At the end of the above steps, a given processor combines all of the Pareto archives in order to create the global Pareto archive. In our first implementation, we only evaluated a model where all islands optimize the same objective functions. However, we believe that increasing the diversity of objective functions may contribute to the quality of the results. For this reason, we plan as future work to implement other parallel schemes (for example, the different islands will optimize different objective functions).

5 Evaluation

This section presents a comparison between several new meta-heuristics offered by our tool and some existing algorithms destined to solve the many-core platform

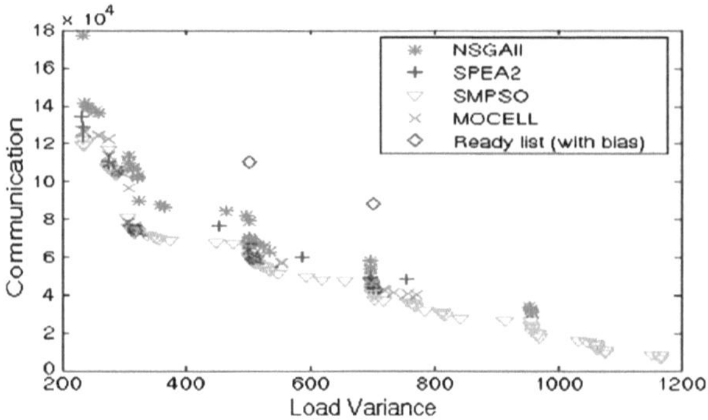

Fig. 10.7 Mapping TNR on an 8 PE platform. Two objective functions are optimized: load variance and communication cost

mapping problem. One of them is based on a graph traversal approach, previously proposed by Hu et al. [25], that we slightly modified to fit in our experimentation environment and called "ready-list". We also considered classical evolutionary algorithms such as *NSGAII* [6] and *SPEA2* [7]. For our experiments, we have used three applications. The first one is a Temporal Noise Reduction (*TNR*) application containing 26 tasks that are interconnected with 40 communication edges. The second application is a *3G WCDMA/FDD* base-station application containing 19 tasks and 18 communication edges. Finally, we have also performed comparisons on random task graphs generated by *TGFF* [26] which was designed to provide a flexible and standard way of generating pseudo-random task-graphs for use in scheduling and allocation research. Several authors used this tool to evaluate their approaches. The target many-core platform that we considered for our experiments was previously presented in Sect. 2. Relevant details about the platform setup for each experiment are given in the following sections.

5.1 TNR

Figure 10.7 presents the comparison of different meta-heuristics for mapping the TNR application on an 8 PE platform. In this experiment, two objective functions are optimized: the load variance and the communication cost. These two objective functions are optimized for a platform containing 8 PEs, 16 hardware communication channels and 128 Kbytes memory associated to each processor. As depicted in Fig. 10.7, each algorithm gives a set of Pareto solutions (each Pareto curve set is noted with the same shape). The solutions given by the ready list algorithm are dominated by all solutions given by the evolutionary algorithms. The solutions given by *SMPSO* [12] and *MOCELL* [11] algorithm dominate all the solutions given by *NSGAII* [6] and *SPEA2* [7]. As mentioned in Sect. 4.2.5, the execution time is mainly depending on the load variance and the communication cost. This means that the solutions given by *SMPSO* [12] (or *MOCELL* [11]) have better chance to give better processing time. Another characteristic of this experiment is that, contrary to other meta-heuristics, the *SMPSO* [12] algorithm gives a uniformly distributed set of solutions.

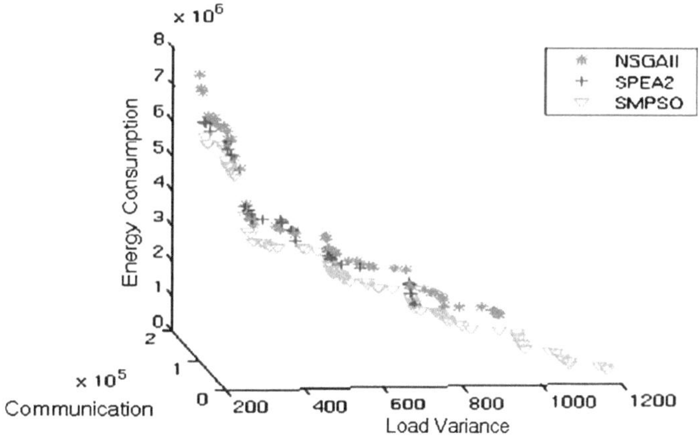

Fig. 10.8 Mapping TNR on an 8 PE platform. Three objective functions are optimized: energy consumption, load variance and communication cost

Figure 10.8 presents the Pareto solutions given by 3 meta-heuristics (*NSGAII* [6], *SPEA2* [7] and *SMPSO* [12]). In this experiment, 3 objective functions are optimized: energy consumption, load variance and communication cost. The same platform as before is used to optimize these objective functions. As depicted in the figure, the evolutionary algorithms propose a set of interesting Pareto solutions. However, the solutions given by *SMPSO* [12] dominate almost all the solutions given by the other algorithms (for figure clarity, we consider only the results given by 3 meta-heuristics; several other new meta-heuristics outperform the results given by *NSGAII* [6] and *SPEA2* [7]). For the channel constraint problem, the ready list heuristic fails to propose solutions. This experiment shows that the communication cost is strongly correlated to the energy consumption and the load variance is in conflict with these two objective functions. To confirm these two assertions, we performed two other experiments. In each of these experiments, we optimized two objective functions:

- Load variance with energy consumption. This experiment is given in Fig. 10.9. As shown in this figure, the load variance is in conflict with the energy consumption.
- Energy consumption with the communication cost. In this experiment, all the algorithms propose one mapping solution. The best solution is given by *SMPSO* [12] and *GDE* [13] which corresponds to the optimal solution. *MOCELL* [11] gives better solutions than *NSGAII* [6] and *SPEA2* [7].

5.2 3G WCDMA/FDD

This section presents the mapping results obtained for the *3G WCDMA/FDD* networking application. This application is composed of 19 tasks connected using 18

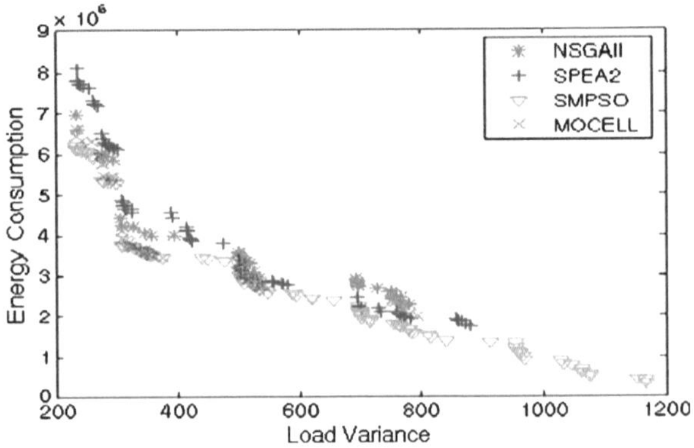

Fig. 10.9 Mapping TNR on an 8 PE platform. Two objective functions are optimized: energy consumption and load variance

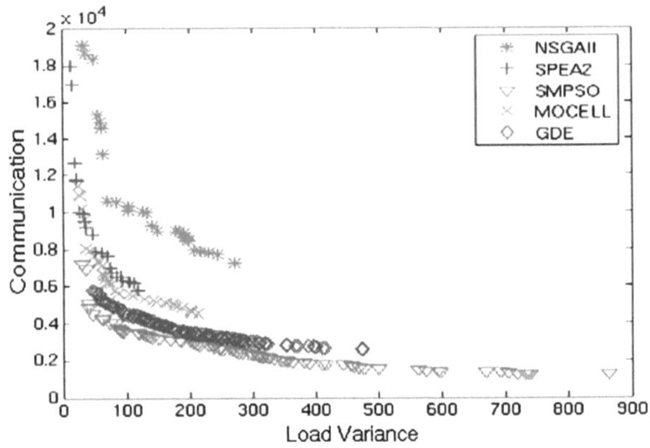

Fig. 10.10 Mapping 3G_V2 on a16 PE platform. Two objective functions are optimized: load variance and communication cost

communication channels. To expose more potential parallel processing, we created a second version of a functionally equivalent application graph of the reference application in which we duplicate each task 3 times. The original reference code and the new one will be called respectively *3G_v1* and *3G_v2*. For all of the following experiments, the results given by the *ready list algorithm* will be not shown because they are dominated by all other solutions.

Figure 10.10 presents the mapping of *3G_V2* on a platform containing 16 PEs. In this platform architecture, there are 16 hardware communication channels and 128 Kbytes memory associated to each processor. This experiment aims at optimizing two objective functions which are the load variance and the communication cost. As depicted in the figure, the *SMPSO* [12] algorithm gives the best results, which means that the solutions given by this algorithm have a better chance to give shorter processing time. As we can also see from this figure, the solutions given by *GDE* [13] and *MOCELL* [11] outperform the solutions given by the classical algorithm such as *NSGAII* [6] and *SPEA2* [7].

Figure 10.11 presents the mapping of *3G_V2* on a 16 PE platform. In this experiment, three objective functions are optimized: energy consumption, load vari-

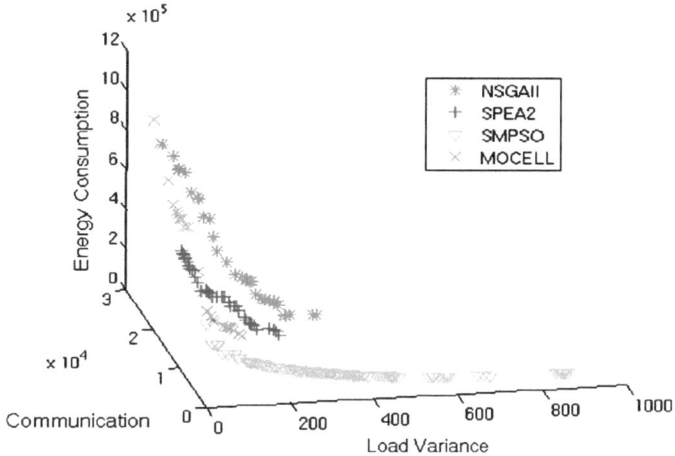

Fig. 10.11 Mapping 3G_V2 on a16 PE platform. Three objective functions are optimized: energy consumption, load variance and communication cost

ance and communication cost. As depicted in the figure, the evolutionary algorithms propose a set of interesting Pareto solutions. However, the solutions given by *SMPSO* [12] dominate all the solutions given by the other algorithms. This experiment also shows that the communication cost is strongly correlated to the energy consumption, and the load variance is in conflict with these two objective functions. We also confirmed these two assertions by optimizing only two objective functions:

- Load variance with energy consumption. This experiment follows the same behavior as the results given in Fig. 10.10. This means that the load variance is in conflict with the energy consumption.
- Communication cost with energy consumption. In this experiment, each evolutionary algorithm gives one mapping solution. This confirms that the load variance is correlated to the communication cost. For all these experiments, the best result is given by *SMPSO* [12] and *GDE* [13].

Figure 10.12 presents the comparison between the two strategies developed for the task limit constraint (Sect. 4.3.3). In this experiment, we supposed that only one task can be mapped to a given PE (task limit constraint = 1). *Genetic1* uses the constraint defined in *jMetal*[10] while *Genetic2* uses the new crossover and mutation operators defined in the second strategy. In this experiment, we considered a 19 PE platform where each task of *3G_V1* is mapped on a processing element. Only one criterion is optimized in this experiment (communication cost). This figure gives the successive optimizations of the communication cost where the iteration number represents the running number of the genetic algorithm. As depicted in Fig. 10.12, *Genetic2* gives a better solution than *Genetic1*.

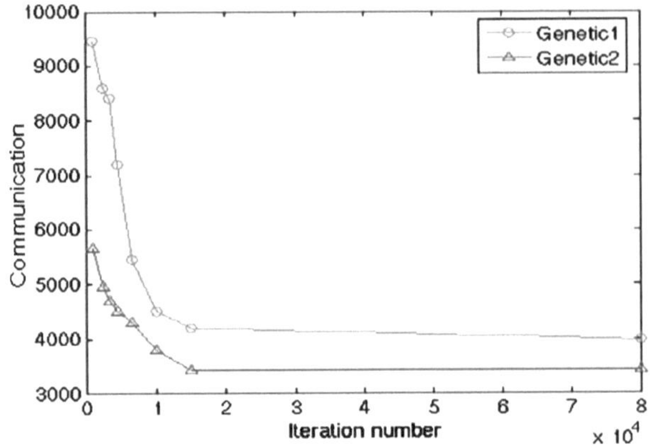

Fig. 10.12 Comparison between the two task mapping strategies developed in Sect. 4.3.3. The experiments are made on a19 PE platform for 3G_v1

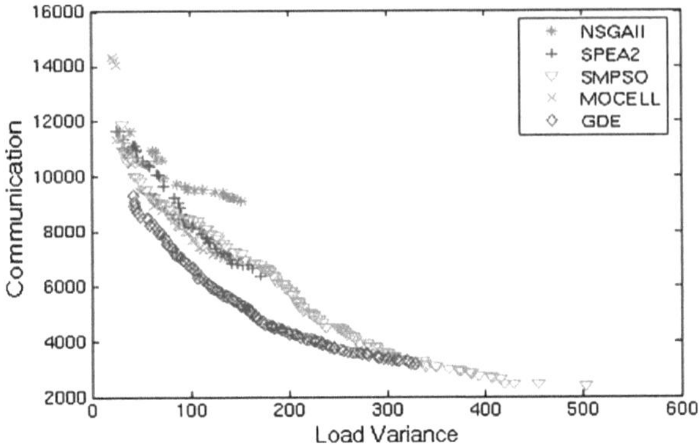

Fig. 10.13 Mapping of randomly generated graph (100 tasks) on a12 PE platform. Two objective functions are optimized

5.3 Experiments on Randomly Generated Graphs

As mentioned previously, *TGFF* [26] is a flexible way of generating pseudo-random task graphs for use in scheduling and allocation research. To evaluate the robustness of our tool, various parameters are used in *TGFF* [26] to generate benchmarks with different topologies and task/communication distributions. Due to a lack of space, only experiments for 3 task graphs will be presented. However, most experiments follow the same trends as the ones presented.

Figure 10.13 gives the comparison between several meta-heuristics where only two objective functions are optimized (load variance and communication cost). In this experiment, a randomly generated graph of 100 tasks is mapped on a 12 PE platform. Contrary to the previous experiments, *SMPSO* [12] outperforms *SPEA2* [7] only in some solutions. On the other hand, *GDE* [13] outperforms *SPEA2* [7] in all solutions. This confirms that a mapping tool must provide several meta-heuristics in order to explore different solution spaces (the best meta-heuristic changes from one task graph to another one).

Figure 10.14 gives the comparison between several meta-heuristics where three objective functions are optimized (load variance, communication cost and energy

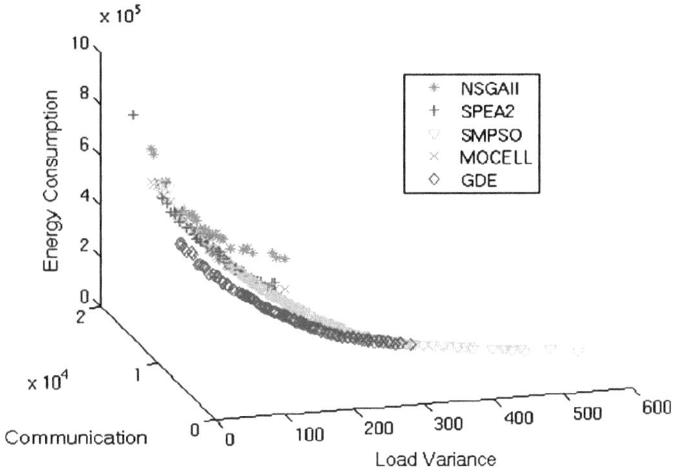

Fig. 10.14 Mapping of randomly generated graph (100 tasks) on a 12 PE platform. Three objective functions are optimized

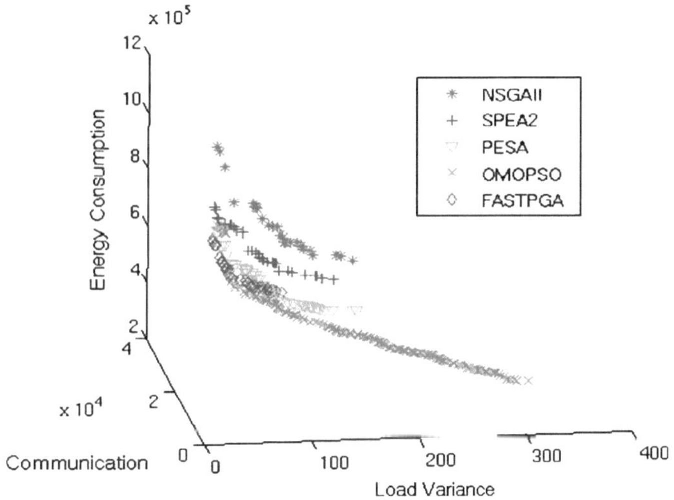

Fig. 10.15 Exploration of other new meta-heuristics offered by our tool. Mapping a randomly generated graph (110 tasks) on a 16 PE platform

consumption). The same task graph used in the precedent experiment is mapped on the same platform. This experiment follows thc same trend as the precedent experiment.

Figure 10.15 explores other new meta-heuristics offered by our tool (*PESA* [14], *OMOPSO* [16], *FASTPGA* [15]). These new meta-heuristics are compared to the classical meta-heuristics such as *NSGAII* [6] and *SPEA2* [7]. In this experiment, 3 objective functions are optimized (load variance, communication cost and energy

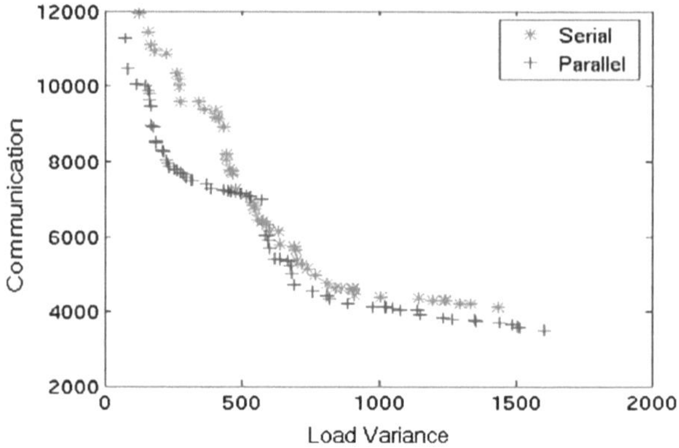

Fig. 10.16 Comparison of serial and parallel multi-objective evolutionary algorithms, mapping a randomly generated graph (120 tasks) on a 12 PE platform

consumption). As shown in this figure, these new meta-heuristics outperform *NSGAII* [6] and *SPEA2* [7]. During our experiments, we have tested more than 100 randomly generated task graphs. For all these tests, the best meta-heuristic changes form one task graph to another one.

Figure 10.16 gives the comparison between the serial *MOCELL* [11] algorithm and the parallel evolutionary algorithm described in Sect. 4.4 that are executed for optimizing a randomly generated application graph with 120 tasks on a 12 PE platform. This figure shows almost all the solutions given by the serial algorithm are dominated by the solutions given by the parallel algorithm. Indeed, we explain this result by the fact that (1) the parallel algorithm explores a wider space than the serial algorithm and (2) the migration operation allows maintaining a steady convergence between the diversity of solutions explored by each computation island.

5.4 Execution Time

As mentioned previously, the execution time depends mainly on load variance and communication cost. To confirm this, we have simulated several categories of solutions for the 3 preceding applications. For the randomly generated task graphs, different topologies and task/communication distributions are tested. The simulated solutions are divided mainly on 3 categories:

- Manually constructed solutions.
- Non Pareto solutions (generated by our tool).
- Pareto solutions. These solutions are generated by our tool by optimizing simultaneously the load variance ant the communication cost.

Our different experiments show that:

1. For some applications (like *TNR* where the number of task input and output is large) it is difficult to construct manually solutions that respect the different architectural constraints (channel constraint for the *TNR* application).

2. The Pareto solutions generated by our tool give better execution time than the non Pareto and the hand constructed solutions.
3. For the *TNR* and *3G WCDMA/FDD* applications, the solutions that give the best execution time are located in the middle of the Pareto front (the solutions having a good compromise between the load variance and the communication cost).
4. For the randomly generated graph, the solutions that give the best execution time are in general located on the Pareto front. However, their positions in the Pareto front change according to the values of task weights and edge communication costs.

6 Related Work

The problem of mapping application on a mesh based NoC has been addressed in several works [24, 25, 27]. Hu and Marculescu [24] presented a branch and bound algorithm for mapping IP cores in a mesh-based NoC architecture. Their technique aims at minimizing the total amount of power consumed by communications with the constraint of performance handled via bandwidth reservation. The same authors introduced an energy-aware scheduling (EAS) algorithm [25], which statically schedules application-specific communication transactions and computation tasks onto heterogeneous network-on-chip (NoC) architectures. The proposed algorithm automatically assigns the application tasks onto different processing elements and then schedules their execution under real-time constraints. Marcon et al. [27] extended the work of Hu and Marculescu by taking into consideration the dynamic behavior of the target application. Bourduas et al. [28] used simulated annealing to perform task assignments. Their algorithm assigns task graph to nodes such as to minimize path length between communicating nodes; however the authors studied a very restrictive case where only one task is assigned to a given node. Lie and Kumar [5] presented an approach that uses genetic algorithms to map an application described as a parameterized task graph, on a mesh based NoC architecture; their algorithm optimize only one criteria (execution time). In [29, 30] mapping methodologies are proposed supporting multi-use case NoCs. In these works, an iterative execution of mapping algorithm increases the network size until reaching an effective configuration.

For Pareto based multi-objective optimization we can cite [2–4, 31, 32]. All these work define only few objective functions, no architecture (or application) constraints are considered and no easy way to extend. Another important limitation of these different works is that only a few meta-heuristics are explored. Ascia et al. [2] use *SPEA2* [7] meta-heuristic to solve the problem of topological mapping of IPs on the tiles of a mesh based network on chip architecture. Their goal is to maximize performance and to minimize the power consumption. Erbas et al. [3] compare two meta-heuristics: *NSGAII* [6] and *SPEA2* [7]. The goal of their work is to optimize processing time, power consumption, and architecture cost. Kumar et al. [4] use *NSGAII* [6] to obtain an optimal approximation of the Pareto-optimal front. Their approach tries to optimize energy consumption and bandwidth requirement. Thiele et al. [31] explored *SPEA2* [7] to solve the mapping problem. They considered only a two dimensional optimization space (computation time and communication time). Zhou et al. [32] treat the NoC mapping problem as a two conflicting objective optimization problem (minimize the average hop and achieving the thermal balance). Their approach is based on the *NSGAII* [6] meta-heuristic.

7 Conclusion

In this chapter, we have studied one of the main challenging problems determining the efficiency of parallel software applications on many-core platforms. We have presented a framework that allows exploring several new meta-heuristics. We have also described many objective and constraint functions for modeling the characteristics of parallel software applications and many-core platforms. We argue that, while our experiments were based on a number of criteria making sense for our application cases, others can extend this framework for their own purposes. Our evaluations based on real life applications have shown that several new meta-heuristics outperform the classical evolutionary algorithms such as *NSGAII* [6] and *SPEA2* [7]. We have also observed that the parallel approach developed in our framework gives better results than the serial meta-heuristics. As future work, we plan to investigate other parameters (e.g. scheduler, NoC contention, etc.) which impact the execution time and the energy consumption objective functions.

References

1. Garey, M.R., Johnson, D.S.: Computers and Intractability: A Guide to the Theory of NP-Completeness. Freeman, New York (1979)
2. Ascia, G., Catania, V., Palesi, M.: Mapping cores on network-on-chip. Int. J. Comput. Intell. Res. **1**(1–2), 109–126 (2005)
3. Erbas, C., Cerav-erbas, S., Pimentel, A.D.: Multi-objective optimization and evolutionary algorithms for the application mapping problem in multiprocessor system-on-chip design. IEEE Trans. Evol. Comput. **10**(3), 358–374 (2006)
4. Jena, R.K., Sharma, G.K.: A multi-objective evolutionary algorithm-based optimisation model for network on chip synthesis. Int. J. Innov. Comput. Appl. **1**(2), 121–127 (2007)
5. Lei, T., Kumar, S.: A two-step genetic algorithm for mapping task graphs to a network on chip architecture. In: DSD (2003)
6. Deb, K., Pratap, A., Agarwal, S., Meyarivan, T.: A fast and elitist multi-objective genetic algorithm: NSGA-II. IEEE Trans. Evol. Comput. **6**(2), 182–197 (2002)
7. Zitzler, E., Laumanns, M., Thiele, L.: SPEA2: improving the performance of the strength Pareto evolutionary algorithm. Technical Report 103, Computer Engineering and Communication Networks Lab (TLK), Swiss Federal Institute of Technology (2001)
8. Coello Coello, C.A., Veldhuizen, D.A.V., Lamont, G.B.: Evolutionary Algorithms for Solving Multi-objective Problems. Kluwer Academic, Dordrecht (2002)
9. Das, I.: Nonlinear multi-criteria optimization and robust optimality. Ph.D. Thesis, Dept. of Computational and Applied Mathematics, Rice University, Houston, TX (1997)
10. http://mallba10.lcc.uma.es/wiki/index.php/Jmetal

11. Nebro, A.J., Durillo, J.J., Luna, F., Dorronsoro, B., Alba, E.: MOCell: A cellular genetic algorithm for multi-objective optimization. Int. J. Intell. Syst. **24**(7), 726–746 (2009)
12. Nebro, A.J., Durillo, J.J., García-Nieto, J., Coello Coello, C.A., Luna, F., Alba, E.: SMPSO: a new PSO-based meta-heuristic for multi-objective optimization. In: 2009 IEEE Symposium on Computational Intelligence in Multi-criteria Decision-Making (2009)
13. Kukkonen, S., Lampinen, J.: GDE3: the third evolution step of generalized differential evolution. In: IEEE Congress on Evolutionary Computation (CEC2005) (2005)
14. Corne, D.W., Jerram, N.R., Knowles, J.D., Oates, M.J.: PESA-II: region-based selection in evolutionary multi-objective optimization. In: GECCO-2001 (2001)
15. Eskandari, H., Geiger, C.D., Lamont, G.B.: FastPGA: a dynamic population sizing approach for solving expensive multi-objective optimization problems. In: 4th International Conference on Evolutionary Multi-Criterion Optimization (2007)
16. Sierra, M.R., Coello Coello, C.A.: Improving PSO-based multi-objective optimization using crowding, mutation and epsilon-dominance. In: EMO (2005)
17. Branke, J., Schmeck, H., Deb, K., Reddy, S.M.: Parallelizing multi-objective evolutionary algorithms: cone separation. In: Proceedings of the 2004 Congress on Evolutionary Computation (2004)
18. Vrenios, A.: Parallel programming in C with MPI and OpenMP (Book Review). IEEE Distrib. Syst. Online **5**(1), 7.1–7.3 (2004)
19. MPI, A Message-Passing Interface Standard. Message Passing Interface Forum, version 2.1 (2008)
20. Thies, W., Karczmarek, M., Amarasinghe, S.: StreamIt: a language for streaming applications. In: 11th International Conference on Compiler Construction (2002)
21. Munshi, A.: The OpenCL Specification version 1.0. Khronos OpenCL Working Group (2009)
22. Coppola, M., Locatelli, R., Maruccia, G., Pieralisi, L., Scandurraa, A.: Spidergon: a novel on-chip communication network. In: Proceedings of the International Symposium on System-on-Chip (2004)
23. Ye, T.T., Benini, L., De Micheli, G.: Packetized on-chip interconnect communication analysis for MPSoC. In: DATE (2003)
24. Hu, J., Marculescu, R.: Energy-aware mapping for tile-based NoC architectures under performance constraints. In: ASP-DAC (2003)
25. Hu, J., Marculescu, R.: Energy-aware communication and task scheduling for network-on-chip architectures under real-time constraints. In: DATE (2004)
26. Dick, R.P., Rhodes, D.L., Wolf, W.: TGFF: task graphs for free. In: Workshop on Hardware/Software Codesign (1998)
27. Marcon, C., Calazans, N., Moraes, F., Susin, A., Reis, I., Hessel, F.: Exploring NoC mapping strategies: an energy and timing aware technique. In: DATE (2005)
28. Bourduas, S., Chan, H., Zilic, Z.: Blocking-aware task assignment for wormhole routed network-on-chip. In: MWSCAS/NEWCAS (2007)
29. Murali, S., Coenen, M., Radulescu, A., Goossens, K., De Micheli, G.: A methodology for mapping multiple use-cases onto networks on chips. In: DATE (2006)
30. Murali, S., Coenen, M., Radulescu, A., Goossens, K., De Micheli, G.: Mapping and configuration methods for multi-use-case networks on chips. In: ASP-DAC (2006)
31. Thiele, L., Bacivarov, I., Haid, W., Huang, K.: Mapping applications to tiled multiprocessor embedded systems. In: Application of Concurrency to System Design (2007)
32. Zhou, W., Zhang, Y., Mao, Z.: Pareto based multi-objective mapping IP cores onto NoC architectures. In: Circuits and Systems, APCCAS (2006)

Chapter 11
Functional Virtual Prototyping for Heterogeneous Systems

Design Flow Evolutions and Induced Methodological Requirements

Yannick Hervé and Arnaud Legendre

1 Introduction

Integration and miniaturization of electronic systems have yielded innovative technical objects that enable many physical disciplines to coexist, in order to achieve increasingly complex functions. Such systems are all denominated under the same neologism of mechatronics (or MOEMS or SoC). While the initial challenge was to associate electronics and mechanics (sensor, control, regulation, etc.) to increase system quality or diminish the final service cost, designers have more recently started to take advantage of the wealth offered by a smart association of other technical disciplines (magnetics, optics, fluidics, etc.).

The coexistence of these sub-systems in the parent system (car, plane, boat, biomedical device, etc.) poses severe problems to system integrators. It is hardly possible to foresee reliability, and to guarantee non-interference between the various subsystems among themselves, or with their environment. The delivery of complex subsystems from independent and external providers raises new challenges to the industrial community. Tools currently used to design and optimize these categories of systems have not evolved fast enough to answer the new needs they induce. This is true for the definition, design and validation steps, but also for their integration in the industrial design flow used by multiple co-workers.

This chapter will cover the 'appropriate' description languages that can be used for implementation. More precisely, we will focus on Functional Virtual Prototyping implemented with the standardized language VHDL-AMS; the process will be

Y. Hervé (✉)
Université de Strasbourg, Strasbourg, France
e-mail: yrv@wanadoo.fr

Y. Hervé · A. Legendre
Simfonia SARL, Strasbourg, France

G. Nicolescu et al. (eds.), *Design Technology for Heterogeneous Embedded Systems*,
DOI 10.1007/978-94-007-1125-9_11, © Springer Science+Business Media B.V. 2012

depicted throughout this chapter with industrial examples extracted from a variety of technical disciplines.

2 Evolution in Design Issues

2.1 Current Industrially Applied Methods

The increasing complexity of modern systems, where many physical domains can interact, raises problems of reliability and R&D cost control. Recent scientific and technological progress, coupled with increasing performance criteria imposed by worldwide competitivity render the industrial design task increasingly complex. Modern systems now make multiple sciences and cultures interact, within space and time scales that outstrip the representational capability of humans. One can witness, as an answer to such globally increasing complexity and its speed of increase, the overspecialization of professions, and as a consequence, a progressive loss of control of the industrial company on its product. In spite of the concatenation of specialists in the firm, one cannot make professions communicate efficiently if they stem from widely separated training. Each culture has its own methodology, languages and tools. In the middle of a complex production chain, between providers with variable quality criteria and customers with perpetually changing expectations, the industrial challenge is to minimize risks and costs, while keeping guarantees on its quality standards, meeting deadlines and market prices, and keeping on innovating intelligently. In this context the usual design flow, called the V-cycle [1], no longer meets industrial needs. Indeed, this cycle is composed of two distinct phases as shown in Fig. 11.1. The first step is a top-down analysis that leads to a progressive and hierarchical definition of the system functions and components down to the lowest level (physical layer). The second step consists in a systematic and recursive prototype test series and corrections until the whole system is validated. An important point to realize is that the later a problem is detected, the more expensive it is to correct it. With complex systems, where multi-physics effects merge together, their early identification becomes almost impossible, and first pass system design success is hardly reachable.

2.2 Industrial Context: What Is Not True Anymore

As already introduced, recent technological progress has rendered traditional design workflows outdated. To be more specific, one can list the following ways of action that used to be applied, and are no longer suitable:

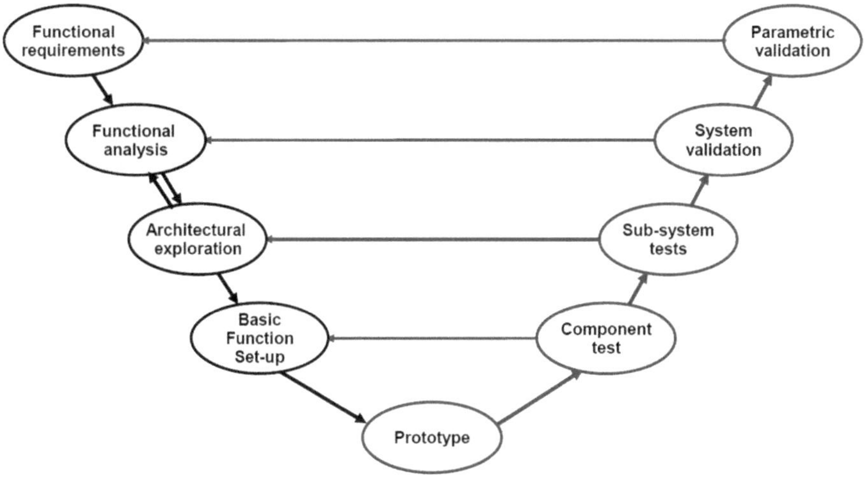

Fig. 11.1 V-cycle: the classical systems' design workflow

- **Mono-disciplinary systems**

The natural evolution of technologies, coupled with the trend towards sophisticated objects, drives companies to propose complex systems. In this flow, mono-disciplinary systems seem to disappear.

- **Centralized firms**

The set of tasks to design a new heterogeneous system is difficult to master inside a single company. Other skills and Industrial Property from other companies are mandatory. Tools and methods have to be shared, though the Industrial Property has to be protected.

- **Unitary tests**

The right branch of the V-cycle is dedicated to the test steps. The rational approach to test a system is to validate the behavior of each of its blocks regarding its specification, and to re-assemble those blocks following the hierarchy. This is one definition of the notion of complexity, and a characteristic of heterogeneous systems is that these sub-blocks interfere with each other, triggering unexpected global behaviors (though each block is validated) and reducing the interest of unitary test plans.

- **Physical prototype validation**

The use is to validate the design with a series of physical tests, on a real prototype. With numerous and sophisticated functions integrated in modern systems, it is impossible for prototype tests to be sufficiently exhaustive so as to entirely guarantee the quality of the system (as revealed by the quality problems that dog the car industry, resulting in massive recall campaigns).

- **Static market and expectations**

Traditional design begins with market identification and the construction of a specification that is coherent with it. The trends of the market are also evolving rapidly. Expectations are increasingly specific and the user requires tailored functions and adaptive systems. The consequence is that the systems must be designed accordingly, i.e. configurable, adaptable and optimized for each usage.

2.3 Systems Design Challenges

These new aspects require specific management so that industrial failures can be avoided, and workflows optimized. Architectural explorations and design decisions have to be achieved early in the design process: new methods and tools have to be properly associated with this flow.

The design flow currently used by the engineers at systems level is the one used for the realization of well-managed mono-disciplinary systems. The input modality is usually based on (inevitably proprietary) graphic tools, or on proprietary languages (such as Matlab® or MAST). Recently, one has seen the birth of definition languages, both freely-defined (Modelica) and standardized (VHDL-AMS, VERILOG-AMS). The designer will define the hardware architecture for each involved discipline and will (if possible) simulate it. Validation tests are established in an empirical way, before the 'validation' on one or several physical prototypes. Most of the complementary studies are achieved with domain-specific tools (mono-disciplinary tools specialized in mechanics, electronics, thermal diffusion, fluidics, etc.) and the final validation is achieved through real prototypes. With these methods, the interactions between disciplines are not studied, and validation steps are extremely expensive. The projects become very long, the industrial risk is not managed and the failures are frequent.

2.3.1 'Time to Market' Stakes

One of the main evolutions throughout the past century has been the contraction of development delays. Market expectations change rapidly (market-pull) and the technological choices are often called into question (technology-push), implying the flexibility of integrating technological breakdowns, while keeping guarantees on the quality requirements, and the expected functions in the specified environmental constraints.

2.3.2 First Pass Validation

The industrial trend concerning the design of systems is to tend to 'first-pass validation', and 'just-by design'. This expectation is pushed to the limit when there is a single testing phase of the reputed-valid physical prototype, before the sale starts. First-pass validation success is verified if all the blocks in the design meet their block-level goals, if the overall system specifications are met and if the yield of the system is at an acceptable level [2].

2.3.3 Technological Breakdown Integration and Management

The rapid evolution of technology, the sophistication of systems and expectations are available at each step of the design chain, for every subsystem provider. Hence (and this is especially true for integrators) it is impossible to understand all the details of the novel technologies included in the numerous subsystems. Nevertheless, these breakthroughs have to be handled in order to fit the final expectations, and to propose competitive products. This aspect of complexity management is also a challenge in itself, which imposes the use of systematic and formal conception methods.

2.3.4 Quality Guarantee: Design Groups' Legal Responsibility

In this oppressive environment, where competition is harsh and delays short, quality is critical, and must be guaranteed by systematic methods. Moreover, the responsibility placed on design teams is still significant, and danger to the end-user is no longer tolerated, especially when the product is massively and internationally distributed. One can cite here Sony® battery explosions, or Renault® cruise controller blockings, among other more or less serious modern system failures. As a result of these problems, responsibility checks and legal proceedings are extremely expensive. In order to avoid negative project publicity and to preserve their corporate image, large companies are likely to slow down innovative developments.

3 Systems Design Workflow Evolutions

3.1 Evolutions of the Design Flow

Figure 11.2 shows how the modeling steps fit into the classical V-design cycle. The modeling steps provide, at each level of the top-down design, a way to verify the global specification. The models are tuned at each step and validated with the previous level. Consequently, when unitary blocks and their models are in accordance with their specifications, the complete design should be correct by construction. In order to reduce cost and time-to-market significantly, it is worth developing modeling methods and tools that automatically provide code and verification.

The Functional Virtual Prototyping (FVP) workflow, depicted in Fig. 11.2, is a set of well-organized tasks used to design or enhance a system through an assembly of "proven" models [1]. By analyzing the functions of a given system, and characterizing them with performance specifications, one can build a set of models of

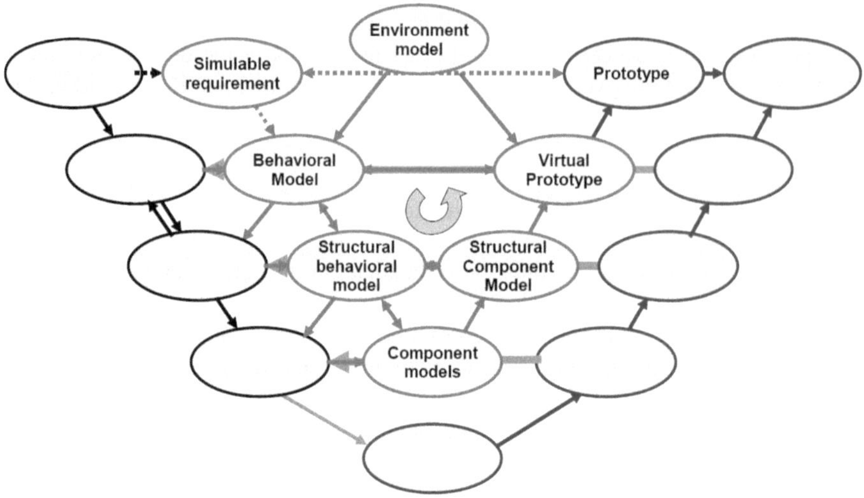

Fig. 11.2 The functional virtual prototyping cycle

increasing complexity that permits a deeper understanding. Studies are based on descriptive or predictive models of objects and parts of the system, and on modeling the environment which represents its operating conditions.

Modeling steps accompany the conventional design steps and are depicted in the following subsections.

3.1.1 Global Optimization vs. Missions

Providing the designer (or the team in charge of defining the goals) with a virtual prototype at multiple abstraction levels allows the description and test of validation scenarios very early in the design cycle. These same tests are used to validate the behavioral model in order for it to be used as a specification and reference for all the other validations (ultimately the validation of the final virtual prototype itself).

Since the virtual prototype can be run surrounded by a modeled and simulated environment, it is possible to evaluate the system subject to environmental conditions representative of its actual final usage. In this approach, one does not test the system as a set of assembled technological components, but as a set of services, which allows the determination of performance metrics, which are the actual perceived requirements of the final user or client.

Using this method, optimization concerns the service and global performance instead of those of each individual subsystem. This approach is called system-level optimization, and indeed optimizing the various sub-systems independently is not the same as optimizing the system itself.

3.1.2 High Level Global Behavioral Description of the System

This important stage consists of translating the characteristics of the system which result from the definition of the project into characteristics that can be simulated. This "simulatable schedule of conditions" reproduces the operation of the system by establishing relations between the inputs and the outputs as in a "black box" approach. This high-level model employs the available data on the system. In the case of functional specifications, which can be supplemented by some normalized standard, the behavioral model produces compatible performance metrics selected by the design group. In this case, financial or technological constraints can lead to the development of a slightly different solution. When data are provided by characteristic sheets and measurement results, the proposed modeling approach can reflect reality with an extreme precision, but without any reference to the physical operation of the device. This type of descriptive model, designed to simulate the operation of the system in a range of "normal" use, cannot handle pathological cases where the system fails to function because of its operating procedure or because it is used outside of its validity range. The advantages of this kind of behavioral model are multiple.

First of all, it provides a reference to validate any later simulation of a lower level system model. Also, a behavioral model naturally simulates faster than a low level model, with a sufficient level of precision. Thirdly, it is useful to simulate the upper level in a hierarchical design, using behavioral models of some blocks in order to test consequences of behavior at standard limits.

3.1.3 Structural Step(s)—System Modeling

It is rare that the project objectives are satisfied with device behavioral simulation. It is then necessary to remove the lid of the "black box" to observe system operation and to isolate physical or functional parts inside. This is a new way to express system behavior rather than a new level of simulation. Structural analysis establishes links between the internal components of the system, independently of their behavioral or physical implementation. The related simulation task consists of building and linking the virtual components, with the relations established in the structural analysis, using the behavioral or the physical models according to the needs. With large or multidisciplinary systems, the identification of the components in the structural model requires specialist knowledge in the various physical fields present in the device. They choose the subsets to be isolated, according to the possible interactions between them and to the required modeling levels. Then, each element of the functional model at its turn can be the object of a structural analysis, and so on, until going down to the physical modeling level of the various components.

3.1.4 Physical Low Level Description of the Components

When a very accurate predictive model of a component is required, it is necessary to delve very deeply into the component's internal mechanisms to emphasize the equa-

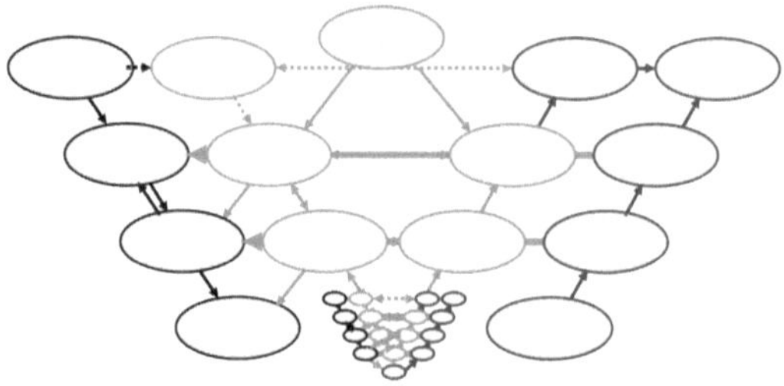

Fig. 11.3 Recursive view of the FVP cycle

tions that control its behavior and make it possible to apprehend the most detailed parasitic effects. These models, which are in general much slower than their behavioral counterparts, have the advantage of "predictive capability". Such models are indeed able to reproduce atypical behavior that would completely escape from modeling at a higher level. The development of these models requires the intervention of all the fields of competences involved in the system. The specialists in physics must develop generic mathematical models, including as far as possible the operational detail of the components, whereas the specialists in the modeling languages must find the computing solution best adapted to process the physical model. Once these stages are explored, it is possible to build a library of models.

While building this set of models, at different levels of abstraction, industrial actors can master and optimize their design flow [1].

In synthesis, this workflow allows one to:

- formalize the expression of specifications,
- take rational decisions, and minimize risks,
- shorten the time to market,
- handle technological breakdowns,
- provide a framework for Intellectual Property capitalization and reuse,
- achieve "correct by design",
- be independent from tools vendors.

It is interesting to note that this workflow naturally includes the top-down, the bottom up and the meet-in-the-middle approaches. It is possible to manage technological choices at a high level of abstraction by taking into account operating conditions thanks to appropriate modeling of the environment. The use of a multi-domain language also allows the modeling of interactions.

3.1.5 FVP Cycle Assets

An accurate analysis of the FVP workflow shows many new assets. We could also develop different points of view about strategic, technical, operational, legal and marketing advantages. In the following paragraph we outline only the industrial assets in collaborative works. This design flow is a recursive process. A sub system may be studied with FVP methodology and may be integrated like a component in another system. Figure 11.3 depicts this capability.

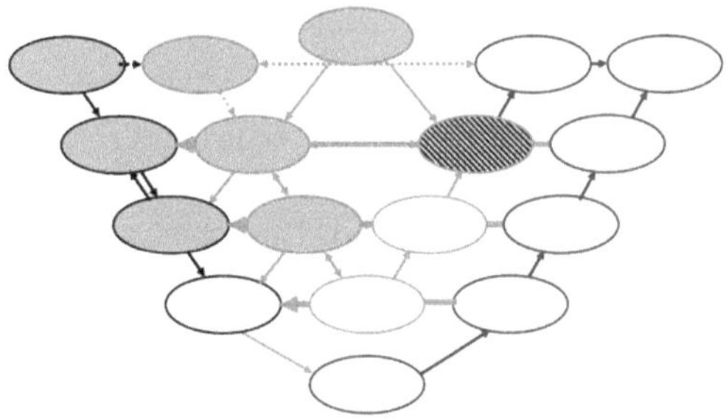

Fig. 11.4 FVP cycle with suppliers

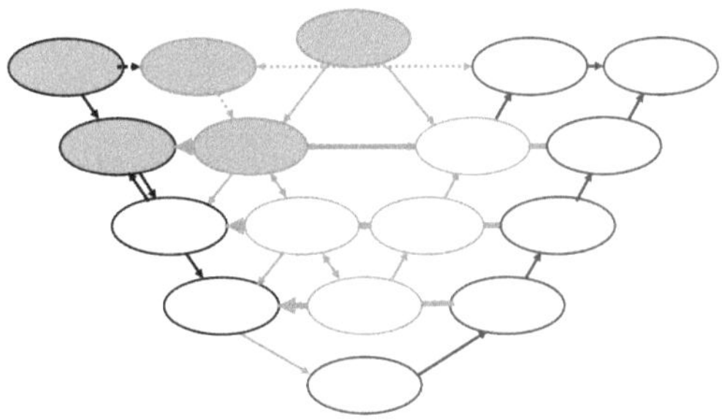

Fig. 11.5 FVP cycle with subcontractors

When a company works with suppliers, it is possible to simplify the workflow including their models. The company defines the behavioral and architectural models, includes supplier models (grey steps) and verifies that they are in agreement in the virtual prototype (dashed block in Fig. 11.4).

If a company works with one or more subcontractors, it is possible to produce high level models (grey steps in Fig. 11.5) and to gather and compare external system models before the final choice and actual developments.

3.1.6 Requirements

The more complex the system is to design, the more the collaboration of many skills in close teamwork is required. Effective teamwork implies constraints that highlight the needs of a shared description formalism brought by methodology, languages and tools, each at a different level. Methodology and project management techniques bring indeed an efficient and standardized partitioning and assignment of tasks and thus improved guarantee of results. Modeling languages, in this context, facilitate communication inside the team. They provide tools to achieve top-down and bottom-up diffusion of information throughout the various business competencies in the design flow. Design and simulation tools have their role to play here. They are at the heart of the process because they can allow models developed with different languages to be simulated together. In this way, they facilitate communication between communities employing different design methodologies and vocabularies, and open the way to the next level of co-design technologies.

3.1.6.1 Collaboration

As mentioned above, teamwork is the cornerstone for prototyping complex systems. Preparative work around a realistic schedule and the distribution of the tasks according to skills are primordial. The design team must be shaped to include the different competencies identified to meet the goals. As an example, design people may not be able to build high-quality models, and also may not be specialists in the various fields of physics representing specific parts in the design. As such, the complementarities of a well-shaped team appear to be essential. This collaboration implies tools and languages that can link together the various layers of the design process.

3.1.6.2 Reusability

As industry is likely to be the main beneficiary of this change in the design process, virtual prototyping must provide significant gain in terms of productivity. One of the aims of virtual prototyping is the construction of model libraries that can be reused in later designs or projects, thus reducing the development time. To achieve this goal, or at least to minimize the model changes to satisfy requirements, modeling rules need to be developed and held to. This modeling methodology first needs standardized languages and adapted tools, as described below. Furthermore, model texts must respect writing conventions to make the most of the method. Here are some examples to highlight:

- The intensive use of datasheet parameters is key to reusability. Datasheet parameters are available for users, as opposed to physical parameters, which are confidential and often extracted to fit in models. Such behavior is not suitable to allow further use of the model in another technology for example.
- Detailed and systematic documentation of models is also very important for understanding and reuse. Considering that many people may access and modify the components in a library, it is primordial to make them comprehensible, and give indications about the model—often comments—and the way it works.
- Naming conventions also provide a way to identify very easily the pins at the interface of a component and then connect it reliably.

All these methodological elements require a strong and common base of languages and tools to grow faster and in a constructive way.

3.1.7 Language Aspects

Programming languages abound in computer science. But not all of them are dedicated to or appropriate for system design. Amongst the ones that may be suitable for modeling, we can distinguish the following categories:

3.1.7.1 Object-Oriented Languages

This class of language (including C++, Visual Basic and Java) is high-level with quasi-infinite possibilities given by the fact that they are the building blocks for the development of applications. It is thus completely possible to use these languages for simulation, especially since dedicated libraries such as SystemC have been developed. Considering that these libraries are not yet ready for analog and mixed-signal purposes and that powerful simulators, which can often interact with C++, already exist, it does not appear necessary to make use of these languages in our case.

3.1.7.2 Digital Modeling Languages

One finds in this category languages like VHDL (IEEE 1076-2000), Verilog (IEEE 1364-2005), SystemC (IEEE 1666-2005), SystemVerilog (IEEE 1800-2005) or SpecC. These very effective languages are associated with tools that allow, with adequate use, the synthesis of complex digital devices and the co-design of advanced heterogeneous platforms such as ASIC/FPGAs with embedded processors. These languages are however unsuited to projects which require mixed-signal circuitry and/or multi-disciplinary simulation.

3.1.7.3 Explicit Formal Mathematical Languages

Languages such as Matlab, associated with Simulink, make it possible to propose to users a graphical representation in the form of block boxes being able to be connected and to have very complex transfer functions thanks to the very thorough mathematical possibilities of Matlab. However implicit equations (Kirchhoff laws) and multidisciplinary aspects are not natively implemented (i.e. in the basic toolset), and links with industrial standard tools (in CAD, CAE, CAM domains) are *ad hoc* and not generalized. These are the two main issues with the use of such platforms to design complex systems.

3.1.7.4 Implicit Modeling Languages for Electronics

In this family of languages, including SPICE as an example, it is only possible to handle analog electrical modeling. For multidisciplinary purposes, this approach requires an explicit analogy from the electrical domain to another. Moreover, this language is not able to carry out mixed-signal modeling alone. In addition, as SPICE components must be in the simulation kernel, it implies a recompilation (or a modification of the simulator) for each new low-level model. This is not very practical and in complete contradiction with the methodological objectives of model reusability.

3.1.7.5 Analog Multi-fields Modeling Languages

This family contains two languages which have now been succeeded: HDLA (which has evolved to VHDL-AMS) and Verilog-A (which has also evolved, to Verilog-AMS). It also includes the MAST language (Synopsys) that remained for a long time a mechatronics analog language before recently evolving to the AMS world. As this language is proprietary and only supported by the Synopsys tool SABER, it does not meet our reusability and standard needs.

3.1.7.6 AMS Multi-fields Modeling Languages

One finds here the Verilog-AMS, VHDL-AMS (IEEE 1076.1) and the future SystemC-AMS standards. These languages are all AMS extensions of their digital predecessors. As their names imply, these languages make it possible to process indifferently logical, analog or mixed-modeling within the same component or system. In addition, these languages are intrinsically multi-fields and natively manage implicit equations [5]. Lastly, the richness of these two languages and the instantiation methods make it possible for tools to approach modeling from several angles and to reach several levels of abstraction, corresponding to the designer needs. These languages are the basis of the complex systems design, but their main weakness is the slowness of low-level simulations. This is why the use of SPICE and FAST-SPICE simulators as a complement for low-level analog simulation is researched.

All these different languages that meet the goals of virtual prototyping do not natively coexist. To allow the complete workflow and the various modeling cultures to communicate efficiently through the methodology points developed previously, specific design and simulation tools are required.

3.1.8 Tools Aspects

Modeling tools differ in their philosophy, and do not provide the same means of accessing information. In the following paragraphs, we develop several qualities that tools should provide in order to facilitate the implementation of the FVP cycle presented in Fig. 11.2.

3.1.8.1 Multi-abstraction

Considering the very wide range and content of projects that virtual prototyping may address, this methodology requires languages and tools able to support such diversity. Within a given project, different views of a component can be created. As an example, a behavioral model can be developed to fit the specifications and create a client integration model or a bottom-up test environment. A fine-grain view could also be achieved by instantiating transistor level models to observe physical details, and to improve behavior or fabrication process. Another possibility could be a black-box model with no feedback effects at a digital level. Higher degrees of abstraction will be required in co-designed systems where a software model will have to run above a hardware model.

3.1.8.2 Multi-designer

Given the increasing number of persons involved in large prototyping projects, many of them do not have the understanding of HDLs. For this reason, the use of the models through the library must be code-transparent, meaning that, with the use of the documentation, it should not be required to understand the model text to use it as a component. People that are not HDL specialists must access and gather models for design, simulation, demonstration or test purposes for example. A way to achieve an easy manipulation is the use of graphical symbols with pins and parameters that users have merely to link and complete. Some simulation or workflow environments allow such a presentation (Simplorer, SystemVision, Cadence), but this does represent a loss of portability. In fact, graphical aspects are proprietary, contrary to such languages that are standardized.

3.1.8.3 Language Independence

As there are currently several HDLs that allow device modeling, users who operate with existing models do not have to take into account which one the designer chose. As a consequence, they expect the software to have the largest standard compliance and to be able to mix possibilities to get access to the widest library and design choices.

3.1.8.4 Easy Management of Complex and Mixed-Fields Systems

VHDL-AMS and some other languages have been designed to allow modeling of systems that do not contain only electrical parts. This becomes increasingly relevant in recent systems that often mix electronics with optics, mechanics or other fields of physics, chemistry or even biology. To achieve such a multi-purpose goal, the software must be able to recognize and connect properly all these different kinds of information in a simple front-end or GUI. This ability to merge results coming out of different fields of expertise, and to address all the various partners of a project should be a quality of system modeling tools (cf. Sects. 3.1.8.2 and 3.1.8.5).

3.1.8.5 Intuitive, Simple and Efficient Graphical User Interface (GUI)

Amongst the conditions that would make software usable by the greatest number, the simplicity and the convenience of the GUI are essential. The difficulty resides in the fact that in the same time it has to cover the huge spectrum of functions that the different blocks of a device may offer. All the details and the compatibility operations—between different languages as an example—must obviously remain as independent as possible from the user, who would ideally just have to click on a button to view the results.

3.1.8.6 Model Creation Tools

Behind the graphical aspect of the components lies the HDL source code. That part of models—hidden because of its needlessness for basic users—remains the most important as it conditions the successful simulation of the device. That is why the software must provide a powerful and convenient way to create and edit model code.

4 A Complete MEMS Example

In order to back up the explanations of the concepts involved in the FVP methodology, an industrial application is presented. The system is a MEMS micro conveyor, and its design implies the management of different domains.

4.1 Air Fluidic MEMS

In this study, we wish to design a distributed smart MEMS fluidic micro-conveyor system with totally distributed control. At the integrated circuit level, this system is composed of several layers. The first layer distributes the pressure to the second, which is a set of electrostatically controlled valves. The air is pushed through a shaped hole, such that the flow of air can either be normal to the surface (for stationary suspension) or at an angle to the normal (for directive suspension). A layer of photo-detectors allows the detection of the presence of an object. The system allows the control of the trajectory of a small object, maintained in suspension within the flow through the management of valves.

4.2 Modeling Design Approach for MEMS

A modeling design approach for MEMS can be viewed from either a top-down or a bottom-up approach, as shown in Fig. 11.6.

The traditional (bottom-up) design approach naturally starts at the device level, with physical level modeling and moving up to system level modeling. In this work however, we focus on the exploration of the system design space to determine our critical system parameters. This is the principal focus of the top-down

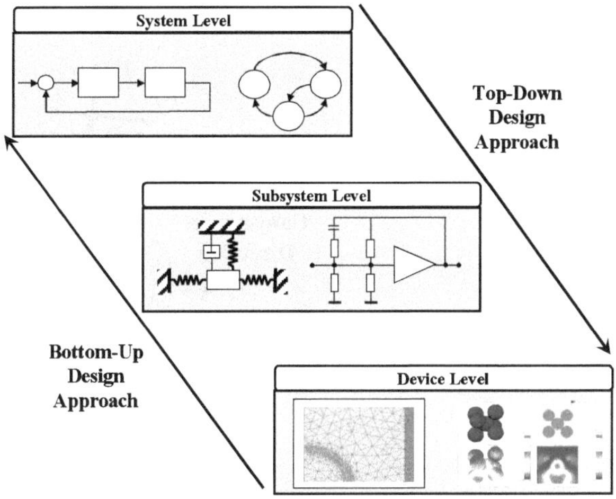

Fig. 11.6 Modeling design approach for MEMS

design method, where developments would proceed from the system to the device level, via one or more intermediate subsystem levels. We start at the highest level of abstraction with specifications (customer needs and associated constraints), formalized through block diagrams or state charts from control engineering and signal processing in order to attain an executable (simulatable) system description. Once the critical system parameter values have been established, more focus can be placed on examining implementation options and specific technologies through the use of reduced-order models at the subsystem level. The term "reduced-order modeling" is used to highlight the fact that the ability to address coupled energy domains such as those involving mechanical and microfluidic components now exists. It should allow a tremendous reduction of model size which becomes important for time-domain simulations with several hundreds of steps needed for MEMS, circuit and control systems.

Finally, we develop the lower abstraction (device) level, with more detailed physical modeling. It is more commonly referred to as "three-dimensional (3-D) modeling" because it usually uses finite element or boundary element solvers, or related methods. Due to their high accuracy, they are well suited to calculate all physical properties of MEMS, but they also cause considerable computational effort.

Each design level, *with the classical approach*, requires a specific language, which is different from one level to another. There is no common language to describe all levels.

With this example, we will explain the principles of the FVP design approach, using solely the VHDL-AMS language to describe a DMASM (Distributed MEMS Air-flow Smart Manipulator) including physical MEMS components.

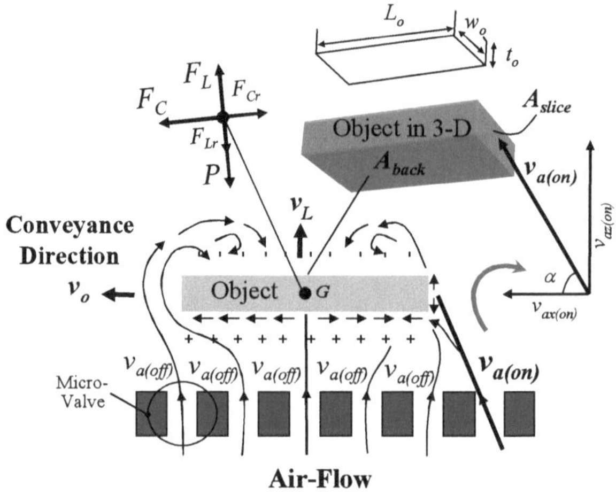

Fig. 11.7 Physical model of the air-flow distributed surface

4.3 Behavioral Model

4.3.1 Modeling Conditions

As introduced previously, the behavioral model is the highest modeling level of the system to be simulated. At this level, the model is based only on the most basic physical effects that occur in or between component modules involved in the system, in order to examine the functional requirements. To model the DMASM, we capture the mathematical description of the physical and informational behavior of the device. The behavior of the DMASM involves multiple interactions between technological and physical elements, such as MEMS-based micro-actuators, optical sensors, IC-based controllers and drivers, and air-fluidic flow effects over a solid body. Firstly, we focus our study on the phenomena of air-flow over a body and the induced fluidic forces. These physical effects have been analyzed by Cengel & Cimbala, who developed a model from experimental data and underlying correlations [8].

4.3.2 Air-Flow Conveyance Model

The DMASM can be described as a fluidic model with interaction between the distributed air-jet surface and an object during a specific sequence of drag and lift forces, as shown in Fig. 11.7.

Here, pneumatic microactuators are replaced by simple micro-valves taking two position states: ON or OFF, respectively when the micro-valve is open or closed. When the micro-valve is closed (OFF), we find an equivalent model of the static model defined with vertical air-flow generated by each micro-valve. The air-flow velocity is then defined by v_a (off). When the micro-valve is open (ON), the air-flow depends on a directional velocity that is defined by v_a (on), and the angle of inclination (α).

All forces of the dynamic model are applied to the center of gravity of the object (G), as shown in Fig. 11.7. Fluidic forces are separated into two tasks: one to maintain the object in levitation, and the second to convey the object in a desired direction. We define two fluidic forces: the levitation force (F_L) and the conveyance force (F_C). The levitation force is due to the combined effects of pressure and wall shear forces, respectively F_{Lp} and F_{Ls}. The pressure force (F_{Lp}) is normal to the object's back-side with an area of A_{back}, whereas the wall shear force (F_{Ls}) is parallel to the object's slice corresponding to area A_{slice}.

The dynamic relationship for a one-dimension conveyance of the object is established in a given axis (z-vertical and x-horizontal), as given respectively by:

$$F_L = F_{Lp} + F_{Ls} \approx F_{Lp}$$

and:

$$F_{Cp} - F_{Cr} = m_o \frac{d^2x}{dt^2} - K v_{ox}$$

with m_o representing the mass of the object, W the weight of the object, K the coefficient of the viscosity in air, and v_{ox} the horizontal component (x-axis) of the velocity v_o.

The two-dimensional representation of the active surface is extracted from the same model established for one-dimensional representation. Indeed, the displacement of the object can be defined as well in the y-horizontal axis.

4.3.3 VHDL-AMS Description

A VHDL-AMS model has the same structure as a VHDL model with two main parts: *entity* and *architecture*. The *entity* declaration describes the interface of the model to external circuitry. Specifically, it describes the physical ports as well as the parameters that are used in the behavioral description of the model, and which influence its performance. The *architecture* declaration contains the actual description of the functionality of the model. It can be a mathematical description describing the physics of the model or it can contain so-called structural constructs. More details on VHDL-AMS can be found in [3] or [4].

Figure 11.8 presents the behavioral model of the DMASM using VHDL-AMS, with a general structure of the description at the header of the figure (entity of design, configuration declaration, component architecture, packages).

4.4 Structural Component Models

4.4.1 Structural Behavioral Model

Decomposition is an essential principle in managing and mastering the complexity of large-scale systems. To establish the structural behavioral model, we first operate

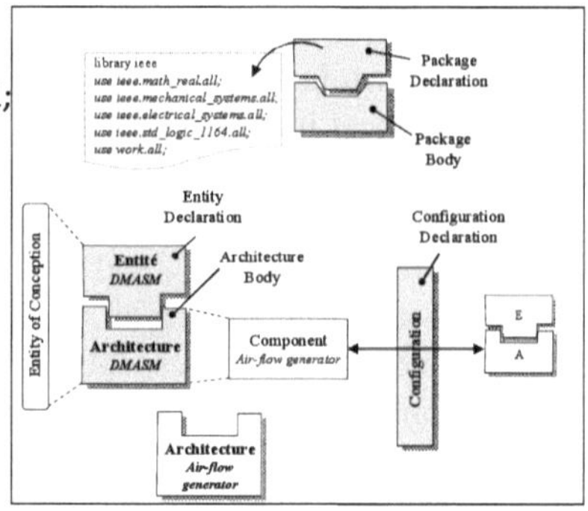


```
Library IEEE; use ieee.math_real.all;
use ieee.std_logic_1164.all;
use ieee.mechanical_system.all;
use ieee.electrical_systems.all;
use work.new_type.all; work.all;

ENTITY DMASM IS
GENERIC (Mo, Dens, Cxp, Cxf, Lo,
Wo, To, Vazoff, DD, DH, DIM,
              Dx0, Vox0, H0: REAL);
PORT (TERMINAL FA, FV: TRANSLATIONAL);
END DMASM;

ARCHITECTURE behavioral OF DMASM IS
CONSTANT g: REAL;
QUANTITY X, Y, Z, Flp, Flf, Fmp, Fmf, P, Vl, Vo, Vox, Voy, K:  REAL;
QUANTITY Vaon ACROSS FV, Theta ACROSS FA;
   BEGIN
     - Submodel: Air-flow generator
        ......................
      - Submodel: N distributed micro-valve
        ......................
      - Submodel: Estimation and detection exposed area
        ......................
      - Submodel: Force and dynamic relations
        ......................
Flp==0.5*Dens*Cxp*Vazoff**2* Lo*Wo;
Fmp==0.5*Dens*Cxp*Vaxon*cos(ang/180.0*MATH_PI)**2*Wo*To;
Flr==0.5*Dens*Cxp*Vl**2*Lo*Wo*To;
Fmr==0.5*Dens*Cxp*Vo**2*Lo*Wo;
P==Mo*g;
Flp-Flr-P==Mo*Vz'dot-KVl;
Vz==Z'dot;
Fmp-Fmr==Mo*Vx'dot-KVox; Fmp-Fmr==Mo*Vy'dot-KVoy;
Vx==X'dot;   Vy==Y'dot;
........................
END ARCHITECTURE behavioral;
```

Fig. 11.8 VHDL-AMS description of the DMASM behavioral model

analyses of the behavioral model which can be composed out of interconnected functions. All extracted functions are independent of their physical descriptions. Decisions about what constitute the functions of the structural behavioral model are usually based on the global behavior of the system and the data/quantities exchanged in it. The parameter of each subfunction gives the values of local performances. The global simulation of this net of interconnected functions has to operate like the behavioral model.

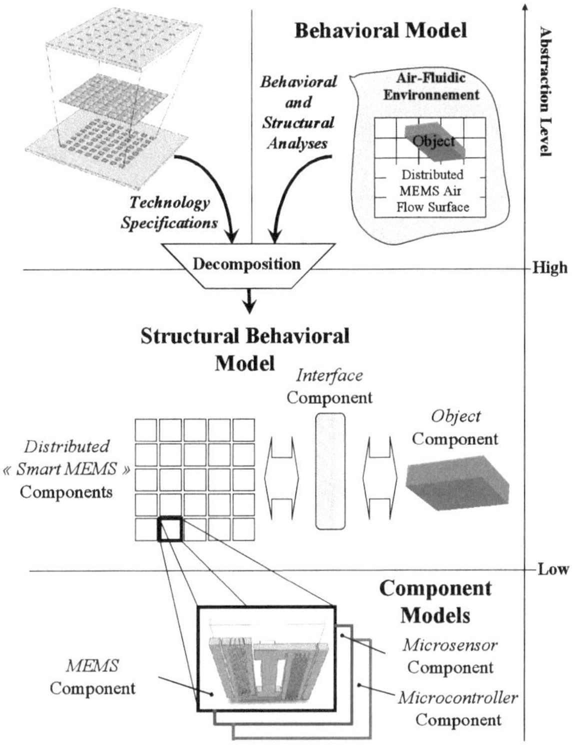

Fig. 11.9 Structural behavioral decomposition flow

Establishing the behavioral model of the DMASM, we carried out a first decomposition with submodels based on forces (actions and reactions) inside the system. However, this decomposition direction does not focus on the substance aspects of the system, e.g. the actual objects and their relations. Such a type of decomposition is defined at the structural level of behavioral models or functional decomposition.

Finally, as represented in Fig. 11.9, we illustrate the three steps of the design flow beginning by the behavioral model, followed by a transformation to the structural behavioral and finishing by an analysis of the model of actual components in order to interconnect them to build a structural technological model. Each structural model (both functional and technological) must behave identically within the performance requirement evaluation setup.

At the component model level, we develop the three-function component based on the *distributed "Smart MEMS"* component. It is composed of three independent component models as given by:

- *MEMS component model* (pneumatic microactuator);
- *Microsensor component model* (micro-photodetector);
- *Microcontroller component model* (decision-making logic).

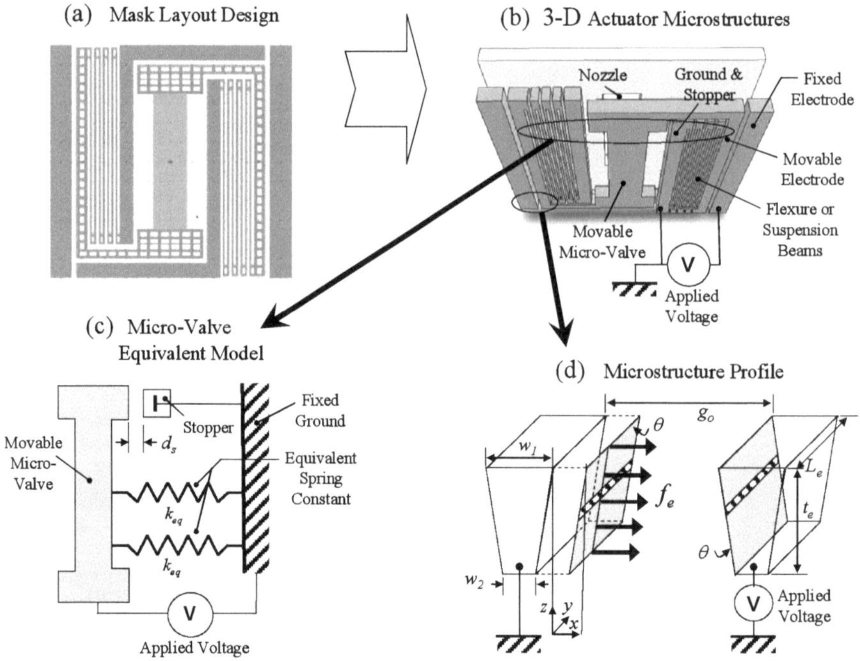

Fig. 11.10 MEMS component model. (**a**) Mask layout design. (**b**) 3-D actuator microstructures. (**c**) Micro-valve equivalent model. (**d**) Microstructures profile

We do not claim this to be the best or only way to decompose the "Smart DMASM", and it is possible to study the case in another way. However, as it will be shown, this approach helps designers to extract component models and analyze their behavior and technology in their functional environment.

4.4.2 MEMS Component Model

At the component model level, the component is described at its lower physical level. In general, the internal constitution of a component can be a behavioral model, or a subsystem consisting of interconnected components, allowing for composable and hierarchical models.

To describe the "MEMS component", we propose four representations of the design, as shown in Fig. 11.10.

Firstly, the mask layout design of the pneumatic microactuator is described in Fig. 11.10(a). The resulting 3-D bulk fabrication of the pneumatic microactuator is illustrated in Fig. 11.10(b). An equivalent model of the micro-valve, based on a movable micro-valve, which depends on electrostatic parallel-plate structures, is shown in Fig. 11.10(c).

```
......................
ARCHITECTURE Structural OF DMASM IS
BEGIN
 Component air-pressure
   generic(...);
    port(...);
END components;
Component MEMS
  ......................
Component microsensor
  ......................
Component microcontroller
  ...................
Component interface
  .....................
Component object
  ......................
Air:      entity air-pressure generic map(...)
             port map (...);
Gen1:     for i in 1 to DIM generate
Actu:     entity MEMS generic map(...)
             port map (...);
          end generate;
Gen2:     for j in 1 to DIM generate
Sens:     entity microsensor generic map(...)
             port map (...);
          end generate;
Gen3:     for k in 1 to DIM generate
Cont:     entity microcontroller generic map(...)
             port map (...);
          end generate;
Inter:    entity interface generic map(...)
             port map (...);
Obj:      entity object generic map(...)
          port map (...);
END ARCHITECTURE structural;
```

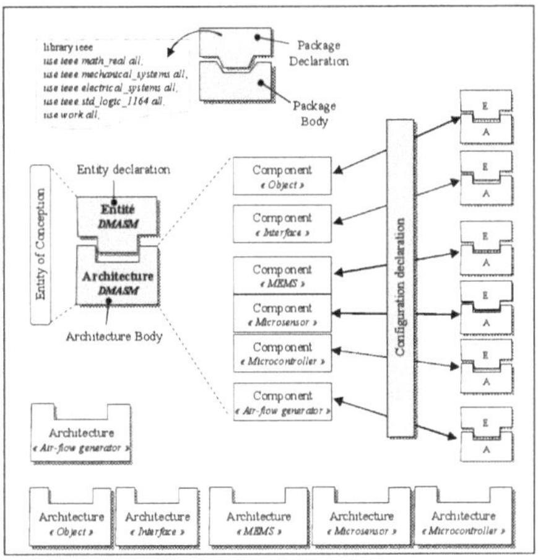

Fig. 11.11 DMASM structural behavioral code

4.4.3 VHDL-AMS Description

The corresponding VHDL-AMS code of the structural behavioral architecture (*structural*) of the DMASM is given in Fig. 11.11. The general structure of the description is also given at the header of the figure (entity of design, configuration declaration, component architectures, packages).

Table 11.1 Model parameters of the DMASM

Parameter	Description	Value	Unit
m_o	Object mass	6.6×10^{-6}	kg
t_o	Object thickness	2.5×10^{-4}	m
w_o	Object width	4.5×10^{-3}	m
L_o	Object length	4.5×10^{-3}	m
C_{xp}	Pressure coefficient	1.11	
C_{xf}	Friction coefficient	0.004	
ρ	Air density	1.3	kg/m^3

The definition and interconnection of *entity* are identical to the behavioral model; only the *architecture* of each entity changes. The structurally defined model uses instantiated components *air-pressure* (*air*), *MEMS* (*actu*), *micronsensor* (*sens*), *microcontroller* (*cont*), *interface* (*inter*), and *object* (*obj*), which have been defined and coded separately. They belong to the work library where they are called in the description. The order of instantiation in the model is not important.

To model the distributed aspect of the DMASM, we use the *GENERATE instruction* (automatic code writing) for each component *MEMS* (*gen1*), *micronsensor* (*gen2*) and *microcontroller* (*gen3*). This instruction generates instructions or instances from a static value known at elaboration time. Here, variables (i, j, k) are defined for respectively *MEMS*, *micronsensor* and *microcontroller* components, where the range of values is from 1 to DIM (which represents the maximum dimension value).

4.5 Simulations

4.5.1 Behavioral Model

To validate the proposed global behavioral model, several simulations have been carried out. In this sub-section, we first performed a 1-D conveyance of the object using a range of five pneumatic microactuators. Air-flow generated by each element is produced when the back-edge of the object is detected at the nozzle entrance. The values of the model parameters we used are listed in Table 11.1. Figure 11.12 presents simulation results of various characteristics of the object (height, acceleration, velocity) according to the physical fluidic conditions (air-flow velocity, drag force, air resistance) for a 1-D conveyance.

The appropriate responses of velocity, acceleration and height of the object according to air-flow velocity, drag force and air resistance are obtained. When the object's end arrives at the exposed area of the micro-valve (1, 2, 3, 4 or 5), an air-flow velocity is applied, which generates a drag force on the object's edge, increasing the velocity of the object. Over five micro-valves, the velocity of the object is approximately 0.05 m/s.

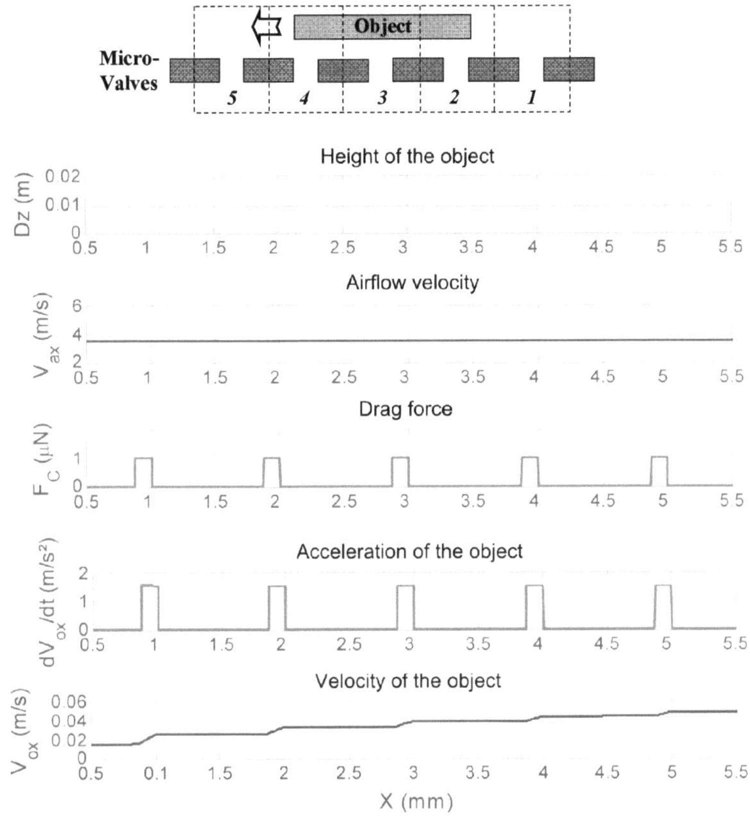

Fig. 11.12 Simulation results using DMASM behavioral model

The models are built with generic parameters, such that all parametric studies can be carried out with the same model. Optimizations using these parameters are also possible.

4.5.2 MEMS Component Model Simulation

Figure 11.13 shows the simulation results of the MEMS-based pneumatic microactuator. In particular, we observe the electrostatic micro-valve behavior by applying a specific profile voltage between 0 and 150 V during 0.03 s. The parameter values of the micro-valve model we used are listed in Table 11.2.

We can observe a displacement response of the micro-valve following a classical pull-in voltage ramp. This displacement is 15 μm, the distance between the rest position and the stopper. We also recorded the contact shock, which appears as a brief variation of the micro-valve displacement. When the voltage is released, we can ob-

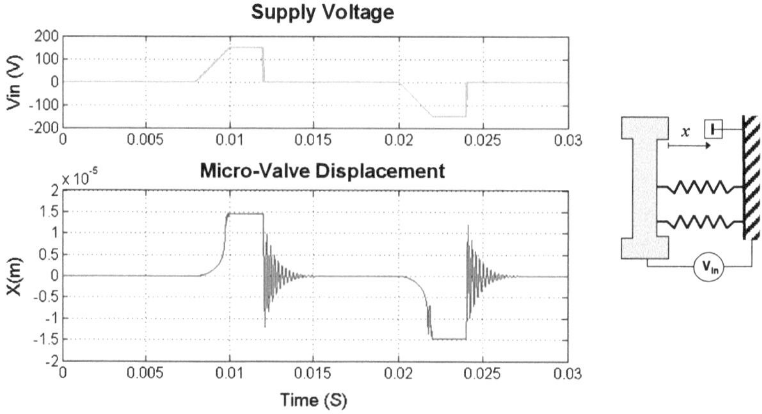

Fig. 11.13 Simulation results based on the MEMS component model

Table 11.2 Model parameters of the MEMS component

Parameter	Description	Value	Unit
m_s	Microstructure mass	6.61×10^{-9}	kg
L_e	Electrode length	900×10^{-6}	m
w_1	Electrode upper width	10×10^{-6}	m
w_2	Electrode lower width	6.5×10^{-6}	m
t_e	Electrode thickness	100×10^{-6}	m
ε_o	Vacuum dielectric constant	8.85×10^{-6}	F/cm
E	Young's modulus	1.3×10^{11}	Pa

serve oscillations of the micro-valve returning to its initial position. All simulations generated with VHDL-AMS present an excellent behavior profile.

4.5.3 Structural Simulation Including the MEMS Component

Figure 11.14 shows simulation results of the DMASM structural behavioral model including the "MEMS component".

Simulation performances such as drag force, acceleration and velocity of the object perfectly match with the previous results based on the behavioral model. This validates the decomposition between two description levels as proposed in the FVP design flow.

At the component model level, we observe the sampling effect on the air-flow velocity signal, which can be identified according to each specific micro-valve. In the behavioral model, the air-flow velocity was simply a continuous signal of the global

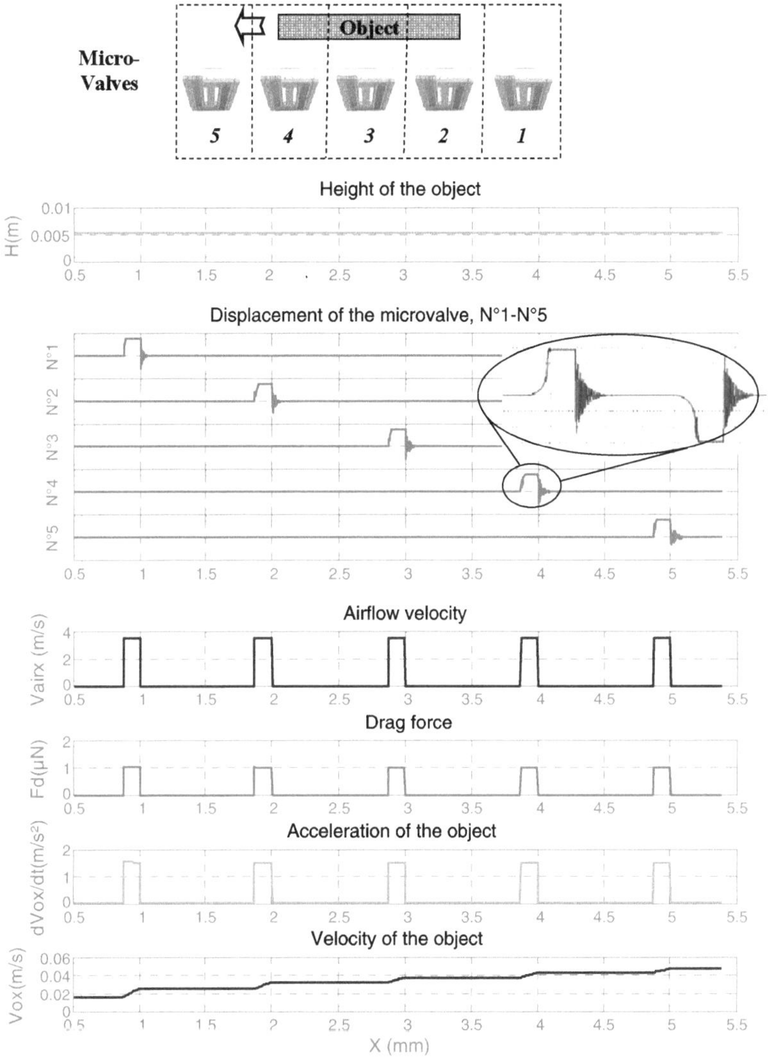

Fig. 11.14 Simulation results of DMASM structural behavioral model including the MEMS component

model—only the drag force, acceleration and velocity of the object were sampled. Finally, we show in Fig. 11.14 that the simulation results of the "MEMS component" are successfully reproduced using the structural behavioral model. Indeed, signals of the "displacement of the micro-valve" shown in Fig. 11.14 are similar in accuracy to those presented in Fig. 11.13.

These results validate the multi-domain and multi-abstraction features of the VHDL-AMS language and confirm the suitability of the FVP design flow approach to develop complex models.

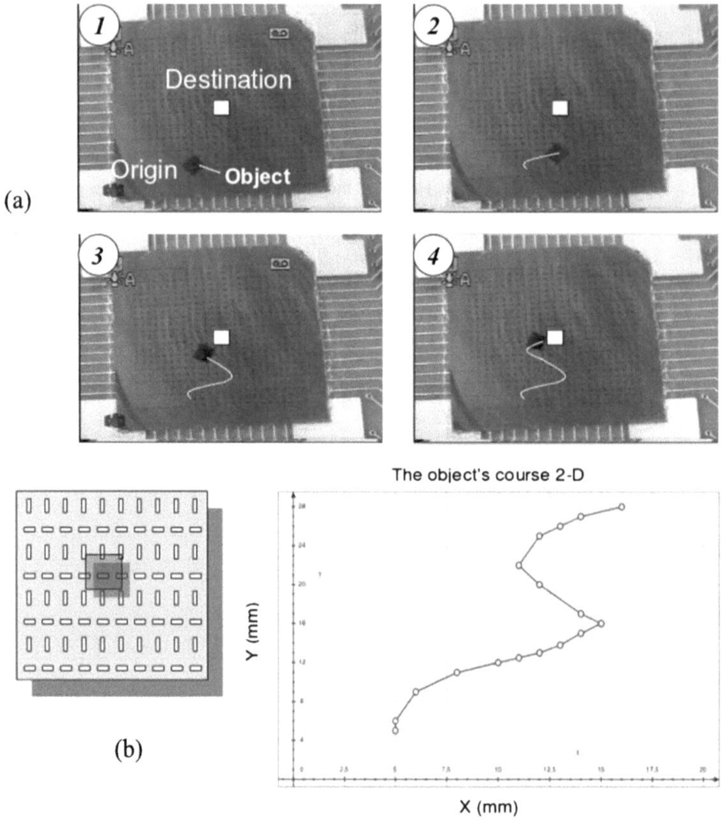

Fig. 11.15 Simulation/experimental comparison for 2-D conveyance. (**a**) Experimental results; (**b**) Simulation results

4.6 Simulation and Experimental Verification

An experimental result of a 2-D micro-manipulation of the real DMASM was done *on the first prototype*, by extracting the conveyance performance of the object with open-loop control. The structural behavioral model of the DMASM was tested under the same conditions with approximate values of the air-flow velocities ($v_{a(\mathrm{off})}$, $v_{a(\mathrm{on})}$). Selected test cases were used to refine the structural behavioral models to match the experimental results.

Good agreement between the two approaches was observed, as shown in Fig. 11.15(a) and Fig. 11.15(b). The experimental object's trajectory is reproduced in detail, along with the simulation results of the structural behavioral model of the DMASM.

These results further illustrate the usefulness and the predictive capabilities of the FVP approach.

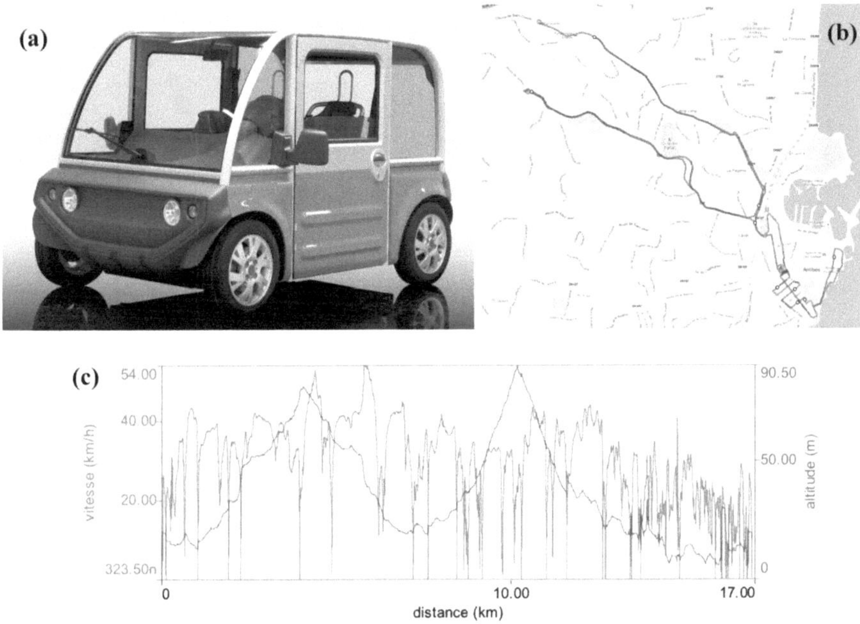

Fig. 11.16 On the basis of a behavioral model of the vehicle (**a**), and a specified mission (**b**), the energy consumption can be determined (**c**), and the battery, inverter and motor technologies and sizes determined

5 Industrial Applications

To show the flexibility of the approach, we now briefly describe several applications in which the FVP methodology has been applied, ranging from the field of transport to that of medicine.

5.1 Electric Vehicle Energetic Optimization

A critical design choice in electric vehicles concerns the battery technology, which essentially depends on the feasibility (particularly in terms of weight), autonomy and cost requirements. A generic behavioral model of an electric vehicle has been developed. It allows the estimation of energy consumption, as plotted in Fig. 11.16(c) on the basis of recorded GPS data for specified-reference trip, shown in Fig. 11.16(b). This study was carried out during the early stages of the design of F-City electrical vehicle, shown in Fig. 11.16(a), by FAM Automobile (Etupes, France), and enabled the determination of the best battery, inverter and motor technologies very early in the design process. With this method, the first working prototype was rolled out 16 months after the first simulation, and the accuracy was observed to be better than 2% (in terms of weight, autonomy, etc.).

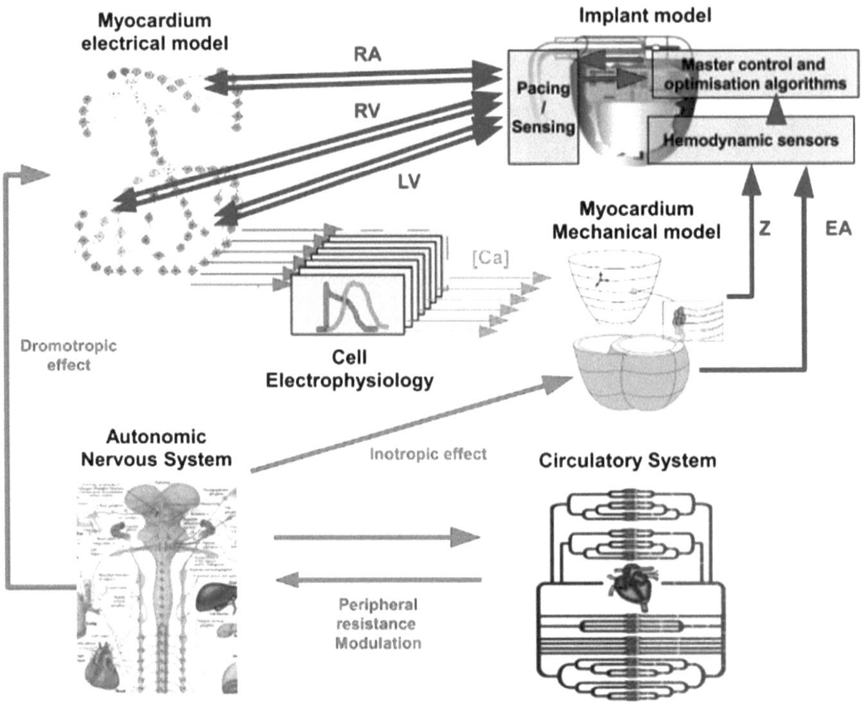

Fig. 11.17 Pacemaker and heart interaction. The cardiovascular system and pacemaker virtual prototype allow the simulation of heart implant behavior

5.2 Heart Modeling and Cardiac Implant Prototyping

In the framework of the Adapter[1] project with ELA Medical, we built a model library describing the heart and main elements of the cardiovascular system. We model and simulate the cardiovascular system, as shown in Fig. 11.17, implanted with a new generation of pacemaker implants, applied to Cardiac Resynchronization Therapy (CRT). By reproducing a specific adaptation phenomenon of the heart to stimulations, illustrated by a bell-shaped curve named the Whinett's curve, presented in Fig. 11.18, the architectural exploration of the pacemaker can be carried out virtually with unequaled simulation performances [6] close to real-time. This allows algorithmic and energy optimizations, leading to vastly improved device performance.

[1]Eureka! #3699.

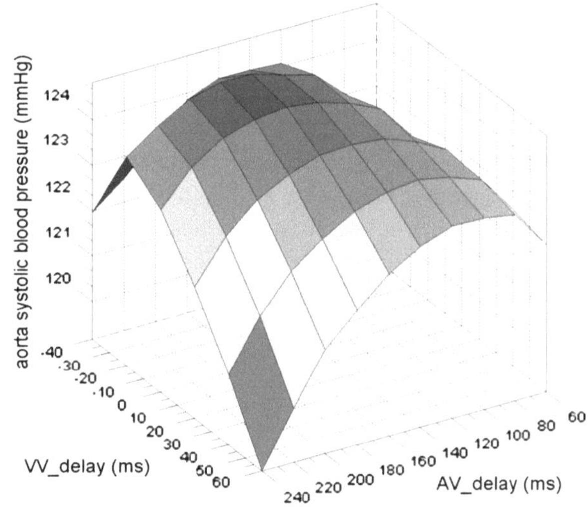

Fig. 11.18 The model allows the generation of Whinett's curve (hemodynamic effects depending on inter-ventricular delays)

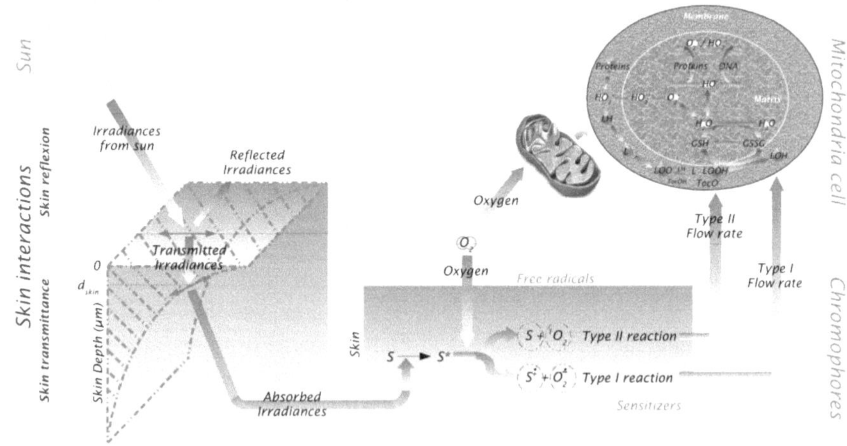

Fig. 11.19 Virtual skin synoptic. The model includes a spectral description of sunlight, the absorption characteristics of the skin, and descriptions of chemical reaction chains leading to the creation of free radicals

5.3 UV-skin Interactions Modeling

Cosmetic firms are in the process of improving R&D methodologies with more formal methods (increasingly based around functional models leading to less laboratory testing). We built, with the Coty-Lancaster Group, a unique and dynamic view of biophysical and chemical phenomena linking human skin to the sun's ultraviolet spectrum [7]. This model, illustrated in Fig. 11.19 was built based on collaboration with cosmetics experts.

5.4 Other Applications

5.4.1 Chemical Process: Inline pH Regulation

The regulation system monitoring and controlling the acidity of the content of a chemical factory's waste pipe leading to the river has to be improved. Experimental tests, which would consist of deliberate pollution experiments, are obviously precluded. A virtual prototype of the installation, including acid/base chemical reactions, flow mixing and regulation automata models, has been developed. This model allowed the characterization and optimization of the installation performance, reaching a higher level of safety.

5.4.2 Mechanics: Dose Pump Modeling and Optimization

An existing precision pump is able to work correctly up to 22 Hz. A functional virtual prototype, including the mechanics of the pump, the electronic control, and results from a magnetic field FEM tool, was built to enable architectural exploration of the device. This approach allowed identification of the limits of the current pump, and solutions to optimize the performances were proposed and implemented, leading to a doubling of the flow rate.

5.4.3 Energy: Magnetocaloric Heat Pump

The performance of the various concepts in the design of innovative cooling systems based on the magnetocaloric effect can be evaluated, and their feasibility validated, before building any prototype. The virtual prototype explores various operating modes and parameter values, so that one can choose the most efficient approach.

6 Conclusion

In this chapter, we presented the Functional Virtual Prototyping methodology and its implementation with the multi-domain, multi-abstraction, standardized language VHDL-AMS (IEEE 1076.1). We illustrated its capabilities and broad range of application with many industrial examples. With these tools, a highly-skilled team of experts can significantly increase efficiency in industrial projects. The next step in the evolution of system design methodology is likely to be the formalization of requirements, for example with the ROSETTA language (IEEE P1699), and graphical expression, for example with SysML (Incose recommendation).

Acknowledgements The example presented in Sect. 4 has been developed with Dr. Lingfeï Zhou and Dr. Yves-André Chapuis (InESS/CNRS, Strasbourg, France).

References

1. Hervé, Y.: Functional virtual prototyping design flow and VHDL-AMS. In: Proc. of Forum on Specification & Design Languages (FDL'06), Darmstadt, Germany, September 19–22, 2006, pp. 69–76 (2006)
2. Hervé, Y., Desgreys, P.: Behavioral model of parallel optical modules. In: Proc. IEEE Int. Workshop on Behavioral, Modeling and Simulation, Santa Rosa, CA, Oct. 2002
3. Design Automation Standards Committee of the IEEE Computer Society: IEEE Standard VHDL Analog and Mixed-Signal Extensions. IEEE Std 1076.1-1999. IEEE Comput. Soc., Los Alamitos (1999). ISBN 0-7381-1640-8
4. Ashenden, P.J., Peterson, G.D., Teegarden, D.A.: The System Designer's Guide to VHDL-AMS. Morgan Kaufman, San Mateo (2003). ISBN 1-55860-749-8
5. Pêcheux, F., et al.: VHDL-AMS and Verilog-AMS as alternative hardware description languages for efficient modeling of multidiscipline system. IEEE TCAD **24**(2) (2005)
6. Legendre, A., Hervé, Y.: Functional virtual prototyping applied to medical devices development: from myocardic cell modeling to adaptive cardiac resynchronization therapy. In: The Huntsville Simulation Conference Proceedings (HSC), Huntsville, AL, USA, October 21–23, 2008
7. Nicolle, B., Ferrero, F., Ferrero, L., Zastrow, L., Hervé, Y.: From the UVA to the lipid chain reaction: archetype of a virtual skin model. In: The Huntsville Simulation Conference Proceedings (HSC), Huntsville, AL, USA, October 21–23, 2008
8. Cencel, Y.A., Cimbala, J.M.: Flow over bodies: drag and lift. In: Fluid Mechanics—Fundamentals and Applications. McGraw-Hill, New York (2006). Chap. 11

Chapter 12
Multi-physics Optimization Through Abstraction and Refinement

Application to an Active Pixel Sensor

L. Labrak and I. O'Connor

1 Multi-physics Design Complexity

Current applications in a wide range of domains such as medicine, mobile communications and automotive, clearly show that future systems on chip will be based on increasingly complex and diversified integration technologies in order to achieve unprecedented levels of functionality. Design methods and tools are lagging behind integration technology development, leading to a limited use of such new functionality. Figure 12.1, showing the V-design cycle (synthesis and verification) projected by the International Technology Roadmap for Semiconductors (ITRS [1]) for heterogeneous system design, as well as data available from the same source, show that the earliest bottlenecks stem from the integration of heterogeneous content. One of the main challenges that clearly appears is to provide efficient Electronic Design Automation (EDA) solutions and associated methods in order to handle system-level descriptions, partitioning and data management through multiple abstraction levels.

1.1 Design Tools and Methods

The field of design methods, in general terms, is a vibrant field of research and is often applied to the management of design, production, logistics and maintenance processes for complex systems in the aeronautics, transport, civil engineering sectors, to name but a few. The micro-electronics industry, over the years and with its spectacular and unique evolution, has built its own specific design methods while focusing mainly on the management of complexity through the establishment of

L. Labrak · I. O'Connor (✉)
CNRS UMR 5270, Lyon Institute of Nanotechnology, Ecole Centrale de Lyon, av. Guy de Collongue 36, Bâtiment F7, 69134 Ecully, France
e-mail: ian.oconnor@ec.lyon.fr

G. Nicolescu et al. (eds.), *Design Technology for Heterogeneous Embedded Systems*,
DOI 10.1007/978-94-007-1125-9_12, © Springer Science+Business Media B.V. 2012

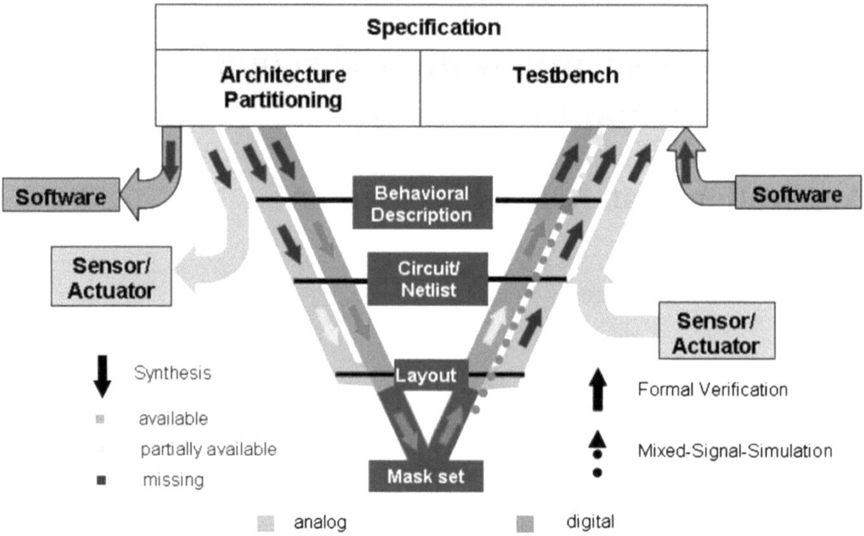

Fig. 12.1 V-cycle for system design [1]

abstraction levels. Today, the emergence of device heterogeneity requires a new approach, and no existing tool has the necessary architecture to enable the satisfactory design of physically heterogeneous embedded systems. Therefore, to design a heterogeneous structure, system level designers must choose from a large number of tools and models leading to application specific design flows (including different simulators and models). The design of heterogeneous systems on chip (SoC), including software components, digital hardware and analog/mixed-signal and potentially mixed-technology hardware, is widely characterized by hierarchical partitioning. Indeed, the heterogeneous nature of its components and the domain-specific design flows associated with each part require that each be processed separately. As shown in Fig. 12.2, we can partition an example system (image sensor) into three main domains. Grouping component blocks into domains is a useful design mechanism to separate concerns and can be broadly defined as grouping component blocks possessing the same physical means of conveying information (optical, fluidics, mechanical, biological, or chemical, etc.), or the same representational means of describing information (analog, digital). Each component of a domain is considered individually, so that a specific synthesis flow can be applied. For example, the electrical part of a design can be synthesized using many different tools to handle electrical simulation (Spice, Cadence Virtuoso® Spectre®, Mentor Graphics Eldo™) as well as layout, placement and routing (such as Cadence Assura® or Mentor Graphics Calibre® to name just a few). For physical component simulation, a wide range of solutions exist, such as Ansys Multiphysics™, Comsol Multiphysics®, Coventorware®, or Ansoft Designer®. Most of these tools can solve physical problems spanning several domains (for example including chemical, thermal or optical phenomena).

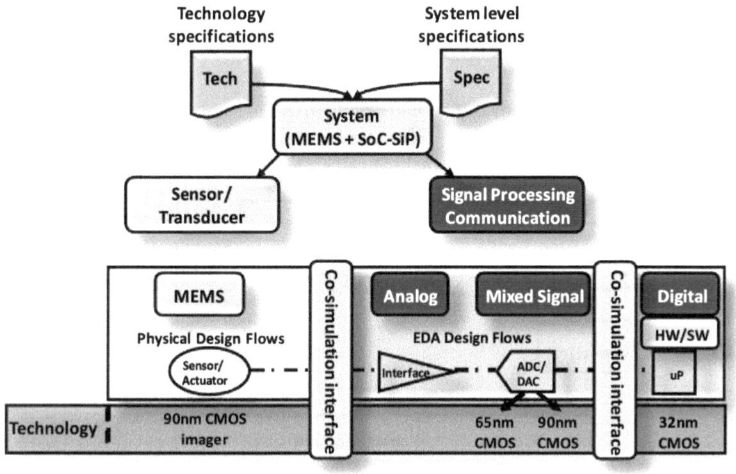

Fig. 12.2 Hierarchical partitioning of heterogeneous SoC/SiP

Figure 12.2 also emphasizes the necessity of simultaneously managing design processes in several domains. For example, in the digital domain, a trade-off must be found between hardware and software implementation. This partitioning of digital functionalities may require reconfigurable capabilities, but also co-simulation with the analog part of the system. Such cross-domain synthesis is ubiquitous in complex design to evaluate the coupling effects between each domain. Coupling a domain-specific simulator and a circuit simulator to carry out co-simulation is possible, but the process of connecting two simulators is challenging and different in each particular case. This is mainly due to the lack of commonly accepted modeling environments, leading to a necessity to know the details of the inner workings of both circuit simulators and domain-specific simulators [2]. Some of the domain-specific simulators (Ansys Multiphysics™, Ansoft Designer®, Comsol Multiphysics®, Matlab/Simscape™, etc.) have included the capability to simulate circuits within physics-based simulations, but it still remains difficult to model complex multi-physics systems.

1.2 Towards High-Level Description Languages

The problem concerning the design specifications for heterogeneous systems can be as detailed and as low-level as the material properties and geometric form of a layer defining a microstructure, to as broad and as high-level as an abstraction of an embedded processor supporting the firmware for a heterogeneous SoC/SiP [3]. The most widely-used solution to handle interaction between different domains is to exploit abstraction. The main idea is to simulate the whole system at a behavioral level using a high level description language (HDL). Each domain-specific part of a

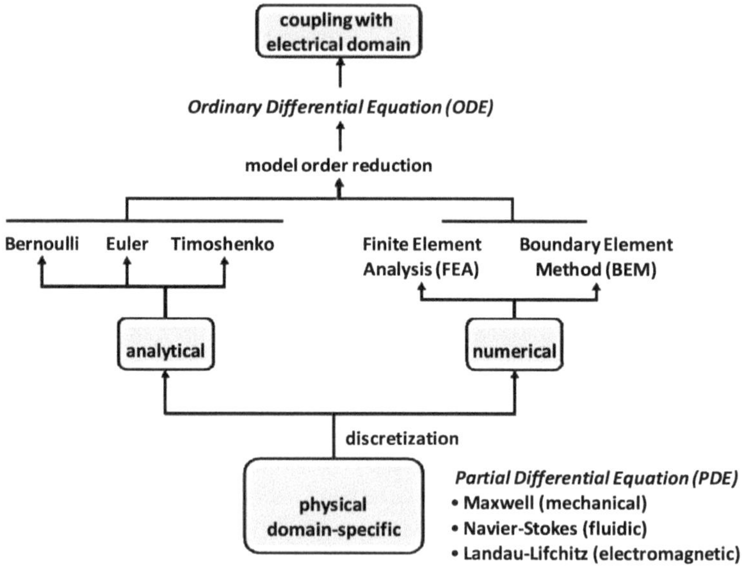

Fig. 12.3 MEMS modeling methods

system can be described at the "behavioral" level (i.e. using a set of equations to describe the functional behavior of a device), or at a more detailed structural level. The system described can then be simulated, regardless of the physical domain and the abstraction level of its various components, using a multi- level and multi-language design strategy. Nevertheless, speed-accuracy trade-offs still exist in the choice of abstraction level: broadly speaking, behavioral levels describe global behavior with analytical equations and target speed at the expense of accuracy, while structural levels describe individual behavior and improve the accuracy of component interaction simulation, while sacrificing speed [4].

Conventional physics-based design flows proceed via an extraction from the physical level to the behavioral level via multiple runs of finite element or boundary element solvers. These solvers are algorithms or heuristics that solve Partial Differential Equations (PDE). Most simulators propose different solvers, depending on the physical characteristics of the problem. In fact, the natural way to simulate sensors and actuators is to use numerical solvers based on finite element analysis (FEA), such as Ansys Multiphysics™ or Simulia Abaqus®. The FEA solver is then used to generate a model based on simplified Ordinary Differential Equations (ODE) that can be integrated into an electrical simulation environment (Fig. 12.3). Some commercial tools propose a solution to automatically generate the HDL code for mixed-signal design and development (e.g. Coventor Mems+® [5] based on Saber®, or SoftMems MemsXplorer [6] based on Verilog-A).

Hence hierarchical design management, using a relatively high-level description language such as Verilog-A, VHDL-AMS or the emerging SystemC-AMS, seems to be the only viable approach to describe and simulate multi-domain systems. Nevertheless, the fact remains that it is difficult to establish relations between two physical domains at the physical level, consequently hindering the determination of, for example, the geometrical constraints on a beam microstructure such that electrical circuits are guaranteed to behave correctly.

To address multi-physics design efficiently, a multi-level and multi-language strategy must be adopted. High-level modeling techniques capable of covering more physical domains should be developed, and multi-level methods and tools should aim to cover more abstraction levels. It is consequently clear that the impact of heterogeneity on design flows is or will be high, and necessary to facilitate heterogeneous device integration in SoC/SiP.

1.3 Proposed Methodology

As mentioned in the previous section, the concept of abstraction levels is key to addressing heterogeneous SoC/SiP design. However, a valid abstraction is difficult to achieve when tightly-coupled physical phenomena are present in the system. Efficient ways must be found to incorporate non-digital objects into design flows in order to ultimately achieve analog and mixed-signal (AMS)/radiofrequency/heterogeneous/digital hardware/software co-design.

While hierarchy in the digital domain is based on levels of signal abstraction, AMS and multi-physics hierarchy is based on a dual abstraction: the structural abstraction and the representational abstraction (Fig. 12.4).

The structural abstraction gives a way to describe the structural partitioning of a system, while the representational abstraction allows the description of each part of a system at different levels [7]. The method we propose is based on four nominal levels: Function, System, Block and Circuit. This capacity to represent any complex system with generic partitioning allows us to take advantage of different model descriptions with different abstraction levels and thus different languages and tools. A loose association between these levels and existing modeling languages can be established: Unified Modeling Language (UML) or SystemC for functional level, Matlab/Simulink for the system level, analog and mixed HDL languages (VHDL-AMS or Verilog-A) for block level, and netlist-based description (SPICE/Eldo/Spectre) for the circuit level. As shown in Fig. 12.4, structural decomposition can be represented by a set of transitions from one block to several (usually, but not necessarily, at the same abstraction level). For example, considering an analog-digital converter (ADC) composed of a digital and an analog part, the analog part can be further decomposed into a comparator and an integrator that are described at a different representational level. Obviously, some structures are not accessible at the functional level; this concerns for example the two stage representation of the integrator, and illustrates the non-representativity of strong physical coupling between blocks at this abstraction level. As a consequence, in a multi-physics design, a refinement process must be defined to update the system-level specifications of the components in each physical domain.

Figure 12.5 shows the representational abstraction levels for a heterogeneous structure. The specifications are defined at the system level in a top-down approach,

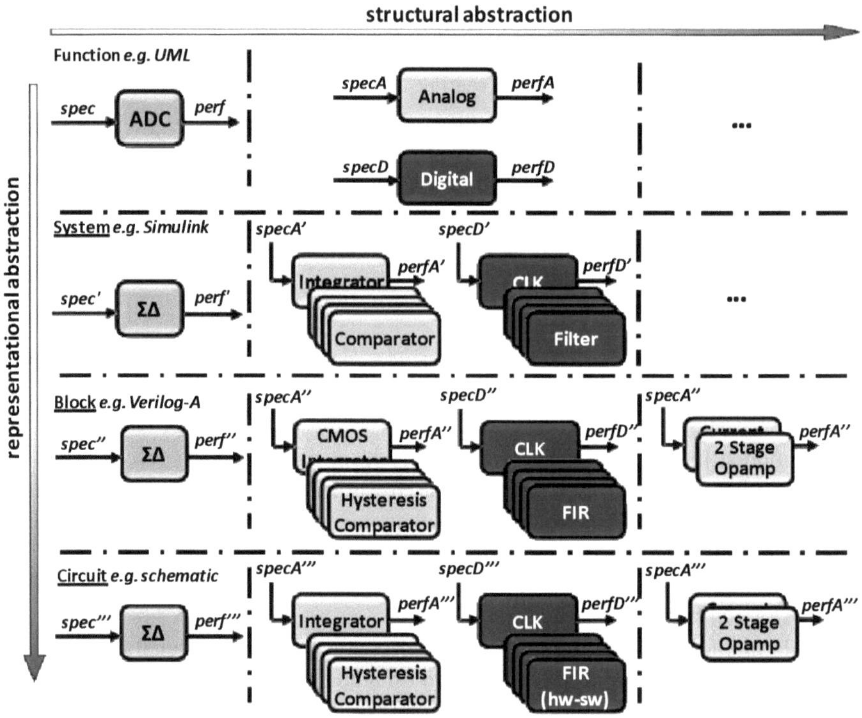

Fig. 12.4 Dual abstraction description of multi-physics systems

and are subsequently propagated through representational levels in each specific domain. To perform this propagation, model specifications at the higher level must be functions of the component model specifications of the lower level [8]. The next step is to perform synthesis using the high-level model for multi-domain simulation (see Sect. 1.2) with updated specifications, where the synthesis process is based on optimization techniques. The representation of a system using different abstraction levels allows us to explore tradeoffs between speed and accuracy, using a circuit-level description for some of the components and a behavioral description for others, carrying out (vertical) multi-level co-synthesis. For the behavioral abstraction level, the spectrum of powerful hardware description languages allows cross-domain (transverse) co-simulation. This multi-directional (transverse-vertical) approach allows us, through refinement at each abstraction level, to propagate and update specifications from different domains. This builds a clear bridge between the system-level and physical-level (or domain-specific) phases of design—in our view, this bridge is critical to setting up a continuum of abstraction levels between system and physical design, enabling detailed analysis of cross-domain tradeoffs and the correct balancing of constraints over the various domains, and hence the achievement of optimally designed heterogeneous systems.

This approach enables the clarification of the available/necessary steps in the design process. Hierarchical representation, as well as the use of a multi-level/multi-

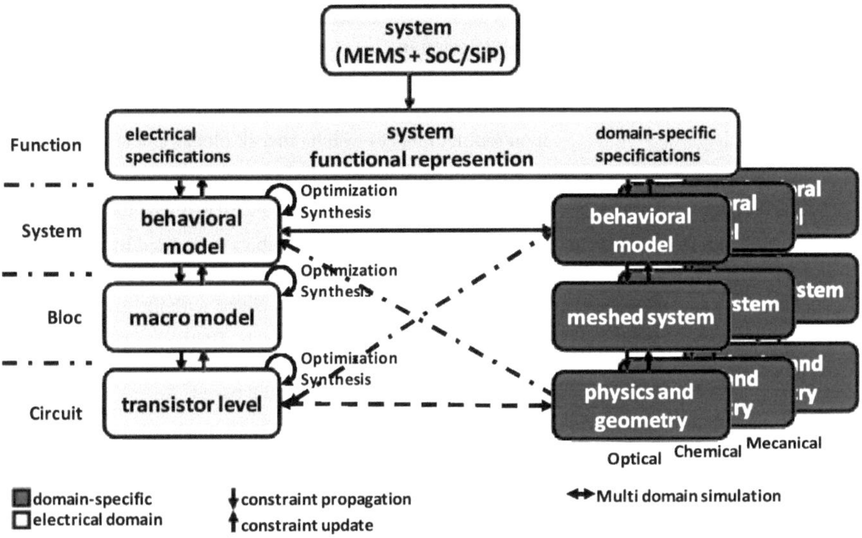

Fig. 12.5 Modeling abstraction and structural hierarchies

language approach allows the handling of heterogeneous system complexity. The synthesis process at each abstraction level is based on optimization, to automate and give predictive information on the system feasibility. The optimization methods are discussed in the next part.

1.4 *Multi-level and Multi-objective Optimization of Synthesizable Heterogeneous IP*

In this section, we will examine how to take advantage of the top-down synthesis method, associated with multi-objective optimization techniques. The main advantage of the top-down methodology is the ability to define the specifications from the system level down to the sub blocks. Thus, we need a high-level partitioning of the system that will bring the complex heterogeneous synthesis problem into the design of a specific domain component. Interaction between blocks and domains is managed through the definition of constraints and their propagation. In particular, in the design of heterogeneous structures, one of the most challenging tasks is to provide AMS IP that can be reused. Indeed, most analog and RF circuits are still designed manually today, resulting in long design cycles and increasingly apparent bottlenecks in the overall design process [9]. This explains the growing awareness in industry that the advent of AMS synthesis and optimization tools is a necessary step to increase design productivity by assisting or even automating the AMS design process. The fundamental goal of AMS synthesis is to generate quickly a first-time-correct sized circuit schematic from a set of circuit specifications. This is critical

Table 12.1 AMS IP block facets

Step	Property	Short description
1	Function definition	Class of functions to which the IP block belongs
	Terminals	Input/output links to which other IP blocks can connect
	Model	Internal description of the IP block at a given abstraction level
	Performance criteria set *S*	Quantities necessary to specify and to evaluate the IP block
	Design variable set *V*	List of independent design variables to be used by a design method or optimization algorithm
	Physical parameter set *P*	List of physical parameters associated with the given model
2	Synthesis method $*m$	Code defining how to synthesize the IP block, i.e. transform performance criteria requirements to design variable values. Can be procedure- or optimization based
3	Evaluation method $*e$	Code defining how to evaluate the IP block, i.e. transform physical parameter values to performance criteria values. Can be equation- or simulation-based (the latter requires a performance extraction method)
4	Performance extraction method	Code defining how to extract performance criteria values from simulation results (simulation-based evaluation methods only)
	Specification propagation method $*c$	Code defining how to transform IP block parameters to specifications at a lower hierarchical level

since the AMS design problem is typically under-constrained with many degrees of freedom and with many interdependent (and often conflicting) performance requirements to be taken into account. Synthesizable (soft) AMS Intellectual Property (IP) [10] extends the concept of digital and software IP to the AMS domain. It is difficult to achieve because the IP hardening process (moving from a technology-independent, structure-independent specification to a qualified layout of an AMS block) relies to a large extent on the knowledge of a designer. It is thus clear that the first step to provide a route to automated system-level synthesis incorporating AMS components is to provide a clear definition. Table 12.1 summarizes the main facets and operations necessary to AMS and heterogeneous IP synthesis. These various facets allow us to distinguish four main steps, and groups of properties. The first consists of the definition and configuration of the IP synthesis problem, while the second concerns solving the formulated problem using either procedural or optimization techniques. The third is the evaluation step which allows the determination of the values of the performance criteria during the synthesis process, and finally the last step consists of propagating the specifications to the next structural level.

An illustration of these steps, brought together in an iterative single-level synthesis loop, is shown in Fig. 12.6. Firstly, the set *S* of performance criteria is used to quantify how the IP block should carry out the defined function. The performance criteria are meaningful measurements with target values composed of functional specifications and performance specifications: for example in an amplifier, *S* will contain gain (the single functional specification), bandwidth, power supply rejection ratio (PSRR), offset, etc. They have two distinct roles, related to the state of the IP block in the design process:

- as block parameters when the IP block is a component of a larger block, higher up in the hierarchy, in the process of being designed. In this case it can be varied and must be constrained to evolve within a given design space, i.e. $s_{\text{low}\_i} < s_i < s_{\text{high}\_i}$;
- as specifications when the IP block is the block in the process of being designed (such as here). In this case the value s_i is a fixed target and will be used to drive the design process through comparison with real performance values s_{ri}.

Thus, specifications are used to write a cost function to formalize the objectives of the design: for the previous example it could be to maximize the gain and bandwidth, while minimizing the area, power and noise. A typical (most common) example of formulation is the normalized squared weighted sum function ε, where n is the size of S, w_i the weight ($w_i \forall i \in \{0, n-1\}$) subject to specification type (cost, condition, etc.):

$$\varepsilon = \sum_{i=0}^{n-1} w_i \left(\frac{s_i - s_{ri}}{s_i} \right)^2$$

The objective function is of great importance, as it must formulate the exact needs of the designer and it must be able to provide all optimal Pareto points (Pareto points represent the best tradeoffs between concurrent performances). Other function types exist and can be used to address a given problem efficiently [11]. This kind of function is at the heart of the multi-objective optimization methods. Indeed, the function established represents the need to achieve several potentially conflicting performance criteria. This function has to be minimized under constraints to solve a multi-objective optimization problem of the form:

$$\begin{aligned} \min_x \; & [\mu_1(x), \mu_1(x), \ldots, \mu_n(x)]^T \\ \text{s.t.} \; & g(x) \leq 0 \\ & h(x) = 0 \\ & x_l \leq x \leq x_u \end{aligned}$$

where μ_i is the i-th objective function, g and h are the inequality and equality constraints, respectively, and x is the vector of optimization or decision variables.

The synthesis method $^*m$ describes the route to determine design variable values. It is possible to achieve this in two main ways:

- through a direct procedure definition, if the design problem has sufficient constraints to enable the definition of an explicit solution;
- through an iterative optimization algorithm. If the optimization process cannot, as is usually the case, be described directly in the language used to describe the IP block then a communication model must be set up between the optimizer and the evaluation method. A direct communication model gives complete control to the optimization process, while an inverse communication model uses an external

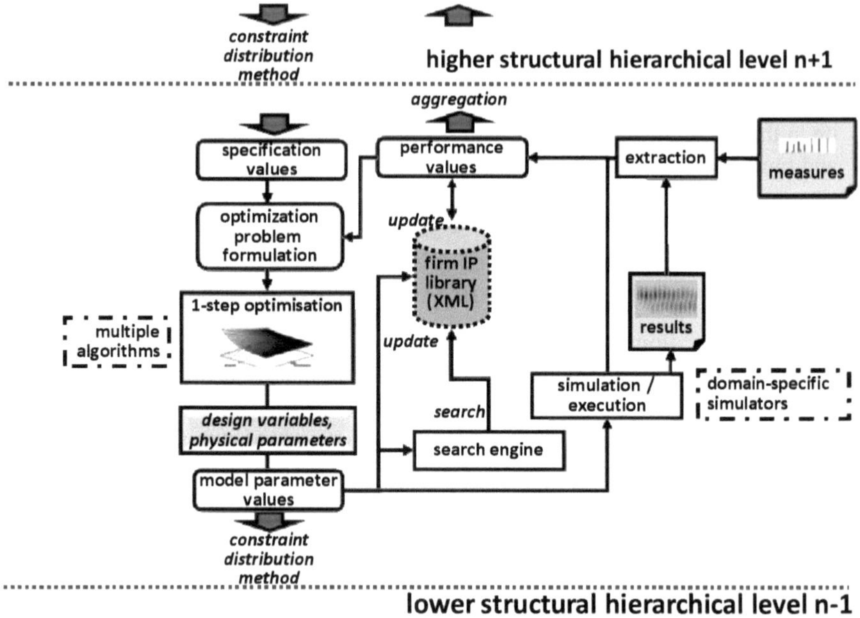

Fig. 12.6 Single-level AMS synthesis loop showing the context of AMS IP facet use

process to control data flow and synchronization between optimization and evaluation. The latter model is less efficient but makes it easier to retain tight control over the synthesis process. The synthesis method then generates a new set V of combinations of design variables as exploratory points in the design space according to $^*m : S \rightarrow V$. The number of design variables, which must be independent, defines the number of dimensions of the design space.

The evaluation method $^*e$ describes the route from the physical variable values to the performance criteria values such that $^*e : P \rightarrow S$. This completes the iterative single-level optimization loop. Evaluation can be achieved in two main ways:

- through direct code evaluation, such as for geometric area calculations;
- through simulation (including behavioral simulation) for accurate performance evaluation (gain, bandwidth, distortion, etc.). If the IP block is not described in a modeling language that can be understood by a simulator, then this requires a gateway to a specific simulator and to a jig corresponding to the IP block itself. For the simulator, this requires a definition of how the simulation process will be controlled (part of the aforementioned communication model). For the jig, this requires transmission of physical variables as parameters, and extraction of performance criteria from the simulator-specific results file. The latter describes the role of the parameter extraction method, which is necessary to define how the design process moves up the hierarchical levels during bottom-up verification phases.

Once the single-level loop has converged, the constraint distribution method $^*c$ defines how the design process moves down the hierarchical levels during top-down design phases. At the end of the synthesis process at a given hierarchical level, an IP block will be defined by a set of physical variable values, some of which are parameters of an IP sub-block. To continue the design process, the IP sub-block will become an IP block to be designed and it is necessary to transform the block parameters into specifications according to $^*c : P_k \rightarrow S_{k+1}$ (where k represents the structural hierarchy level). This requires a definition of how each specification will contribute to the cost function ε for the synthesis method in the new block.

This description gives the general framework of our multi-level and multi-objective optimization method. It is based on the hierarchical management of a complex system to distribute the synthesis process. The synthesis is then performed with optimization methods that can be combined with several evaluation procedures. It is implemented in a Java-based application called Rune.

2 Rune, a Framework for Heterogeneous Design

The Rune framework aims at researching novel design methods capable of contributing to the management of the increasing complexity of the heterogeneous SoC-SiP design process due to growth in both silicon complexity and in system complexity. Current design technology is at its limits and is in particular incapable of allowing any exploration of high- and low-level design tradeoffs in systems comprising digital hardware/software components and multi-physics devices (e.g. instruction line or software arguments against sensor or device characteristics). This functionality is required to design (for example) systems in which power consumption, temperature issues and, with the advent of 3D integration, vertical communication cost, are critical.

2.1 Main Objectives of the Framework

The ultimate overall goal of the platform is to enable the concurrent handling of hardware/software and multi-physics components in architectural exploration. Specifically, the objectives include:

- the development of hierarchical synthesis and top-down exploration methods, coherent with the design process model mentioned above, for SoC-SiP comprising multiple levels of abstraction and physical domains. Synthesis information for AMS components is formalized and added to behavioral models as a basis for synthesizable AMS IP. Developed tools exploit this information and are intended to guarantee the transformation of the system specifications into a feasible set of components specified at a lower (more physical) hierarchical level. Since multiple levels of structural abstraction are implied in the process, it is necessary to clearly specify bridges between the levels (through performance-based partitioning and synthesis). Technology-independence is a key point for the establishment of a generic approach, and makes it possible to generate predictive information when the approach is coupled with device models at future technology nodes.

- the definition and development of a coherent design process for heterogeneous SoC-SiP, capable of effectively managing the whole of the heterogeneous design stages—through multiple domains and abstraction levels. A primary objective is to make clear definitions of the levels of abstraction, the associated design and modeling languages and the properties of the objects at each level, whatever their nature (software components, digital/AMS/RF/multi-physics hardware). This makes it possible to establish the logistics of the design process, in particular for actions that could be carried out in parallel, and to take a first step towards a truly holistic design flow including economic and contextual constraints.
- the heterogeneous specification of the system by high-level modeling and co-simulation approaches to allow the analysis of design criteria early in the design cycle.
- the extension of current hardware/software partitioning processes to non-digital hardware. Methods to formalize power, noise, silicon real estate and uncertainty estimation in AMS and multi-physics components need to be developed, thus allowing the estimation of feasibility as critical information for the partitioning process. Although this information is intrinsically related to the implementation technology, efforts need to be made to render the formulation of information as qualitative as possible (thus circumventing the need to handle, in the early stages of the design process, the necessary numerical transposition to the technology). This formulation is employed to enrich the high-level models in the system.
- the validation of design choices using model extraction and co-simulation techniques. This relates to a bottom-up design process and requires model order reduction techniques for the modeling of non-electronic components (including the management of process and environmental variability), as well as the abstraction of time at the system level. This opens the way to the development of formal verification methods for AMS to supplement the design flow for "More than Moore" systems.

These concepts are at the heart of our vision of a high-level design flow embodied in an experimental design framework for heterogeneous SoC-SiP.

2.2 Rune Key Features

Rune is an existing in-house AMS synthesis framework. As shown in Fig. 12.7, the main inputs are the hierarchical description of the system and associated system level performances. From the user's point of view, there are two main phases leading to the synthesis of an IP block:

1. definition of AMS *soft-IP*, described in the Extended Markup Language (XML) format (directly into an XML file or through the graphical user interface, GUI).

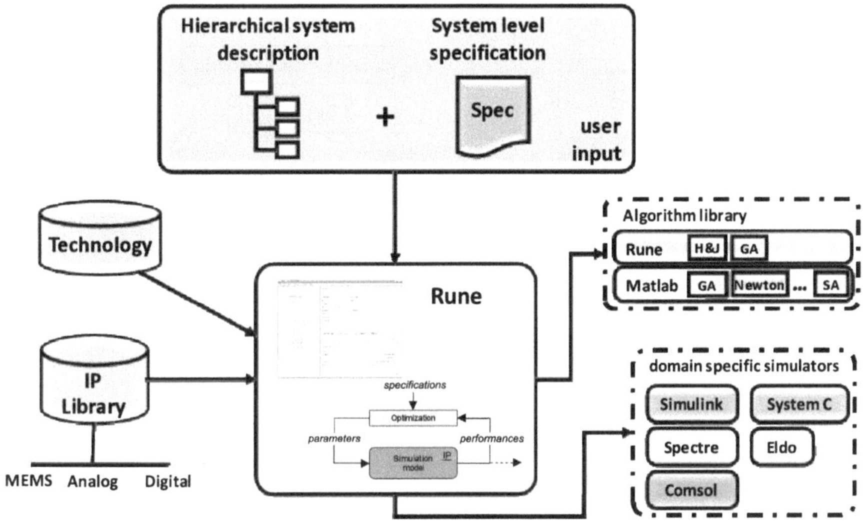

Fig. 12.7 Rune block diagram functions

In this step, all information related to the system must be provided (hierarchy, models, variables, performances specifications, etc.).

2. configuration of the AMS *firm-IP* synthesis method. In this step, the user must define an optimization strategy, i.e. a numerical method or algorithm and the formulation of the problem according to the specifications.

As explained in the previous section, the hierarchical description of the system is key to heterogeneous synthesis. In Rune, different kinds of models describing the whole or part of the system at a given representational abstraction level can be entered. These models are stored in a database allowing each soft-IP to be used as part of a system. Also, in order to evaluate the performance of these domain-specific models, a simulation Application Programming Interface (API) has been developed in order to plug in several external simulators. In this way, the user can select the external simulators to use in the specification evaluation phase.

At the system level, in order to enable the satisfactory partitioning of system-level performance constraints among the various digital, software and AMS blocks in the system architecture, top-down synthesis functionality needs to be added. This can actually be done by providing models at a given abstraction (structural) level with parameters corresponding to specifications of blocks of the lower level. With such models, optimization at the system level allows the balancing of specifications on each sub-block, such that the optimization of each individual block is guided to correspond to an optimization of the system. The goal of this approach is to enable accurate prediction of AMS architectural specification values for block-level synthesis in an optimal top-down approach by making reasoned architectural choices about the structure to be designed. Having established how we have applied the hierarchical management of heterogeneous system, we can see how it is used in the optimization process [12].

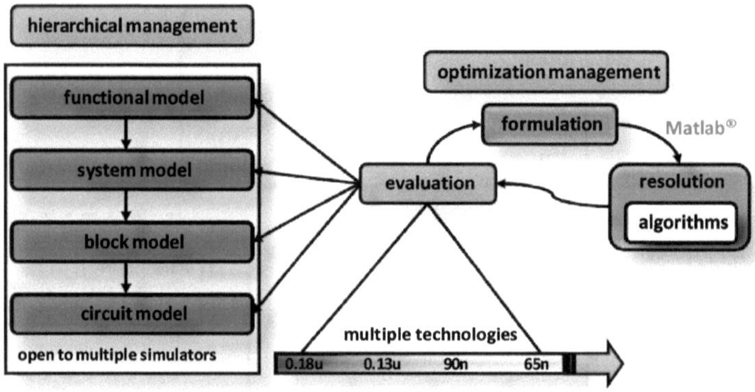

Fig. 12.8 Rune optimization steps

The optimization process can be used at each abstraction level and for every structural (sub-) component. Three main steps are followed:

- a cost function is formulated from specifications and design parameters set and stored in XML files.
- a design plan is set to define which optimization algorithms will be used to perform synthesis.
- a model at a given abstraction level for each specification must be defined for the performance evaluation during optimization process.

From the set of information provided by the designer, a multi-objective optimization problem is automatically formulated and run (see Fig. 12.8). This is the formulation step, which consists of defining the objectives and the constraints of the problem, as well as the variables and parameters, their ranges and initial values. The implementation of this step is set up to use either Matlab® or an algorithm directly implemented in Rune. The evaluation method called during the optimization process can use a model from any abstraction level, since Rune can call various simulators to perform an evaluation through its standard API. For example in the electrical domain, a given block can be described at circuit level (schematic representation) and its performance metrics can be evaluated with electrical simulation tools such as Spectre or Eldo, with various target technologies.

The ability to use different models and tools, and to manage heterogeneity, plays an important role in the definition of multi-physics design, as will be seen in the following section describing an example application.

2.3 Active Pixel Sensor Application

Rune has been used to explore integrated active pixel sensor (APS) design tradeoffs, both (i) to automatically size circuits according to image sensor specifications and

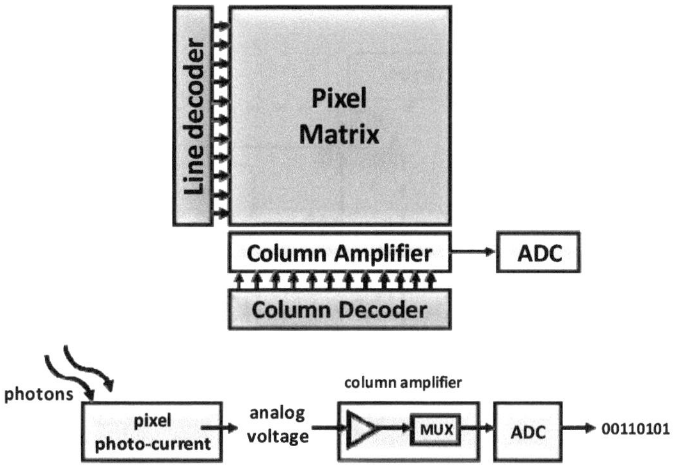

Fig. 12.9 Conventional CMOS imager structure

technology characteristics, thus enabling a complete sizing of the APS, and (ii) to explore the impact of physical device characteristics on photodiode performance metrics, thus leading to the quantitative identification of bottlenecks at the device level. Due to the very diverse nature of the exploration space variables, and the level of detail required in the investigations and analyses, this work could only be carried out using an automated and predictive simulation-based synthesis approach. In this section, we will describe how the Rune synthesis flow was applied to this design problem. This consists of the establishment of models required for the simulation and synthesis of the pixel sensor, the top-down specification- and technology-driven synthesis method; and the definition of the performance metrics and specification sets to be used in the investigation program.

2.3.1 Models for the Simulation and Synthesis of an APS

Most CMOS image sensors dedicated to consumer applications (such as cell phone cameras or webcams) require roughly the same characteristics. The conventional architecture, shown in Fig. 12.9, consists of (i) a pixel sensor matrix, (ii) decoders to select lines and columns, (iii) readout circuit consisting of a column amplifier (with correlated double sampling (CDS) correction), and (iv) an ADC. The luminous flux is measured via the pixel sensor which converts the photo-generated current into a voltage, and subsequently transferred to the column readout circuit and ultimately to the ADC (see conversion chain Fig. 12.9). To extract the data from each pixel sensor, every pixel sensor integrates the photocurrent either at the same time (global shutter imaging) or line by line (rolling shutter imaging).

This short description allows us to highlight that optimized pixel sensor design is critical to a high performance image sensor. Indeed, the smaller the pixel sensor, the higher the resolution, and consequently image quality, for a given circuit size. The

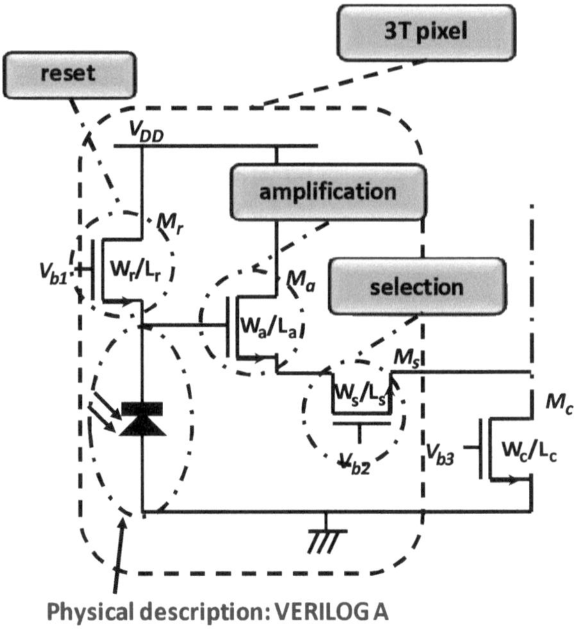

Fig. 12.10 Conventional CMOS 3T pixel sensor structure

trade-off here is of course that the signal to noise ratio of the complete signal acquisition chain must be maintained, while the luminous flux is reduced proportionally to the photodiode size (assuming constant illuminance).

There are many types of active pixel sensors, and one of the most used architectures in the design of CMOS image sensors is based on a three-transistor (3T) pixel sensor design. A typical 3T pixel sensor consists of a photodiode (PD), Reset Gate (RG), Row Select (RS), and source follower, as shown in Fig. 12.10.

The heterogeneous nature of this structure means that the determination of good tradeoffs between area and other performance metrics requires the management of variables from several physical domains. Indeed, to extract meaningful physical data from analyses where advanced CMOS technologies are involved and accurate device models are key to the relevance of investigation, it is essential to work towards design technology including the simulation of a complete pixel sensor in an Electronic Design Automation (EDA) framework. A direct consequence of this is that it is necessary to develop behavioral models for the optoelectronics devices for concurrent simulation with the transistor-level interface circuit schematics. For all behavioral models, the choice of an appropriate level of description is prerequisite to developing and using the models in the required context. In this work, we consider the system level to be composed of the whole imager structure, which we can split according to the conversion chain of Fig. 12.9 into three main smaller blocks: the pixel sensor, the column amplifier and the ADC (the digital part is not discussed). To focus on the multi-physics aspects, we will consider the two former elements for optimization, i.e. the pixel sensor structure and the column amplifier.

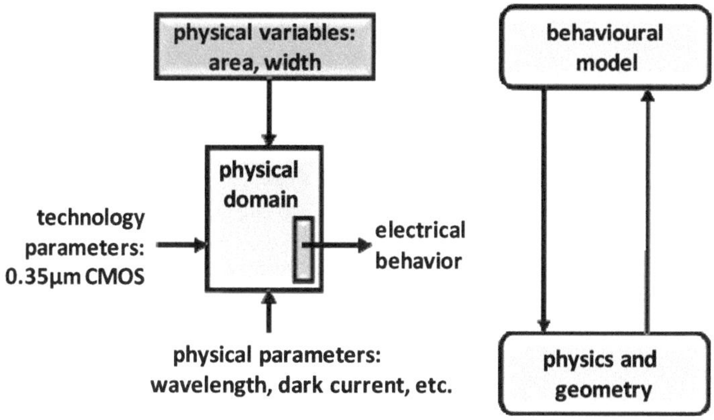

Fig. 12.11 Photodiode model (VerilogA)

2.3.2 Pixel Sensor Model and Specifications for Automated Synthesis

In order to model the physical behavior of the photodiode, and to take into account the strong coupling between the electrical elements (i.e. transistors) and the photodiode, the Verilog-A language has been used. This model describes the behavior using variables that belong to both optical and electrical domains, without defining the structure of the device. The device can thus be parameterized depending on the target specifications, and cross-domain variables can be changed to model a given interaction between the optical and electrical domains. Figure 12.11 illustrates the fixed parameters, related to the target technology (0.35 µm CMOS in this example), and the optical characteristics such as light wavelength and dark current. The active area and the depth of the depletion zone of the photodiode represent the cross-domain variables. It is important to bear in mind that this high-level model should be linked to a more detailed physical description to refine the behavioral model according to physical simulations. Conversely, a physically detailed model of the photodiode can be used to refine the specifications at a higher abstraction level.

This model allows us to design the pixel sensor transistor size, taking into account the effect of the physical dimension of the photodiode. Table 12.2 presents the main specifications of the 3T pixel sensor.

In a full CMOS imager system design flow, these specifications would be inherited from the system-level description. For example the readout time of the circuit, generally limited by the decoders and/or the ADC [13] in the overall CMOS imager, would lead to a highly relaxed pixel readout speed specification. In this work there is no such dependency, since we only consider the active pixel sensor and the column amplifier, and we have chosen to apply a more stringent set of specifications than is generally necessary to demonstrate our approach.

Table 12.2 CMOS 3T pixel sensor specifications

Technology	0.35 μm CMOS
Supply voltage	3.3 V
Fill factor	>0.65
Area	<10 μm^2
Photodiode discharge slope	>200 kV/s
Amplifier input voltage @ end of read	>0.45 V
IR drop reset	–
IR drop select	–
Power	<8 μA
Responsivity	<0.3 A/V

2.3.3 Optimization Based Pixel Sensor Synthesis Using Rune

Setting the pixel sensor synthesis problem has been done through the Rune platform interface (shown in Fig. 12.12) and can be broadly divided into three main steps:

1. The first step consists of **identifying the problem variables**, thereby fixing the number of degrees of freedom. All transistor dimensions are taken as variables. The area and depletion depth are the cross-domain variables for the photodiode.
2. The second step concerns the **identification of the specifications or performance metrics**. In this step, the designer must obviously identify meaningful specifications for which a trade-off has to be found, but also define a means by which to extract this information automatically from the simulation results. Pattern files are provided in Rune in order to help the designer to write scripts in the appropriate language, depending on the simulator used (Spectre or Eldo). The performance values can be directly extracted from simulation results or can also be the results of an arithmetic operation defined in the XML description file. This allows us to fix the specifications on the fill factor of the photodiode, and to extract the slopes of the signals to find a trade-off between the area and speed of the photodiode.
3. The last step consists of **setting up the optimization**. A mathematical formulation of the optimization problem is carried out automatically within Rune; the designer has merely to select between several available algorithms such as Hooke and Jeeves, Newton-based, genetic algorithm, pattern search and simulated annealing. In this application, we adopted a hybrid approach [12] consisting of combining (i) a global search algorithm, to map out a rough evaluation of the design space and provide a good starting point, with (ii) a direct search technique that can converge quickly to a solution. This combination naturally leads to a good solution with reasonable computational cost.

Synthesis of the 3T pixel sensor has been performed, targeting a 0.35 μm CMOS technology. The results obtained are summarized in Table 12.3. The formulated problem has 10 variables and is solved using the Newton-based solver from Matlab (fmincon), after 32 iterations (16 minutes) with a 65% fill factor and an area of 4.5 μm^2.

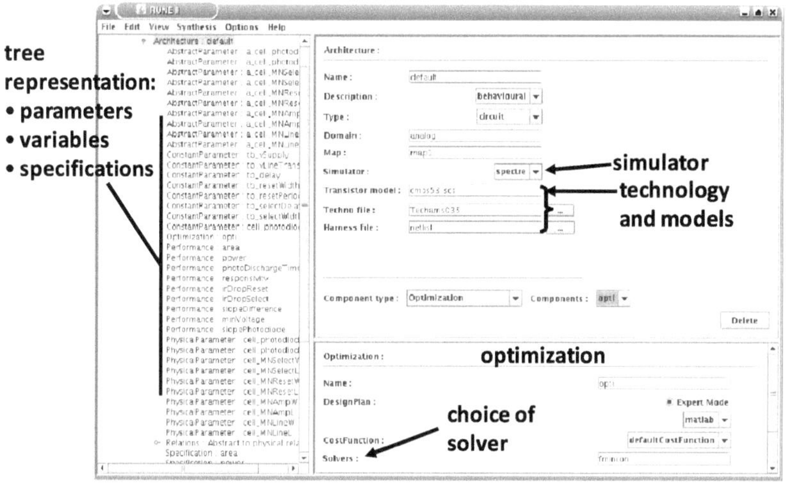

Fig. 12.12 Setting up the pixel sensor design problem with Rune

Table 12.3 CMOS 3T pixel sensor synthesis results

Technology	0.35 μm CMOS	
Supply voltage	3.3 V	
Fill factor	>0.65	0.65
Area	<10 μm^2	4.5 μm^2
Photodiode discharge slope	>200 kV/s	200 kV/s
Amplifier input @ end of select	>0.45 V	0.9 V
IR drop reset	–	1.4 V
IR drop select	–	53 mV
Power	<8 μA	1.2 μA
Responsivity	<0.3 A/V	0.3 A/V

The shape of the slopes (shown in Fig. 12.13) can then be displayed using a post processing script developed for the Spectre® simulator. Furthermore, a major feature of the Rune framework is the ability to reconfigure a synthesis problem formulation easily (i.e. modifying variables, constraints, objectives and/or solver) such that one can evaluate the pixel behavior in different contexts (e.g. a new target technology process or a new target performance).

3 Conclusion

In this chapter, we have outlined our vision for the synthesis of multi-physics components in the context of complex systems on chip design methods. We believe that hierarchical design management, with a coherent multi-level and multi-language strategy, is the only viable approach to achieve this goal. A key observation is

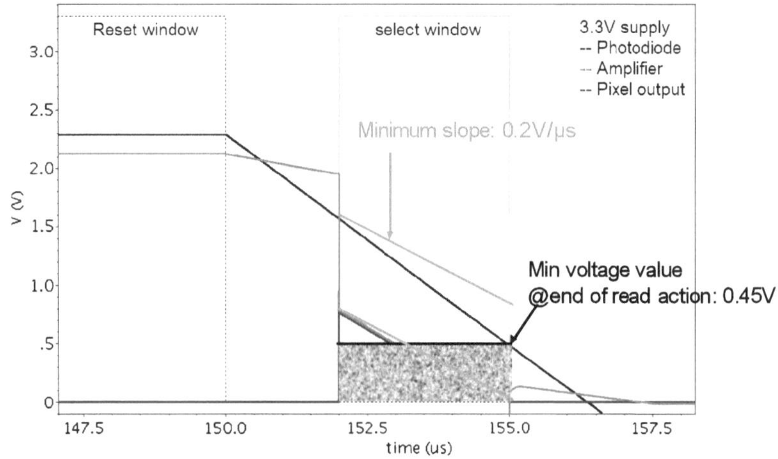

Fig. 12.13 Simulation of the synthesized 3T pixel sensor (specification window overlaid)

the multiple nature of abstraction, taking into account both structural abstraction and representational abstraction. Exploiting such abstraction, a top-down synthesis method, associated with multi-objective optimization techniques, enables a high-level partitioning of the system that brings the complex heterogeneous synthesis problem into the design scope of specific domains, which can then be processed individually. The interaction between blocks and domains is managed through the definition of constraints and their propagation.

We have described our implementation of these concepts in the Rune framework, an ongoing work aimed at researching novel design methods for complex integrated heterogeneous systems. We used Rune to explore integrated active pixel sensor (APS) design tradeoffs which, due to the very diverse nature of the exploration space variables, and the level of detail required in the investigations and analyses, could only be achieved using an automated and hierarchical synthesis approach. In the example, the aim was to determine the width and area of the photodiode to find a trade-off between the fill factor (depending on the area of each component) and response speed (represented by the discharge slope). The aim of the example application was to demonstrate the capabilities of multi-physics optimization through abstraction and refinement. By describing a system including blocks from several physical domains at a high abstraction level, this method enabled multi-domain design. Moreover, it allowed the formulation of an optimization problem including cross-domain variables, and enabled the exploration of trade-offs at an early design stage.

References

1. International Technology Roadmap for Semiconductors. http://www.itrs.net/

2. Nikitin, P.V., Shi, C.R.: VHDL-AMS based modeling and simulation of mixed-technology microsystems: a tutorial. Integr. VLSI J. **40**(3), 261–273 (2007)
3. McCorquodale, M.S., Gebara, F.H., Kraver, K.L., Marsman, E.D., Senger, R.M., Brown, R.B.: A top-down microsystems design methodology and associated challenges. In: Design, Automation and Test in Europe Conference and Exhibition, pp. 292–296 (2003) (suppl.)
4. Levitan, S.P., Martinez, J.A., Kurzweg, T.P., Davare, A.J., Kahrs, M., Bails, M., Chiarulli, D.M.: System simulation of mixed-signal multi-domain microsystems with piecewise linear models. IEEE Trans. Comput.-Aided Des. Integr. Circuits Syst. **22**(2), 139–154 (2003)
5. Coventor. http://www.coventor.com/
6. Softmems. http://www.softmems.com/
7. O'Connor, I., Tissafi-Drissi, F., Revy, G., Gaffiot, F.: UML/XML-based approach to hierarchical AMS synthesis. In: Vachoux, A. (ed.) Advances in Specification and Design Languages for SoCs. Kluwer Academic, Dordrecht (2006)
8. Fellah, Y., Labrak, L., Abouchi, N., Tixier, T., Condemine, C.: Synthèse automatique de convertisseur analogique numérique de type sigma delta $\Sigma\Delta$. In: 4th Int. Conf. Science of Electronics, Information Technologies and Telecommunications (SETIT 2007), Hammamet, Tunisia, 25–29 March 2007, pp. 277–282 (2007). ISBN: 978-9973-61-475-9
9. Gielen, G., Dehaene, W.: Analog and digital circuit design in 65 nm CMOS: end of the road. In: Proceedings of Design, Automation and Test in Europe, 7–11 March 2005, vol. 1, pp. 37–42 (2005)
10. Hamour, M., Saleh, R., Mirabbasi, S., Ivanov, A.: Analog IP design flow for SoC applications. In: Proceedings of the 2003 International Symposium on Circuits and Systems, ISCAS '03, 25–28 May 2003, vol. 4, pp. IV-676–IV-679 (2003)
11. Marler, R.T., Arora, S.T.: Survey of multi-objective optimization methods for engineering. Struct. Multidiscip. Optim. **26**(6), 369–395 (2004)
12. Labrak, L., Tixier, T., Fellah, Y., Abouchi, N.: A Hybrid approach for analog design optimization. In: 50th Midwest Symposium on Circuits and Systems (MWSCAS 2007), Montreal-Quebec, 5–8 August 2007, pp. 718–721 (2007). ISBN 978-1-4244-1176-9
13. El Gamal, A., Eltoukhy, H.: CMOS image sensors. IEEE Circuits Devices Mag. **21**(3), 6–20 (2005)

Part II
Design Contexts

Chapter 13
Beyond Conventional CMOS Technology: Challenges for New Design Concepts

Costin Anghel and Amara Amara

1 Introduction

The International Technology Roadmap for Semiconductors (ITRS) evaluates the progress of technologies for Beyond CMOS era and judges their readiness for development [1]. Today the time horizon prospected is 2022—the end of CMOS scaling. ITRS 2007 makes clear distinction between memories and logic devices. The 2008 ITRS update shows only the candidate technologies chosen for evaluation. Six candidates are proposed for memory and logic circuits, together with two hybrid architectures. The last mentioned architectures, the CMOS/Molecular hybrid (CMOL) and the Field Programmable NanoWire Interconnect (FPNI) are designed to interconnect the nano-molecular structures and CMOS gates. From a broader perspective, the research activity in memory domain largely exceeds the research activity in the logic devices domain. This is also reflected in this chapter where new concepts that are placed beyond the conventional CMOS technologies are discussed. To sum up, one must remember that the promises of the new technologies are tremendous but so are the difficulties of these technologies.

2 Near Term Alternatives

This section discusses the near-term alternative considering a practical case—the Static Random Access Memory (SRAM). SRAM cells are high-speed memories usually used inside the processors. Today the SRAM is faster that the other technologies, but pays a prohibitive surface price. The 2007 ITRS roadmap and the 2008 update predict that the gap in terms of speed is going to widen between the SRAM and its counterparts [1]. With no potential replacement, the design and the

C. Anghel (✉) · A. Amara
Institut Superieur d'Electronique de Paris (ISEP), 21 rue d'Assas, 75270 Paris, France
e-mail: costin.anghel@isep.fr

G. Nicolescu et al. (eds.), *Design Technology for Heterogeneous Embedded Systems*,
DOI 10.1007/978-94-007-1125-9_13, © Springer Science+Business Media B.V. 2012

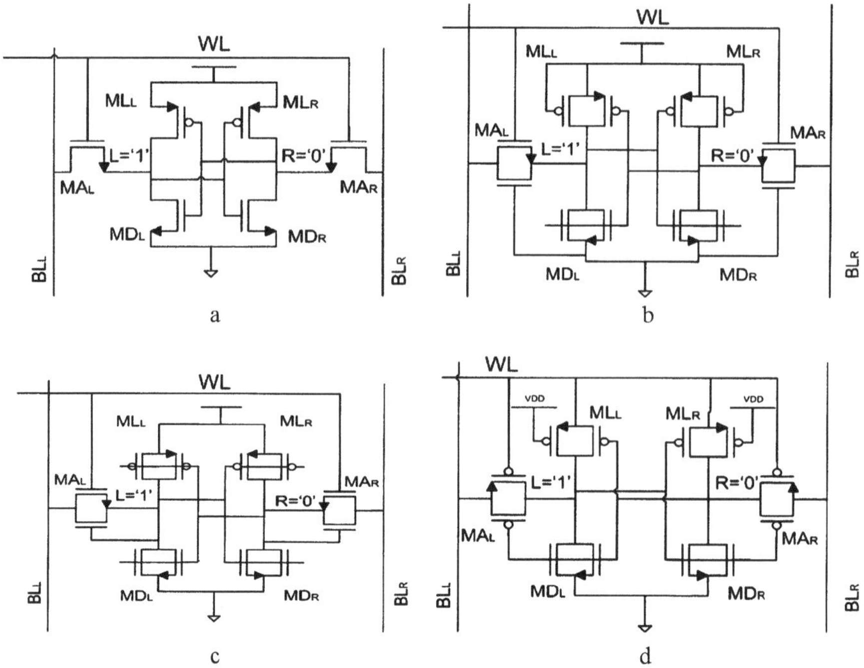

Fig. 13.1 Schematic representation of standard 6T SRAM cell (**a**); Hitachi DG static 6T cell (**b**); Berkeley dynamic DG 6T SRAM cell (**c**); and Hybrid cell (**d**)

improvement of SRAM memory remains of large interest for stand alone and embedded memory applications. The continuous size scaling imposes, beyond 45 nm node, the migration of the MOS transistor architecture from bulk towards double gate (DG) or multi-gate (MG) structures. This migration also leads to a change in the design of the circuits in particular of the SRAM cell. As the SRAM memory cell continues to be implemented using the same basic design structure (Fig. 13.1a), some research groups proposed to take advantage of the DG device and to redesign the cell in order to improve the cell ratio or the pull-up ratio, the signal to noise margin (SNM) or the read access time. Due to the lack of space, we present in the following only a small review of some of these designs. The extended overview proposed by Amara and Rozeau covers in detail all the design aspects for SRAM memory using the double-gate transistor and the advantages of each cell [2].

Among the first designs proposing the improvement of the SRAM, the Hitachi group has been oriented on a static approach, in which the access and the load transistors are weakened [3] (Fig. 13.1b). This improves the stability making the design adapted for low power applications. Nonetheless, the reduced read cell current limits the performance as the design space curves show in Fig. 13.2a.

Several research groups proposed dynamic approaches that improve the performance of SRAM. Figure 13.1c. presents the 6T Berkeley Pass-Gate Feed-Back

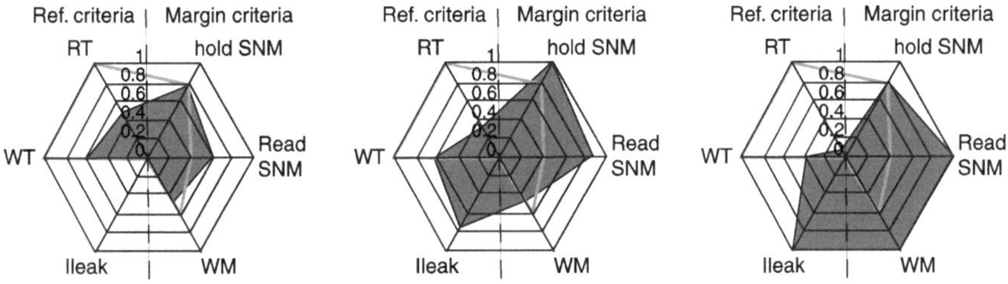

Fig. 13.2 (Color online) Design space curves for standard 6T cell (*red curve*) versus: (**a**) Hitachi cell; (**b**) Berkeley DG 6T SRAM cell; (**c**) Hybrid cell—*blue surface*; SNM—signal to noise margin, WM—write margin, Ileak—leakage current, WT—write time, RT—read time

cell [4]. This technique consists of connecting the second gate of the access transistors to the storage node. The cell shows a read SNM dramatically improved, without leakage or area penalty thanks to the feedback connection that controls the access transistor (Fig. 13.2b). A second approach that uses the dynamic configuration proposes a hybrid cell that consists of two cross-coupled inverters and two DG PMOS access transistors [5] (Fig. 13.1d). The write margin of this architecture is significantly improved in comparison to the other cells (Fig. 13.2c).

Figure 13.2 illustrates a comparison between the optimized SRAM cells and the reference 6T cell. The same transistor dimensions are used to design the different cells for this comparison. A predictive DG-MOSFET model based on ITRS forecast is used for Spice simulation.

The main conclusion highlighted by this section is that the design of the cell can be adapted as a function of the targeted application and by exploiting the increased flexibility offered by the new devices (i.e. the second gate for the DG MOSFETs).

This section presented some design aspects for SRAM memories that will be used as near-term solutions. The comparison with classical 6T SRAM structure highlighted that design adapted cells may improve the SNM margin in applications while keeping the area and leakage constant. Even if the 6T SRAM represents a "classical" design for the CMOS technology, it was shown in this section that the utilization of the new devices influences the topology of the cell and higher flexibility with respect to the cell performance is obtained.

3 Future Alternative—Devices Based on Nanotube, Nanowire and Graphene Nanoribbon

Carbon Nanotubes (CNTs), Nano-Wires (NWs) and Graphene Nano-Ribbons are one-dimensional or two-dimensional objects that successfully proved their ability to replace silicon (Si) devices in electronic applications. CNTs were discovered about twenty years ago by Iijima from NEC Corporation [6]. CNTs are regular structures of purely carbon atoms. Two types of CNTs can be encountered [7]: (i) Single Wall Nanotubes (SWNTs) and (ii) Multi Wall Nanotubes (MWNTs). SWNTs may have metallic or semiconductor properties depending on the chirality of the tube [8–10]. Nowadays, the three major processes (arc discharge, laser ablation and chemical vapor deposition) that are used to fabricate the SWNTs are not selective and mixtures of metallic and semiconductor CNTs are produced. The distribution of the SWNTs is given by the statistics as 1/3 metallic and 2/3 semiconductor. The separation between metallic and semiconductor CNTs is the major limiting factor for the CNTs applications in electronics, as only a few methods for CNTs sorting are known today [11–14]. From the charge transport point of view, the MWNTs are metallic. MWNTs found applications as interconnects and electrodes, because they are roughly ten times more conductive than copper and, do not suffer from electromigration at equivalent field [15–18].

Nano-Wires (NWs) are, like the CNTs, one dimensional nano objects, used for development of electronics applications. A large variety of semiconducting NWs exists ranging from amorphous to crystalline and from single element to compound semiconductors. These are obtained by several techniques like Epitaxial Growth, Chemical Vapor Deposition (CVD), etc. [19–22]. From the vast NWs family we distinguish Si NWs with the particular interest for the electronics applications, as they are compatible with the Si technology. A major limitation, common for NWs and CNTs based electronics is the variability issue, as these objects are, in many cases, deposited from solutions. Both CNTs and NWs are one-dimensional objects that do not provide, intrinsically, large regular structures. Therefore, methods that perform the alignment and control the surface distribution of the CNTs or NWs are needed in order to use these nano-objects for electronic applications. Recent progress in the research field demonstrate that NWs and CNTs devices are intrinsically adapted for flexible electronics applications [23–25] an unaffordable domain for bulk Si CMOS due to cost reasons.

In the last five years, Graphene Nano-Ribbons (GNRs) grabbed the attention of research community. Graphene represents only one plane of sp^2 carbon atoms from the 3D structure of graphite. The existence of Graphene was revealed by the School of Physics and Astronomy, University of Manchester [26]. Like the CNT-FETs, GNR-FETs exhibit transistor properties. However, unlike the NWs or the CNTs, GNRs are two-dimensional objects. In theory, Graphene Nano-Ribbons should circumvent the spread in the device characteristics of the CNT-FETs and NW-FETs due to the diameter variability. GNR-FETs should provide also the width scaling if increased drive currents are needed in applications. The ITRS 2008 update included Graphene as material for “Beyond CMOS information processing paradigms”, reflecting clearly the interest of electronics community for this material. Additionally, some groups demonstrated the epitaxial growth of Graphene on several substrates, narrowing the research-application gap [27–30]. It is, however, premature to claim that Graphene will take the lead in nano-electronics for Beyond CMOS applications and replace 1D nano-objects.

Carbon Nanotubes (CNTs), Nano-Wires (NWs) and Graphene Nano-Ribbons (GNRs) represent families of nano-objects that can be used to build transistors, and they may replace Si FETs for Beyond CMOS roadmap. All these nano-objects present their advantages when compared with classical bulk devices, and all of them

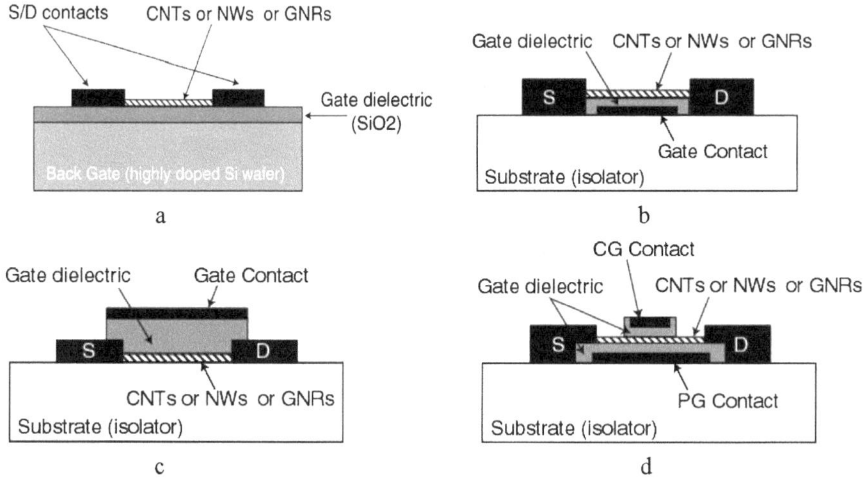

Fig. 13.3 Schematic representation of different device architectures: (**a**) Bottom gate wafer contact; (**b**) Bottom gate local contact; (**c**) Top gate local contact; (**d**) DGSB transistor

have to surpass some drawbacks in developing their specific technology. The first CNTs transistors were fabricated more than 10 years ago [31, 32] and a lot of research was performed in the field of nanotransistors and nanocircuits since that time. The fabricated transistors present high current densities mainly due to the high carrier mobility of these materials. Aggressive scaled devices using eBeam lithography show ballistic transport for channel lengths below 200 nm [33, 34]. However, due to the separation issue, mentioned at the beginning of this section, it is only recently that the CNTs circuit demonstrators appeared [23–25].

Many studies performed on CNTs, NWs and GNRs use a bottom gate back wafer contact structure (Fig. 13.3a). This structure, adapted for material characterization and performance estimation, is prohibitive for circuit applications as no individual gate is available. Independent gate structures were built, using both bottom gate configuration and top gate configuration (Fig. 13.3b and c). Each structure has its own advantage and can be used for specific, dedicated applications. The bottom contact structure is easily used as a sensor since only the lower part of the semiconductor is in contact with the dielectric, the rest being exposed to the environment. The top contact configuration is analogous to a gate all-around device and, therefore, is more adapted to the logic circuit applications.

An important difference between the FETs built using nano-objects and the classical Si MOSFETs is the operation principle. The CNT-FETs and some of the NW-FETs are Schottky barrier devices i.e. the current in the device is controlled by the injection of the charge through the Schottky barriers at the contacts [35, 36].

Some works in the literature took advantage of the special characteristics of the Schottky barrier FETs and presented several design deviations from the classical top or bottom contact architectures [37–40]. One structure referred as Double Gate Schottky-Barrier (DGSB), uses a polarity gate (PG) to control the device polar-

a b

Fig. 13.4 Schematic representation of the (**a**) Crossbar architecture and (**b**) one single memory element built at the intersection of two NWs

ity and a control gate (CG) to control de device operation. (Fig. 13.3d). PG influences the charges injection at the contacts by modulating the width of the Schottky barriers. CG modulates the current through the device being similar to the gate of a classical MOS. Three different biasing conditions are possible for DGSB: PG in negative—the transistor behaves as a p-type device, PG zero—the transistor is blocked and PG positive—the device behaves as a n-type device. With this design, two benefits can be noted: first, the control of the polarity offers the compatibility with the requirements for the circuit design, and, second, the polarity control at the device level expands the control of the circuit adding reconfigurable features (see Sect. 5).

The CNTs and NWs are also interesting for memory applications. One example of special design for the nanoelectronic memories is the Crossbar. The design of Crossbar consists in one plane of parallel NWs on which a second plane of parallel NWs is superposed. The direction of the NWs in the second plane is orthogonal with respect to the NWs in the first plane (Fig. 13.4a). Each intersection of a vertical and a horizontal wire forms a single cell element (Fig. 13.4b). The Crossbar memories theoretically offer very high density of cells per surface area [41–43].

The arrangement the NWs in parallel formation at nano level represents a difficulty in constructing the Crossbar matrices. Top-Down techniques, i.e. photolithography, or Bottom-Up techniques like fluidic assisted alignment [44, 45], or Langmuir-Blodgett [46, 47] overcome this problem. One example presented in Sect. 4 of this chapter shows that the feature size and function of the basic device can be defined during the NWs synthesis process when Bottom-Up techniques are used. Therefore, Bottom-Up techniques are extremely powerful, when used to design the systems, the main concern being the systems assembly and not the device functionalization. At the same time, the complexity is shifted from the upper functional level down-to the device level. Hence, Bottom-Up techniques offer more flexibility when compared to the classical lithographic processes.

In the following sections some possible applications of the nano-objects in electronics are discussed. The design aspects are considered taking as examples the future memory architectures and simple logic circuits.

4 Memory Alternatives

Many candidates are present in the race for the ideal memory that will replace the Dynamic Random Access Memory (DRAM). One major goal is to combine the compactness and the performance of DRAM with the non-volatility feature of Flash memory [1]. In this section we discuss the design of the memory elements that compete for Crossbar implementation.

Two types of Crossbar architectures can be found in the literature: passive [48–50] and active matrices [51, 52]. The passive matrices are built by the direct superposition of orthogonal NWs as shown in Fig. 13.4. In active matrices a transistor is used at each node to decouple the inactive storage cell. In contrast with the passive matrices, the active matrices do not suffer from crosstalk and disturbance in signals in the matrix. However, the integration of one transistor at each NW junction is a difficult solution especially for a self-aligned process at the nano level. From this perspective, passive matrices are intrinsically adapted for applications and offer the highest density of integration, $4F^2$, with F being the minimum feature size.

In passive matrices, the cells size is defined by the diameter of the NWs (or CNTs) and the way these 1-D objects are assembled [44, 53–57]. Tera-cells per square centimeter can be achieved if direct assembly [55] or advanced nanofabrication techniques [58–61] are used to scale the NWs (or CNTs) pitch down to 10 nm.

For a successful design, the individual memory elements have to provide special features when placed in the Crossbar architecture [62]. First, the element should be scalable, ideally down to, or comparable to the molecule size. Second, high ON/OFF current ratio is a prerequisite. This is needed to prevent signal degradation in the matrix. Third, reduced write energy is required for programming. Fourth, reduced programming time is desired. Ideally, the memory element should show some rectifying character (Fig. 13.4b), to reduce the crosstalk between the lines and prevent the signal loss [62, 63].

Keeping in mind that almost all future candidates in ITRS for the memory applications target the passive Crossbar configuration, all the constrains presented above show that a careful design is needed in large passive matrices. Three of the ITRS 2007–2008 candidates for beyond CMOS applications are analyzed in the following from a designer perspective. A broader review on the memory devices can be found in Chap. 18 of this book.

4.1 Passive Matrix—Implementations

Starting with the 00's new ways appeared to treat the information with new materials, outside the Si sphere [64–69]. The "Molecular Memory" [57] developed by Hewlett-Packard (HP) in collaboration with the University of California, Los Angeles (UCLA), represents an interesting case in line with the development produced in the last 10 years. The Rotaxane molecules form the memory element. The name comes from Latin words "Rota" that means wheel and "Axis" meaning axle. As the

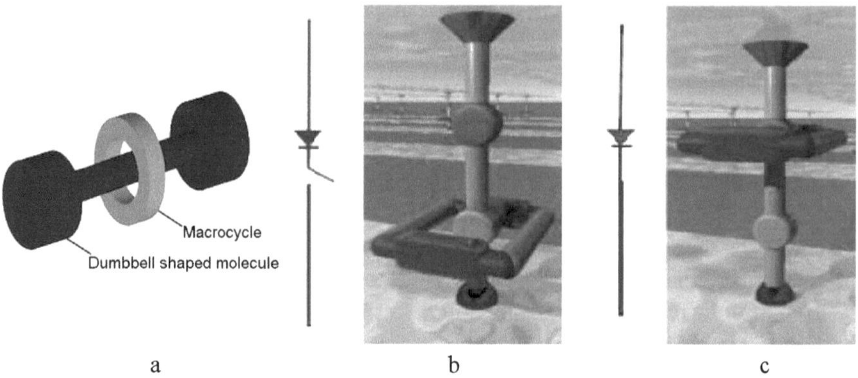

Fig. 13.5 Schematic representation of the Rotaxane molecular system (**a**). Illustration of the electrical switching mechanism with the corresponding displacement of the macrocycle along the axle: (**b**) OFF case, (**c**) ON case. Figures 13.5b and c reproduced with permission from [73]. Copyright Wiley-VCH Verlag GmbH & Co. KGaA

name suggests, Rotaxane is a molecular system composed by the wheel—a molecular macrocycle and the axle—a dumbbell shaped molecule (Fig. 13.5a). The ends of the dumbbell molecule act as stoppers preventing the macrocycle to escape. Rotaxane is interesting for applications as the macrocycle can be moved, in a controlled way, between two positions along the axle (Fig. 13.5b and c). The change of the position of the macrocycle corresponds to a change in the conductivity of the axle. The position of the macrocycle can be controlled electrically, chemically or optically [70–72].

In 2002 UCLA proposed the use of Rotaxane as two-dimensional electronic circuit, demonstrating both memory and logic functions [73]. The Rotaxane molecules were sandwiched between two titanium-platinum NWs, that serve as electrodes. This was only the first demonstrator with modest performances, but the authors foresaw already the interest of exploiting the properties of the Rotaxane in the Crossbar architecture.

In 2003 the group from UCLA joined forces with HP to extend the study of the Rotaxane system in the Crossbar architecture [57], taking advantage of the patent portfolio of the latter [74–76]. The two groups used the nano-imprint lithography to demonstrate that memory devices with a pitch as low as 40 nm can be obtained. The association between the two laboratories was extremely beneficial for the molecular electronics domain, as HP advertised the concept in an active way. However, the results reported by HP in 2003 appeared too early, as even now the switching mechanisms in metal/organic/metal systems are not completely understood. Several of the earlier reported experimental results on electron transport through molecules were found to be due to formation of metal filaments along the molecules attached between the two metal electrodes [68, 77–88]. Consequently, other effects may often mask the intrinsic behavior of molecular switches. Even so, the research in this domain continued and in 2007, UCLA group improved the architecture by lowering the pitch down to 33 nm and demonstrated a 160 kbit molecular-memory [50].

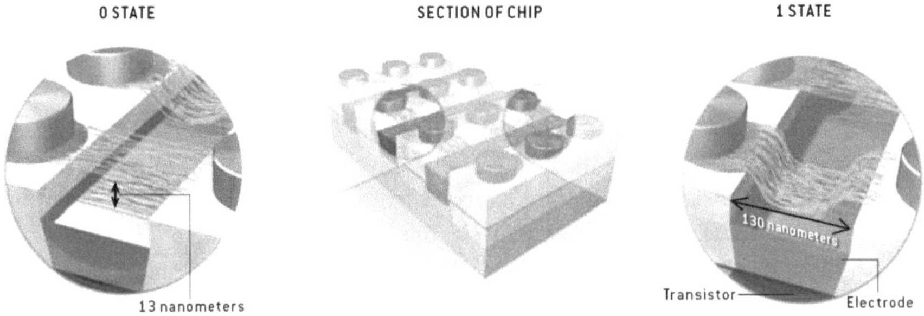

Fig. 13.6 Three-dimensional representation of NRAM®; (**a**) 0 state position; (**b**) NRAM® 3D representation; (**c**) 1 state position. Courtesy of J.F. Podevin [89] (Copyright 2005 J.F. Podevin)

The statistics show that many devices present poor switching or are shorted/not-connected, thus some improvement to reduce the device variability is still needed. In another report, the Superlattice NAnowire Pattern transfer (SNAP) technique was used to reduce the feature size down-to 15 nm [80]. These results represent important achievements, as dimensions like 15 nm offer high densities memories (10^{11} bits/cm^2), which correspond to the 2020 node on ITRS roadmap. Even with all the effort put in the development of the hybrid molecular/NWs system and regardless of the elegance of the solution provided for nonvolatile memories, the Rotaxane cell presents one weakness: from a circuit designer perspective the cell is not intrinsically rectifying. This represents a serious drawback, as it requires special addressing in order to avoid crosstalk between the cells [63]. Therefore, the Rotaxane cell is not the ideal candidate for the Beyond CMOS memory applications.

Another approach in the design of the elementary cell of the Crossbar memory came from C. Lieber's group, at Harvard University, one of the main drivers in the field of NWs and CNTs research [55]. Early in 2000, T. Rueckes invented the Nano-Electro-Mechanical Memory (NEMM) known also as Nanotube-RAM [55] (NRAM®). After the initial proof of the concept, NEMM was developed by Nantero Company, which holds the NRAM® patent [86]. NRAM® represents a Nano Electro Mechanical System (NEMS) device that is electrically actuated. From the electrical point of view, NRAM® belongs to the class of resistive memories that can be fabricated by using CNTs or NWs [55, 81–85]. The device consists of one or several CNTs or NWs that are stretched between two lands and suspended over a conductive electrode (EL) (Fig. 13.6). The CNTs (or NWs) for this application are conductive. One of the lands is metallic too, and represents the first contact of the memory cell. The second contact is the electrode over which the CNTs (or NWs) are suspended (Fig. 13.6a). The write operation is performed by applying a voltage on the cell contacts. Due to the electrostatic forces the suspended CNTs are attracted and collapse against the EL surface (1 logic—Fig. 13.6c). When the actuation voltage is removed, the Van der Waals forces between the CNTs and the EL surface maintain the CNTs in the collapsed position against the mechanical strain that exists in the CNTs. To reset the memory, an opposite field is applied. This generates repulsive forces that overcome the Van der Waals forces and release the CNTs back to the original position (0 logic—Fig. 13.6a). To read the cell a small voltage is applied between the contacts. If the resistance is low (CNTs collapsed) 1 is read, otherwise 0 is read (CNTs are suspended).

The design of the NRAM® cell pushes this memory cell into the spot as possible replacement of DRAM for the following five reasons. First, the NRAM® is intrinsically highly dense as the geometrical features of the CNTs are below the lithographic limits. Therefore, theoretically, the limit of the NRAM® is exclusively imposed by the lithography. In contrast, for the DRAM technology the scaling limits are given by the charge that can be effectively stored and read [62]. A second advantage in comparison with DRAM is that NRAM® does not need "refresh". The NRAM® cell conserves its status even when the voltage is removed. The third advantage comes with the energy that is needed to write the device, which is reduced for NRAM® as compared to the DRAM. Fourth, the NRAM® cell is intrinsically insensitive to the radiation. This property comes from the fact that NRAM® is based on a mechanical effect and not on the stored charges. Fifth, NRAM® offers the portability feature on different substrates including the flexible ones. This last advantage expands the perspective of the use of this memory for new applications beyond the actual use of the SRAM and DRAM.

Experimental results on NRAM® demonstrated the proof of the concept. The cell is fully compatible with the CMOS technology. It is important to mention that this cell is not intrinsically rectifying. From this point of view, NRAM® cell is similar with Rotaxane cell, and some additional cell engineering is needed to use NRAM® for memory applications in Crossbar architecture [90]. Little is known to date on the system reliability and device variability, and therefore it is hard to predict that this cell will take the lead in the development of future memory applications. However, advanced tests were performed on NRAM® on NASA shuttle mission recently [91]. According to the news, the NRAM® modules performed the same before, during and after the completion of the mission. This represents a step forward towards the development of fast, nonvolatile, radiation hardened memory.

The last example we discuss here for the passive Crossbar architecture is the Crystalline Si/Amorphous Si/Metal cell (c-Si/a-Si/M). Lieber's group recently revealed the c-Si/a-Si/M cell [47]. The device is formed at the intersection of a core crystalline Si NW, wrapped in amorphous Si shell, with a metallic NW (Fig. 13.7a). The resistance switching is attributed to the metal filament formation (retraction) inside the a-Si matrix that yields high (low) conductance. The device matrix is assembled by using the Langmuir-Blodgett method for the semiconductor NWs and lithography for the metallic NWs. This demonstrates that it is possible to assemble the devices by combining both Bottom-Up and Top-Down techniques. The memory effect is present independently of the device dimensions. The authors show the presence of the memory effect for devices scaled between 1000 nm and 20 nm. The result gives a promising pitch size of 100 nm for highly dense memory applications. The writing time is less than 100 ns and the retention time is more than 2 weeks. The devices can be switched more than 10^4 cycles without degradation of the performance. Additionally the authors demonstrate that the devices can be correctly packed both on rigid and flexible substrates.

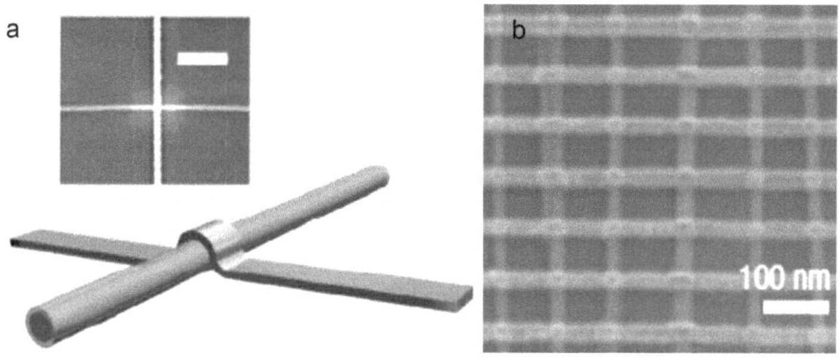

Fig. 13.7 (**a**) Single cell memory element built at the cross-point of a c-Si/a-Si core/shell NW and a metallic NW. (**b**) SEM image of the 2D highly dense Crossbar memory. Reprinted with permission from [47] and [62], copyright 2008 American Chemical Society and 2007 Macmillan Publishers Ltd.

Another group from University of Michigan, inspired by Lieber's work [47], developed the same type of devices on Si platform demonstrating that this device is fully compatible with the CMOS technology [87]. The smallest device realized (50×50 nm) gives a density of 10 Gbit/cm^2 using a cell size of 4F^2. Preliminary tests showed improved speed and endurance with respect to Lieber's demonstrator [87].

Overall, the c-Si/a-Si/M memory cell is compatible or better than the other prototypes present in ITRS 2007. The two contributions [47, 87] demonstrated the proof of concept and the benefits in terms of performance that can be obtained by using a well-established technology. In contrast to Rotaxane and NRAM® cells, the c-Si/a-Si/M cell is intrinsically rectifying. The rectification combined with the compatibility with the Si platform, places the c-Si/a-Si/M cell in favorable position in the race for the future nonvolatile memory cell. We note with this example that the special design of the elementary cell delivers a simple solution for system assembly. In addition, the c-Si/a-Si/M technology can be extended with the cost of some additional technological steps to obtain logic devices. This approach is of high interest for the future technologies as it is among the rare examples that have the potential to integrate both logic and high-density memory circuits. In conclusion, the c-Si/a-Si/M cell represents a serious candidate for the beyond CMOS roadmap for both memory and logic applications.

4.2 Active Matrix

Active matrices are built by using a transistor for each memory element. To keep the memory density high, the Crossbar active matrices need special implementation that requires particular design of the cell. In the following we present a short description of the devices adapted for the active matrices together with their implementation.

In early 00's new organic functionalized devices, which combine the electrical properties of the nano-objects like CNTs or semiconductor NWs with the function-

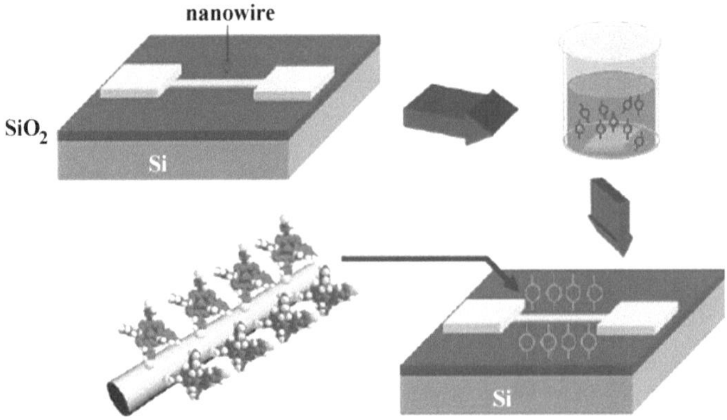

Fig. 13.8 Schematic representation of the process steps in the fabrication of a hybrid molecular-NW Thin Film Transistor (TFT). Reprinted with permission from [92], copyright 2004 American Chemical Society

alities of the molecules appeared [45, 88]. The design of these hybrid systems can be modified and adapted as a function of the targeted application. In most of the cases, these combined systems are operated as Thin-Film Transistors (TFT) in which the CNTs or NWs represent the semiconductor material. The molecules complete the system being either covalently grafted or simply deposited on CNTs or NWs. The operation of these combined devices is straightforward: the molecules respond to an electrical [92, 93] or optical [94, 95] excitation and influence the conduction through the CNTs or NWs. All these hybrid systems bring new functionalities to the classical Thin Film Transistors (TFTs) expanding the application domain outside the classical platforms.

In 2002 Liber's group presented the first hybrid non-volatile memory device called "molecular gated nanowire" [45]. Following Liber's idea, the group from the University of Southern California in collaboration with the Center for Nanotechnology, NASA, revealed the importance of complex inorganic-organic molecule design for applications in electronics [92]. They used In_2O_3 NWs covered by a Self Assembly Monolayer (SAM) of porphyrin molecules (Fig. 13.8). The presence of the memory effect is revealed only for the samples covered by metal core porphyrin molecules while no memory effect is obtained when protio-porphyrin is used. The direct link between the metal nanoparticle in the porphyrin coating and the memory effect highlights the importance of the organometallic chemistry engineering for the electronic applications. In a second contribution, the two groups presented a multilevel non-volatile molecular memory [93]. Redox active molecules (Fe-terpyridine) are used to store discrete charge multilevels. The molecule comes from terpyridine family, and, once again, is functionalized with different inorganic elements (Fe, Zn or Ru). The metal can be chosen as a function of the targeted application. The active part of the device is realized by In_2O_3 NWs. The fabricated system can store up to 8 levels, which corresponds to a three-bit memory device (Fig. 13.9). The demonstrator is nano in diameter (10 nm) and has a long channel (2 μm). Improvements can

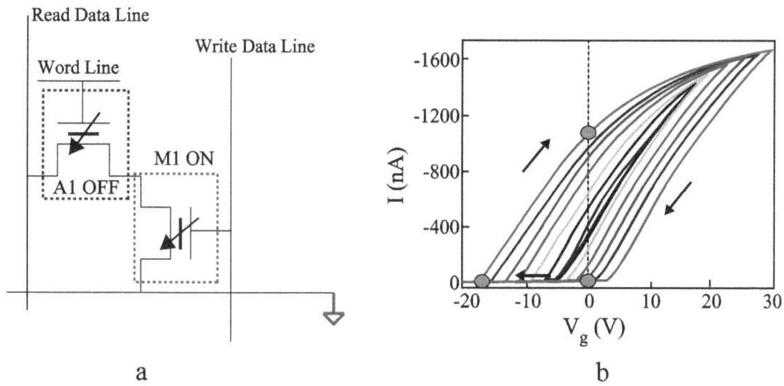

Fig. 13.9 Schematic illustration of the write cycle. (**a**) Initial conditions: A1—OFF, M1—ON; (**b**) electrical characteristics that illustrate the write sequence applied on M1 (A1 is kept OFF): $V_{G\,\mathrm{M1}} = 0$ V $\rightarrow$ $V_{G\,\mathrm{M1}}$ negative $\rightarrow$ $V_{G\,\mathrm{M1}} = 0$ V. Figure 13.9b reprinted with permission from [93]. Copyright 2004, American Institute of Physics

be done in order to bring the length size in the nano-domain. However, it is not sure that all the 8 levels will be correctly reproduced if scaling is performed.

Figure 13.9b presents the device operation as memory based on the static electrical. Sweeping the gate in the negative domain while keeping the source and drain grounded programs the device. The V_G sweep is performed with a step of −2.5 V up to −20 V, yielding 8 distinct levels. The memory is erased in the same conditions but with the positive gate voltage (30 V). The read operation is performed at low V_D (10 mV) in order to do not affect the status of the cell.

Endurance tests performed by repeating tens of write cycles do not highlight significant degradation. The retention time is as long as several days. Retention tests show that the memory levels were stable within the 80% for the first 120 h.

We note, however, that the multilevel molecular gated transistor presented above cannot be used as a classical transistor to be implemented in standard memory architecture like EEPROM, NAND or NOR Flash as writing or reading the data in a point cell will affect the status of the other neighboring cells. Therefore, the French groups from "Commissariat à l'Énergie Atomique" CEA-LETI and "Institut Superieur d'Electronique de Paris" gathered their efforts and used the above-presented device in transformed Crossbar active matrix architecture [96, 97]. In the new design, transistor A1 acts as a programmable switch and is used to control the memory element M1 (Fig. 13.9).

The design benefits are: (i) Compact cells due to the reduced dimensions of the NWs; (ii) Increased density of integration due to the ability to store multivalued information in the memory components; (iii) Reduced number of photolithographic steps due to the chemical functionalization of the cell elements.

The cell is operated by voltage control. To perform an erase operation, the access transistor A1 is kept open while a positive pulse is applied on the gate of M1. This clears the charge stored in the molecules leading the transistor in a nonconductive state (Fig. 13.9). The same conditions are applied in order to write the memory, with the exception that the gate of the memory transistor is pulsed in the negative domain

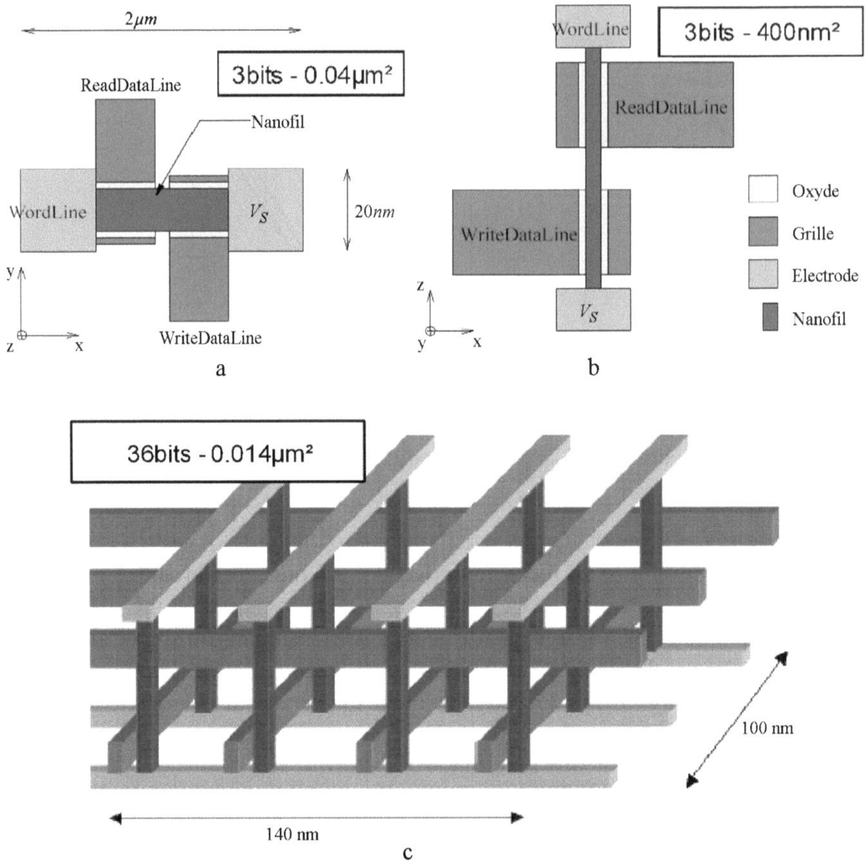

Fig. 13.10 (**a**) and (**b**) 2D and 3D layout of the multivalued memory cell consisting in two molecular-gated NWFET transistors in series; (**c**) 3D integration of 12 cells in an area of 0.014 mm^2

(Fig. 13.9b). As in the description of the memory element above, the negative pulse amplitude gives the level of the memory. To read the cell the access transistor (A1) is closed and biased in the most conductive state, while the gate of the memory transistor is kept at 0 V. Once the read operation is finished, the access transistor is turned back OFF by applying a positive pulse on its gate.

The authors presented a first layout adapted for standard photolithographic process (Fig. 13.10a), for which they estimate an area of 0.04 μm^2 needed for a 3 bit-cell [97]. A more compact design, which includes three-dimensional integration (Fig. 13.10b), results in a cell area of only 400 nm^2. Figure 13.10c illustrates the integration of twelve 3 bit-cells in an area 0.014 μm^2. The authors show that the proposed cell occupies 48 times less surface that the 1 bit, 50 nm NAND Flash cell.

The above contribution shows the interest in having compact active matrix memory cells. We note however the challenges and the difficulties that this design has to surpass when scaled down to the NWs diameter, together with the technological challenges that a 3D integration faces. The read operation in two steps also adds additional pressure on this design. Regardless of the difficulties that this idea confronts today in terms of fabrication, the design remains interesting for its compactness. With the progress in the nano fabrication techniques, the design might inspire other groups to develop the concept further.

5 Logic Circuits Alternative

While several alternatives already exist for future nano memory circuits, we witness only a few reports on the development of logic circuits. This situation is striking as research in Beyond CMOS domain exploded in the last years. However, only a few groups concentrated their efforts to exploit the particular features of the new available devices and provide original solutions at the circuit level. The following section presents two innovative solutions that use the modified DG CNT transistors and a third solution that uses the passive matrices for the implementation of simple logic circuits.

5.1 Logic Circuits—Modified DG Devices

Recently, the special features of the DGSB device presented in Fig. 13.3d were exploited in an original work proposed by O'Connor et al. [98] It is shown that by using DGSB devices, reconfigurable circuits can be designed. The equivalent implementation in CMOS delivers the same functionality with the price of increased circuit complexity that results in significant system power penalties. The authors propose two dynamically reconfigurable two-input logic gates. The first provides 8 logic functions—mainly AND and OR based (Fig. 13.11). The other has 6 logic functions—mainly AND and XOR based. Both complementary and non-complementary functions can be provided. The gates are organized in two logic stages the logic function and the follower/inverter. These types of gates can be successfully used to build compact circuits that need less clock cycles for operation.

Another interesting idea is to use the DGSB devices as pass gates in the Programmable Logic Arrays (PLAs). A group from Ecole Polytechnique Fédérale de Lausanne (EPFL) used this idea to demonstrate the ability of the Generalized NOR gates (GNOR, Fig. 13.12) to be integrated and used in array based architectures [99]. The three polarization cases explained in the description of the DGSB were used in this circuit study. The authors proposed a compact interconnect array realized by using ambipolar DGSB devices that operate as pass transistors.

The EPFL group compared three PLA implementations—Flash, EEPROM and GNOR CNT-FETs. They showed that the GNOR CNTFET cell is 50% larger when compared with the Flash, but 40% smaller than the EEPROM basic cell. Therefore, the DGSB PLA is always more compact than the EEPROM PLA. In comparison with the Flash, DGSB PLA can save area if the number of inputs is large enough

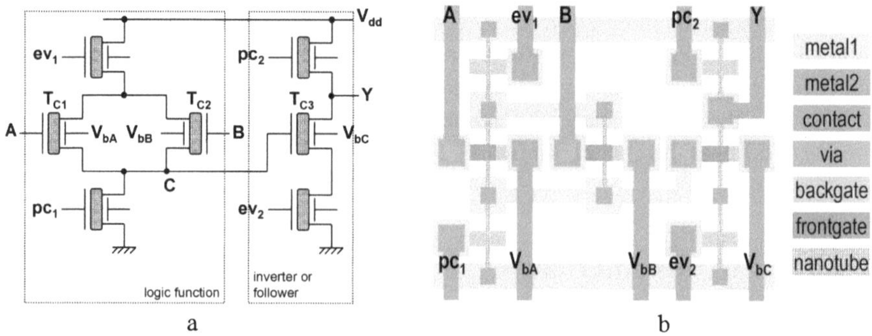

Fig. 13.11 (**a**) Schematic of the dynamically reconfigurable (DR) 8-function logic gate; (**b**) Circuit layout using arbitrary design rules. Reprinted with permission from [98]. Copyright 2007, IEEE

Fig. 13.12 Schematic representation of a GNOR gate configured as $Y = \mathrm{NOR}(A, \bar{B}, D)$. Reprinted with permission from [99]. Copyright 2008, IEEE

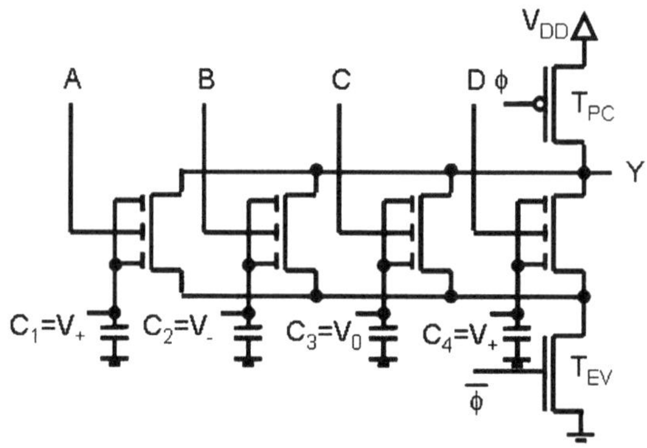

taking advantage that fewer inputs are needed for the DGSB. It is also estimated that the number of signals to route is reduced by almost a factor of 2 because the inverted signals are not routed but generated internally. This has a dramatic impact on the performance by almost doubling the frequency.

These examples show once more the crucial influence of design on circuit applications. The device architecture reduces the complexity at the circuit level and provides reconfigurable features to the circuit.

5.2 Logic Circuits—Passive Matrix

A second class of logic applications appears in the domain of the molecular logic. Molecules like Rotaxane or ionic redox devices like c-Si/a-Si/metal (already presented in Sect. 4.1 of this chapter) represent switches and can be associated as basic elements in the construction of the logic cells [73, 100]. Figure 13.13 presents a small demonstrator highlighting the realization of the logical functions using exclusively this type of switches. The architecture promises, like in the case of Crossbar memory, a high density of integration. However two major limitations have to be

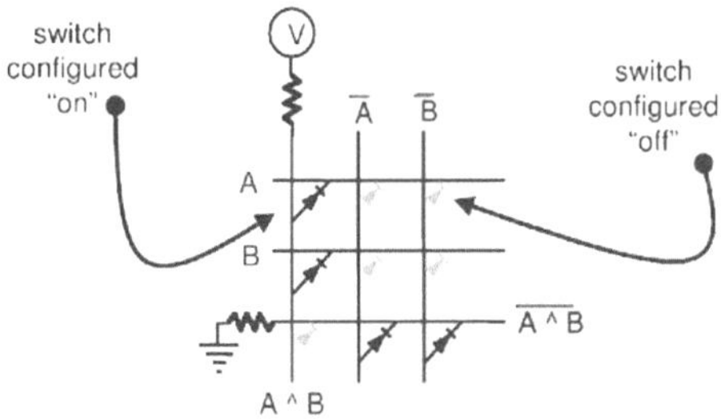

Fig. 13.13 AND gate implemented using rectifying switches. Reprinted with permission from [100]. Copyright 2001, IEEE

faced in order to make the design practical: (i) The diode-based logic does not compensate for the signal degradation and therefore additional blocks for signal restoration are needed; (ii) The proposed logic blocks dissipate power in the static regime when the switch is closed. Even with these limitations, the devices may be used for logic applications by carefully designing the power management and the layout of the logic blocks. From this point of view, the particularity of these switches to recall their status even after a power-down—power-up cycle makes them extremely interesting for future applications.

We witness today key changes in the design of the elementary devices. One example in this direction comes with the rotaxane molecule. Besides the development recorded recently by the group from UCLA (in Sect. 4.1), a new breakthrough was reported in 2009. Two associated British groups—one from the University of Edinburgh, the other from the University of Manchester modified the structure of Rotaxane by transforming it into a hybrid organic-inorganic system [101]. Basically they replaced the aromatic rings with metals converting the macrocycle into a metal-based ring. This change may have a spectacular impact on the utilization of Rotaxane. As it was shown in this chapter for the porphyrin case, the design of organic/metallic molecular systems results in particularly interesting applications in electronics (Sect. 4.2). The metal-based ring Rotaxane could be eventually used by tuning the electrical, magnetical or even the catalytical properties of the metals that form the ring. At this moment the British groups demonstrated the motion of the metallic ring along the axle. The next step in the development of these systems is to control the position of the ring along the axle—a significant issue that faces some challenges up to now. Even so, we note that this realization may open the door for the quantum computation, since a direct application of this new-engineered rotaxane structure may be the quantum bit (q-bit)—the fundamental unit for the quantum computation.

6 Hybrid Molecular-CMOS Architectures

Scientists have already proposed Hybrid Micro/Nano architectures that integrate together the Crossbar NWs (already presented in Sect. 3) and the CMOS [102–104]. These compact structures take advantage of the high device density provided by Crossbar and the flexibility offered by CMOS. The Hybrid architectures operate in

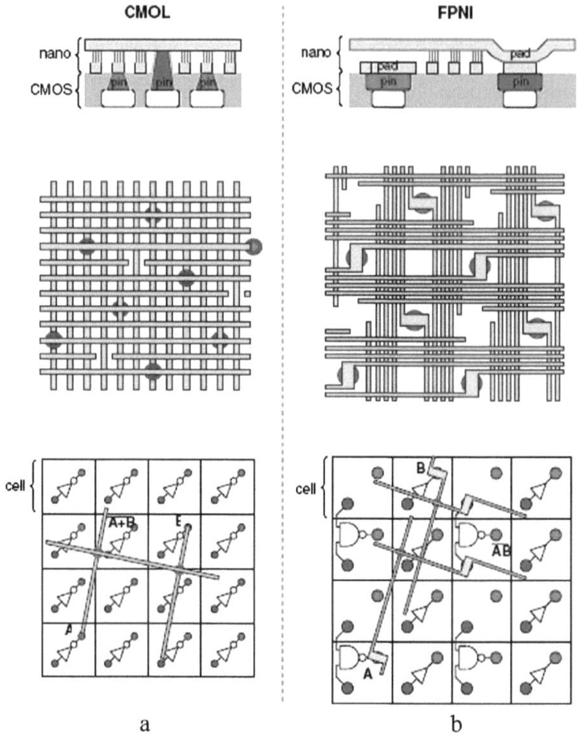

Fig. 13.14 Schematic diagram of hybrid circuits: (**a**) CMOL, (**b**) FPNI. *Upper part*: cross-section; *Middle part*: top view; *Lower part*: superposition CMOS/molecular. Reprinted with permission from [104]. Copyright 2007, IOP Publishing Ltd.

a way similar to FPGAs and present high tolerance to defects. The same criteria have to be established in order to construct a hybrid CMOS/molecular architecture [62, 104]. The functionality split between the CMOS and the molecular level has to be specified. The interconnection of the CMOS to the molecular layer has to be solved. Strategies for device variability and defect handling in the Crossbar architecture have to be defined, too.

The first hybrid architecture that we highlight here is the CMOS/Molecular hybrid—CMOL (Fig. 13.14a), proposed by Likharev's group from the Department of Physics and Astronomy, Stony Brook University [102, 103]. CMOL uses the functionality segregation: the CMOS is used for logic, gain and demultiplexing while NWs are used for wired-OR logic and signal routing. The signal transfer between the molecular plane and CMOS is performed using pins uniformly distributed in the CMOS plane (Fig. 13.14a). The Crossbar plane is slightly rotated so that each NW connects electrically only one pin from CMOS (Fig. 13.14a lower part). Despite the impact that CMOL produced on the research community, the architecture faces some challenges for the practical use [104]. First, it requires ultra low supply voltage (0.3 V) that is well below the end of the CMOS roadmap. Second, the device variability may limit the performance. Third, the CMOS pins designed to connect the molecular layer are just a few nanometers in diameter and represent a fabrication challenge.

An architecture inspired from CMOL is the Field Programmable Nanowire Interconnect (FPNI—Fig. 13.14b). FPNI was proposed by Williams' group[104] from HP and tries to overcome some of the limitations of CMOL. The main differences between the CMOL and FPNI are in terms of segregation logic-interconnects, physical alignment and operation voltage. FPNI architecture uses the standard CMOS voltages. In FPNI the logic operations are performed only in CMOS and the signal routing is performed only by the NWs. The alignment of the NWs Crossbar on CMOS is performed with the CMOS accuracy and not by nano-pins like in CMOL case (Fig. 13.14).

The simulations performed on both CMOL and FPNI show that these architectures are extremely tolerant to high defect rates [102–104]. However, in the case of high defect rates, both CMOL and FPNI present a major economic challenge from the compilation point of view, as the chip manufacturers cannot perform sequentially, chip by chip, the defect mapping and place and route around defects. We also note that the ITRS 2008 update takes into account these architectures as "incremental extension of CMOS beyond the 2022 time horizon". According to the ITRS, these architectures "do not appear to offer an information processing technology addressing "beyond CMOS" domain" [1].

7 Summary

One analysis of the technology alternatives carried out by ITRS in 2007 concluded by saying: "Nothing beats MOSFETs for performing Boolean logic operations at comparable risk levels" [105]. This statement is true and stands well today. Technologies that were developed over the last 20 years like Fully-Depleted SOI, DG-FET or FinFET will be progressively deployed and put in production. An example in this direction discussed in this chapter is the SRAM cell. However, application areas in which the CMOS cannot compete exist already (one example is the flexible electronics domain). These areas, outside the CMOS sphere, represent perfect yards in which new technologies can be developed and applications will find their ways to the market. Therefore, the development of the CNT-FETs, NW-FETs and GNR-FETs may impact "today" in the "outside CMOS domains" and "tomorrow" in the "Beyond CMOS domain". At the same time, new applications emerged aside with the CMOS platform. One good example that was analyzed in this chapter is the Crossbar architecture that provides highly dense memories. It was shown that the performance of the Crossbar memory is directly influenced by the design of the individual cell. Not only memories are the subject of reshaping in the Beyond CMOS era. Phaedon Avouris, a leading scientist from IBM stated in 2005 "We have no problems in finding choices for memory, but we're running out of choices for logic" [89]. In this light, the fifth part of this chapter presented some design solutions for reconfigurable cells. It was shown how the DGSB devices provide an attractive solution to reduce the complexity of the design. Finally, hybrid molecular-CMOS architectures were presented in the sixth part of the chapter. These hybrids may represent the next solution for memory and computation in the early "beyond CMOS" era.

References

1. ITRS homepage: http://www.itrs.net/
2. Amara, A., Rozeau, O.: Planar Double-Gate Transistors. Springer, Berlin (2009)
3. Yamaoka, M., Osada, K., Tsuchiya, R., Horiuchi, M., Kimura, S., Kawahara, T.: In: Technical Digest of VLSI Circuits Symposium, pp. 288–291 (2004)
4. Guo, Z., Balasubramanian, S., Zlatanovici, R., King, T.-J., Nikolić, B.: In: Proceedings of the International Symposium on Low Power Electronics and Design, pp. 2–7 (2005)
5. Giraud, B., Thomas, O.: French Patent No. 08511027, 2008
6. Iijima, S.: Nature **354**, 56–58 (1991)
7. Deleonibus, S. (ed.): Electronic Device Architectures for the Nano-CMOS Era: From Ultimate CMOS Scaling to Beyond CMOS Devices. World Scientific, Singapore (2008). ISBN 9814241288
8. Jorio, A., Saito, R., Hafner, J.H., Lieber, C.M., Hunter, M., McClure, T., Dresselhaus, G., Dresselhaus, M.S.: Phys. Rev. Lett. **86**, 1118–1121 (2001)
9. Telg, H., Maultzsch, J., Reich, S., Hennrich, F., Thomsen, C.: Phys. Rev. Lett. **93**, 177401 (2004)
10. Strano, M.S., Doorn, S.K., Haroz, E.H., Kittrell, C., Hauge, R.H., Smalley, R.E.: Nano Lett. **3**, 1091–1096 (2003)
11. Arnold, M.S., Green, A.A., Hulvat, J.F., Stupp, S.I., Hersam, M.C.: Nat. Nanotechnol. **1**, 60–65 (2006)
12. Chen, Z., Du, X., Du, M.-H., Rancken, C.D., Cheng, H.-P., Rinzler, A.G.: Nano Lett. **3**, 1245–1249 (2003)
13. Zheng, M., Jagota, A., Strano, M.S., Santos, A.P., Barone, P., Chou, S.G., Diner, B.A., Dresselhaus, M.S., Mclean, R.S., Onoa, G.B., Samsonidze, G.G., Semke, E.D., Usrey, M., Walls, D.J.: Science **302**, 1545–1548 (2003)
14. Tanaka, T., Jin, H., Miyata, Y., Kataura, H.: Appl. Phys. Express **1**, 114001 (2008)
15. Li, J., Meyyappan, M.: United States Patent No. 7094679
16. Nihei, M., Horibe, M., Kawabata, A., Awano, Y.: Jpn. J. Appl. Phys. **43**, 1856–1859 (2004)
17. Kong, J., Soh, H.T., Cassell, A.M., Quate, C.F., Dai, H.: Nature **395**, 878–881 (1998)
18. Kreupl, F., Graham, A.P., Duesberg, G.S., Steinhögl, W., Liebau, M., Unger, E., Hönlein, W.: Microelectron. Eng. **64**, 399–408 (2002)
19. Gudiksen, M.S., Lauhon, L.J., Wang, J., Smith, D.C., Lieber, C.M.: Nature **415**, 617–620 (2002)
20. Lauhon, L.J., Gudiksen, M.S., Wang, D., Lieber, C.M.: Nature **420**, 57–61 (2002)
21. Yang, P., Yan, H., Mao, S., Russo, R., Johnson, J., Saykally, R., Morris, N., Pham, J., He, R., Choi, H.J.: Adv. Funct. Mater. **12**, 323–331 (2002)
22. Goldberger, J., He, R., Zhang, Y., Lee, S., Yan, H., Choi, H.-J., Yang, P.: Nature **422**, 599–602 (2003)
23. Cao, Q., Kim, H.-S., Pimparkar, N., Kulkarni, J.P., Wang, C., Shim, M., Roy, K., Alam, M.A., Rogers, J.A.: Nature **454**, 495–500 (2008)
24. Ju, S., Li, J., Liu, J., Chen, P.-C., Ha, Y.-G., Ishikawa, F., Chang, H., Zhou, C., Facchetti, A., Janes, D.B., Marks, T.J.: Nano Lett. **8**, 997–1004 (2008)
25. Sekitani, T., Nakajima, H., Maeda, H., Fukushima, T., Aida, T., Hata, K., Someya, T.: Nat. Mater. **8**, 494–499 (2009)
26. Novoselov, K.S., Geim, A.K., Morozov, S.V., Jiang, D., Zhang, Y., Dubonos, S.V., Grigorieva, I.V., Firsov, A.A.: Science **306**, 666–669 (2004)
27. de Heer, W.A., Berger, C., Wu, X., First, P.N., Conrad, E.H., Lia, X., Li, T., Sprinkle, M., Hass, J., Sadowski, M.L., Potemski, M., Martinez, G.: Solid State Commun. **143**, 92–100 (2007)
28. Sutter, P.W., Flege, J.-I., Sutter, E.A.: Nat. Mater. **7**, 406–411 (2008)
29. Coraux, J., N'Diaye, A.T., Engler, M., Busse, C., Wall, D., Buckanie, N., Meyer zu Heringdorf, F.-J., van Gastel, R., Poelsema, B., Michely, T.: New J. Phys. **11**, 023006 (2009)
30. Geim, A.K.: Science **324**, 1530–1534 (2009)

31. Tans, S.J., Verschueren, A.R.M., Dekker, C.: Nature **393**, 49–52 (1998)
32. Martel, R., Schmidt, T., Shea, H.R., Hertel, T., Avouris, Ph.: Appl. Phys. Lett. **73**, 2447 (1998)
33. Javey, A., Guo, J., Wang, Q., Lundstrom, M., Dai, H.: Nature **424**, 654–657 (2003)
34. Javey, A., Guo, J., Farmer, D.B., Wang, Q., Yenilmez, E., Gordon, R.G., Lundstrom, M., Dai, H.: Nano Lett. **4**, 1319–1322 (2004)
35. Martel, R., Derycke, V., Lavoie, C., Appenzeller, J., Chan, K.K., Tersoff, J., Avouris, Ph.: Phys. Rev. Lett. **87**, 256805 (2001)
36. Avouris, Ph.: Chem. Phys. **281**, 429–445 (2002)
37. Wind, S.J., Appenzeller, J., Avouris, Ph.: Phys. Rev. Lett. **91**, 058301 (2003)
38. Appenzeller, J., Lin, Y.-M., Knoch, J., Avouris, Ph.: Phys. Rev. Lett. **93**, 196805 (2004)
39. Koo, S.-M., Li, Q., Edelstein, M.D., Richter, C.A., Vogel, E.M.: Nano Lett. **5**, 2519–2523 (2005)
40. Chen, B., Wei, J., Lo, P., Wang, H., Lai, M., Tsai, M., Chao, T., Lin, H., Huang, T.: Solid-State Electron. **50**, 1341–1348 (2006)
41. Heath, J.R., Kuekes, P.J., Snider, G.S., Williams, R.D.: Science **280**, 1716–1721 (1998)
42. Stan, M.R., Franzon, P.D., Goldstein, S.C., Lach, J.C., Ziegler, M.M.: Proc. IEEE **91**, 1940–1957 (2003)
43. DeHon, A.: IEEE Trans. Nanotechnol. **2**, 23–32 (2003)
44. Huang, Y., Duan, X., Cui, Y., Lauhon, L.J., Kim, K.-H., Lieber, C.M.: Science **294**, 1313–1317 (2001)
45. Duan, X.F., Huang, Y., Lieber, C.M.: Nano Lett. **2**, 487–490 (2002)
46. Jin, S., Whang, D., McAlpine, M.C., Friedman, R.S., Wu, Y., Lieber, C.M.: Nano Lett. **4**, 915–919 (2004)
47. Dong, Y., Yu, G., McAlpine, M.C., Lu, W., Lieber, C.M.: Nano Lett. **8**, 386–391 (2008)
48. Kaeriyama, S., Sakamoto, T., Sunamura, H., Mizuno, M., Kawaura, H., Hasegawa, T., Terabe, K., Nakayama, T., Aono, M.: IEEE J. Solid-State Circuits **40**, 168–176 (2005)
49. Wu, W., Jung, G.-Y., Olynick, D.L., Straznicky, J., Li, Z., Li, X., Ohlberg, D.A.A., Chen, Y., Wang, S.-Y., Liddle, J.A., Tong, W.M., Williams, R.S.: Appl. Phys. A **80**, 1173–1178 (2005)
50. Green, J.E., Choi, J.W., Boukai, A., Bunimovich, Y., Johnston-Halperin, E., DeIonno, E., Luo, Y., Sheriff, B.A., Xu, K., Shik Shin, Y., Tseng, H.-R., Stoddart, J.F., Heath, J.R.: Nature **445**, 414–417 (2007)
51. Baek, I.G., Lee, M.S., Seo, S., Lee, M.J., Seo, D.H., Suh, D.-S., Park, J.C., Park, S.O., Kim, H.S., Yoo, I.K., Chung, U.-In., Moon, J.T.: In: Technical Digest of IEEE International Electron Devices Meeting, pp. 587–590 (2004)
52. Dietrich, S., Angerbauer, M., Ivanov, M., Gogl, D., Hoenigschmid, H., Kund, M., Liaw, C., Markert, M., Symanczyk, R., Altimime, L., Bournat, S., Mueller, G.: IEEE J. Solid-State Circuits **42**, 839–845 (2007)
53. Cui, Y., Lieber, C.M.: Science **291**, 851–853 (2001)
54. Whang, D., Jin, S., Wu, Y., Lieber, C.M.: Nano Lett. **3**, 1255–1259 (2003)
55. Rueckes, T., Kim, K., Joselevich, E., Tseng, G.Y., Cheung, C.-L., Lieber, C.M.: Science **289**, 94–97 (2000)
56. Collier, C.P., Wong, E.W., Belohradský, M., Raymo, F.M., Stoddart, J.F., Kuekes, P.J., Williams, R.S., Heath, J.R.: Science **285**, 391–394 (1999)
57. Chen, Y., Jung, G.-Y., Ohlberg, D.A.A., Li, X., Stewart, D.R., Jeppesen, J.O., Nielsen, K.A., Stoddart, J.F., Williams, R.S.: Nanotechnology **14**, 462–468 (2003)
58. Zankovych, S., Hoffmann, T., Seekamp, J., Bruch, J.U., Torres, C.M.S.: Nanotechnology **12**, 91–95 (2001)
59. Chou, S.Y., Krauss, P.R., Renstrom, P.J.: Science **272**, 85–87 (1996)
60. Melosh, N.A., Boukai, A., Diana, F., Gerardot, B., Badolato, A., Petroff, P.M., Heath, J.R.: Science **300**, 112–115 (2003)
61. Brueck, S.R.J.: In: Guenther, A.H., Holst, G.C. (eds.) International Trends in Applied Optics, pp. 85–110. SPIE, Bellingham (2002)
62. Lu, W., Lieber, C.M.: Nat. Mater. **6**, 841–850 (2007)

63. Scott, J.C., Bozano, L.D.: Adv. Mater. **19**, 1452–1463 (2007)
64. Reed, M.A., Zhou, C., Muller, C.J., Burgin, T.P., Tour, J.M.: Science **278**, 252–254 (1997)
65. Dimitrakopoulos, C.D., Malenfant, P.R.L.: Adv. Mater. **14**, 99–117 (2002)
66. Horowitz, G.: J. Mater. Res. **19**, 1946–1962 (2004)
67. Singh, T.B., Sariciftci, N.S.: Annu. Rev. Mater. Res. **36**, 199–230 (2006)
68. Waser, R., Aono, M.: Nat. Mater. **6**, 833–840 (2007)
69. Bogani, L., Wernsdorfer, W.: Nat. Mater. **7**, 179–186 (2008)
70. Collier, C.P., Mattersteig, G., Wong, E.W., Luo, Y., Beverly, K., Sampaio, J., Raymo, F.M., Stoddart, J.F., Heath, J.R.: Science **289**, 1172 (2000)
71. Credi, A., Ferrer, B.: Pure Appl. Chem. **77**, 1051–1057 (2005)
72. Serreli, V., Lee, C.-F., Kay, E.R., Leigh, D.A.: Nature **445**, 523–527 (2007)
73. Luo, Y., Collier, C.P., Jeppesen, J.O., Nielsen, K.A., DeIonno, E., Ho, G., Perkins, J., Tseng, H.-R., Yamamoto, T., Stoddart, J.F., Heath, J.R.: ChemPhysChem **3**, 519–525 (2002)
74. Snider, G.S.: United States Patent No. 7203789, 2001
75. Snider, G.: United States Patent No. 7359888, 2003
76. Kuekes, P.J.: United States Patent No. 6586965, 2003
77. Stewart, D.R., Ohlberg, D.A.A., Beck, P.A., Chen, Y., Williams, R.S.: Nano Lett. **4**, 133–136 (2004)
78. Blackstock, J.J., Stickle, W.F., Donley, C.L., Stewart, D.R., Williams, R.S.: J. Phys. Chem. C **111**, 16–20 (2007)
79. Lau, C.N., Stewart, D.R., Williams, R.S., Bockrath, M.: Nano Lett. **4**, 569–572 (2004)
80. Dichtel, W.R., Heath, J.R., Stoddart, J.F.: Philos. Trans. R. Soc. Lond. A **365**, 1607–1625 (2007)
81. Ward, J.W., Meinhold, M., Segal, B.M., Berg, J., Sen, R., Sivarajan, R., Brock, D.K., Rueckes, T.: In: Non-Volatile Memory Technology Symposium, pp. 34–38 (2004)
82. Jang, J.E., Cha, S.N., Choi, Y., Amaratunga, G.A.J., Kang, D.J., Hasko, D.G., Jung, J.E., Kim, J.M.: Appl. Phys. Lett. **87**, 163114 (2005)
83. Badzey, R.L., Zolfagharkhani, G., Gaidarzhy, A., Mohanty, P.: Appl. Phys. Lett. **85**, 3587–3589 (2005)
84. Tsuchiya, Y., Takai, K., Momo, N., Nagami, T., Mizuta, H., Oda, S., Yamaguchi, S., Shimada, T.: J. Appl. Phys. **100**, 094306 (2006)
85. McClelland, G.M., Atmaja, B.: Appl. Phys. Lett. **89**, 161918 (2006)
86. Rueckes, T., Segal, B.M., Vogeli, B., Brock, D.K., Jaiprakash, V.C., Bertin, C.L., United States Patent No. 6944054, 2003
87. Jo, S.H., Lu, W.: Nano Lett. **8**, 392–397 (2008)
88. Cui, Y., Wei, Q.Q., Park, H.K., Lieber, C.M.: Science **293**, 1289–1292 (2001)
89. Stix, G.: Sci. Am., February 2005, 82–85
90. Confidential data provided to the authors by G. Schmergel, President and CEO of Nantero, Inc
91. http://www.lockheedmartin.com/news/press_releases/2009/1118_ss_nanotubes.html
92. Li, C., Ly, J., Lei, B., Fan, W., Zhang, D., Han, J., Meyyappan, M., Thompson, M., Zhou, C.: J. Phys. Chem. B **108**, 9646–9649 (2004)
93. Li, C., Fan, W., Lei, B., Zhang, D., Han, S., Tang, T., Liu, X., Liu, Z., Asano, S., Meyyappan, M., Han, J., Zhou, C.: Appl. Phys. Lett. **84**, 1949 (2004)
94. Borghetti, J., Derycke, V., Lenfant, S., Chenevier, P., Filoramo, A., Goffman, M., Vuillaume, D., Bourgoin, J.-P.: Adv. Mater. **18**, 2535–2540 (2006)
95. Anghel, C., Derycke, V., Filoramo, A., Lenfant, S., Giffard, B., Vuillaume, D., Bourgoin, J.-P.: Nano Lett. **8**, 3619–3625 (2008)
96. Jalabert, A., Clermidy, F., Amara, A.: In: Proc. of IEEE International Conference on Electronics, Circuits and Systems, pp. 1034–1037 (2006)
97. Jalabert, A., Clermidy, F., Amara, A.: Molecular Electronics Materials, Devices and Applications. Springer, Berlin (2008). ISBN 978-1-4020-8593-2
98. O'Connor, I., Liu, J., Gaffiot, F., Pregaldiny, F., Lallement, C., Maneux, C., Goguet, J., Fregonese, S., Zimmer, T., Anghel, L., Dang, T.-T., Leveugle, R.: IEEE Trans. Circuits Syst. I, Regul. Pap. **54**, 2365–2379 (2007)

99. Ben Jamaa, M.H., Atienza, D., Leblebici, Y., De Micheli, G.: In: Proc. of ACM/IEEE Design Automation Conference, DAC, pp. 339–340 (2008)
100. Budiu, M., Goldstein, S.C.: In: Proc. of The 28th Annual International Symposium on Computer Architecture (2001)
101. Lee, C.-F., Leigh, D.A., Pritchard, R.G., Schultz, D., Teat, S.J., Timco, G.A., Winpenny, R.E.P.: Nature **458**, 314–318 (2009)
102. Strukov, D.B., Likharev, K.K.: Nanotechnology **16**, 888–900 (2005)
103. Strukov, D.B., Likharev, K.K.: Nanotechnology **16**, 137–148 (2005)
104. Snider, G.S., Williams, R.S.: Nanotechnology **18**, 035204 (2007)
105. Hutchby, J.: In: ITRS Public Conference, Dec. 2007

Chapter 14
Through Silicon Via-based Grid for Thermal Control in 3D Chips

José L. Ayala, Arvind Sridhar, David Atienza, and Yusuf Leblebici

1 Introduction

The traditional chip fabrication technology in 2D is facing lots of challenges in utilizing the exponentially growing number of transistors on a chip. The wire delay and power consumption is increasing dramatically and achieving interconnect design closure is becoming a challenge. Vertical stacking of multiple silicon layers, referred to as 3D stacking, is emerging as an attractive solution to continue the pace of growth of Systems on Chips (SoCs). The 3D technology results in smaller footprint in each layer and shorter vertical wires that are implemented using Through Silicon Vias (TSVs) across the layers.

Despite the advantages of 3D ICs over 2D ICs, thermal effects are expected to be significantly exacerbated in 3D ICs due to higher power density and greater thermal

J.L. Ayala · A. Sridhar · D. Atienza
Embedded Systems Laboratory (ESL), Faculty of Engineering, Ecole Polytechnique Fédérale de Lausanne (EPFL), Lausanne, Switzerland

J.L. Ayala
e-mail: jose.ayala@epfl.ch

A. Sridhar
e-mail: arvind.sridhar@epfl.ch

D. Atienza
e-mail: david.atienza@epfl.ch

J.L. Ayala (✉)
Department of Computer Architecture (DACYA), School of Computer Science, Complutense University of Madrid (UCM), Madrid, Spain
e-mail: jayala@fdi.ucm.es

Y. Leblebici
Microelectronic Systems Laboratory (LSM), Faculty of Engineering, EPFL, Lausanne, Switzerland
e-mail: yusuf.leblebici@epfl.ch

G. Nicolescu et al. (eds.), *Design Technology for Heterogeneous Embedded Systems*,
DOI 10.1007/978-94-007-1125-9_14, © Springer Science+Business Media B.V. 2012

Fig. 14.1 View of the 3D chip

resistance of the insulating dielectric, and this can cause greater degradation in device performance and chip reliability which have already plagued 2D ICs. Thus, it is essential to develop 3D-specific design tools that take a thermal co-design approach so as to address the thermal effects and generate reliable and high performance designs. This work considers the design of a nano-structure to help on the thermal dissipation in 3D chips.

In 3D ICs, devices are fabricated on a number of active layers, which are separated by silicon dioxide and joined by an adhesive material. Within each device layer, interconnections among devices can be achieved with traditional interconnect wires and vias. Connections between active layers are facilitated by vertical interconnect vias that span through multiple layers, providing a means for electrically connecting wires in those layers. This type of via is different from a regular 2D via: in particular, it is significantly taller than conventional vias, and has a larger landing pad to maintain a viable aspect ratio. We refer to such vias as TSVs.

Due to the ultra short lengths of the TSVs (50–100 μm), they easily overcome RC delays of long, horizontal circuit traces in conventional 2D circuits, and they also provide a higher density of connections. The TSVs are good conductors of heat, and hence they can be effective in dissipating some of the temperature of the devices.

The design and placement of TSVs can be proposed as an effective mechanism for thermal dissipation in 3D chips. The thermal resistivity of the via diffuses the heat and can create a homogeneous thermal distribution in the stack if placed carefully. The aim of this work is to propose a nano-structure, built as a grid of TSVs, for the thermal dissipation and optimization in 3D stacks. The capability of selecting a higher density grid of TSVs in specific areas of the system will enable to cool down selectively those zones with a bigger power density.

The experimental work of this paper is carried out through a novel thermal analysis of a real 5-tier 3D stack (see Fig. 14.1). Then, the material layers and TSVs are modeled mathematically, and the effect of a non-homogeneous distribution of the vias for thermal control is analyzed and effective inclusion of localized TSVs conforming a grid of nano-structures for thermal control is proposed. Also, the effect of specific interface materials used as inter-layer glue is considered. These interfaces will expose unique characteristics due to the presence of aluminum dopants.

The thermal model is also validated against the on-silicon implementation shown in Fig. 14.1, where the extensive measurements of several existing heaters and sensors per layer can be used to study the horizontal and vertical heat diffusion. The obtained results show interesting conclusions in the area of thermal modeling and optimization for 3D chips, as well as bringing new opportunities in the design of nano-structures based on TSVs for thermal balancing and control.

The paper structure is as follows: next section reviews the related work on this area, Sect. 3 presents the configuration of the 3D stack developed for the experimental work, and the developed thermal model is explained in Sect. 4. Then, the validation of the model and the rest of the experimental work are covered in Sects. 5 and 6, respectively. Finally, the conclusions of the work are drawn.

2 Related Work

Three-dimensional (3D) integration consists of the vertical placement and interconnections of several layers of active circuits. The main interests of this technology are to reduce global interconnect lengths, to increase circuit functionality and to enable new 3D circuit architectures [1–3].

Recently 3D die stacking is drawing a great deal of attention, primarily in embedded processor systems. Some previous works analyze the application of 3D stacks in system-on-chip designs: [4–8]. Others explore cache implementations: [9–11]. Some other works design 3D adder circuits: [12, 13]. While others evaluate wire benefits in full microprocessors: [4, 5, 14–16]. This emerging 3D technology is considered as a very attractive method for integrating complex systems by the embedded industry. Furthermore, existing 3D products from Samsung [17] and Tezzaron [18] corporations demonstrate that the silicon processing and assembly of structures in 3D stacks are feasible in large scale industrial productions.

In the literature, the "1D" approximation is often assumed to evaluate the thermal behavior of 3D integration [2, 19–21]. This means that the power is uniformly produced on "active levels" (or on part of it), one per stratum. This assumption may lead to strongly underestimated maximum temperature. Some authors [22] use this simplification but perform detailed simulation of 3D thermal effects due to the presence and localization of vias. Other works [23] analyze the local (3D) and global (1D) modeling contribution to the maximum temperature, showing that thermal resistance can be higher than 1D thermal resistance due to local 3D effects.

Numerical thermal simulations have been carried out to convert power dissipation distribution into a temperature distribution in a 3D IC [24]. Based on the past work, the development of a fundamental analytical model for heat transport in 3D integrated circuits is highly desirable. Such an analytical model will provide a framework in which to analyze the general problem of heat dissipation in 3D ICs, and will offer simple thermal design guidelines.

A key component of 3D technology is a Through-silicon via (TSV) that enables communication between the two dies as well as with the package. Some work has been reported on optimizing the problem of placement of vias for heat dissipation

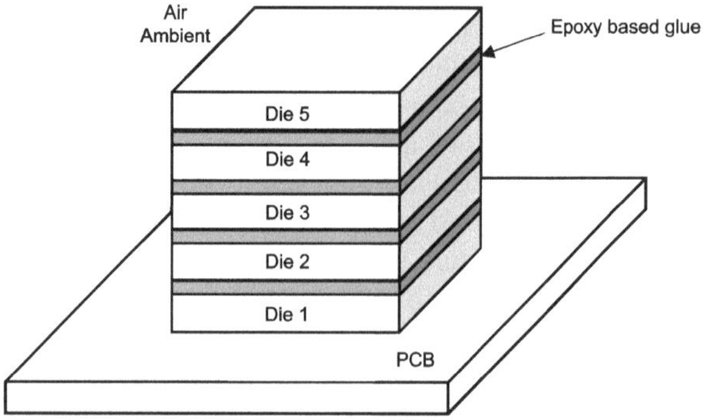

Fig. 14.2 The test 3D stacked structure

in 3D ICs [22, 25]. Other works [26] propose analytical and finite-element models of heat transfer in 3D electronic circuits and use this model to analyze the impact of various geometric parameters and thermophysical properties (through silicon vias, inter-die bonding layers, etc.) on thermal performance of a 3D IC.

This is the first time that a nano-structure of TSVs is proposed on purpose as an effective way to optimize the thermal profile in 3D stacks. The closest work to our proposal is [27], where the authors analyze the impact of thermal through silicon vias (TTVs) in vertically integrated die-stacked devices. However, while the work presented in [27] performs a theoretical analysis, our approach proposes an accurate thermal modeling of the through-silicon vias and it is validated against measurements collected in a real chip. Finally, the thermal effect of the nano-structure of the TSVs will be examined.

3 Configuration of the 3D Stack

The 3D chip manufactured for our experimental set-up is created as a multi-level chip, built by stacking silicon layers and fixed with an interface glue. In this configuration, we can find five silicon layers (Die 1–Die 5), the epoxy-based interface glue, and a bottom PCB layer (see Fig. 14.2). Each stack has an area of 1 cm^2.

This 3D stack resembles the thermal effects that can be found in a 3D multiprocessor systems on chip by the use of heaters that create the power dissipation. As the power dissipated in a chip is not uniform on its surface (microprocessors can dissipate between 200 to 300 W/cm^2 while memories only dissipate about 10 W/cm^2) each layer contains several microheaters located at different points to simulate the heat dissipated by the integrated components.

These microheaters are built as a serpentine wire created with thin-film technologies. The material used for the heaters is Platinum, due to its capability to operate at very high temperature and its long stability.

Some thermal sensors are also placed in specific places as detector devices to monitor the temperature inside of the stack and check the heat dissipated and the heat interactions between neighboring microheaters. Platinum has also been selected as the material to build the sensors; therefore, sensors and microheaters can be manufactured at the same time in a single step of the technology process. These sensors are Resistance Temperature Detectors (RTDs). In this way, the temperature of the heater creates a variation in the resistance of the sensor. The absolute value of the tolerance on RTD resistance is not an issue because the experimental setup will employ relative (differential) measurements. Then, the temperature can be obtained by the observation of the voltage drop at both extremities of the sensor (with a fixed current) and applying the resistivity temperature dependence of Platinum:

$$R_T = R_0(1 + \alpha T + \beta T^2) \tag{14.1}$$

with R_T the resistance at temperature T, R_0 the nominal resistance at 0°C, $\alpha = 3.9083e^{-3}$°C^{-1} and $\beta = -5.77e^{-7}$°C^{-2}.

Heaters and sensors are connected to the PCB by wire bonding, allowing the direct access to perform the measurements. All pads are located in one side of the chip.

Each layer comprises 10 heaters of 1 mm^2 each, very similar to the area of common processing elements. These microheaters have been designed to resemble a hot-spot on the surface of the chip of 300 W/cm^2; therefore, each heater dissipates 3 W. The heaters are aligned in three vertical lines. The 5 layers of the stack have the same configuration so the alignment of the heaters appears also out of the plane.

In our configuration, RTDs are placed around the heaters. These sensors are designed for a value of 100 Ω and are driven with a current of 1 mA. A view of the placement of heat and sensors in each layer will of the 3D stack will be analyzed later in the text (see Fig. 14.7 in Sect. 5).

The fabrication process is schematically shown in Fig. 14.3 and outlined here. A 200 nm wet oxide as an insulating layer for heaters and resistance temperature detectors (RTDs) is deposited over a double side polished wafer with a 1000.5 mm diameter and 52525 μm depth. If these components were deposited directly on the silicon surface some current may flow in the semiconductor choosing non-appropriate way through the material (not following the metallic path) with an energy loss at the desired place. Other influences like Schottky diode effects may appear.

For micro heaters and sensors, all the design and dimensioning have been done for a 500 nm platinum evaporation on the surface. Platinum has been selected as the fabrication material due to its capability to operate at very high temperature and its long stability. These structures are created using a lift-off process which involves a low-pressure vapor deposition (evaporation must be done instead of sputtering). To do this, a large distance between source and substrates is needed for a lift-off process and then a larger amount of material is used (what is not welcome here due to high cost of platinum). As 500 nm is quite large for platinum deposition, then for first trial aluminum has been evaporated instead of platinum to study the feasibility of the process.

Aluminum is suitable for wire bonding, if platinum would have been evaporated as it was planed at the beginning, a soft metal must be evaporated on it in the pad location. Aluminum metalization was done using titanium adhesion layer, 10 nm

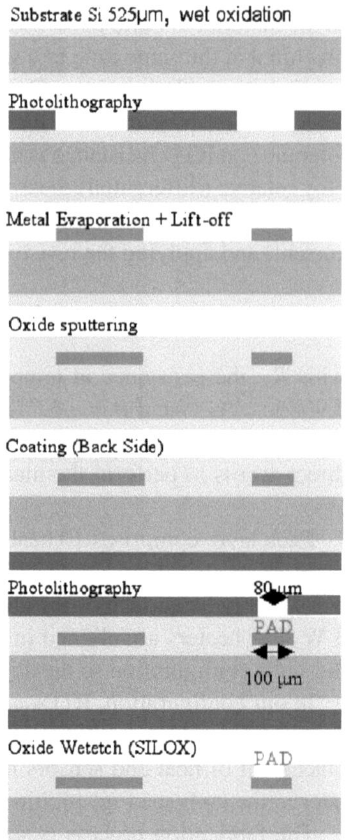

Fig. 14.3 Schematic view of the fabrication process

Ti +500 nm Al with lift-off process. Wet etching in SILOX bath with activation is done for pad etches to open contacts for heaters and sensors for wire bonding.

To build the stack, single layers have been glued on top of each other. To have a thin uniform thickness, glue has been deposited by screen printing. To do this, a semiautomatic machine (EKRA) dedicated to thick layers fabrication and suitable for gluing has been used. An epoxy resist material doped with alumina particles (Al_2O_3) has been used to avoid the creation of thermal insulating layer at the interface. Thickness deposited is around 30 μm. Stacking itself has been done manually. For alignment, layers have been leaned against a corner angle.

4 Thermal Model

The test five layered 3D stack structure considered in this work is shown in Fig. 14.2. As seen in this figure, five silicon dies, stacked one on the top of another fixed with an interface epoxy glue, are placed on the printed circuit board (PCB). The bottom surface of the 3D stack attached to the PCB is assumed to be adiabatic; therefore,

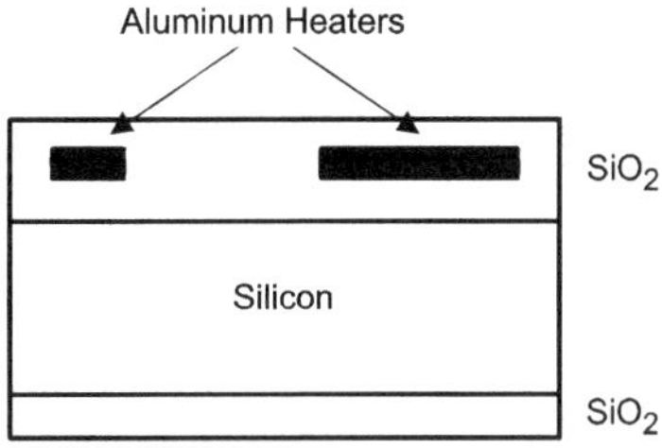

Fig. 14.4 Cross sectional view of the layers in a single die

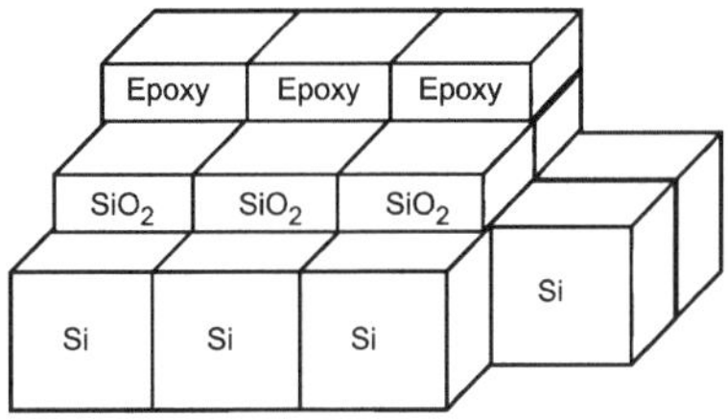

Fig. 14.5 The unitary thermal cells of the 3D stack

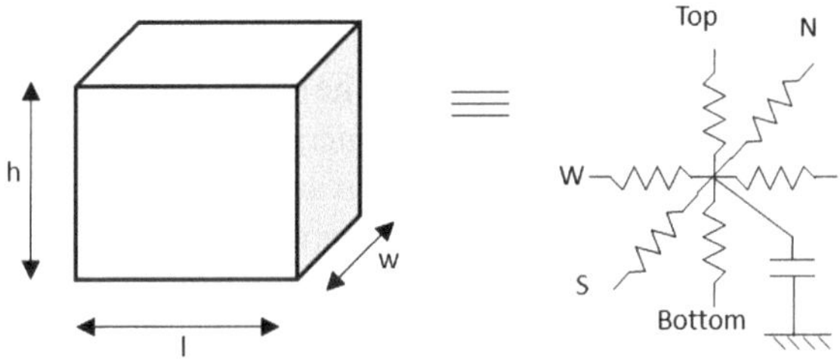

Fig. 14.6 Equivalent RC circuit of a single cell

the heat will be exchanged through the vertical active and interface layers in the system.

Within each die, the aluminum resistor-based heaters are fabricated in the silicon dioxide layer on the top of the substrate, as shown in the cross sectional view in Fig. 14.4. These heaters model the thermal effects of the hot-spot cores in an actual 3D MPSoC. The heat generated by these heaters flows through the body of the 3D stack, and ends at the environment interface (ambient) where it is spread through natural convection.

The heat flow inside this structure is diffusive in nature and hence, is modeled by its equivalence to an electronic RC circuit [28–30]. This is done by first dividing the entire structure into small cubical thermal cells as shown in Fig. 14.5. Each cell is then modeled as a node containing six resistances that represent the conduction of heat in all the six directions (top, bottom, north, south, east and west), and a capacitance that represents the heat storage inside the cell, as shown in Fig. 14.6. The conductance of each resistor and the capacitance of the thermal cell are calculated as follows:

$$g_{\text{top/bottom}} = k_{\text{th}} \frac{lw}{h/2}, \tag{14.2}$$

Table 14.1 Thermal properties of materials

Silicon thermal conductivity	$295 - 0.491T$ W/mK
Silicon specific heat	1.659×10^6 J/m^3 K
SiO_2 thermal conductivity	1.38 W/mK
SiO_2 specific heat	4.180×10^6 J/m^3 K
Aluminum electrical resistivity	$2.82 \times 10^{-8}(1 + 0.0039\Delta T)$ Ωm $\Delta T = T - 293.15$ K

$$g_{\text{north/south}} = k_{\text{th}} \frac{lh}{w/2}, \tag{14.3}$$

$$g_{\text{east/west}} = k_{\text{th}} \frac{wh}{l/2}, \tag{14.4}$$

$$c_{\text{top}} = sc_{\text{th}}(lwh). \tag{14.5}$$

Here, the subscripts top, east, south etc. indicate the direction of conduction, k_{th} and sc_{th} are the thermal conductivity and the specific heat capacity per volume unit of the material, respectively.

Current sources, representing the sources of heat, are connected to the cells in the regions where the Aluminum heaters are present. The entire circuit is grounded to the ambient temperature at the top and the side boundaries of the 3D stack through resistances, which represent the thermal resistance from the chip to the air ambient.

The behavior of the resulting RC circuit can be described using a set of first order differential equations via nodal analysis [31] as follows:

$$\boldsymbol{G}\boldsymbol{X}(t) + \boldsymbol{C}\dot{\boldsymbol{X}}(t) = \boldsymbol{B}\boldsymbol{U}(t), \tag{14.6}$$

where $\boldsymbol{X}(t)$ is the vector of cell temperatures of the circuit at time t, $\boldsymbol{G}$ and $\boldsymbol{C}$ are the conductance and capacitance matrices of the circuit, $\boldsymbol{U}(t)$ is the vector of input heat (current) sources and $\boldsymbol{B}$ is a selection matrix. $\boldsymbol{G}$ and $\boldsymbol{C}$ present a sparse block-tridiagonal and diagonal structure respectively, due to the characteristics and definition of the thermal problem.

In addition, $\boldsymbol{G}$ and $\boldsymbol{U}(t)$ are functions of the cell temperatures $\boldsymbol{X}(t)$, making the behavior of the circuit non-linear. This is because of the temperature-dependent thermal conductivity of silicon and the temperature-dependent electrical resistance of the aluminum heaters respectively. In this work, a first-order dependence of these parameters on temperatures around 300 K is assumed. Some of these parameters are shown in Table 14.1 [32].

For the validation of the thermal library, profuse temperature measurements on the 3D stack were performed with DC current inputs for the heaters. Hence, an efficient way to calculate the steady state response of the circuit described by (14.6) is required. The DC solution for the circuit can be found by solving the corresponding steady state equations,

$$\boldsymbol{G}\boldsymbol{X} = \boldsymbol{B}\boldsymbol{U}. \tag{14.7}$$

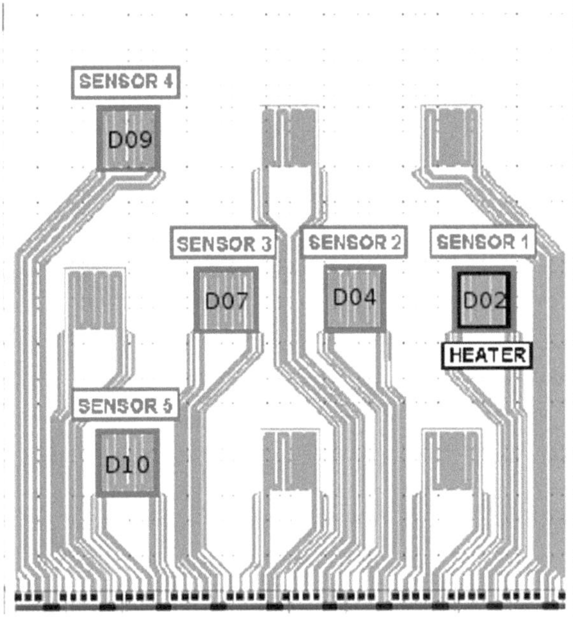

Fig. 14.7 Lateral heat transfer measurement in layer 2 (test case A)

The above set of equations are solved by the inversion of the matrix $\boldsymbol{G}$ using the sparse LU decomposition method [33]. Because of the non-linearity of the circuit and the input sources, these equations were solved repeatedly, by updating the matrices after each iteration of solving, until convergence is reached. The description of this iterative algorithm is described in the following pseudocode. In most of the test cases, 5–6 iterations were found to be sufficient to reach convergence within an error of 10^{-6}.

---

1 Define:
  $\boldsymbol{X}^r$ = vector of cell temperatures during the rth iteration,
  $\boldsymbol{G}^r$ = conductance matrix during the rth iteration,
  $\boldsymbol{U}^r$ = input vector during the rth iteration.
2 Set $r = 0$. Generate an initial-guess for $\boldsymbol{X}^0$
3 Calculate $\boldsymbol{G}^0$ and $\boldsymbol{U}^0$ using the initial-guess $\boldsymbol{X}^0$
4 $\boldsymbol{X}^{r+1} = (\boldsymbol{G}^r)^{-1}\boldsymbol{B}\boldsymbol{U}^r$
5 Calculate $\boldsymbol{G}^{r+1}$ and $\boldsymbol{U}^{r+1}$ using the updated temperatures $\boldsymbol{X}^{r+1}$
6 If $\|\boldsymbol{X}^{r+1} - \boldsymbol{X}^r\| <$ (a predetermined error criterion), exit. Else set $r = r + 1$ and go to step 4.

---

5 Electrical Measurements and Validation of the Model

Two different test cases were chosen for experimental measurements and validation of the Thermal model: A. Lateral heat transfer measurement and B. Vertical heat

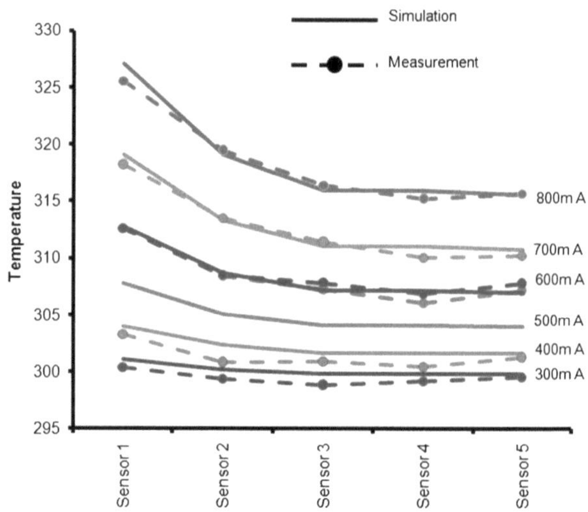

Fig. 14.8 Simulation results from lateral heat transfer measurement (test case A)

transfer measurement. The two cases are described in detail in the ensuing of this section.

5.1 *Lateral Heat Transfer Measurement (Layer 2)*

In this test case, the lateral heat diffusion in a given layer of the 3D stack is characterized. For this, the heater in device D02 in Die 2 (as described in Sect. 3) is excited with different current levels. For each current level temperature several measurements are made at the sensors in devices D02, D04, D07, D09 and D10 within the same die (since the vertical heat diffusion flows from the bottom die to the top, only the heater of D01 is used). This test case is illustrated in Fig. 14.7. These measurements provide information on the behavior of the lateral temperature distribution as function of the distance of the sensor from the heat source within a layer.

Figure 14.8 shows the comparative results between the measurements and the simulation for various heater current levels in the lateral heat transfer test case. The solid lines here indicate the simulation results and dashed lines indicate measurement results from the different sensors respectively. Heater current levels from 300 mA (1.25 W/mm^2) to 800 mA (9 W/mm^2) were used in the experiments. As can be seen from this figure, the simulation results from the thermal model accurately predict the experimental results for all sensors. The percentage of average errors between measurement and simulation results for each sensor are tabulated in Table 14.2, being these errors less than 1% for all the sensor-heater configurations.

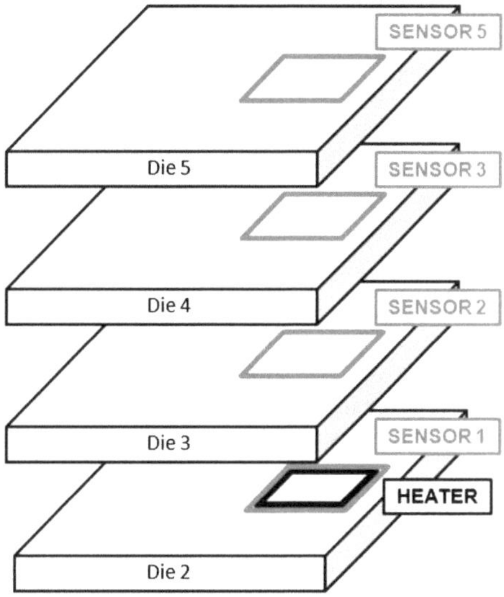

Fig. 14.9 Vertical heat transfer measurement in device D02 (test case B)

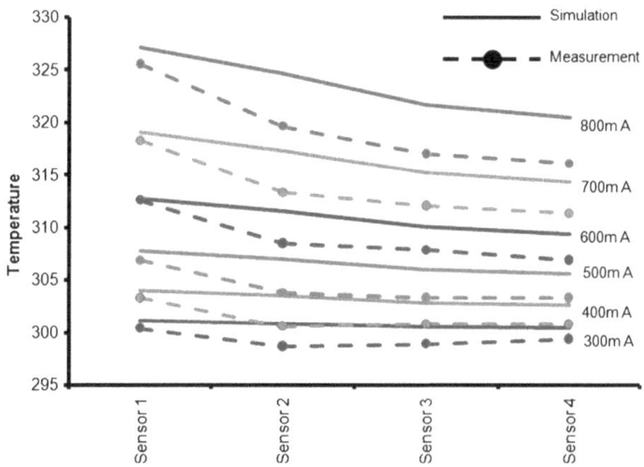

Fig. 14.10 Simulation results from vertical heat transfer measurement (test case B)

5.2 *Vertical Heat Transfer Measurement (Device D02)*

In this test case, the vertical heat flow from one layer to another in the 3D stack is characterized. For this, the heater in device D02 in Layer 2 is again excited with different current levels. For each current level, temperature measurements are made at the sensors in devices D02 of Die 2, Die 3, Die 4 and Die 5. This test case is illustrated in Fig. 14.9. These measurements provide information about the behavior of the temperature distribution as function of the vertical distance of the sensor from the heat source in different layers of the 3D stack.

Figure 14.10 shows the comparative results for the vertical heat transfer between measurements and the simulation, obtained for the same heater current levels as in test case A. Again, as in test case A, the solid lines indicate the simulation results and dashed lines indicate the measurement results for the different sensors respec-

Table 14.2 Average error for lateral heat transfer measurement (test case A)

Sensor	Average error between simulation and measurement
1	0.669%
2	0.823%
3	0.240%
4	0.426%
5	0.314%

Table 14.3 Average error for vertical heat transfer measurement (test case B)

Sensor	Average error between simulation and measurement
1	0.669%
2	1.344%
3	1.556%
4	1.783%

tively. As can be seen in this figure, the simulation results from the thermal model (implemented with a cubic grid where the size of the unitary cell is 100 μm) accurately predicts the experimental results for all sensors. The percentage of average errors between measurement and simulation results for each sensor are tabulated in Table 14.3, being in this case always less than 2%. This result is accurate enough for the purpose of this work (in-chip thermal modeling and thermal management) and responds to the tolerances in the material coefficients.

6 Thermal TSVs and Thermal Grids

During the last years, many fabrication-based solutions for the thermal management in 3D integrated circuits have been proposed. Thermal through silicon vias (TTSVs) have a prominent place among these solutions. TTSVs can also serve to transfer data vertically in the 3D stack, and this configuration could also be analyzed by the proposed approach because the thermal model would not be impacted.

Many times, it is more desirable to reduce the difference in the temperatures between various parts of the IC, rather than the reduction of the absolute temperature of the chip. This is because variations in operating temperatures affects performance of different parts of the IC (e.g. processor and memory) differently, leading to timing errors and chip failures. Moreover, thermal gradients have been observed as a determinant negative factor on system reliability.

To overcome the above mentioned challenges and to simulate the effects of on-chip metalizations on the thermal behavior of the 3D stack, thermal through silicon vias and thermal grids were introduced in the thermal model developed in the previous section. For the ensuing experiments, a 3-layered 3D stack was used instead of

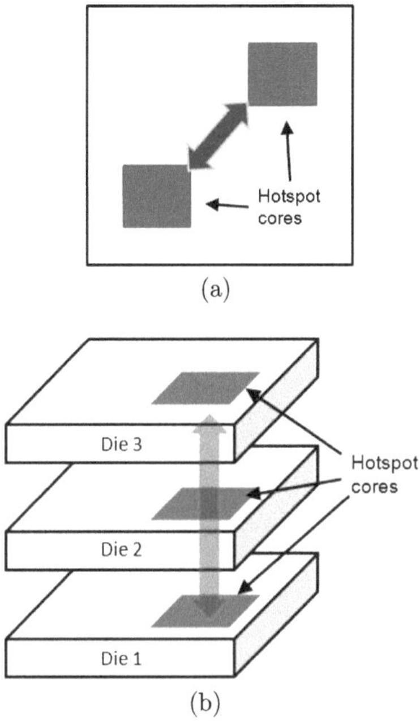

Fig. 14.11 Communication between active cores in a 3D IC: (**a**) within one layer, (**b**) between different layers

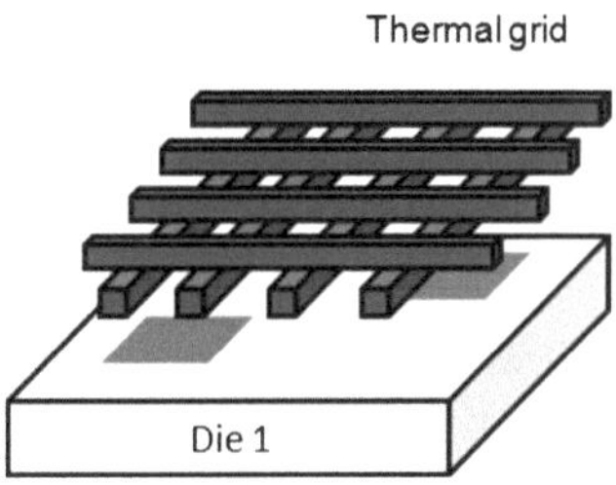

Fig. 14.12 Thermal grid for reducing temperature variation within a single layer

the 5-layered stack. Figure 14.11 shows two test cases- (a) with 2 hot-spot cores in the same die of the 3D stack and (b) with 3 cores, one on the top of another, communicating each other through different layers (from the performance-enhancement perspective, it is desirable to place the most frequently communicating cores of a 3D IC one on the top of the other to reduce communication delay).

For case (a), to reduce the temperature variations within the same layer, thermal grid networks- dedicated metalizations as well as existing metalizations for the electronic design, are proposed. These thermal grid networks lower the effective thermal conductivity of the dielectric material within the layer and hence, reduce the temperature variations in the layer. This is illustrated in Fig. 14.12 shows the schematic configuration of the horizontal grid.

For case (b), to address the temperature variations between different layers in regions where the communicating cores exist, TTSVs are placed around the active cores as shown in Fig. 14.13. This placement of TTSVs, in addition with the metalizations that naturally exist between the cores meant for electronic routing, reduces

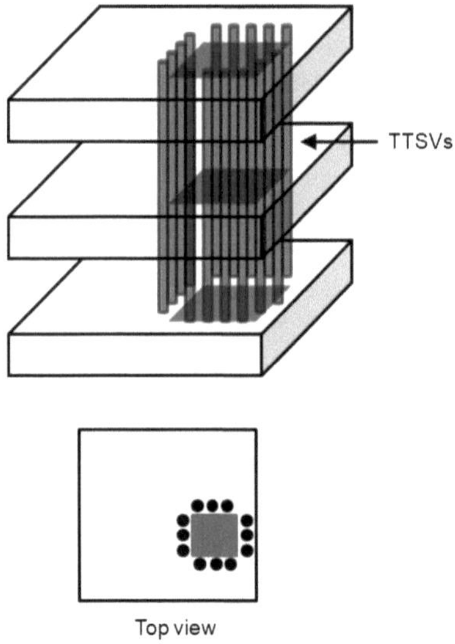

Fig. 14.13 TTSVs for reducing temperature variations along the different layers in a 3D IC

the effective thermal conductivity of this region. This, in turn, brings the temperature of different parts of this region closer to each other because of the favored thermal flow.

To incorporate both the thermal grid and the TTSVs in the thermal model, effective thermal conductivity was calculated for the cells in the region containing these metalizations, using the following relation:

$$k_{\mathrm{eff}} = k_{\mathrm{cu}}\omega + k_{\mathrm{th}}(1-\omega), \tag{14.8}$$

where, k_{cu} is the thermal conductivity of copper (the metal used for all metalizations in the IC), k_{th} is the thermal conductivity of the surrounding material and ω is the wiring/via density in the region. In the case of TTSVs, a slight modification was made for the effective thermal conductance in the lateral direction. This parameter was calculated by computing the equivalent thermal resistance of the cells depending upon the path of heat flow while traversing it along north-south and east-west direction (a series/parallel combination of vias and surrounding material). Hence, anisotropic cells were created in order to more accurately capture the effects of TTSVs.

Figure 14.14 shows the devised nano-grid[1] of horizontal interconnects and vertical TTSVs. The TTSVs that integrate the nano-structure improve the overall thermal conductivity of the active layer, provided the thermal coupling is good between the TTSVs. To improve the thermal coupling, these TTSVs must be placed as close as possible to each other but electrically isolated. On the other hand, the horizontal grid helps to spread the temperature along the die and also improve the thermal conductivity of the Inter-Layer Material.

[1] The term *nano* is used here to emphasize the size of this structure, in the same order of magnitude as the TSVs used in the device.

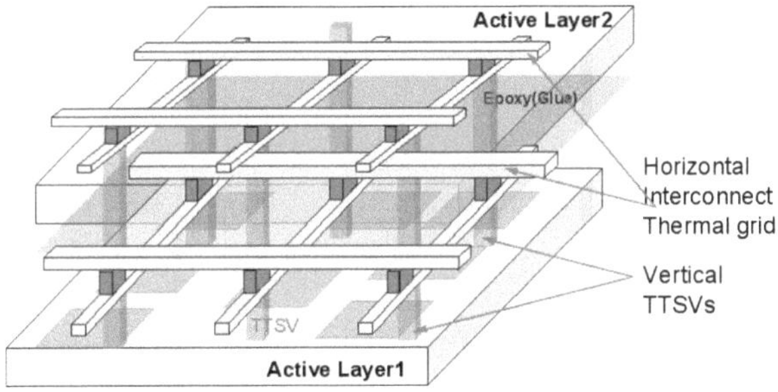

Fig. 14.14 Vertical and horizontal thermal grid

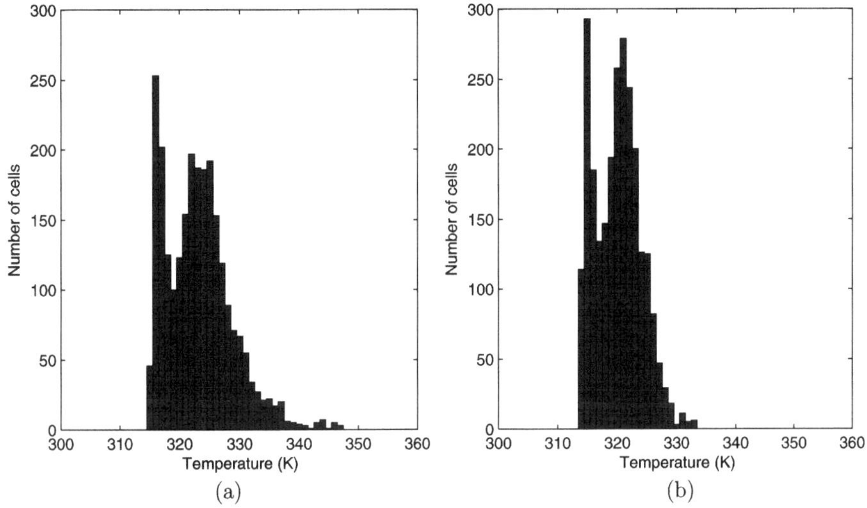

Fig. 14.15 Lateral temperature distribution profile for Die 1: (**a**) without thermal grid and (**b**) with thermal grid

Two experiments were performed to measure the performance of these two strategies. In the first experiment, 4 heaters (in devices D02, D04, D07 and D10) in Die 1 of the 3-layered 3D stack were excited, each with a current of 300 mA (1.25 W/mm^2). First, this experimental set up was simulated without any thermal grid. Next, thermal grid was added to Die 1 (with 50% wiring density) in the same experimental set up and the resulting model was simulated again. The temperature distribution profile was drawn for each case. These histograms are shown in Fig. 14.15. As can be seen from this figure, the temperature spread within this layer has been reduced by the effect of the thermal grid, that easies the diffusion of the extra heat.

In the next experiment, the same set up was used. TTSVs were laid around each of the active heaters in Die 1 as shown in Fig. 14.13. The resulting thermal circuit was then simulated, once without the TTSVs and then once with the TTSVs. Temperatures in the region covered by the TTSVs of one of the heaters (the region enclosed by the TTSVs encompassing all the 3 dies as shown in Fig. 14.13) were

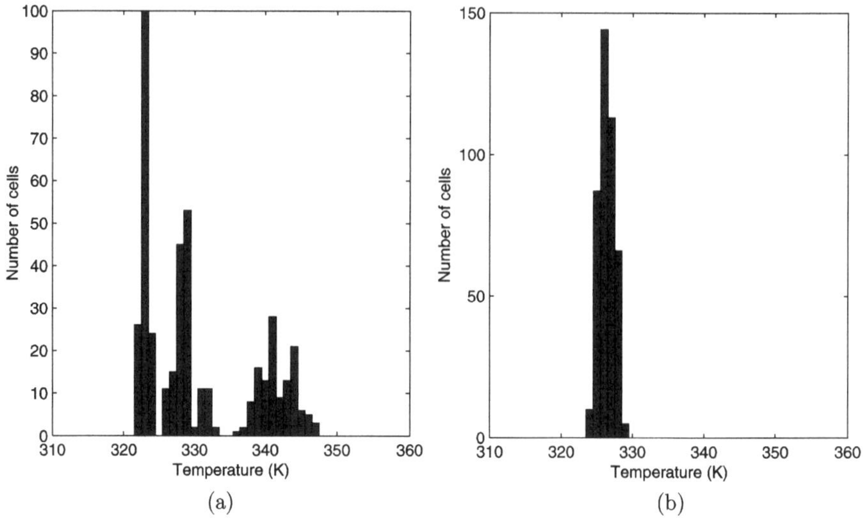

Fig. 14.16 Vertical temperature distribution profile for region around D06: (**a**) without TTSVs and (**b**) with TTSVs

recorded in each case. The corresponding temperature distribution profiles for one such active heater regions are shown in Fig. 14.16. We find that the temperature spread was considerably reduced in this region along the vertical direction. Therefore, the grid of TTSVs can be considered as an effective mechanism to optimize the thermal profile in 3D stacks, both in the vertical and lateral direction.

7 Conclusion

This paper presents a nano-grid of TSVs as an effective mechanism to optimize the thermal profile in 3D integrated systems. In this work, an accurate modeling of the thermal effects that appear in these structures has been developed, and a profuse validation process has been carried out. The measurements performed in a real 5-layered 3D chip manufactured on purpose confirm the validity of the model with an error lower than a 2% in all the cases (lateral and vertical heat diffusion).

The proposed thermal model has then used to evaluate the capability of a nanostructure of thermal through-silicon vias to improve the thermal response of a complex 3D system. The nano-grid is configured to reduce the impact of high-density temperature hot-spots, providing very positive results in the optimization and homogenization of the vertical and lateral diffusion of heat.

Acknowledgements This research has been partially funded by the Nano-Tera.ch RTD Project CMOSAIC (ref. 123618), which is financed by the Swiss Confederation and scientifically evaluated by SNSF. This work is also funded by the Spanish Ministry under contract TIN2008-508.

References

1. Das, S., Chandrakasan, A., Reif, R.: Design tools for 3-D integrated circuits. In: Proceedings of the 2003 Asia and South Pacific Design Automation Conference, pp. 53–56 (2003)
2. Banerjee, K., Souri, S.J., Kapur, P., Saraswat, K.C.: 3-D ICs: a novel chip design for improving deep-submicrometer interconnect performance and systems-on-chip integration. In: Proceedings of the IEEE, vol. 5, pp. 602–633 (2001)
3. Topol, A.W., et al.: Three-dimensional integrated circuits. IBM J. Res. Dev. **4–5**, 494–506 (2006)
4. Deng, Y., Maly, W.P.: Interconnect characteristics of 2.5-D system integration scheme. In: Proceedings of the 2001 International Symposium on Physical Design, pp. 171–175 (2001)
5. Deng, Y.S., Maly, W.: 2.5D system integration: a design driven system implementation schema. In: Proceedings of the 2004 Asia and South Pacific Design Automation Conference, pp. 450–455 (2004)
6. Kgil, T., D'Souza, S., Saidi, A., Binkert, N., Dreslinski, R., Mudge, T., Reinhardt, S., Flautner, K.: Picoserver: using 3D stacking technology to enable a compact energy efficient chip multiprocessor. Oper. Syst. Rev. **40**(5), 117–128 (2006)
7. Mysore, S., Agrawal, B., Srivastava, N., Lin, S.-C., Banerjee, K., Sherwood, T.: Introspective 3D chips. Comput. Archit. News **34**(5), 264–273 (2006)
8. Rahman, A., Reif, R.: System-level performance evaluation of three-dimensional integrated circuits. IEEE Trans. Very Large Scale Integr. **8**(6), 671–678 (2000)
9. Morrow, P., et al.: Wafer-level 3D interconnects via CU bonding. In: Proceedings of the 2004 Advanced Metalization Conference (2004)
10. Tsai, Y.-F., Xie, Y., Vijaykrishnan, N., Irwin, M.J.: Three-dimensional cache design exploration using 3DCacti. In: Proceedings of the 2005 International Conference on Computer Design, pp. 519–524 (2005)
11. Zeng, A.Y., Lu, J., Gutmann, R., Rose, K.: Wafer-level 3D manufacturing issues for streaming video processors. In: Proceedings of the Advanced Semiconductor Manufacturing Conference, pp. 247–251 (2004)
12. Mayega, J., Erdogan, O., Belemjian, P.M., Zhou, K., McDonald, J.F., Kraft, R.P.: 3D direct vertical interconnect microprocessors test vehicle. In: Proceedings of the 13th ACM Great Lakes Symposium on VLSI, pp. 141–146 (2003)
13. Puttaswamy, K., Loh, G.: The impact of 3-dimensional integration of the design of arithmetic units. In: Proceedings of the IEEE International Symposium on Circuits and Systems, pp. 4951–4954 (2006)
14. Black, B., Nelson, D.W., Webb, C., Samra, N.: 3D processing technology and its impact on IA32 microprocessors. In: Proceedings of the IEEE International Conference on Computer Design, pp. 316–318 (2004)
15. Nelson, D.W., et al.: A 3D interconnect methodology applied to IA32-class architectures for performance improvement through RC mitigation. In: Proceedings of the 21st International VLSI Multilevel Interconnection Conference (2004)
16. Xie, Y., Loh, G.H., Black, B., Bernstein, K.: Design space exploration for 3D architectures. ACM J. Emerg. Technol. Comput. Syst. **2**(2), 65–103 (2006)
17. http://www.samsung.com/us/business/semiconductor/newsView.do?news_id788.0
18. http://www.tezzaron.com/technology/FaStack.htm
19. Im, S., Banerjee, K.: Full chip thermal analysis of planar (2-D) and vertically integrated (3-D) high performance ICs. In: IEDM Technical Digest. International Electron Devices Meeting (2000)
20. Rahman, A., Reif, R.: Thermal analysis of three-dimensional (3-D) integrated circuits (ICs). In: IITC Conference (2001)
21. Chiang, T.-Y., Souri, S.J., Chui, C.O., Saraswat, K.C.: Thermal analysis of heterogeneous 3-D ICs with various integration scenarios. In: IEEE International Electron Devices Meeting (2001)
22. Goplen, B., Sapatnekar, S.: Placement of thermal vias in 3-D ICs using various thermal objectives. IEEE Trans. Comput.-Aided Des. Integr. Circuits Syst. **25**(4), 692–709 (2006)

23. Leduca, P., et al.: Challenges for 3D IC integration: bonding quality and thermal management. In: IEEE International Interconnect Technology Conference (2007)
24. Puttaswamy, K., Loh, G.H.: Thermal analysis of a 3D die-stacked high-performance microprocessor. In: Proceedings of the 16th ACM Great Lakes Symposium on VLSI, pp. 19–24 (2006)
25. Cong, J., Zhang, Y.: Thermal via planning for 3-D ICs. In: Proceedings of the 2005 IEEE/ACM International Conference on Computer-Aided Design, pp. 745–752 (2005)
26. Jain, A., Jones, R., Chatterjee, R., Pozder, S., Huang, Z.: Thermal modeling and design of 3D integrated circuits. In: Intersociety Conference on Thermal and Thermomechanical Phenomena in Electronic Systems (2008)
27. Natarajan, V., Deshpande, A., Solanki, S., Chandrasekhar, A.: Thermal and power challenges in high performance computing systems. In: International Symposium on Thermal Design and Thermophysical Property for Electronics (2008)
28. Heo, S., Barr, K., Asanovic, K.: Reducing power density through activity migration. In: Proceedings of the ISPD (2003)
29. Skadron, K., Stan, M.R., Sankaranarayanan, K., Huang, W., Velusamy, S., Tarjan, D.: Temperature-aware microarchitecture: modeling and implementation. ACM Trans. Archit. Code Optim. **1**, 94–125 (2004)
30. Su, H., Liu, F., Devga, A., Acar, E., Nassif, S.: Full chip leakage estimation considering power supply and temperature variations. In: Proceedings of the ISPD, pp. 78–83 (2003)
31. Vlach, J., Singhal, K.: Computer Methods for Circuit Analysis and Design. Springer, Berlin (1983)
32. Incropera, F.P., Dewitt, D.P., Bergman, T.L., Lavine, A.S.: Fundamentals of Heat and Mass Transfer. Wiley, New York (2007)
33. Davis, T.A., Duff, I.S.: An unsymmetric-pattern multifrontal method for sparse LU factorization. SIAM J. Matrix Anal. Appl. **1**, 140–158 (1997)

Chapter 15
3D Architectures

From 3D integration technologies to complex MPSoC architectures

Walid Lafi and Didier Lattard

1 Introduction

The electronics industry has achieved unusual development over the last decades, mainly due to the rapid progresses in integration technologies. As more and more complex functions are required in present electronic devices such as video processing and telecommunication, the need to integrate these functions in a small system/package and to reduce their power consumption is also increasing.

To address this issue, current technological solutions consist of continuing system scaling according to Moore's Law (the number of transistors on a chip doubles every 18 to 24 months). An important consequence of this trend is the emergence of System-on-Chip (SoC) which refers to integrating many functions into a single integrated circuit (chip). Another alternative direction of today's integrated circuit is the "More than Moore" that consists of using existing CMOS processes to develop new micro-devices with extended functionality such as sensors, MEMS, RF or analog. At heterogeneous level, System in Package (SiP), which is composed of a set of several integrated circuits in a single package, is a classic solution for reducing system form factor.

But the standard packaging technologies used for SiP production are still limited in terms of form factor and wire length. Another issue with SiP is the decrease in the manufacturing yield since any defective module in the package will result in a non-functional final circuit, even if all other chips in that same package are functional. At the same time, advanced integration technologies are facing several problems such as the exponential increase of leakage and the not-so-exciting gain in performance. CMOS technology seems to reach its physical limits with the 22 nm technological node, and further scaling of SoC becomes uncertain and very expensive.

3D die stacking is emerging as a promising solution allowing convergence of SiP and SoC, by combining higher performance (More Moore) with complex and ex-

W. Lafi · D. Lattard (✉)
CEA-LETI, MINATEC, 17 avenue des Martyrs, 38054 Grenoble Cedex 9, France
e-mail: didier.lattard@cea.fr

G. Nicolescu et al. (eds.), *Design Technology for Heterogeneous Embedded Systems*,
DOI 10.1007/978-94-007-1125-9_15, © Springer Science+Business Media B.V. 2012

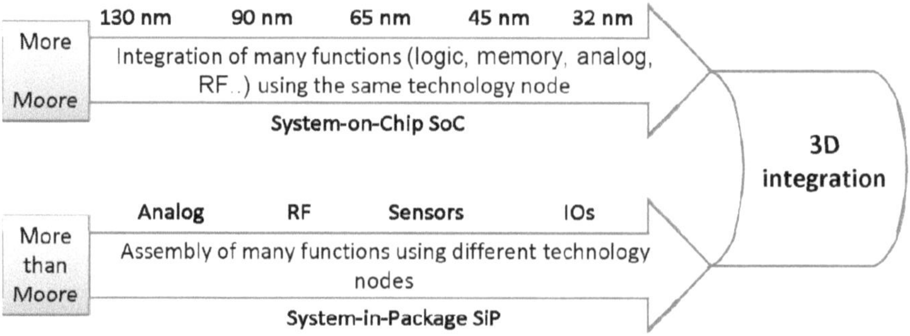

Fig. 15.1 Convergence of the More Moore and the More than Moore trends

tended functionalities (More than Moore) (see Fig. 15.1). 3D integration technology consists of stacking many circuits vertically and connecting those using Through-Silicon-Vias (TSVs). This results in smaller circuit footprint and shorter vertical interconnections, which improves system performance and power. Besides, heterogeneous systems can be built easily, since each layer can support diverse technology.

The remainder of the chapter is organized as follows. In Sect. 2, 3D integration solutions for heterogeneous systems are described. An overview of 3D manufacturing technologies and related issues such as yield is presented in Sect. 3. Section 4 introduces some potential applications. Section 5 focuses on 3D MPSoC architectures.

2 3D Integration Solutions for Heterogeneous Systems

Three-dimensional ICs, as shown in Fig. 15.2, consist of planar device layers stacked one atop another and interconnected by short, vertical wires.

Present system-on-chips (SoCs) usually integrates heterogeneous functions. These functions are initially designed for different manufacturing technologies. This further complicates the process and increases manufacturing cost. For example, in a RF-CMOS process, the total price of a final wafer exceeds that of pure CMOS by more than 15% [1]. Although it is possible to manufacture digital logic, memories, DSPs, analog and RF devices on a single die using one technology, this is suboptimal in terms of performance, area, and power. Advanced digital technologies are not well adapted to realize functions such as analog or RF circuits. Past attempts to converge these different functions onto a single monolith resulted in many issues related mainly to cost and performance. Therefore, it is preferable to make each type of circuit in its own mature technology node in order to get higher performance and lower cost. Even within the same process, it may be advantageous to have layers with different power and performance requirements or clocking domains.

At the same time, with increasing per-die circuit size, the delay generated by the long interconnections (global wires connecting opposite ending of the chip) dominates gate delay and becomes the limiting factor in the performance of today's in-

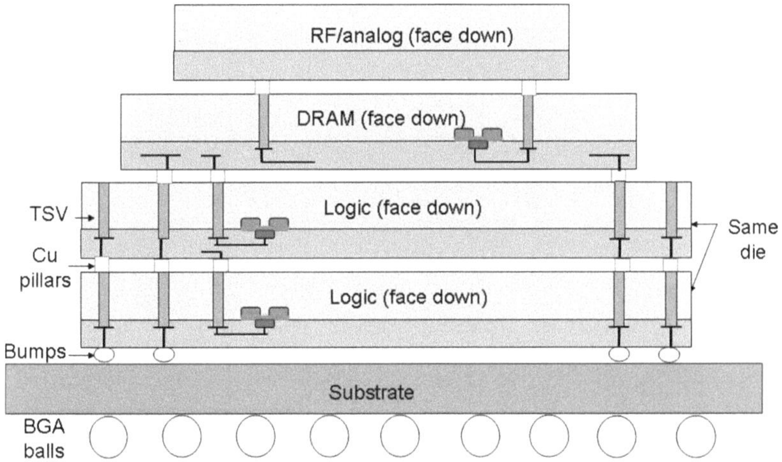

Fig. 15.2 3D heterogeneous system example

tegrated circuits. Furthermore, long interconnects result in high power consumption and significant coupling noise between signal wires.

Vertical integration enables both heterogeneous technologies integration, interconnects' length shortening and chip size reduction. These opportunities have been shown to provide a number of distinct advantages in terms of cost, performance and form factor.

2.1 Reducing Cost

A significant advantage of 3D integration is the ability to integrate heterogeneous technologies built with different processes on the same stacked chip. This means independent manufacturing of different functions such as analog, digital, or memory and their subsequent integration in the same final system. It is then possible to manufacture each type of circuit using the most adapted technology [2].

Currently, a general VLSI application without regular system architecture requires multiple sets of masks. This can be extremely expensive since mask prices for cutting-edge processes have been increasing steadily (see Fig. 15.3). According to the ITRS, the cost of just one mask set and its corresponding probe already exceeds 1 M€. For this reason, reusing mask and reducing the number of mask layers is becoming critical.

Given that devices in the non-monolithic integration approaches may require a smaller number of mask layers, 3D die stacking technology enables reducing the number of masks for each stacked application. Reducing the number of mask layers for expensive technologies (such as logic circuit) allows significant manufacturing cost savings.

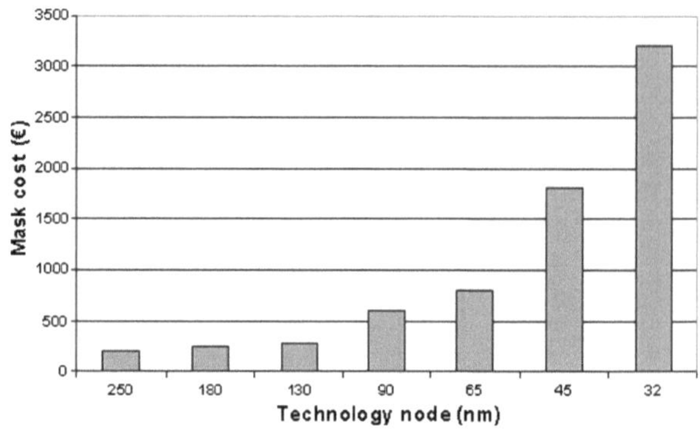

Fig. 15.3 Increase in the cost of a set of masks according to technological nodes

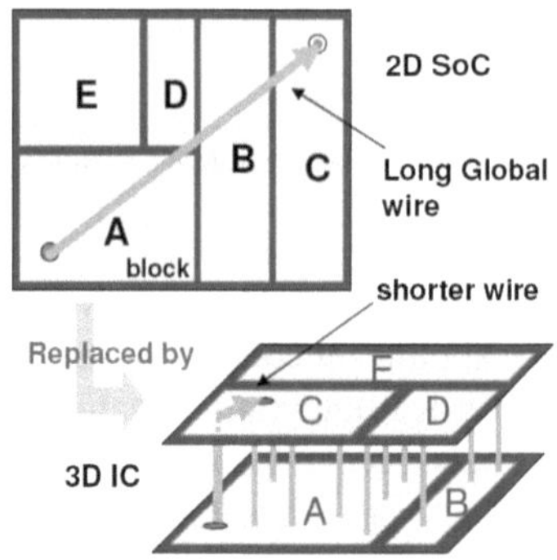

Fig. 15.4 Interconnects' length shortening in 3D ICs

Moreover, 3D integration technologies enable assembling circuits from a grouping of existing masks rather than designing a new mask set. While timing-critical digital masks are persistently updated for newer processes, existing masks for certain components, especially analog devices, could be reused for many years.

In conclusion, 3D stacking enables the integration of heterogeneous technologies on the same chip. Therefore, it is possible to make each type of circuit using its ideal technology. This enables using the best manufacturing process, to reduce the number of mask layers for each stacked circuit and to reuse masks.

2.2 Enhancing Form Factor and Performance

One of the most obvious advantages of 3D integration is to replace long horizontal wires with short vertical interconnects (TSV—Through-Silicon Via). Figure 15.4 illustrates the overall reduction of interconnections. The global inter-blocks wiring in 2D circuits (the longest wires) here are replaced by short vertical interconnections.

These shorter wires will decrease the average load capacitance and resistance (wire's capacitance and resistance are proportional to wire length) and reduce the number of repeaters needed for long wires.

Since interconnect wires with their supporting repeaters consume a significant portion of total active power, the average interconnect length reduction in 3D IC will significantly reduce overall power consumption. Looking ahead, die stacking would allow the upper layer to be dedicated to low-skew clock distribution, enabling significant reduction in clocking power.

Moreover, the shorter interconnects in 3D ICs with consequent reduction of load capacitance and reduced numbers of repeaters will reduce the noise resulting from simultaneous switching events and coupling between signal lines. This should provide better signal integrity.

Another major consequence of the reduced wire resistance and capacitance is the significant reduction of signal propagation delay (proportional to the product resistance times capacitance), resulting in significant system performance gain.

As shown in Fig. 15.4, 3D integration technologies allow chip surface reduction and better form factor. Thus, it would be possible to continue chip miniaturization without necessarily following Moore's Law.

In conclusion, 3D stacking allows to have shorter global interconnects, and thus reduce total active power, switching and coupling noise, and signal delay.

In spite of all these benefits relevant to cost and performance, vertical integration is currently facing a major problem that hinders its industrial application -that is the high temperature increase. The next subsection deals with the causes of the thermal issue and presents some potential solutions.

2.3 Open Issue in 3D Systems: Thermal Management

One of the most critical challenges of 3D design is heat dissipation. The thermal issue in the vertical stacking is caused by two main reasons. Firstly, in the 3D circuits more devices are packed into a smaller volume, resulting in a rapid increase of power density, especially if stacked tiers are highly active (typically logic circuit). In addition, heat from the top-most core of a 3D chip has to travel through several layers, inserted between device layers for insulation, to reach a heat sink. The thermal conductivity of these dielectric layers is very low compared to metal layers or even silicon. High temperature has two main drawbacks: it can limit the operating frequencies of vertically-stacked chip and degrades chip reliability. These two factors have made thermal management to be identified as a critical issue in 3D devices.

Many academic and industrial researches propose several promising methods to solve thermal issues. An important technological solution consists of inserting thermal vias (see Fig. 15.5) that are used to create thermal paths helping heat dissipation from a core on a stacked chip to the heat sink [3]. Other researches deal with the thermal problem from a design point of view. They present thermal-driven 3D floor planning algorithms which take a thermal model of the 3D circuit into account in order to achieve temperature optimization [4].

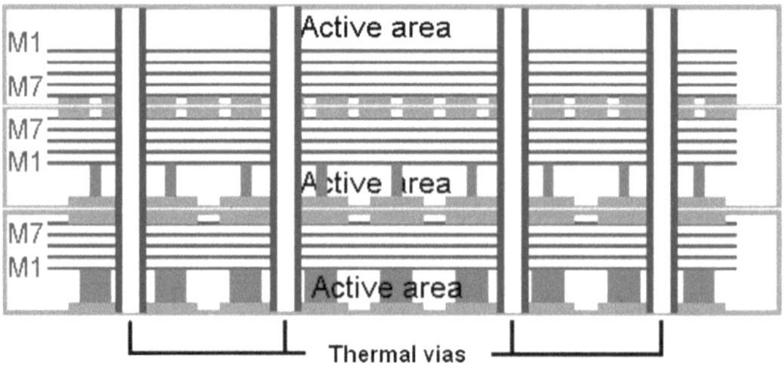

Fig. 15.5 Thermal vias in a 3D chip

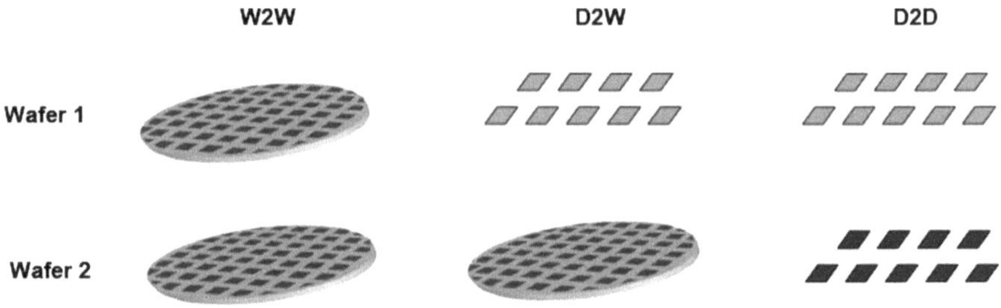

Fig. 15.6 Three ways to assemble the circuits vertically

3 3D Circuit Manufacturing Technologies

3D integration consists of stacking many integrated circuits so that the final 3D device has the same thickness as a classic planar device. There are many technological options to set up a 3D integrated circuit [5].

A critical issue is to choose the way to assemble chips (see Fig. 15.6):

- Die-to-die (D2D): This approach requires a stringent alignment effort since it seems difficult to handle small dies. Besides, chips' assembly is time consuming (Pick and Place).
- Wafer-to-wafer (W2W): In this case, the time necessary for chips' assembly is much shorter. Further, alignment is easier since assembly is performed on bigger objects.
- Die-to-wafer (D2W): The time needed for stacking is less significant than in the D2D option. Alignment issue is also less critical than in the die-to-die approach.

Another important topic is to decide on how active levels are oriented (see Fig. 15.7):

- Face-to-face (F2F): This option gives a high integration density, since there is no TSV (Through Silicon Via) in the active area. However, it is not possible to stack more than two layers this way.

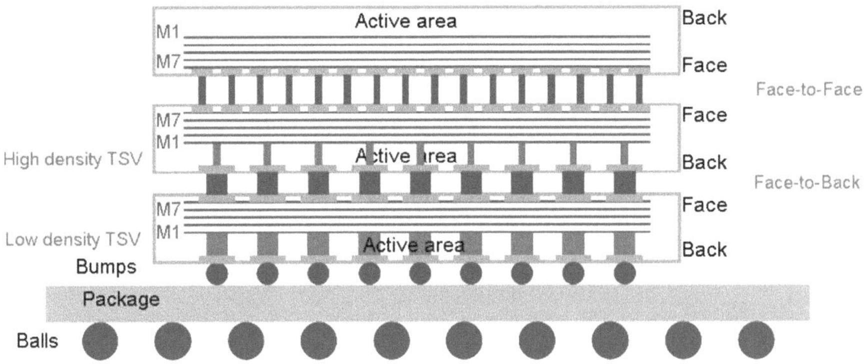

Fig. 15.7 F2F and F2B approaches in 3D stacking

- Face-to-back (F2B): In this case, integration density is lower, but stacking more than two layers is possible.

To achieve 3D stacking, some key technological steps must be fully involved [5, 6]:

- Wafer thinning: the chips have to be thinned in order to have a good form factor and reduce interconnections' length;
- Bonding;
- Alignment (associated with the bonding step): alignment precision affects integration density;
- Vertical interconnections' (TSVs) formation.

3.1 Wafer Thinning

The wafer can be either a bulk Si or SOI (Silicon on Insulator). Although more expensive, the buried oxide layer (BOX) in SOI wafers provides a selective etch stop for the uniform removal of the Si substrate. This step is more critical on bulk Si wafer. Substrate thinning is mandatory to fabricate high density TSVs with an aspect ratio from 1 to 10. It may be performed in two steps:

- Grinding step consists of removing a large amount of material at high speed using an abrasive wheel.
- Polishing step allows getting rid of roughness and obtaining a good quality uniform surface. This is indispensable to enable the subsequent technological process. This step is often carried out by combining chemical–mechanical polishing (CMP) and wet chemical surface treatment.

Current technologies can achieve wafer thicknesses as low as 10 μm.

3.2 Bonding

A variety of bonding methodologies are being considered. Nevertheless, there is no rework solution so far. For all types of bonding methods, the quality of the bonded interface strongly depends on surface roughness and cleanliness. Bonding may be achieved two different ways: metallic bonding or molecular (direct) bonding.

3.3 Alignment

The alignment step may be performed before or during bonding using either optical or infrared microscopy. Depending on the chosen integration scheme (F2F or F2B), the alignment will be performed according to the substrate face or the substrate back. On the other hand, the higher the integration density will be the greater alignment accuracy will be required. Alignment precision depends not only on equipment but also on substrates' deformation generated by the technological steps preceding alignment. If the bonding interface is not smoothed sufficiently, alignment can be degraded by the following process. Currently, obtained alignment precision may reach 1 μm.

3.4 TSVs' Formation

The development of TSVs may be performed at different stages of the manufacturing process:

- Via first—mid-process: after the front-end (transistors);
- Via first—post-process: after the back-end (interconnections);
- Via last: after wafers' bonding.

TSVs may be filled with copper (most often used), tungsten or even poly-silicon. Depending on applications, the TSV diameter may vary from 1 to 100 μm and their form factor (height to width ratio) from 1 to 30 (see Fig. 15.8). Smaller TSV sizes give the highest integration density. Therefore, reducing vias' size is a critical issue for both manufacturer and designer in order to reduce 3D circuit cost.

3.5 3D IC Manufacturing Yield

For a technology to be economically viable, cost and yield should be the most important issues for chip designer and producer.

In general, the larger chip size provides the lower yield. Hypothetically, we can suppose that a wafer has a known number of fatal defects that are spread randomly

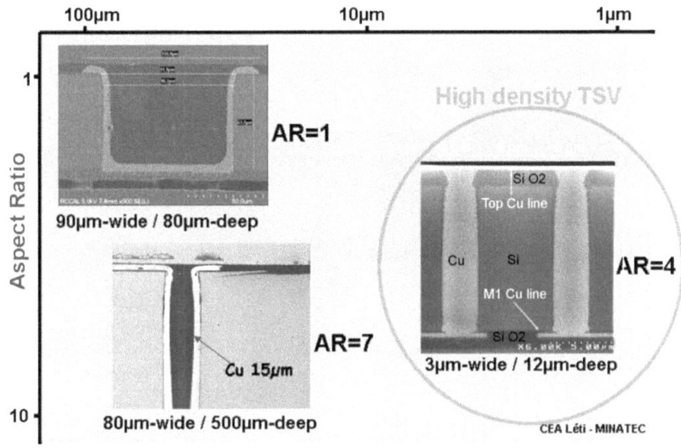

Fig. 15.8 Different TSV sizes and form factors

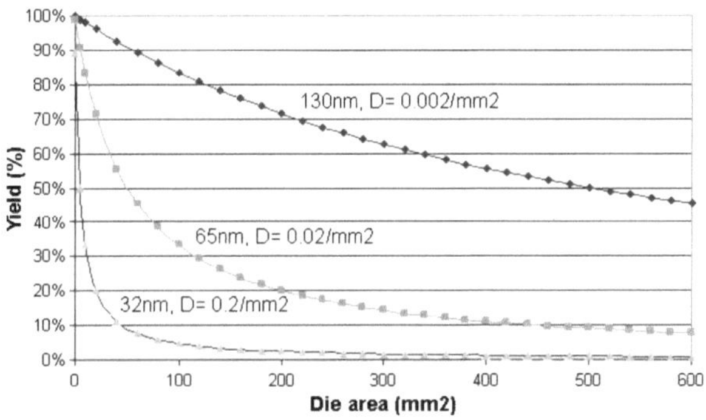

Fig. 15.9 Yield according to area

over the wafer surface. Thus, the average number of defects per chip would be AD, where A is the chip area and D is the number of defects on the wafer divided by the surface of the wafer. The yield can be defined as the Poisson yield model: $Y = \exp(-AD)$. Figure 15.9 shows the yield variation according to chip area for different technology nodes; when the chip size increases, yield falls with rates that depend on the technology maturity. As seen is Sect. 2, 3D integration allows reducing total chip area and thus enables significant enhancement in individual chip yield.

In reality, total manufacturing yield in 3D stacking technology depends not only on the yield of individual die production but also on the chosen assembly scheme: die-to-die, wafer-to-wafer or die-to-wafer. Indeed, when stacking n dies made with technology yield Y_d, the final yield Y_f of the 3D system enclosing these n stacked chips becomes: $Y_f = Y_s \cdot Y_d^n$, with Y_s being the 3D integration (interconnections and assemblies) technology yield. For example, when stacking 4 wafers having a yield of 80%, using an assembly technology with a yield of 90%, the final yield drops to 37% with the W2W approach. In the case of assembly schemes D2D or D2W, it

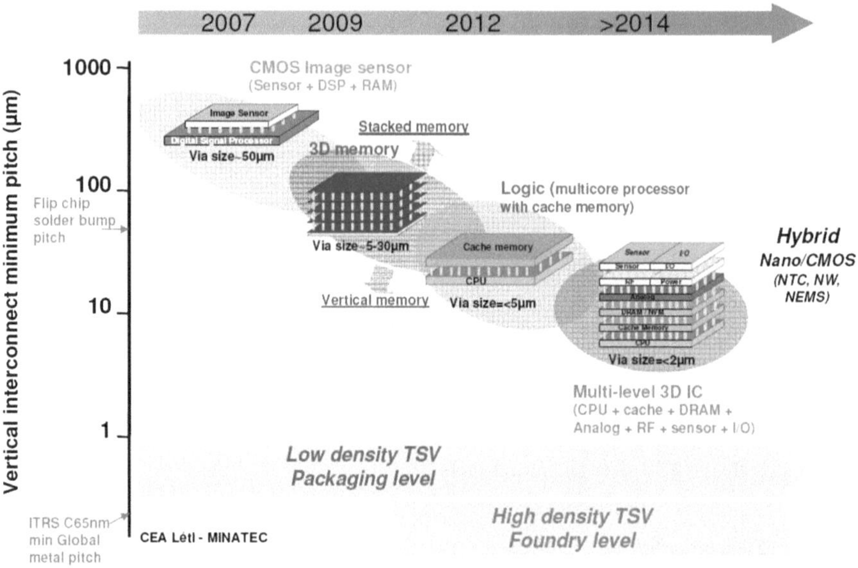

Fig. 15.10 3D applications roadmap

is possible to test the chip in advance and select only the functional chips (known good die); the final yield reaches 72%.

4 Potential Applications

Many manufacturers and research laboratories are working on 3D integration. Several demos have been made to prove the validity of TSV and bonding technologies. Figure 15.10 depicts a roadmap of 3D applications depending on advances in reducing TSV size.

The first achievements looked at homogeneous technologies, especially memory. Since 2006, Samsung Electronics has been the first to announce the stacking of 8 wafer-level processed 2 Gb NAND flash memory, for a total height of 0.56 mm using TSV interconnection technology. This miniaturization of memory presents several potential benefits especially for mobile applications. In 2006, Intel published the wafer-level stacking of 4-MB SRAM memories [7]. The bonding was done using 330 mm Si bulk wafers processed in 65 nm technology and face to face assembly. The top wafers were thinned to different thicknesses ranging from 5 to 28 µm.

As shown in Fig. 15.10, first 3D heterogeneous technologies integration began in 2006 with Tohoku University proposing a novel retinal prosthesis system consisting of several LSI (Large-Scale Integration) chips vertically stacked and electrically connected using 3D integration technology [8]. The retinal prosthesis chip is made including photo-detectors. In 2007, MIT and Yale University presented the design

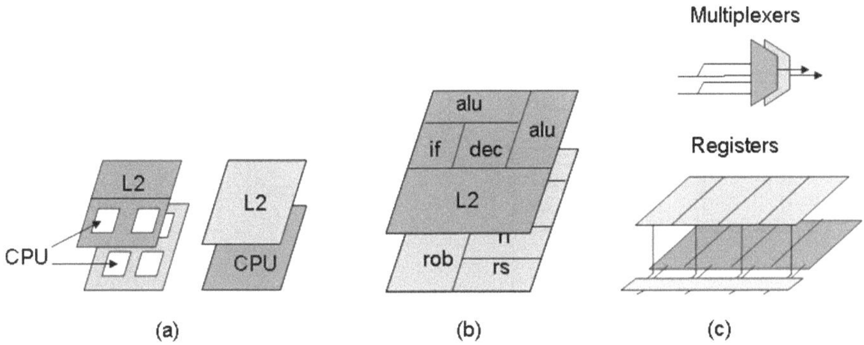

Fig. 15.11 Partitioning granularities: (**a**) macroscopic blocks; (**b**) functional units; (**c**) basic operators

and measurement results of 3D photo-detectors made using 0.18 μm SOI CMOS technology [9]. From the year 2009 on, advances in 3D integration technologies would allow drastic reduction of TSV size, enabling several other applications especially 3D MPSoC. First versions of such system consist of a vertically stacked IO, analog and digital chips. Future works will focus on stacking IO, digital and memory. 3D MPSoC is a very representative subsystem of a completely 3D heterogeneous circuit incorporating also RF, MEMS, sensors, etc. For this reason, Sect. 5 is entirely dedicated to 3D MPSoC (stacking memory on top of processor).

With further progresses in integration technologies, it would be possible to make a complete 3D heterogeneous system; this depends especially on advances in terms of TSV size reduction and thermal management.

5 3D MPSoC Architectures

5.1 Challenges in Partitioning Aspects

Circuit partitioning between layers may be carried out according to different granularities [10]. As shown in Fig. 15.11(a), the designer may choose to stack macroscopic blocks such as memory and processor. Examples include staking L2 cache or main memory on top of a CPU. It is also possible to partition a circuit according to functional units such as stacking the arithmetic and logical unit directly above the register file (Fig. 15.11(b)). This means splitting the processor itself across two or more layers. The finest granularity consists of stacking basic operators such as multiplexers and logic gates. This enables splitting functional units like the instruction scheduler or the register file across several layers (Fig. 15.11(c)). With current technologies, the last two approaches seem not to be viable given that current TSV size is huge compared to the size of stacked blocks. Besides, the power density in such circuits is considerable, and temperature increase is prohibitive.

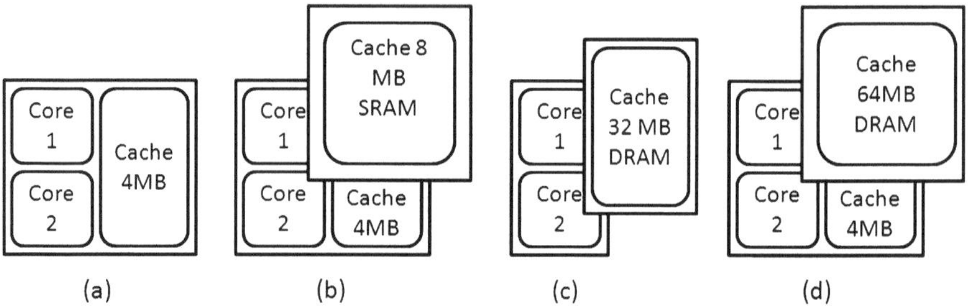

Fig. 15.12 Memory stacking schemes: (**a**) original 2D circuit; (**b**) stacking more SRAM above the initial circuit; (**c**) stacking L2 DRAM cache on top of the CPU; (**d**) stacking more DRAM on top of the initial 2D circuit

For this reason, almost all scientific publications dealing with 3D heterogeneous circuits look at stacking macroscopic blocks (memory/processor), since it seems to be the most viable approach with present technology limitations. The most obvious advantage of this type of partitioning is reusing existing 2D IPs, thereby reducing redesign efforts. Although this approach allows shortening the length of global inter-block wires, each component (processor or memory) keeps the same performance since it is identical to its original 2D version. Therefore, the power and performance improvements concern only global interconnects.

5.2 Examples of Memory on Processor Stacking

Several studies are conducted to explore this type of partitioning in order to reveal its contributions in terms of performance, power and thermals, as well as to identify the most promising architectures. We focus on some research carried out in universities and industrial laboratories; we present the proposed architectures and the obtained results. These studies were carried out based on simulations, not on real physical implementations.

5.2.1 Stacking L2 Cache on Top of a Processor

A survey performed by a team of researchers from Intel Corporation [11] compares three schemes to improve the performance of a planar circuit composed of a Core 2 Duo processor, with an L1 cache for each core and a shared L2 SRAM cache (see Fig. 15.12). The first 3D scheme consists of retaining the original planar circuit and expanding the L2 cache by stacking more SRAM on top. The second option proposes to replace the L2 SRAM cache by a denser (therefore larger) L2 DRAM cache. In this case, cache memory is stacked on top of the circuit which is composed only of the CPU and L1 cache. The last design consists of stacking DRAM memory (as L2 cache) above the initial circuit.

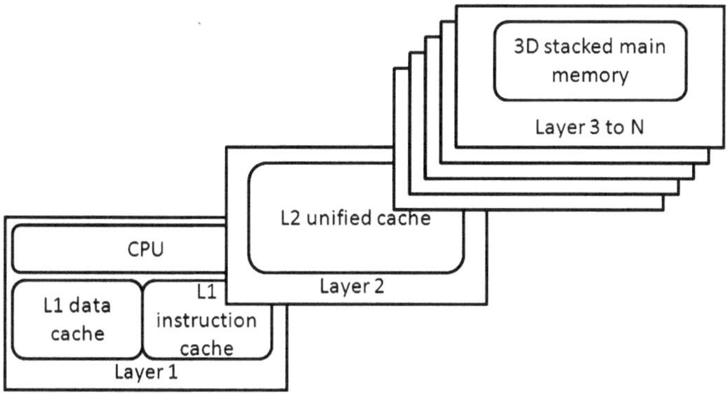

Fig. 15.13 The 3D stacked processor-cache-memory system

Performances of the proposed circuits are measured in terms of cycles per memory access CPMA and off-die bandwidth BW. According to simulations, the three proposed schemes outperform the original planer circuit. In particular, the second 3D system (L2 DRAM cache on top of the CPU) can reduce CPMA by 13% and off-die BW by 66% on average compared to the original 2D circuit. Simulations also register an average 66% power reduction in average bus power. In the same time, none of the stacking schemes considerably impacts on the thermals. Indeed, temperature increase is not prohibitive and varies between 0.08°C (stacking DRAM on top of the CPU) and 4.5°C (stacking more SRAM above the initial circuit). This can be explained by the higher power density of SRAM compared to DRAM. Source: Intel Corporation.

5.2.2 Stacking Main Memory and L2 Cache on Top of Processor

A group of researchers from the University of California [12] present a 3D system composed of a processor, a cache and a main memory, all integrated on the same chip (see Fig. 15.13). The CPU and L1 cache are implemented on the first layer in 130 nm technology, while the L2 cache is on the second layer. SDRAM main memory, implemented in 150 nm technology, is partitioned across 16 layers above the first two layers.

According to simulations, the considerable increase in memory bus frequency and bus width contribute to a significant decrease in execution time in the 3-D system. Indeed, for a typical 1 MB L2 cache configuration, execution time improvement over 2-D system is found to reach 57.7% with a 16-byte bus width. However, temperature increase in this case is higher than that of the previously presented system; it reaches 12°C for a 3 GHz frequency. Consequently, thermal constraint in this 3D design imposes a maximum allowed operating frequency lower than that of 2-D designs. In spite of this, the overall system performance can still be significantly better than conventional planar designs, especially for memory intensive applications.

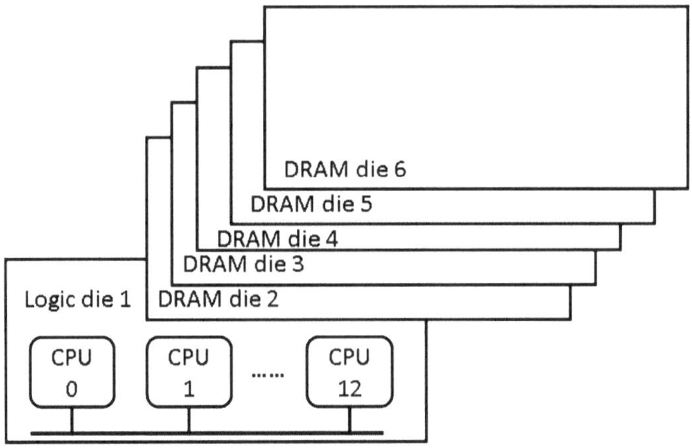

Fig. 15.14 PicoServer: a multi-processor architecture connected to a conventional DRAM using 3D integration technology

5.2.3 Stacking main memory on top of a processor without L2 cache

Another architecture called PicoServer is proposed by researchers from Michigan University and ARM Company [13]. As illustrated by Fig. 15.14, the proposed circuit is composed of a first layer containing several slow processor cores, and several other layers forming the DRAM main memory. This architecture has no L2 cache and main memory is connected directly through a shared bus to the L1 cache of each core. This approach is justified by the fact that latency and bandwidth obtained using stacked DRAM are comparable to those of the L2 cache. Therefore, it is more advantageous to remove the L2 cache and replace it with additional processor cores. The additional cores allow the clock frequency to be lowered without affecting performance. Lower clock frequency in turn reduces power and thus allows decreasing thermal constraints which are a concern with 3D stacking.

Simulations confirm that for a similar logic die area, a 12 CPU system with 3D stacking and no L2 cache outperforms an 8 CPU system with a large on-chip L2 cache by about 14% while consuming 55% less power. Besides, PicoServer shows the same performance as a Pentium 4-like class device while consuming only about 1/10 of the power. According to simulations, temperature increase is not a major limitation in the PicoServer platform since the power density is relatively low (it does not exceed 5 W/cm^2).

In conclusion, in spite of the diversity of the previously exposed architectures, they all confirm the role that 3D integration can play to improve heterogeneous system performance (in terms of cycles per memory access, bandwidth and execution time), compared to classic planar architectures. Stacking memory on top of logic leads to significant reduction in the overall circuit power by decreasing global interconnects' length. Given that memory is not a highly active layer, reached temperatures in this type of architecture are imperceptibly higher than those of planar circuits.

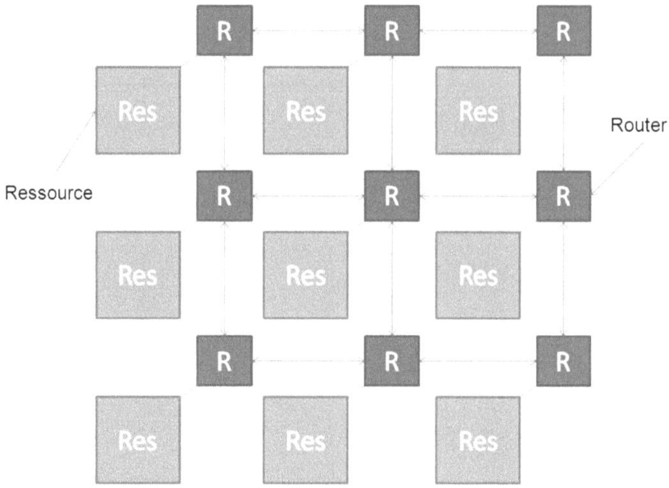

Fig. 15.15 2D mesh NoC

In parallel with advances in 3D architectures, inter-layer interconnects' design requires an equal attention in order to ensure effective communication between the cores on different tiers. Recently, several studies carried out in academic and industrial laboratories look at developing efficient interconnects' design, based mainly on the network-on-chip (NoC) approach. Next subsection will present some researches dealing with 3D NoC.

5.3 Communication Infrastructure Based on 3D NoC

A typical 2D NoC is composed of a set of nodes (also called switches or routers) which are connected by links (or channels). Each node includes a range of ports that allow it to connect to other nodes and also to a functional element in the network (resource) such as processor or memory. The topology of a NoC specifies the way its nodes are organized: ring tree, mesh, torus, fat-tree [14], etc. Currently, the 2D mesh networks (see Fig. 15.15) are recognized to be the most used since they provide the best compromise in terms of wire length and routing complexity.

Nowadays, NoCs are commonly used in 2D circuits by many companies and research centers such as CEA-LETI [15]. However, 3D NoC is still an emerging research topic. Here are some interesting researches dealing with expanding the NoC paradigm in the third dimension.

5.3.1 3D Symmetric NoC

A simple way of extending the 2D NoC router in the vertical dimension consists of adding one port for upward traffic and one port for downward traffic [16] (see

Table 15.1 Area and power comparison of different crossbars assessed in [16]

Crossbar type	Area (μm^2)	Power with 50% switching activity (mW)
5 × 5	8523.65	4.21
6 × 6	11579.10	5.06
7 × 7	17289.22	9.41

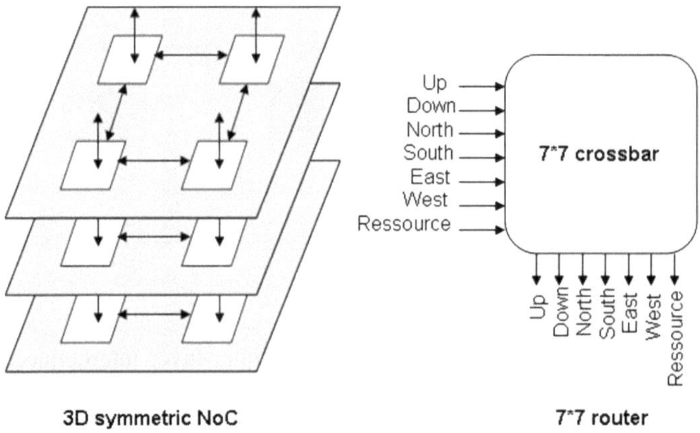

Fig. 15.16 3D symmetric router

Fig. 15.16). Of course, it would be necessary to extend also buffers, arbiters, and crossbar. This approach is called 3D Symmetric NoC given that communications in all directions (up, down, north, south, east and west) present the same characteristics. Although this model is simple to design and implement, it has two main disadvantages. The first problem is that each dataset moving from one layer to another should be buffered and arbitrated at each router: this is a misuse of the advantage of the insignificant distance between layers in 3D chips. The second major issue is that adding two ports to each router would require a bigger crossbar which leads to significant area and power overheads compared to a 2D router (see Table 15.1).

5.3.2 3D NoC-bus

To solve the two limitations of the symmetric 3D NoC, a promising solution is to use 2D router for intra-layer communication and vertical bus for inter-layer upward and downward communication [17] (see Fig. 15.17). This approach requires adding only one physical port to each router to connect it to the vertical bus. Consequently, router area and power decrease compared to 3D symmetric architecture. Moreover, flits wishing to move to another layer do not need to be buffered since vertical communication is supported by bus. Therefore, this hybrid system also provides performance benefits. Nevertheless, given that the bus is a shared medium, it cannot

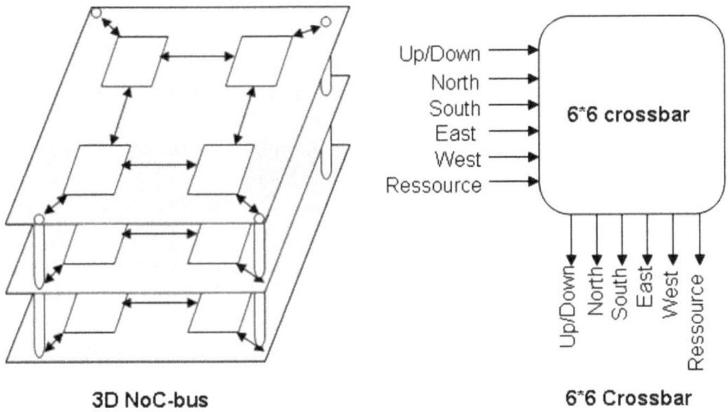

Fig. 15.17 3D NoC-bus router

carry more than one flit at a time. This drastically increases contention and blocking probability in the network.

6 Conclusion

3D stacking technology is a promising solution to deal with cost and performance issues in current heterogeneous systems. Circuit manufacturing cost may be reduced considerably since each application can be implemented using the most adapted technology. Moreover, global system form factor, performance and power could be enhanced drastically by reducing overall circuit area and shortening global interconnects. Power density in 3D stacked circuits is high and leads to a prohibitive temperature increase that should be studied carefully.

Current manufacturing technologies of 3D circuits offer many options and consist of 4 main steps: wafer thinning, alignment, bonding and TSV formation. Each of these steps and the chosen technological options may affect final yield.

Potential applications of 3D stacking are numerous and varied such as 3D photodetectors and 3D processor-memory system. A complete 3D heterogeneous circuit incorporating also RF, MEMS, sensors, IOs could be achieved in the few years to come, when manufacturing processes become more mature, and thermal issues and TSV size are managed efficiently.

Several academic and industrial researches look into 3D memory-processor systems such as stacking SRAM cache or DRAM main memory on top of the CPU. These advances in 3D architectures require equal progresses in inter-layer communication. Recently, some studies have looked at developing efficient interconnects' design, based mainly on the network-on-chip (NoC) approach.

In order to roll out the full potential of 3D integration technology, new EDA (Electronic Design Automation) tools are more and more needed to address challenges in physical design, thermal analysis and system level design, and to find out precise performance-cost estimates for a 3D system. At present, these tools are not sufficiently available, and much effort should be made to fill this gap.

References

1. Maly, W.P., Yangdong, D.: 2.5-dimensional VLSI system integration. IEEE Trans. Very Large Scale Integr. (VLSI) Syst. **13**, 668–677 (2005)
2. Yu, C.H.: The 3rd dimension—More Life for Moore's Law. In: Microsystems, Packaging, Assembly Conference Taiwan, Oct. 2006, pp. 1–6 (2006)
3. Wong, E., Lim, S.K.: 3D floorplanning with thermal vias. In: Proceedings of the Conference on Design, Automation and Test in Europe, March 2006, vol. 1, pp. 1–6 (2006)
4. Cong, J., Wei, J., Zhang, Y.: A thermal-driven floorplanning algorithm for 3D ICs. In: IEEE/ACM International Conference on Computer Aided Design, Nov. 2004, pp. 306–313 (2004)
5. Leduc, P., et al.: Enabling technologies for 3D chip stacking. In: International Symposium on VLSI Technology, Systems and Applications, April 2008, pp. 76–78 (2008)
6. Topol, A.W., et al.: Three-dimensional integrated circuits. IBM J. Res. Dev. **50**, 491–506 (2006)
7. Morrow, P.R., Park, C.-M., Ramanathan, S., Kobrinsky, M.J., Harmes, M.: Three-dimensional wafer stacking via Cu–Cu bonding integrated with 65-nm strained-Si/low-k CMOS technology. IEEE Electron Device Lett. **27**, 335–337 (2006)
8. Watanabe, T., Kikuchi, H., Fukushima, T., Tomita, H., Sugano, E., Kurino, H.: Novel retinal prosthesis system with three dimensionally stacked LSI chip. In: Proceedings of the 36th European Solid-State Device Research Conference, Sept. 2006, pp. 327–330 (2006)
9. Culurciello, E., Weerakoon, P.: Vertically-integrated three-dimensional SOI photodetectors. In: IEEE International Symposium on Circuits and Systems, May 2007, pp. 2498–2501 (2007)
10. Loh, G.H., Yuan, X., Bryan, B.: Processor design in 3D die-stacking technologies. IEEE MICRO **27**, 31–48 (2007)
11. Black, B., Annavaram, M.M., Brekelbaum, E., DeVale, J., Loh, G.H., Jiang, L., McCauley, D., Morrow, P., Nelson, D., Pantuso, D., Reed, P., Rupley, J., Shankar, S., Shen, J.P., Webbn, C.: Die stacking (3D) microarchitecture. In: IEEE International Symposium on Microarchitecture, pp. 469–479 (2006)
12. Loi, G., Agrawal, B., Srivastava, N., Sheng-Chih, L., Sherwood, T., Banerjee, K.: A thermally-aware performance analysis of vertically integrated (3-D) processor-memory hierarchy. In: Design Automation Conference, July 2006, pp. 991–996 (2006)
13. Kgil, T., D'Souza, S., Saidi, A., Binkert, N., Dreslinski, R., Reinhardt, S., Flautner, K., Mudge, T.: Picoserver: Using 3D stacking technology to enable a compact energy efficient chip multiprocessor. Oper. Syst. Rev. **40**, 117–128 (2006)
14. De Michelli, G., Benini, L.: Networks on Chip: Technology and Tools. Morgan Kauffman, San Mateo (2006)
15. Lattard, D., Beigné, E., Clermidy, F., Durand, Y., Lemaire, R., Vivet, P., Berens, F.: A reconfigurable baseband platform based on an asynchronous network-on-chip. IEEE J. Solid-State Circuits **43**(1), 223–235 (2008)
16. Kim, J., Nicopoulos, C., Park, D., Das, R., Xie, Y., Vijaykrishnan, N., Yousif, M.S., Das, C.R.: A novel dimensionally-decomposed router for on-chip communication in 3D architectures. Comput. Archit. News **35**, 138–149 (2007)
17. Li, F., Nicopoulos, C., Richardson, T., Yuan Xie, M.K., Narayanan, V.: Design and management of 3D chip multiprocessors using network-in-memory. Comput. Archit. News **34**, 130–141 (2006)

Chapter 16
Emerging Memory Concepts

Materials, Modeling and Design

Christophe Muller, Damien Deleruyelle, and Olivier Ginez

1 Overview of Nonvolatile Memories

Currently, the industry of microelectronics must overcome new technological challenges to continuously improve performances of memory devices in terms of access time, storage density, endurance, retention and power consumption. The main difficulty to be surmounted is the downsizing of memory cell dimensions necessary to integrate larger number elementary devices and subsequently more functionalities on a constant silicon surface. This drastic size reduction is restricted, in particular, by the intrinsic physical limits of materials integrated in the memory architecture.

Beside volatile memories such as Dynamic (DRAM) or Static (SRAM) RAM, nonvolatile memory (NVM) technologies may be subdivided in two categories depending upon the mechanism used to store binary data. A first group of solid state devices is based on *charge storage* in a poly-silicon floating gate (Fig. 16.1). In this family, gathering usual EPROM, EEPROM and Flash technologies, new concepts or architectures are currently developed to satisfy CMOS "More Moore" scaling trends. For instance, $Si_{1-x}Ge_x$ "strained silicon" technology enables boosting the charge carrier mobility and discrete charge trapping (SONOS, TANOS, etc.) improves data retention and allows extension towards "multi-bits" storage with silicon nanocrystals. Beside, multiple level cell (MLC) architecture enables increasing the density of NAND Flash. Presently, Flash technology still remains the undisputed reference [1], regardless of its NAND (dense and cheap) or NOR (fast) architecture. However, as the scaling of the conventional floating-gate cell becomes ever more complicated below 32 nm technological node (Fig. 16.2), opportunities for alternative concepts are emerging. In the last decade, several major Companies (Samsung, IBM, Micron, Everspin, etc.) have explored new solutions integrating a *functional material*, whose fundamental physical property enables data storing (Fig. 16.1).

C. Muller (✉) · D. Deleruyelle · O. Ginez
IM2NP, Institut Matériaux Microélectronique Nanosciences de Provence, UMR CNRS 6242, Aix-Marseille Université, IMT Technopôle de Château Gombert, 13451 Marseille Cedex 20, France
e-mail: christophe.muller@im2np.fr

G. Nicolescu et al. (eds.), *Design Technology for Heterogeneous Embedded Systems*,
DOI 10.1007/978-94-007-1125-9_16, © Springer Science+Business Media B.V. 2012

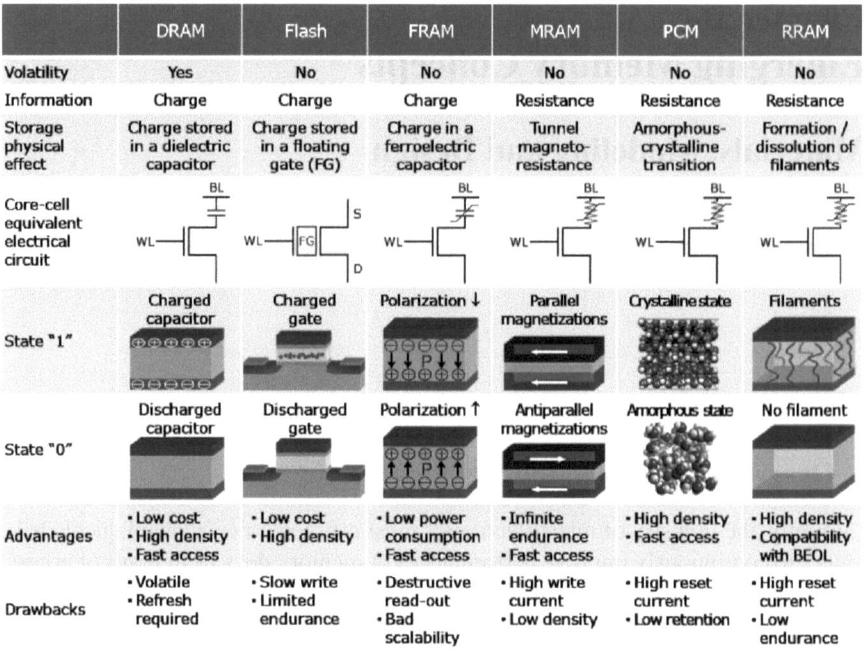

	DRAM	Flash	FRAM	MRAM	PCM	RRAM
Volatility	Yes	No	No	No	No	No
Information	Charge	Charge	Charge	Resistance	Resistance	Resistance
Storage physical effect	Charge stored in a dielectric capacitor	Charge stored in a floating gate (FG)	Charge in a ferroelectric capacitor	Tunnel magneto-resistance	Amorphous-crystalline transition	Formation / dissolution of filaments
Core-cell equivalent electrical circuit	WL, BL	WL, FG, S, D	WL, BL	WL, BL	WL, BL	WL, BL
State "1"	Charged capacitor	Charged gate	Polarization ↓	Parallel magnetizations	Crystalline state	Filaments
State "0"	Discharged capacitor	Discharged gate	Polarization ↑	Antiparallel magnetizations	Amorphous state	No filament
Advantages	• Low cost • High density • Fast access	• Low cost • High density	• Low power consumption • Fast access	• Infinite endurance • Fast access	• High density • Fast access	• High density • Compatibility with BEOL
Drawbacks	• Volatile • Refresh required	• Slow write • Limited endurance	• Destructive read-out • Bad scalability	• High write current • Low density	• High reset current • Low retention	• High reset current • Low endurance

Fig. 16.1 Comparison of few characteristics in volatile (DRAM) and nonvolatile (Flash, FRAM, MRAM, PCM and RRAM) memory technologies

The oldest technology, called FRAM for Ferroelectric RAM, is in small volume production for several years (Fujitsu, Texas Instruments, etc.). FRAM memories with a DRAM-like architecture integrate ferroelectric capacitors that permanently store electrical charges [2, 3] (Fig. 16.1). Thanks to their low voltage operation, fast access time and low power consumption, FRAM devices mainly address "niches" such as contactless smart cards, RFID tags, ID card, and other embedded applications. Nevertheless, due to scaling issues, conventional FRAM cell cannot be extended below 65 or 45 nm node (Fig. 16.2). For future generations, three-dimensional ferroelectric capacitors [4–6] will certainly enable increasing effective polarization charge and innovative architectures for the memory cell (e.g. FeFET, Ferroelectric Field Effect Transistor) will probably overcome the destructive read-out encountered in conventional FRAM cells.

Excluding FRAM technology, the R&D efforts also led to a new class of disruptive technologies based on a *resistive switching*. These memories, presenting two stable resistance states controlled by an external current or voltage, attract a lot of attention for future high density nonvolatile memorization devices. The physical origin of the resistance switching straightforwardly depends upon the *nature* and fundamental *physical properties* of materials integrated in the memory core-cell (Fig. 16.1).

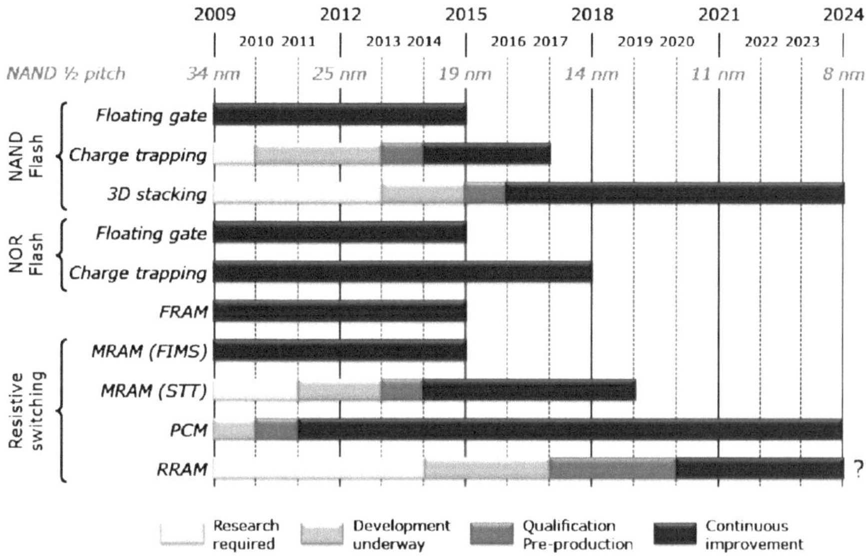

Fig. 16.2 Comparison of memory technologies: legend indicates the time during which research, development, and qualification/pre-production should be taking place for each technological solution. Inspired from ITRS, "Process Integration, Devices, and Structures", 2009

As a result, a broad panel of new concepts is currently emerging: magneto-resistive memories (MRAM) and derived concepts ("Toggle", "Thermally-Assisted" or "Spin Torque Transfer" MRAM); phase change memories (PCM); resistive memories (RRAM) including oxide resistive OxRRAM and "nanoionic" memories CBRAM and PMC.

As compared to other technologies, PCM and RRAM memories are more favorably positioned as alternatives to Flash for sub-32 or sub-22 nm nodes (Fig. 16.2). Nevertheless, it must be stressed out that these two technologies have not the same maturity level: while PCM prototypes already exist (cf. Sect. 3), RRAM memories are still at the early R&D stage for a time which is difficult to estimate (Fig. 16.2).

To summarize, alternative NVM concepts, *either evolutionary or revolutionary*, are being explored to satisfy the need for continuously higher storage capacity and better performances (lower power consumption, smaller form factor, longer data-retention, etc.). Within the broad panel of emerging solutions, this chapter will specifically address MRAM, PCM and RRAM *resistive switching technologies*. After a description of integrated "memory materials" and their ability to withstand a downscaling of their critical dimensions, a focus will be done on the models describing memory cell operations and their implementation in electrical simulators for design purposes or to apprehend robustness of innovative architectures.

2 MRAM: Magneto-Resistive Memory

2.1 Operations Based on Tunnel Magneto-Resistance

Recent developments in the physics of *spin electronics, or spintronics*, have enabled emergence of a new class of nonvolatile memories based on the interaction of the electron spin (seen as the electron rotation on itself) with the magnetization of a ferromagnetic layer. Nonvolatility, fast access time and compatibility with CMOS process make magneto-resistive RAM (MRAM) a very well suited candidate for applications where Flash-like nonvolatility is combined with SRAM-like speed and DRAM unlimited endurance (e.g. computer cache memory). Thus, in MRAM devices, data are no longer stored by electrical charges, as in semiconductor-based memories, but by a resistance change of a complex magnetic nanostructure [7].

MRAM cells integrate a magnetic tunnel junction (MTJ) consisting of a thin insulating barrier (i.e. tunnel oxide) separating two ferromagnetic (FM) layers. Using lithographic processes, junctions are etched in the form of sub-micron sized pillars connected to electrodes. MTJ resistance depends upon the relative orientation of magnetizations in the two FM layers: the magnetization in FM reference layer is fixed whereas the one in FM storage layer is switchable. *Tunnel magneto-resistance* (TMR) is due to tunneling of electrons through the thin oxide layer sandwiched between two FM films having either antiparallel (high resistance state "0") or parallel (low resistance state "1") magnetizations (Fig. 16.1). In conventional FIMS (Field Induced Magnetic Switching, called Stoner-Wohlfarth switching), each MTJ is located at the intersection of two perpendicular metal lines (Fig. 16.3a). Magnetization reversal in FM storage layer is controlled by two external magnetic fields produced by currents injected in the surrounding metal lines (Fig. 16.3b).

In MRAM technology three major problems are clearly identified: (i) low resistance discrimination between "0" and "1" states (i.e. small sensing margin); (ii) high sensitivity to disturb during writing ("bit fails"); (iii) high currents (few mA) necessary to reverse magnetization in FM storage layer.

To overcome these issues, different solutions have been proposed:

- To enlarge the sensing margin, the tunnel barrier in amorphous aluminum oxide AlO_x of first generations of devices was progressively replaced by crystallized magnesium oxide MgO. TMR was subsequently increased from 50 to 200% for stabilized processes.
- Everspin Company (former Freescale), world's first volume MRAM supplier, already sells in 2011 a 16 Mb MRAM chip based on 180 nm CMOS technology and using "Toggle" concept to limit disturbs during writing [8, 9]. Thanks to a new free layer magnetic structure, bit orientation and current pulse sequence, MRAM cell bit state is programmed via "Savtchenko switching toggle" mode [10]. Savtchenko switching relies on the behavior of a synthetic antiferromagnet (SAF) free layer that is formed from two FM layers separated by a non-magnetic coupling spacer layer. To exploit the unique field response of this free layer, a two-phase programming current pulse sequence in metal lines is required to efficiently

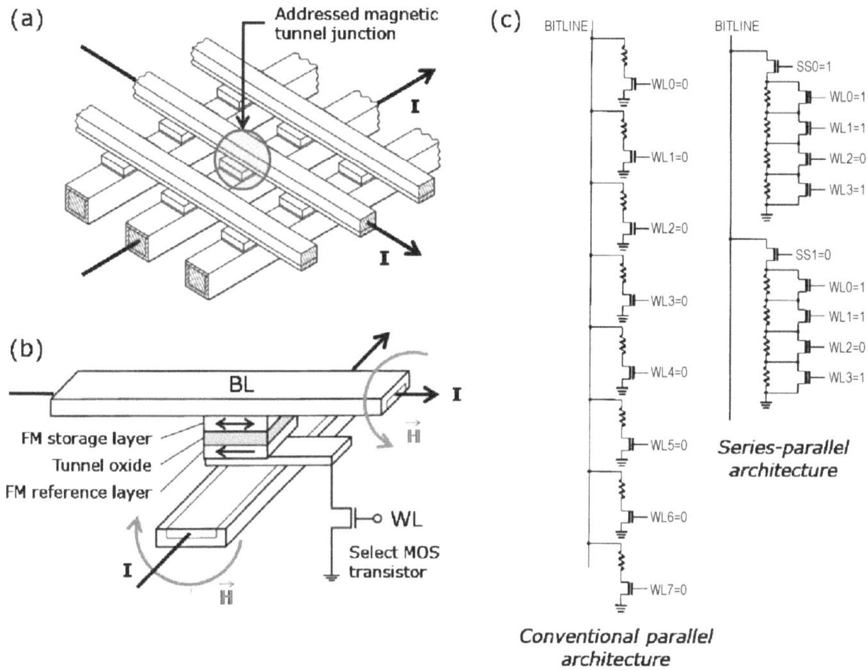

Fig. 16.3 (**a**) To write an MRAM bit, currents are passed through perpendicular metal lines surrounding the magnetic tunnel junction (MTJ) to produce magnetic fields. (**b**) The resulting magnetic field enables programming the bit in reversing magnetization of the FM storage layer (select MOS transistor is OFF). To read a bit, a current is passed through the MTJ and its resistance is sensed (select MOS transistor is ON). (**c**) Schematic diagrams of either conventional parallel or alternative series-parallel architectures for 1T/1MTJ cells. Inspired from US Patents no. 6,331,943, 6,806,523 and 7,411,816 [11–13]

rotate SAF magnetization [9, 14, 15]. Because of the inherent symmetry, this sequence toggles the bit to the opposite state regardless of existing state. In that configuration, a single metal line alone cannot switch the bit, providing greatly enhanced selectivity over the conventional FIMS MRAM.

- In June 2009, French startup Crocus Technology announced an agreement to transfer TA-MRAM process into Tower Semiconductor's 130 nm foundry with a migration path to 90 nm. In "Thermally-Assisted MRAM" cell, jointly developed by Crocus and CEA/CNRS Spintec Laboratory, a current injected in MTJ during writing induces a Joule heating in FM layers, the temperature increase facilitating the magnetization reversal [16, 17]. TA-MRAM concept enables (i) shrinking the memory cell size with only one metal line to produce magnetic field; (ii) reducing the power consumption thanks to limited writing current; (iii) improving bit fails immunity. Moreover, thermally-assisted writing mode requires a tight control of effective junction temperature and heating dynamics. Both drastically depend on heat leakages between the magnetic memory cell and CMOS level. This issue may be solved by introducing "thermal barriers" below and above MTJ with materials simultaneously exhibiting low thermal conductivity and high electrical conductivity [18]. This approach, developed in TIMI [19] European project, enables (i) concentrating heat flux close to the addressed junction; (ii) significantly reducing power consumption during writing; (iii) limiting heat spread towards adjacent cells.

In conventional FIMS MRAM technology, the memory cell remains quite large around 30–40 F^2 (*the term F refers to the smallest lithographic feature size of the respective lithographic technology node*). With TA-MRAM concept based on a low voltage 130 nm CMOS technology, cell size should be reasonably shrunk from 35 down to 20 F^2 with timing comparable to low power SRAM (35 ns for read/write cycle) and writing currents 5 times smaller than those of the first generation of "Toggle" MRAM. More recently, STT ("Spin Torque Transfer") switching was proposed to solve aforementioned issues and make MRAM memory compatible with more aggressive technological nodes. Big Companies (Renesas, Hynix, IBM, Samsung, etc.) and few startups (Grandis, Avalanche, Everspin, Crocus, etc.) are actively working on this new concept and state they can push the integration limits beyond 45 and 65 nm nodes for standalone and embedded memories respectively [20]. Few prototypes have already been achieved: 1 Mb chip from Grandis/Renesas consortium; 2 Mb SPRAM chip from Hitachi Ltd. in collaboration with Tohoku University. In STT writing mode a spin-polarized current, which flows through MTJ, exerts a spin torque on the magnetization of FM storage layer due to the interaction between the conduction electron spin and the local magnetization [21]. Technologically, STT concept requires the integration of a polarizer to select the spin of electrons injected into the magnetic nanostructure and ensuring the magnetization reversal. As compared to a conventional FIMS MRAM cell, STT-RAM solution is said to consume less power, to improve bit selectivity and to reduce the memory cell size to approximately 6–9 F^2. In summary, STT-RAM concept appears as a promising and disruptive technology that combines all the benefits of SRAM, DRAM and Flash memories, as well as offering scalability to leading-edge technological nodes.

2.2 Modeling of Magnetic Tunnel Junctions

Spintronics takes advantage of the spin of the electrons in addition to their electrical charge to design innovative electronic functions (memory, FPGA, logical devices, etc.). Devices integrate magnetic thin films (used as polarizer or analyzer) together with standard semiconductor and insulator materials. Consequently, to investigate new architectures, it is of primary importance to develop Design Kits (DK) taking into account magnetic components in microelectronics design suites (EDA) like CADENCE or MENTOR. Design kits dedicated to memories require (i) a compact modeling describing MTJ's electrical and magnetic properties and (ii) verification tools that enable designing and checking the circuit layout.

Targeting optimized architectures for TA-MRAM cells, Spintec and IEF Laboratories in close collaboration with Crocus Technology, developed a MTJ compact model taking into account (i) static and dynamic behavior of magnetization under magnetic field; (ii) thermal dependence of parameters; (iii) heat propagation in the vicinity of addressed junction (Fig. 16.4a); (iv) spin transfer torque effect for an extension towards STT-RAM devices [22]. This model, containing equations describing physical behavior of magnetic tunnel junction, is compatible with a verification environment developed in the framework of CILOMAG [23] ANR project.

In the 2000s, Das and Black [24] were the first to propose a generalized circuit macro-model for Spin-Dependent-Tunneling (SDT) devices. The macro-model was done as a four terminal sub-circuit emulating nonlinear and hysteretic behavior of SDT device over a wide range of sense and word line currents. HSPICE simulations of memory circuits using this model showed expected outcomes.

MTJ may be also modeled as a resistor and a capacitance in parallel. Using this approach, a team from InESS [25, 26] Laboratory proposed a MTJ compact model taking into account magnetic and non-linear electronic transport phenomena suitable for circuits simulation. Authors employed a graphical (i.e. vectorial) method to model the magnetization hysteresis loops for a Stoner-Wohlfarth switching. This MTJ compact model was used to design a conventional FIMS MRAM array and a two-axis magnetic sensor.

On the other hand, K.J. Hass [27] from the University of Idaho developed another SPICE simulation model using a pair of MTJs and including the junction capacitance of MTJ itself. To simplify the model, one of the MTJs was assumed to be in the antiparallel state, with a higher resistance than the other MTJ. The bias dependence of the magnetoresistance ratio was modeled by varying the resistance of the antiparallel junction as a function of the voltage across it.

2.3 MRAM Circuit Design Overview

Similarly to DRAM, EEPROM or NOR Flash memories and whatever the writing mode (Toggle, TA or STT), MRAM core-cell is essentially based on the association of 1 select MOS transistor and 1 magnetic tunnel junction (i.e. 1T/1MTJ cell). To individually access each memory cell, magnetic tunnel junctions are located at each intersection of a bottom "Word Line" (WL) connected to the transistor gate and an upper "Bit Line" (BL) (Fig. 16.3b). As shown in Fig. 16.3c, either conventional parallel or series-parallel architectures are achievable. Nevertheless, in contrast to multi-layer crossbar structure proposed for RRAM technology (cf. Sect. 4), such 1T/1MTJ cells does not enable stacking memory layers due to disturbs between neighboring cells during magnetic field-assisted writing. Consequently, MRAM architecture is restricted to two-dimensional memory arrays.

Polarization of "Word", "Bit" and "Source" (SL) lines used to select one core-cell in the memory matrix straightforwardly depends on the writing mode. In Thermally-Assisted switching [16], writing requires a heating current injection in MTJ obtained in applying a positive voltage on the gate of the select transistor (WL) of the addressed core-cell whereas the source line is grounded. Concomitantly, a current is

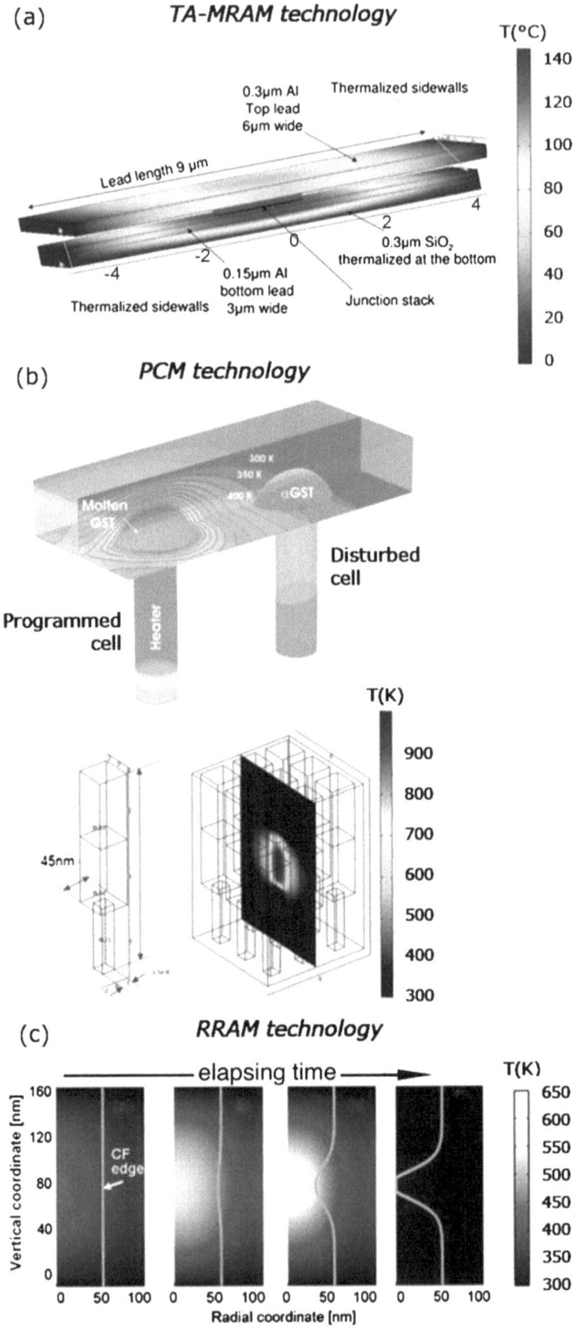

Fig. 16.4 Illustration of different thermal simulations performed on each memory technology. (**a**) 3D thermal simulation of TA-MRAM cell showing temperature profile around MTJ during programming [16]. (**b**) 3D temperature map simulation of an elementary 2-bit PCM array showing the impact of programming on a neighboring cell [49]. (**c**) 2D temperature map obtained on NiO-based RRAM showing the continuous shrinkage of the conductive filament during the self-accelerated thermal process occurring during reset operation [28, 29]

injected in single metal line (BL) to create a magnetic field that enables switching of FM storage layer. In STT writing mode [30], as for TA-MRAM, a positive voltage is applied on the select transistor gate (WL) of the addressed cell. Since writing does not require magnetic field, only bipolar voltage is applied between BL and SL to inject a top-down or bottom-up spin polarized current enabling magnetic switching.

From a design point of view, write drivers used for Toggle, TA or STT concepts are not very aggressive in terms of complexity, surface, consumption and timing [31], these circuits being similar to those used for SRAM, DRAM or Flash technologies. In contrast, read circuit for MRAM memories is crucial since a reliable read operation of MRAM core-cell (whatever the writing mode) requires an unambiguous discrimination of R_{High} and R_{Low} resistances. However, as previously mentioned (cf. Sect. 2.1), TMR, proportional to $\Delta R = R_{\mathrm{High}} - R_{\mathrm{Low}}$, does not exceed few hundreds of percents. As a consequence, it is of primary importance to limit process variability with subsequent narrower bit resistance distributions [32]. In stabilized Toggle MRAM process, a ratio $\Delta R/\sigma$ of about 25 was reported at room temperature, with σ the standard deviation of resistance distribution [15]. To perform a reliable read, $R_{\mathrm{Ref}} = \Delta R/2$ median resistance is generally used to sense “0” and “1” logical states, the sense amplifier being able to discriminate low and high resistances below and above R_{Ref} respectively. In the literature, different sense amplifier circuits are proposed. Differential amplifier working in current measurement mode was developed for Toggle-MRAM devices [33, 34]. In that case, the select transistor is ON and the source line is grounded. After BL is polarized, the current going through MTJ is compared to a reference I_{Ref}. Another innovative solution was developed by a research team from LIRMM [35, 36] Laboratory: a sense amplifier using a SRAM cell (i.e. 6 transistors) dedicated to reconfigurable FPGAs. The principle is based on the imbalance of two bit lines (BL_x and $/BL_x$) caused by the resistance difference ($|R_{\mathrm{Ref}} - [R_{\mathrm{Low}}$ or $R_{\mathrm{High}}]|$) between a reference and the core-cell to read. These authors proposed to integrate nonvolatile and fast FIMS or TA-MRAM cells in FPGA: using this heterogeneous design there is no need to load the configuration data from an external nonvolatile memory as required in SRAM-based FPGAs. To summarize, a lot of design solutions [35, 36], essentially based on conventional memories, can be transposed to MRAM concept even if tuning and adjusting operations are mandatory (especially for read operation).

3 PCM: Phase Change Memory

3.1 Operations Based on Crystalline–Amorphous Transition

In February 2008, Numonyx Company, joint venture formed by STMicroelectronics and Intel, announced the shipment to customers of “Alverstone” 128 Mb PCM prototypes based on a 90 nm CMOS technology (in 2011, Micron, former Numonyx, sells Omneo^{TM} 128 Mb PCM chip). One year earlier, Samsung [37–40] showed a 512 Mb PRAM chip developed on the same CMOS technological node. Do these two prototypes correspond to different memory technologies? The answer is no. PCM, PRAM or PCRAM are different acronyms for a single technology which uses phase change material to store data.

In the 1970s, S.R. Ovshinsky [41] reported in some chalcogenide materials (i.e. containing chalcogen chemical elements such as S, Se, Te, etc.) reversible transition between amorphous and crystalline phases, the rearrangements at atomic scale inducing huge change in optical and electrical properties. In the crystalline state, the atoms present a long range order and the material exhibits high optical reflectivity and low electrical resistivity. In contrast, in the amorphous state characterized by a short range atomic order, chalcogenide alloys demonstrate low reflectivity and high resistivity. Historically, chalcogenide materials were first used for optical storage in re-writable CD then in DVD and recently in Blue-Ray disks. Currently, they are progressively integrated in nonvolatile memory devices [42, 43].

PCM concept exploits the resistance change (few orders of magnitude) due to reversible amorphous (state "0") to crystalline (state "1") phase transitions (Fig. 16.1). Technologically, a layer of chalcogenide alloy (e.g. $Ge_2Sb_2Te_5$, GST) is sandwiched between top and bottom electrodes: the phase change is induced in GST programmable volume through an intense local Joule effect caused by a current injected from bottom electrode contact acting as a "heater" (Fig. 16.5a). To reach amorphous state (i.e. reset operation), GST material is heated above the melting temperature T_M (typically 600–700°C) and then rapidly cooled. In contrast, GST material is placed in the crystalline state (i.e. set operation) when "slowly" (few tens of ns) cooled from a temperature in between melting point T_M and crystallization temperature T_X (Fig. 16.5b).

PCM materials face three main challenges: (i) lower melting temperature that enables reducing reset current which remains quite high in existing prototypes (typically 0.5 mA); (ii) higher crystallization temperature to improve data retention (spontaneous amorphous-to-crystalline transition over a time range as long as possible); (iii) higher crystallization rate to decrease switching time. In terms of device, it is crucial to control cooling rates and to achieve heat confinement in the region of programmed cell.

As already mentioned, reset current in existing prototypes is still quite high, but unlike MRAM memory, it should be scaled down by shrinking programmable volume. Recently, IBM/Qimonda/Macronix consortium [44] reported phase changes in *scaling demonstrator PCB structures* ("Phase Change Bridge") consisting of a narrow line of ultrathin doped GeSb phase change material (width W, thickness H) bridging two underlying TiN electrodes. Unlike in earlier line-device concepts [42], the electrodes are formed very close together in order to obtain a reasonable threshold voltage and are separated by a small oxide gap that defines the bridge length L. At the smallest cross-sectional areas of 60 nm^2 ($H \times W = 3 \times 20$ nm^2), the reset current decreases below 100 μA. GeSb integration in small dimensions PCB structures enabled demonstrating several promising characteristics [45]: (i) very fast switching with 40 ns set pulses; (ii) resistance ratio $R_{\mathrm{High}}/R_{\mathrm{Low}}$ between amorphous and crystalline states larger than 10^4; (iii) high crystallization temperature improving data retention; (iv) capability of operating using phase change material thicknesses as small as 3 nm.

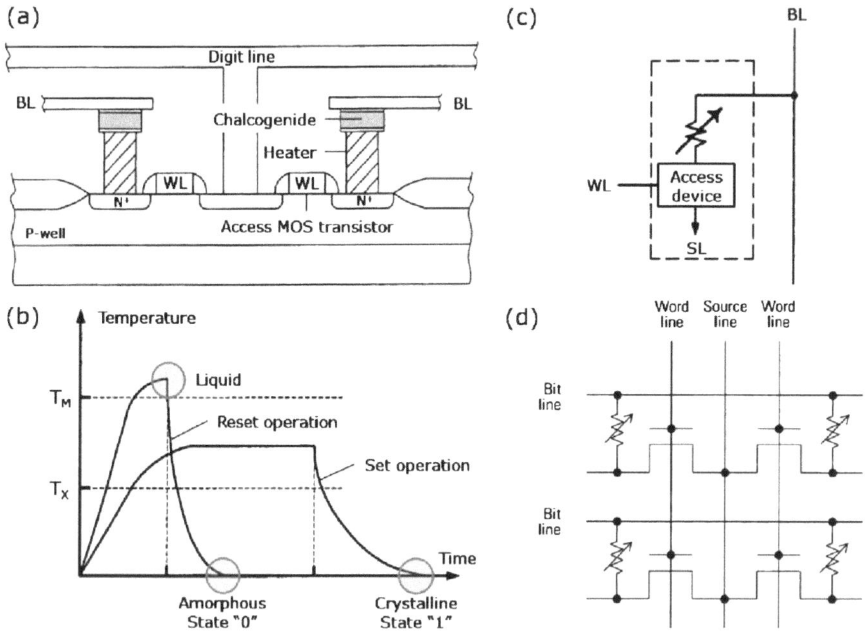

Fig. 16.5 (**a**) Conventional PCM architecture in which two adjacent memory cells are coupled to a common digit line. Access MOS transistor driven by a word line WL is connected to chalcogenide-based phase change element through a conductive plug acting as a "heater". (**b**) Temperature profiles used for set and reset operations. (**c**) 1 access device/1 resistor memory cell. (**d**) Electrical schematic arrangement of a PCM memory array based on core-cell illustrated in (**a**). Inspired from US Patents no. 6,791,859, 7,238,994 and 7,656,719 [38, 46, 47]

As compared to MRAM technology, PCM memories have a better ability to withstand size reduction. Indeed, although the endurance is still limited (less than 10^{10} cycles), the memory cell is much more compact (typically 10 F^2) and downsizing should be extended beyond 25 nm node.

To summarize, resistor-based PCM technology appears as an advanced alternative memory concept which offers faster read and write speeds at lower power than conventional NOR and NAND Flash memories.

3.2 Models for PCM Cells

As early reported in the original work of S.R. Ovshinsky [41], electrical switching was observed in amorphous GST alloy when the applied bias is over a threshold voltage. So far, this phenomenon, known as "threshold switching", was described by models involving purely electrical mechanisms such as (i) balance between a recombination process involving Valence Alternation Pairs (VAP) and high-field avalanche-like process [48, 49] or (ii) a non-uniformity of the electrical field in a trap-limited conduction process [50]. Other works also mentioned the possibility of a field-induced nucleation process resulting in the growth of conductive clusters [51]. However, since amorphous GST layers spontaneously recover their low conductivity state after a phase transition, another kind of switching, involving changes at atomic scale (i.e. crystallization), namely "memory switching", is mandatory to obtain an exploitable memory effect. Therefore, set and reset operations [49] rely on an electrothermal model (coupling electronic transport and self-heating equations in a semiconductor-like material) with a thermodynamical model (describing phase transition and crystallization process within the GST region situated just above the heater). Radaelli et al. [49] used this simulation approach to map 3D temperature profile within a portion of 2 bits PCM array (Fig. 16.4b). They showed that the programming operation on a cell induces heat spread towards the neighboring cells, in which GST material may crystallize if it was initially in amorphous state (with subsequent data loss). These simulations stress out that thermal diffusion between adjacent cells is a limiting factor to the maximum density.

3.3 Architecting Phase Change Memory

Due to its fast access time, PCM technology could be envisaged to replace volatile RAM and NOR Flash memories in microcontrollers. This enables the code execution directly from the memory, without intermediate copy to RAM. Nevertheless, PCM access latencies, although tens of nanoseconds, are several times slower than those of DRAM [52]. At present, considering limitations in terms of high reset current and low performances in retention, PCM is rather positioned as a NOR Flash substitute. On this market segment, PCM technology presents significant improvements in terms of endurance, access time and bit alterability (i.e. switching from "1" to "0" or "0" to "1" without an intermediate erase required in floating gate technologies). To exploit PCM's scalability as a DRAM alternative, PCM memories should be architected to manage long latencies, high energy writing and finite endurance [52].

PCM technology basically uses 1X/1R core-cells (Fig. 16.5c), X being one access device associated with one phase change resistor R. As shown in Fig. 16.5d, each core-cell is addressed thanks to WL connected to access device, BL and SL. Considering the high reset current required to switch chalcogenide material from crystalline to amorphous phase, access device must be able to drive high current during this operation. Access is typically controlled by one of these three devices: bipolar junction transistor (BJT), diode or MOS transistor:

- BJTs are faster, expected to scale more robustly and able to drive high current. Gill et al. proposed to use BJT in 0.18 μm technology to develop a PCM memory array [53]. Nevertheless, the big hurdle of this solution is the BJT leakage current under bias polarization.
- Diodes occupy smaller areas and potentially enable greater cell densities, but require higher operating voltages. At present, a diode or a transistor is typically

used as the access device. While a diode can provide a current-to-cell size advantage over a planar transistor at sizes as low as those used at 16 nm node, the diode scheme is more vulnerable to errors induced by writing data to adjacent cells because of bipolar turn-on of the nearest-neighbor cells. A 5.8 F^2 PCM cell was demonstrated using a 90 nm technology in which the diode supplied 1.8 mA at 1.8 V [39].
- Finally, access may be also achieved using a MOS transistor (Figs. 16.5a and 16.5c): Oh Hyung-rok et al. [54] proposed a 64 Mb PCM array based on a 120 nm CMOS technology. Using this structure, the read access time and set write time were 68 and 180 ns, respectively.

Dedicated peripheral circuits have to be designed to write and read data in PCM memory cell: during writing, access device injects current into the storage material and thermally induces phase change, which is detected during reading. Contrary to MRAM concept, a special emphasis is required on write drivers. As previously mentioned, the time-dependent temperature profiles used to switch chalcogenide material have to be carefully controlled through the monitoring of set and reset currents injected in the programmable volume. Woo Yeong Cho et al. [37, 38] proposed a solution based on a current mirror source belonging to a local column constituted of 1NTMOS/1R cells. The main advantage of such a structure is the reuse of the current source for all columns through the entire memory chip.

For read operations, PCM memory circuits require sense amplifier able to discriminate the two resistance states. As compared to MRAM devices (cf. Sect. 2.3), the resistance margin ΔR is larger (at least 1 decade) and constraints on sense amplifier are slightly relaxed. Bedeschi et al. [55] proposed a solution based on current measurement mode implemented in a 8 Mb PCM memory matrix using a BJT as selector. The current passing through the cell is measured thanks to a mirror biased structure and the current is compared to a reference given by a 1BJT/1R dummy cell.

In summary to write/read 1T/1R PCM cells, WL of the selected element is polarized at V_{dd}, BLs are driven by I_{set}, I_{reset} or I_{read} (according to requested operation) and unselected WLs are grounded.

4 RRAM: Resistive Memory

4.1 Operations Based on Resistance Switching

Last ten years have seen the emergence of new memories labeled with the acronym RRAM for "Resistive RAM", based on various mechanisms of resistance switching *excluding phase change* as used in PCM (cf. Sect. 3). In its simplest form, RRAM device relies on MIM structures (Metal/Insulator/Metal) whose conductivity can be electrically switched between high (state "0") and low (state "1") resistive states (Fig. 16.6a). RRAM memory elements are gaining interest for (i) their intrinsic scaling characteristics compared to the floating gate Flash devices; (ii) their potential

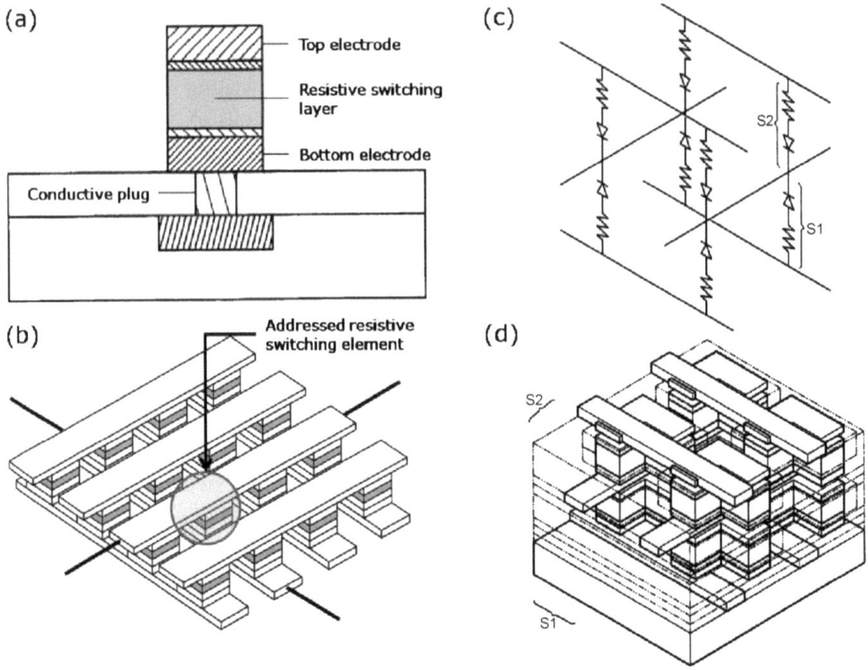

Fig. 16.6 (**a**) Scheme of a typical RRAM memory element: resistive switching layer is sandwiched between top and bottom electrodes to form simple MIM (Metal/Insulator/Metal) structures. (**b**) Crossbar-type memory architecture in which memory elements are located at the intersections of perpendicular metal lines. (**c**) & (**d**) Schematic and three-dimensional view of a two-layer crossbar memory array integrating 1 diode/1 resistor (1D/1R) core-cells. Inspired from US Patents no. 6,849,891, 2009/0184305 and 2009/0026434 [56–58]

small size; (iii) their ability to be organized in dense crossbar arrays (Fig. 16.6b). Hence, RRAM concept is seen as a promising candidate to replace Flash memories at or below 22 nm technological nodes (Fig. 16.2).

Three different "flavors" of RRAM memory elements were disclosed depending on the nature of bi-stable material and mechanism involved in the resistance change [59, 60] (this classification must be considered with caution since switching mechanisms are not yet fully elucidated in many systems):

- *Thermal effect*: in Oxide Resistive RAM (OxRRAM), reversible switching is achieved thanks to reproducible formation/dissolution of conductive filaments (CF) within resistive oxide [61] (Fig. 16.1). A typical resistive switching based on a thermal effect shows a *unipolar current-voltage* characteristic (i.e. switching does not depend upon the sign of voltage or current). During set operation a partial dielectric breakdown occurs in the material and conductive filaments are formed. In contrast, they are thermally disrupted during reset because of high power density generated locally, similar to a traditional house fuse. Although several simple transition metal oxides exhibit resistive switching (TiO_2, CuO_x, FeO_x, etc.), NiO seems to be currently the most promising RRAM material due to its high resistance ratio between "0" and "1" states, simple constituents and compatibility with CMOS process. Different routes for NiO formation are reported in literature: either oxidation of parent Ni metallic layer [62–64] or deposition on top of pillar bottom electrode [65, 66].

- *Ionic effect*: CBRAM [67] ("Conductive Bridge RAM") and PMC [69] ("Programmable Metalization Cells") belong to "nanoionic" memories. MIM-like memory elements consist of an inert electrode (W, Pt, etc.), an ionic conductor used as solid electrolyte (WO_3, MoO_3, GeSe, AgGeSe, etc.) and an active electrode (Ag, Cu, etc.) producing, through an electrochemical reaction, ions (Ag^+, Cu^+, etc.) diffusing within electrolyte. For this type of mechanism, a polarity change of applied voltage is absolutely required, or in other words the *switching is bipolar* [59]. The same mechanism may be invoked for charged oxygen vacancies in few systems reported in the literature. Field-induced oxygen diffusion plays a predominant role in few specific oxide-based memory elements: TiO_2-based "Memristors" [70] recently demonstrated by Hewlett-Packard Company; CMO_x™ elements [71] developed by Unity Semiconductor; NiO layers obtained from nickel oxidation in small interconnect structures [62–64].
- *Electronic effect*: electronic charge injection and/or charge displacement effects can be considered as another origin of resistive switching. One possibility is the charge-trap model, in which charges are injected by Fowler–Nordheim tunneling at high electric fields and subsequently trapped at sites such as defects or metal nanoparticles in insulator [59].

Compared to floating gate technologies, resistive memories (especially OxRRAM) are very promising with attractive expected performances: scalable memory cells; switching time around 10 ns; programming voltages below 2 V; resistance ratio between "0" and "1" states ranging from 10 to 10^6. Moreover, the first demonstrators showed read endurance of 10^{12} cycles and about 10^6 cycles for set/reset operations. However, as mentioned in introduction, RRAM memories are still at the R&D stage and many issues are still waiting for solutions: high reset currents of few mA; retention capability of 10 years at 85°C not yet proven; controversial switching mechanisms; models of reliability not established etc.

4.2 Various Modeling Approaches of RRAM Cells

To achieve the design of RRAM memory circuit and apprehend its robustness, an electrical model of the resistive switching memory element is mandatory. For this purpose, three ways are possible: (i) models relying on physical/technological parameters; (ii) electrical models based on dedicated components and functions available in electrical simulators; (iii) analytical models compliant with simulators like SPICE or ELDO.

As there is no consensus on a universal physical mechanism explaining resistance switching phenomena, several approaches were developed to model the electrical characteristics of various RRAM devices.

For "nanoionic" CBRAM and PMC memories, drift-diffusion models [67, 69] of mobile ions in solid electrolyte layer were deployed to describe the self-limited growth of conductive filaments (CF). This approach was also satisfactory to explain local electrical switching (set operation) measured by conductive AFM in CuTCNQ [72] metal organic-based memory elements.

Percolative models based on a network of random circuit breakers were proposed to describe both set and reset operations in binary oxides OxRRAM devices [73, 74]. In this model, the active layer, i.e. nickel or titanium oxide layer, is treated as a two-dimensional network of resistances representing local conduction paths. By switching the elementary resistances from high to low values when the voltage drop is over a threshold voltage, percolative paths grow and qualitatively reproduce the sharp current increase associated with set operation. This model can be employed to describe either set or reset operations [73].

In the framework of EMMA [75] European project, a research team from Politecnico di Milano [28, 29, 76] proposed a new physics-based model for RRAM reliability and programming. These authors built a model based on the assumption that reset operation is due to a self-accelerated thermal dissolution of conductive filament (CF) in which conductive elements diffuse out of CF where they anneal or react with other elements [28, 29]. In this thermoelectrical model a temperature-dependent resistivity is used to calculate the current flow inside a columnar CF. The resulting Joule dissipation is used as a source term in the Fourier heat-flow equation. The CF shrinkage is computed by introducing a temperature-dependent filament dissolution velocity as a fitting parameter (Fig. 16.4c). Russo et al. [28, 29] showed that, inside the CF, temperature may reach more than 500 K. When the temperature is over a critical value (around 530 K), CF starts to shrink. Current density subsequently increases and a self-accelerated process is initiated leading to the CF rupture. The model was shown to be able to reproduce experimental data by extracting physical parameters such as resistivity thermal coefficient or the activation energy involved during reset. In complement, Bocquet et al. [68] proposed in 2011 a self-consistent physical model accounting for both set and reset operations in unipolar switching memories. This model was then simplified and implemented into circuit simulators for design purpose.

A substitute to physical modeling is the elaboration of an electrical model based on switch circuits which can be easily implemented in electrical simulator such as ELDO. Unfortunately, this type of model presents a severe lack of accuracy, since resistive switching memory element cannot be considered as a basic switch between two resistances (i.e. non-ohmic behavior over the whole voltage range). As a consequence, this 1st order model does not fully satisfactorily describe the actual electrical behavior.

For that reason, a solution based on an analytical modeling can be envisaged. In that case, the elaboration process of such models is done with the help of current-voltage characteristics measured on actual memory elements. Resulting curves are decomposed in several parts and a polynomial interpolation is done to model each part separately. Once the interpolations are satisfactory, the set of equations is inserted in ELDO black-box component with two electrodes. In this approach, codes in C language are developed to be implemented in ELDO simulator and models are validated through different simulations of set and reset operations in a crossbar array configuration. Ginez et al. [77] deployed this approach to apprehend the impact of resistive defects in NiO-based memory arrays.

4.3 Innovative Architectures for RRAM

As already stated (cf. Sect. 1), RRAM memories are still in a R&D phase and only few memory circuits can be found in the literature. As RRAM memory elements can be integrated into "Back-End Of Line" (BEOL), this technology is of particular interest for high density storage with possibly multi-levels three-dimensional architectures. For this promising class of memories, two cell architectures are generally proposed depending on the nature of RRAM element: 1T/1R, a transistor associated with a resistive element; 1R enabling crossbar-type memory matrix. The first demonstrators integrated memory cells with size of 4–8 F^2 for active matrix and $(4/n)$ F^2 for a passive matrix (i.e. crossbar) with n storage layers [59, 67].

4.3.1 Crossbar Architecture for OxRRAM Elements

Samsung Company recently proposed a "non-CMOS" solution with a two-layer architecture based on a 1D/1R memory cells [57, 78] associating a diode with an oxide-based resistive element exhibiting unipolar switching. OxRRAM crossbar memory matrix was easily designed with the help of BL and WL to access each core cell individually (Figs. 16.6c and 16.6d).

Regarding the design of peripheral circuits, a big effort is devoted to set and reset operations. In many unipolar switching systems, the reset operation is performed through a voltage sweep between top and bottom electrodes of memory element whereas set operation is achieved thanks to a current-controlled sweep. From a circuit point of view, the use of two distinct write sources is mandatory: one voltage source for reset and one current source for set. To reduce the surface of write circuit, Hosoi et al. [79] suggested using only one voltage source with a multiplexer scheme that enables connecting or not the voltage source to a shunt resistance.

Even if the write driver circuit is easy to design, reset operation remains an issue to overcome. As previously mentioned (cf. Sect. 4.1), reset current may reach few mA due to 10–100 Ω resistance in low resistive state and reset voltage around 1 V. Moreover, several works [28, 29, 66] showed that the reset current does not scale down with decreasing size of memory element. Nevertheless, Kinoshita et al. [80] demonstrated that the reset current is mainly due to a dynamic phenomenon of parasitic capacitance charge. By reducing this capacitance, these authors succeeded to reduce dynamic current and subsequently reset current by a factor 100. Such a work open a new way for designers and smart write circuits could be developed in a near future.

Regarding read operations, sense amplifier circuits that enable discriminating two resistance states are similar to those used in PCM and MRAM technologies.

4.3.2 Memory Circuit for CBRAM

As previously detailed (cf. Sect. 4.1), CBRAM derives from parent technology PMC [69], which was developed by Axon TC in collaboration with Arizona State University. In 2007, Qimonda/Altis/Infineon [81] consortium demonstrated a 2 Mb CBRAM test chip with read-write control circuitry implemented in a 90 nm technological node with read/write cycle time less than 50 ns. The corresponding CBRAM circuit was developed using 8 F^2 core-cells associating 1 MOS transistor with 1 Conductive Bridging Junction (i.e. 1T/1CBJ). The chip design was based on a fast feedback regulated CBJ read voltage and on a novel program charge control using dummy cell bleeder devices. The low power resistive switching operation at voltages below 1 V, the ability to scale to minimum geometries below 20 nm [82] and the multi-level capability make CBRAM a very promising nonvolatile emerging memory technology.

From a circuitry design viewpoint, main industrial actors involved in CBRAM technology development remain attentive. Few patents [83, 84] deposited by Infineon/Qimonda deal with technical aspects on how to design a CBRAM memory circuit. To read the state of resistive element, a current is fixed between the two CBRAM electrodes (I to V conversion) and the resulting potential drop is compared to a reference V_{ref}. This method can be also applied in current measurement mode.

As compared to other emerging technologies, CBRAM concept requires a refresh operation due to the poor retention capability of low resistance state. In contrast, the high resistive state is usually stable in time. Thus, a refresh voltage is applied to the CBRAM memory element at a predetermined time to strengthen and stabilize low resistance state. This smart refresh circuit enables preserving the resistance margin ΔR necessary to unambiguously discriminate high and low resistance states during read. This refresh is performed without destroying data stored in CBRAM element whereas in DRAM a rewriting of respective state is mandatory.

To conclude this section, RRAM technology is very promising and once again many design solutions exist for implementation. However, depending on the RRAM concept, specific peripheral circuits are required to guaranty reliable memory operations.

5 Summary

After a brief overview of conventional charge storage-based technologies, this chapter was devoted to *resistive switching memories* exhibiting attractive and disruptive performances as compared to conventional devices. To penetrate the markets currently covered by SRAM, DRAM and Flash memories, these emerging concepts are facing the stringent requirements of *process compatibility* and *scalability* at material, device and circuit levels. Status and outlook may be proposed as follows:

- MRAM arising from *spintronics* may be viewed as a credible candidate to replace existing technologies in many applications requiring standalone or embedded solutions combining Flash-like nonvolatility, SRAM-like fast nanosecond switching and DRAM-like infinite endurance. For future, STT-RAM concept appears as a promising technology able to merge aforementioned advantages and scalability with aggressive technological nodes.
- PCM is probably the most advanced alternative memory concept in terms of process maturity, storage capacity and access time. Furthermore, the *phase change* memory element exhibits an excellent ability to withstand a downsizing of its critical dimensions. Considering limitations in terms of high reset current and low performances in retention, PCM is rather positioned as a NOR Flash substitute. Nevertheless, it may be also envisaged to rethink PCM subsystem architecture to bring the technology within competitive range of DRAM. To exploit PCM's scalability as a DRAM alternative, new design solutions could be proposed to balance long latencies, high energy writing and finite endurance.
- RRAM concept relying on a *reversible resistance change* is in an earlier stage of development compared with MRAM and PCM. Since RRAM memory elements can be integrated into "Back-End Of Line" (BEOL), this technology is of particular interest for high density storage with possibly multi-levels three-dimensional architectures. Nickel oxide-based OxRRAM and chalcogenide-based CBRAM appear as the most promising candidates. However, retention and endurance still remain to be demonstrated for CBRAM devices, while the high reset current is an issue in NiO-based memory devices. As a consequence, RRAM concepts, still in their infancy, requires further academic and industrial investigations (i) to validate integration and scalability capabilities; (ii) to uncover the origin of resistance switching which remains sometimes controversial; (iii) to model the reliability.

Even without consolidation of memory technology choices and despite different maturity levels, there are two *common guidelines* for designing *generic resistive switching memory circuits* (MRAM, PCM or RRAM):

- The first guideline is related to the implementation of bi-stable resistive elements in a memory array. In most of cases, the core cell relies on the association of a select/access device with a resistive element (formally 1T/1R cell). Depending on the resistive concept (magnetic, phase change or filamentary), the select/access device can be a BJT, a MOS transistor or a diode. Even if MOS transistor is often preferred, access device is adapted to electrical characteristics of resistive element (set and reset currents and voltages, resistance levels, memory window etc.). As shown in Fig. 16.7a, universal and robust memory array is experienced for many years. To access each 1T/1R core-cell individually, suited bias conditions are applied to Bit (BL), Source (SL) and Word (WL) Lines of the addressed cell, other lines being grounded and/or floating. For illustration, Hush and Baker [46] have proposed an architecture for sensing the resistance state of a programmable conductor random access memory element so-called PCRAM (Fig. 16.7b). The memory design is based on complementary PCRAM elements, one holding the resistance state being sensed and the other holding a complementary resistance

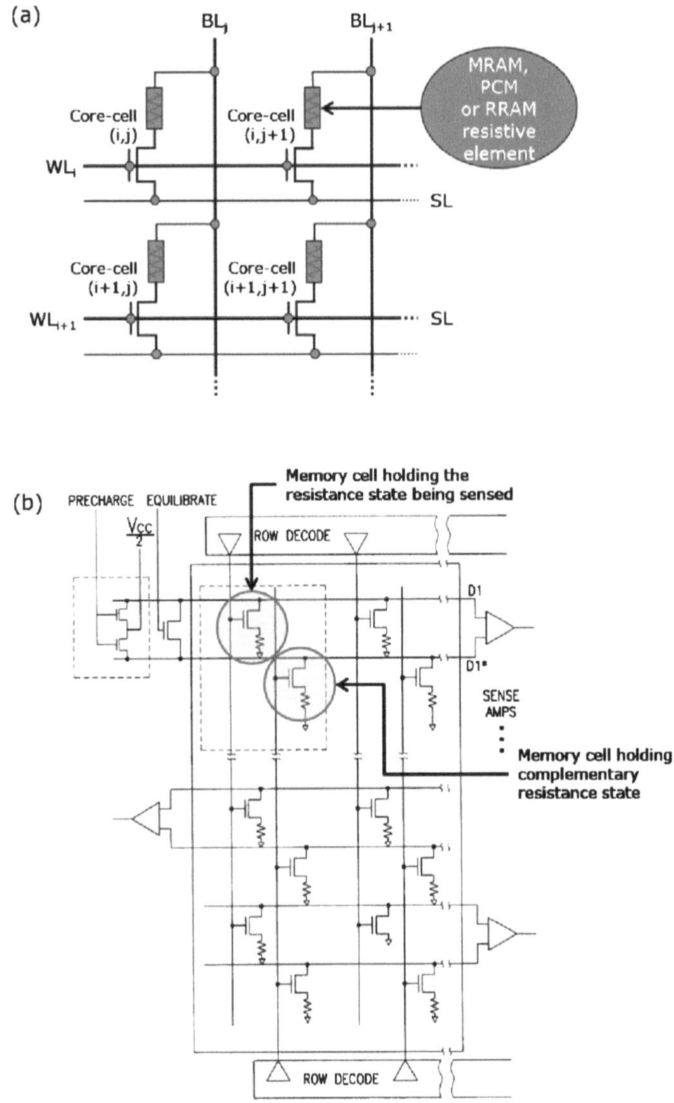

Fig. 16.7 (**a**) Schematic of 2 × 2 bits memory array in which each 1T/1R core-cell associates an access device (e.g. transistor) with a resistive switching element. Each cell is addressed through World (WL), Bit (BL) and Source (SL) lines. (**b**) Architecture that enables sensing the resistance state of a programmable conductor random access memory element using complementary resistive elements, one holding the resistance state being sensed and the other holding a complementary resistance state (dashed square). Inspired from US Patent no. 6,791,859 [46]

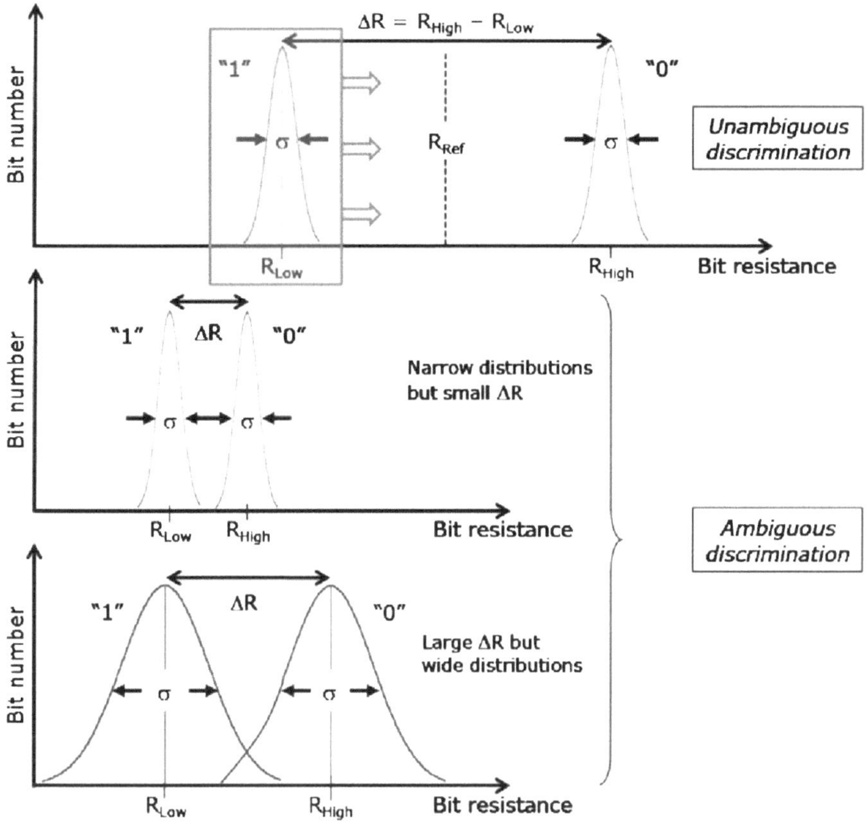

Fig. 16.8 Bit cell resistance distributions enabling or not reliable discrimination between "0" and "1" resistance states. Unambiguous discrimination relies on large ΔR margin and narrow bit cell resistance distributions σ leading to a large $\Delta R/\sigma$ ratio. Moreover, to reduce power consumption, R_{Low} value must be as large as possible

state. A sense amplifier detects voltages discharging through high and low resistance elements to determine the resistance state of an element being read.

- The resistance values in "0" and "1" states represent the second common guideline (Fig. 16.8). Indeed, reliable read operations require an unambiguous discrimination of low and high resistances, below and above the median resistance R_{Ref} defined as $\Delta R/2 = [R_{\mathrm{High}} - R_{\mathrm{Low}}]/2$. Moreover, the bit cell resistance distributions must be narrow to avoid any overlap between "0" and "1" states (i.e. large $\Delta R/\sigma$ ratio, with σ the distribution standard deviation). Finally, as the maximum current consumed during read operation is linked to R_{Low}, the resistance in "1" state must be as large as possible to decrease overall power consumption.

In summary, reliable memory operations (Fig. 16.8) require a close matching between *electrical characteristics* of resistive elements and *design rules*. In other words, to reach suitable sensing margin (i.e. large ΔR) shifted toward high resistances altogether with narrow bit cell resistance distributions (i.e. low σ), it is crucial (i) to control materials microstructure, critical dimensions, process variability, etc.; (ii) to understand physical switching mechanisms; (iii) to develop relevant models implemented in electrical simulators. Hence, the development of *emerging memory concepts* emphasizes the necessity to make stronger links between *memory cell materials and processes, modeling and circuit design.*

References

1. Van Houdt, Y., Wouters, D.J.: Memory technology: where is it going? Semicond. Int. **29**(13), 58–62 (2006)
2. Scott, J.F.: Ferroelectric Memories. Springer, Berlin (2000)
3. Böttger, U., Summerfelt, S.R.: In: Waser, R. (ed.) Nanoelectronics and Information Technology. Wiley-VCH, Weinheim (2003)
4. Nagai, A., et al.: Conformality of $Pb(Zr,Ti)O_3$ films deposited on trench structures having submicrometer diameter and various aspect ratios. Electrochem. Solid-State Lett. **9**(1), C15–C18 (2006)
5. Goux, L., Russo, G., Menou, N., Lisoni, J.G., Schwitters, M., Paraschiv, V., Maes, D., Artoni, C., Corallo, G., Haspeslagh, L., Wouters, D.J., Zambrano, R., Muller, Ch.: A highly reliable 3-dimensional integrated SBT ferroelectric capacitor enabling FeRAM scaling. IEEE Trans. Electron Devices **52**(4), 447–453 (2005)
6. Menou, N., Turquat, Ch., Madigou, V., Muller, Ch., Goux, L., Lisoni, J.G., Schwitters, M., Wouters, D.J.: Sidewalls contribution in integrated three-dimensional $Sr_{0.8}Bi_{2.2}Ta_2O_9$-based ferroelectric capacitors. Appl. Phys. Lett. **87**(7), 073502 (2005)
7. Tehrani, S.: Status and outlook of MRAM memory technology. In: IEEE Proc. of Int. Electron Device Meeting, pp. 1–4 (2006)
8. Engel, B.N., et al.: A 4-Mbit Toggle MRAM based on a novel bit and switching method. IEEE Trans. Magn. **41**, 132–136 (2005)
9. Andre, T.W., Nahas, J.J., Subramanian, C.K., Garni, B.J., Lin, H.S., Omair, A., Martino, W.L.: A 4 Mb 0.18 μm 1T1MTJ Toggle MRAM with balanced three input sensing scheme and locally mirrored unidirectional write drivers. IEEE J. Solid-State Circuits **40**(1), 301–309 (2005)
10. Savtchenko, L., et al.: Method of writing to scalable magnetoresistive random access memory element. US Patent 6,545,906 B1, 8 April 2003
11. Naji, P.K., DeHerrera, M., Durlam, M.: MTJ MRAM series-parallel architecture, US Patent 6,331,943 B1, 18 December 2001
12. Mattson, J.: Magnetoresistive memory devices, US Patent 6,806,523 B2, 19 October 2004
13. Leung, E.T.: Enhanced MRAM reference bit programming structure, US Patent 7,411,816 B2, 12 August 2008
14. Nahas, J.J., Andre, T.W., Garni, B., Subramanian, C., Lin, H., Alam, S.M., Papworth, K., Martino, W.L.: A 180 Kbit embeddable MRAM memory module. IEEE J. Solid-State Circuits **43**, 1826–1834 (2008)
15. Slaughter, J.M.: Recent advances in MRAM technology. In: IEEE Proc. of Device Research Conference, pp. 245–246 (2007)
16. Prejbeanu, I.L., Kula, W., Ounadjela, K., Sousa, R.C., Redon, O., Dieny, B., Nozières, J.P.: Thermally assisted switching in exchange-biased storage layer magnetic tunnel junctions. IEEE Trans. Magn. **40**(4), 2625–2627 (2004)
17. Sousa, R.C., Kerekes, M., Prejbeanu, I.L., Redon, O., Dieny, B., Nozières, J.P., Freitas, P.P.: Crossover in heating regimes of thermally assisted magnetic memories. J. Appl. Phys. **99**(8), 08N904 (2006)

18. Cardoso, S., Ferreira, R., Silva, F., Freitas, P.P., Melo, L.V., Sousa, R.C., Redon, O., MacKenzie, M., Chapman, J.N.: Double-barrier magnetic tunnel junctions with GeSbTe thermal barriers for improved thermally assisted magnetoresistive random access memory cells. J. Appl. Phys. **99**(8), 08N901 (2006)
19. TIMI, Thermally Insulating MRAM Interconnects, EURIPIDES project no. EUR-06-204; Partners: Crocus Technology (leader), Singulus, Tower Semiconductor, and IM2NP
20. Nagai, H., Huai, Y., Ueno, S., Koga, T.: Spin-transfer torque writing technology (STT-RAM) for future MRAM. IEIC Tech. Rep. **106**(2), 73–78 (2006)
21. Diao, Z., et al.: Spin-transfer torque switching in magnetic tunnel junctions and spin-transfer torque random access memory. J. Phys., Condens. Matter **19**(16), 165209–165221 (2007)
22. Prenat, G., El Baraji, M., Wei, G., Sousa, R., Buda-Prejbeanu, L., Dieny, B., Javerliac, V., Nozières, J.-P., Zhao, W., Belhaire, E.: CMOS/magnetic hybrid architectures. In: IEEE Proc. of Int. Conf. on Electronics, Circuits and Systems, pp. 190–193 (2007)
23. CILOMAG, Circuits Logiques Magnétiques, ANR project no. ANR-06-NANO-066; Partners: IEF (leader), Spintec, Crocus, CEA-LETI, CMP, and LIRMM
24. Das, B., Black, W.C.: A generalized HSPICE macro-model for pinned spin-dependent-tunneling devices. IEEE Trans. Magn. **35**(5), 2889–2891 (1999)
25. Kammerer, J.B., Hebrard, L., Hehn, M., Braun, F., Alnot, P., Schuhl, A.: Compact modeling of a magnetic tunnel junction using VHDL-AMS: computer aided design of a two-axis magnetometer. In: Proceedings of IEEE Sensors 2004, vol. 3, pp. 1558–1561 (2004)
26. Madec, M., Kammerer, J.B., Pregaldiny, F., Hebrard, L., Lallement, C.: Compact modeling of magnetic tunnel junction. In: IEEE Proc. of Northeast Workshop on Circuits and Systems and TAISA Conf., pp. 229–232 (2008)
27. Hass, K.J.: Radiation-tolerant embedded memory using magnetic tunnel junctions. Ph.D. Thesis, University of Idaho (2007)
28. Russo, U., Ielmini, D., Cagli, C., Lacaita, A.L.: Filament conduction and reset mechanism in NiO-based resistive-switching memory (RRAM) devices. IEEE Trans. Electron Devices **56**(2), 186–192 (2009)
29. Russo, U., Ielmini, D., Cagli, C., Lacaita, A.: Self-accelerated thermal dissolution model for reset programming in unipolar resistive-switching memory (RRAM) devices. IEEE Trans. Electron Devices **56**(2), 193–199 (2009)
30. Hosomi, M., et al.: A novel nonvolatile memory with spin torque transfer magnetization switching: spin-RAM. In: IEEE Proc. of Int. Electron Device Meeting, pp. 459–462 (2005)
31. Gogl, D., et al.: A 16-Mb MRAM featuring bootstrapped write drivers. IEEE J. Solid-State Circuits **40**(4), 902–908 (2005)
32. Nicolle, E.: Caractérisations et fiabilité de mémoires magnétiques à accès aléatoires (MRAM). Ph.D. Thesis, Université Paris Sud (2008)
33. Durlam, M., et al.: 90 nm toggle MRAM array with 0.29 μm^2 cells. In: IEEE Proc. of VLSI Technology Symp., pp. 186–187 (2005)
34. Liaw, J.-J., Tang, D.: High speed sensing amplifier for an MRAM cell. US Patent 7,286,429 B1, 23 October 2007
35. Bruchon, N., Torres, L., Sassatelli, G., Cambon, G.: Magnetic tunnelling junction based FPGA. In: Proc of Int. Symp. on Field Programmable Gate Arrays, pp. 123–130 (2006)
36. Guillemenet, Y., Torres, L., Sassatelli, G., Bruchon, N.: On the use of magnetic RAMs in field-programmable gate arrays. Int. J. Reconfigurable Comput. **2008**, 1–9 (2008)
37. Cho, W.-Y., et al.: A 0.18 μm 3.0 V 64 Mb nonvolatile Phase transition Random Access Memory (PRAM). IEEE J. Solid-State Circuits **40**(1), 293–300 (2005)
38. Cho, B.-H., Cho, W.-Y., Park, M.-H.: Phase change memory device generating program current and method thereof. US Patent 7,656,719 B2, 2 February 2010
39. Oh, J.H., et al.: Full integration of highly manufacturable 512 Mb PRAM based on 90 nm technology. In: IEEE Proc. of Int. Electron Device Meeting, pp. 49–52 (2006)

40. Kang, S., et al.: A 0.1-μm 1.8-V 256-Mb Phase-Change Random Access Memory (PRAM) with 66 MHz synchronous burst-read operation. IEEE J. Solid-State Circuits **42**(1), 210–218 (2007)
41. Ovshinsky, S.R.: Reversible electrical switching phenomenon in disordered structures. Phys. Rev. Lett. **21**(20), 1450–1453 (1968)
42. Lankhorst, M.H.R., Ketelaars, B.W., Wolters, R.A.: Low-cost and nanoscale non-volatile memory concept for future silicon chips. Nat. Mater. **4**(4), 347–352 (2005)
43. Castro, D.T., Goux, L., Hurkx, G.A.M., Attenborough, K., Delhougne, R., Lisoni, J., Jedema, F.J., Wolters, R.A.M., Gravesteijn, D.J., Verheijen, M.A., Kaiser, M., Weemaes, R.G.R., Wouters, D.J.: Evidence of the thermo-electric Thomson effect and influence on the program conditions and cell optimization in phase-change memory cells. In: IEEE Proc. of Int. Electron Device Meeting, pp. 315–318 (2007)
44. Chen, Y.C., et al.: Ultra-thin phase change bridge memory device using GeSb. In: IEEE Proc. of Int. Electron Device Meeting, pp. 777–780 (2006)
45. Raoux, S., et al.: Phase-change random access memory: a scalable technology. IBM J. Res. Dev. **52**(4/5), 465–479 (2008)
46. Hush, G., Baker, J.: Complementary bit PCRAM sense amplifier and method of operation, US Patent 6,791,859 B2, 14 September 2004
47. Chen, S.-H., Lung, H.-L.: Thin film plate phase change RAM circuit and manufacturing method, US Patent 7,238,994 B2, 3 July 2007
48. Adler, D., Shur, M.S., Silver, M., Ovshinsky, S.R.: Threshold switching in chalcogenide-glass thin films. J. Appl. Phys. **51**(6), 3289–3309 (1980)
49. Radaelli, A., Pirovano, A., Benvenuti, A., Lacaita, A.: Threshold switching and phase transition numerical models for phase change memory simulations. J. Appl. Phys. **103**(11), 111101 (2008)
50. Ielmini, D., Zhang, Y.: Analytical model for subthreshold conduction and threshold switching in chalcogenide-based memory devices. J. Appl. Phys. **102**(5), 054517 (2007)
51. Karpov, I.V., Mitra, M., Kau, D., Spadini, G., Kryukov, Y.A., Karpov, V.G.: Fundamental drift of parameters in chalcogenide phase change memory. J. Appl. Phys. **102**(12), 124503 (2007)
52. Lee, B.C., Ipek, E., Mutlu, O., Burger, D.: Architecting phase change memory as a scalable dram alternative. Comput. Archit. News **37**(3), 2–13 (2009)
53. Gill, M., Lowrey, T., Park, J.: Ovonic unified memory—a high-performance nonvolatile memory technology for stand-alone memory and embedded applications. In: IEEE Proc. Int. Solid State Circuits Conf., vol. 1, pp. 202–204 (2002)
54. Hyung-rok, O., et al.: Enhanced write performance of a 64 Mb phase-change random access memory. IEEE J. Solid-State Circuits **41**(1), 122–126 (2006)
55. Bedeschi, F., Resta, C., Khouri, O., Buda, E., Costa, L., Ferraro, M., Pellizzer, F., Ottogalli, F., Pirovano, A., Tosi, M., Bez, R., Gastaldi, R., Casagrande, G.: An 8 Mb demonstrator for high-density 1.8 V phase-change memories. In: IEEE Proc. of VLSI Circuits Symp., pp. 442–445 (2004)
56. Hsu, S.T., Pan, W., Zhang, F., Zhuang, W.-W., Li, T.: RRAM memory cell electrodes, US Patent 6,849,891 B1, 1 February 2005
57. Lee, C.-B., Park, Y.-S., Lee, M.-J., Wenxu, X., Kang, B.-S., Ahn, S.-E., Kim, K.-H.: Resistive memory devices and methods of manufacturing the same, US Patent 2009/0184305 A1, 23 July 2009
58. Malhotra, S.G., Kumar, P., Barstow, S., Chiang, T., Phatak, P.B., Wu, W., Shanker, S.: Non-volatile memory elements, US Patent 2009/0026434 A1, 29 January 2009
59. Waser, R., Aono, M.: Nanoionics-based resistive switching memories. Nature **6**, 833–840 (2007)
60. ITRS, International Technology Roadmap for Semiconductors: Emerging Research Devices; Process Integration, Devices and Structures. http://www.itrs.net/ (2009)
61. Sawa, A.: Resistive switching in transition metal oxides. Mater. Today **11**(6), 28–36 (2008)
62. Courtade, L., Lisoni-Reyes, J., Goux, L., Turquat, C., Muller, Ch., Wouters, D.J.: Method for manufacturing a memory element comprising a resistivity-switching NiO layer and devices obtained thereof. US Patent 7,960,775 B2, 14 June 2011

63. Courtade, L., Turquat, Ch., Muller, Ch., Lisoni, J.G., Goux, L., Wouters, D.J., Goguenheim, D., Roussel, P., Ortega, L.: Oxidation kinetics of Ni metallic film: formation of NiO-based resistive switching structures. Thin Solid Films **516**(12), 4083–4092 (2008)
64. Courtade, L., Turquat, Ch., Lisoni, J.G., Goux, L., Wouters, D.J., Deleruyelle, D., Muller, Ch.: Integration of resistive switching NiO in small via structures from localized oxidation of nickel metallic layer. In: IEEE Proc. of European Solid State Device Research Conf., pp. 218–221 (2008)
65. Spiga, S., Lamperti, A., Wiemer, C., Perego, M., Cianci, E., Tallarida, G., Lu, H.L., Alia, M., Volpe, F.G., Fanciulli, M.: Resistance switching in amorphous and crystalline binary oxides grown by electron beam evaporation and atomic layer deposition. Microelectron. Eng. **85**(12), 2414–2419 (2008)
66. Dumas, C., Deleruyelle, D., Demolliens, A., Muller, Ch., Spiga, S., Cianci, E., Fanciulli, M., Tortorelli, I., Bez, R.: Resistive switching characteristics of NiO films deposited on top of W and Cu pillar bottom electrodes. Thin Solid Films **519**(11), 3798–3803 (2011)
67. Symanczyk, R., Dittrich, R., Keller, J., Kund, M., Muller, G., Ruf, B., Albarede, P.-H., Bournat, S., Bouteille, L., Duch, A.: Conductive bridging memory development from single cells to 2 Mbit memory arrays. In: IEEE Proc. of Nonvolatile Memory Technology Symp., pp. 71–75 (2007)
68. Bocquet, M., Deleruyelle, D., Muller, Ch., Portal, J.-M.: Self-consistent physical modelling of set/reset operations in unipolar resistive-switching memories. Appl. Phys. Lett. **98**(26), 263507 (2011)
69. Kozicki, M.N., Park, M., Mitkova, M.: Nanoscale memory elements based on solid-sate electrolytes. IEEE Trans. Nanotechnol. **4**(3), 331–338 (2005)
70. Strukov, D.B., et al.: The missing memristor found. Nature **453**, 80–83 (2008)
71. Meyer, R., Schloss, L., Brewer, J., Lambertson, R., Kinney, W., Sanchez, J., Rinerson, D.: Oxide dual-layer memory element for scalable non-volatile cross-point memory technology. In: IEEE Proc. of Nonvolatile Memory Technology Symp., pp. 54–58 (2008)
72. Deleruyelle, D., Muller, Ch., Amouroux, J., Müller, R., Electrical nano-characterization of copper tetracyanoquinodimethane layers dedicated to resistive random access memories. Appl. Phys. Lett. **96**(26), 263504 (2011)
73. Chae, S.C., et al.: Random circuit breaker network model for unipolar resistance switching. Adv. Mater. **20**, 1154–1159 (2008)
74. Liu, C., et al.: Abnormal resistance switching behaviours of NiO thin films: possible occurrence of both formation and rupturing of conducting channels. J. Phys. D, Appl. Phys. **42**(1), 015506 (2009)
75. EMMA, Emerging Materials for Mass-storage Architectures: IST project no. 33751. Partners: IMEC, Numonyx, MDM, IUNET, RWTH-Aachen, and IM2NP, http://www.imec.be/EMMA
76. Cagli, C., Ielmini, D., Nardi, F., Lacaita, A.L.: Evidence for threshold switching in the set process of NiO-based RRAM and physical modeling for set, reset, retention and disturb prediction. In: IEEE Proc. of Int. Electron Device Meeting, pp. 301–304 (2008)
77. Ginez, O., Portal, J.-M., Muller, Ch.: Design and test challenges in resistive switching RAM (ReRAM): an electrical model for defect injections. In: IEEE Proc. of European Test Symp., pp. 61–66 (2009)
78. Lee, M.-J., et al.: 2-stack 1D-1R cross-point structure with oxide diodes as switch elements for high density resistance RAM applications. In: IEEE Proc. of Int. Electron Device Meeting, pp. 771–774 (2007)
79. Hosoi, Y., et al.: High speed unipolar switching resistance RAM (RRAM) technology. In: IEEE Proc. of Int. Electron Device Meeting, pp. 1–4 (2006)
80. Kinoshita, K., et al.: Reduction in the reset current in a resistive random access memory consisting of NiO_x brought about by reducing a parasitic capacitance. Appl. Phys. Lett. **93**(3), 033506 (2008)
81. Dietrich, S., et al.: A nonvolatile 2-Mbit CBRAM memory core featuring advanced read and program control. IEEE J. Solid-State Circuits **42**(4), 839–845 (2007)

82. Kund, M., et al.: Conductive bridging RAM (CBRAM): an emerging non-volatile memory technology scalable to sub 20 nm. In: IEEE Proc. of Int. Electron Device Meeting, pp. 754–757 (2005)
83. Liaw, C., Symanczyk, R.: A Method for operating a PMC memory and CBRAM memory circuit, European Patent 1727151 B1, 27 February 2008
84. Hoenigschmid, H., et al.: Resistive memory device and method for writing to a resistive memory cell in a resistive memory device, US Patent 7,518,902 B2, 14 April 2009

Chapter 17
Embedded Medical Microsystems

Neural Recording Implants

Benoit Gosselin and Mohamad Sawan

1 Introduction

Medical smart implantable devices are among the most critical embedded systems in term of safety and reliability. These devices must operate with a low energy budget, which has to be harvested from external power sources such as electromagnetic techniques [1–5]. These implants are, generally, divided into two main categories:

(1) Sensors to assess several biophysical parameters such as O_2, pH, impedances, pressures, temperature, various types of biopotentials [6–8], as well as biosensing and imaging at the level of molecules, cells, ions, etc. [9].
(2) Actuators which include drug delivery microsystems and several types of electrical stimulators [4, 10, 11]. The latter devices inject charge to excite tissue and then produce action potentials necessary to recover lost neuromuscular functions.

Implantable medical devices are supplied by a broad variety of increasingly complex microsystems. The requirement of implantability places severe restrictions on size, weight, power consumption, and bandwidth: (i) Implants must dissipate very low power in order to limit heat and temperature rises in surrounding tissues. (ii) The small capacity of implantable batteries and suitable wireless power supplies also contribute to restrict available power in implants. (iii) Current low-power telemetry links exhibit narrow bandwidths that cannot sustain more than a few Mbit per second [4, 12].

In contrast, emerging medical implants necessitate very-high densities. For instance, recent advances in neuroscience have motivated a plethora of research efforts

B. Gosselin (✉)
Université Laval, Quebec, Canada
e-mail: benoit.gosselin@gel.ulaval.ca

M. Sawan
École Polytechnique de Montréal, Montreal, Canada

G. Nicolescu et al. (eds.), *Design Technology for Heterogeneous Embedded Systems*,
DOI 10.1007/978-94-007-1125-9_17,

Table 17.1 Common measured neural signal variables

Signal/variable	Sampling rate	Frequency of events	Event duration	Duty cycle (%)
Extracellular APs	30 kHz	10 to 150/s	1–2 ms	2 to 30
EMG	15 kHz	0 to 10/s	0.1–10 s	0 to 100
EEG, LFP	200 Hz	0 to 1/s	0.5–1 s	0 to 100
O_2, Ph, Temp.	0.1 Hz	0.1/s	N/A	Very low

towards building sensors incorporating extraordinarily large channel counts. High-density recording microsystems are intended to simultaneously record from several hundreds of neurons. The number of achievable recording channels is rapidly growing, and it is anticipated that thousands of channels will be required for use of such microsystems in clinical prosthetic applications. With as much as hundreds of independent processing channels to supply, and around 0.5 Mbit/s per channel to handle, such an application is very challenging in terms of power and throughput. Thus, dedicated on-chip management of data and power is becoming essential.

Indeed, manageable data rates can be achieved by applying event detection and selecting only the relevant portions in the measured input. Such a strategy can provide substantial data reduction in biophysical sensing applications in general, because most biological signals are intermittent in nature, and have low duty cycles. Moreover, biological variables have usually very low rates of change. Table 17.1 presents a list of low-duty cycle biophysical signals and biological variables suitable for such a management scheme. For example, the main mechanism of information transmission in extracellular neural recording waveforms is a change in action potential (AP) generation rate by individual neurons [13]. Because APs are short waveforms of a few milliseconds of duration whose rate of occurrence ranges between ten and a few hundred per second, automatic detectors built with integrated circuits can be used in implantable microsystems to discriminate neural events from the background noise and achieve on-line data reduction [8, 14–16].

Besides, various dedicated techniques have been proposed to reduce power consumption in neural recording implants. Adaptive biasing schemes [17] and high-efficiency circuit design approaches [13, 18, 19] are also used to save power in analog building blocks. Likewise, several system-level approaches are employed for such a purpose. Parallelism and multiplexing/sampling schemes using both the positive and negative clock edges (as in [14]) enable for lower clock speed and for reduced power consumption, whereas clock gating is applied to save dynamic power in sampled circuit blocks when no biophysical events occur (as in [16]).

Existing system-level power management techniques are predominantly digital-oriented, focusing on reducing dynamic power and leakage power in switching circuits. However, data acquisition circuits are mostly built from analog and mixed-signal blocks, such as low-noise amplifiers, filters, sample-&-hold (S/H) circuits, data converters, etc., predominately draining static power. In this case, low-power analog design practices can be applied to minimize static power as much as possible.

However, it is expected that much more efficiency will be needed to empower the next generation of neuroengineering microdevices, dedicated to tackle some tremendously complex biological structures like vast cortical neural networks.

On the other hand, achieving the required density with a high resolution involves several design challenges in terms of size, I/O interfacing, and system architecture. Devices featuring up to hundreds of recording channels have been implemented so far [14, 15, 20]. Several approaches to build microsystems dedicated to such a purpose have been put forward. Monolithic devices integrating the signal processing electronics with the electrodes on a common silicon substrate are built using IC fabrication technology [14, 16]. This approach generates planar structures, but devices in 3D are obtained by means of complex assemblies. Hybrid devices have been assembled using flip-chip approaches where one IC was connected to a silicon microelectrode [15]. This strategy enables for more flexibility because the array and mounted components can be fabricated separately; but these works have only demonstrated the integration of a single IC on top of the base of an array. Other techniques using polymer substrate are used for integrating the probe, the circuits and the components. Although such an approach enables for more complexity, it can result in bulky devices.

In this chapter, we present interfacing circuits and on-chip management strategies to improve efficiency in dense neural recording implants which are probably one of today's most challenging and critically constrained applications of embedded systems. We discuss design approaches for low-power operation and optimization of specific circuits. We cover on-chip data management strategies to decrease the quantity of data to be handled in such implants. We introduce power management strategies to reduce dynamic and static power dissipation in analog and mixed-signal circuits. Finally, we discuss system integration strategies to achieve high-density in such critical embedded devices.

2 Low-Power Neural Interfacing Microsystems

The emission of APs by neurons form the basis of the neural activity recorded in the central nervous system. APs are emitted when a neuron suddenly depolarizes after being stimulated by the combined neural activity of several other cells in a network [21]. APs can be recorded in the extracellular fluid with a sharp biocompatible microelectrode placed in the vicinity of a neuron [13]. The recorded bioelectrical signals have weak amplitude in the order of 50–200 μV and lie in the 100 Hz–10 kHz frequency range.

Figure 17.1 reports an AP waveform extracted from a neural signal previously recorded *in vivo* in a small rodent. Recording such biopotentials requires suitable interfacing circuits for providing a sufficient gain, an appropriate bandwidth, a high signal-to-noise ratio (SNR) and an excellent linearity in the recording channel. First, APs must be amplified and filtered to remove offset and noise and to avoid aliasing. Then, they must be sampled at a sufficient rate, and finally outputted in a digital

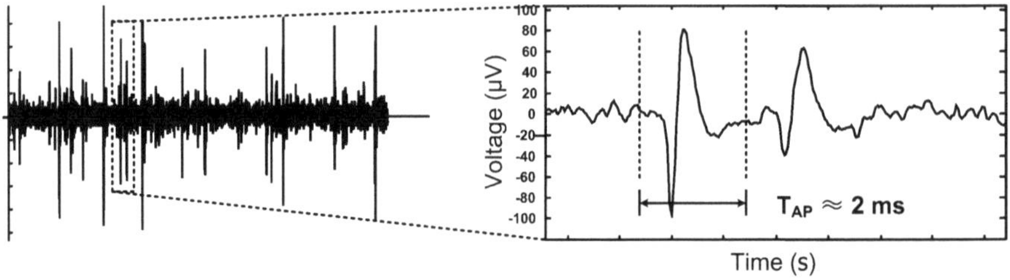

Fig. 17.1 An action potential waveform extracted from a neural signal recorded *in vivo* from the cortex of an anesthetized rat. The duration of a typical AP is of about 2 ms

Fig. 17.2 Conceptual representation of the 16-channel multi-chip neural interface. The ICs implementing the active part are stacked and bonded on the base of a micromachined stainless-steel microelectrode array

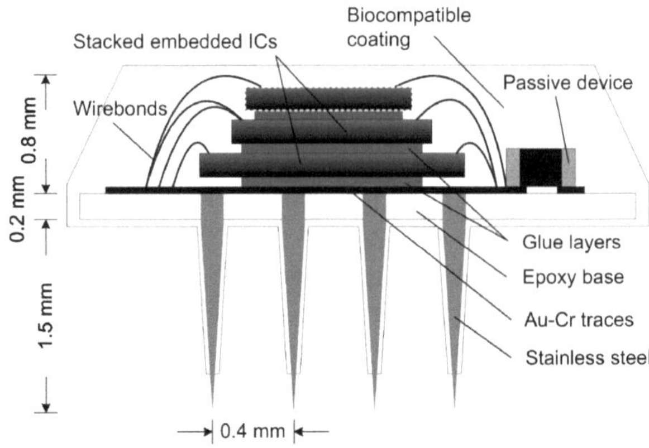

format for further processing or storage. We developed a multichannel microsystem capable of recording simultaneously from several neurons in a miniature format [8]. This microscale implantable device is composed of several ICs assembled with a microelectrode array in a vertical structure to acquire, process and transmit the measured neural waveforms to a host controller [8]. A prototype that includes 16 neural recording channels was built based on this multichip integration strategy. It employs dedicated mixed-signal circuit blocks with low-power dissipation to perform data acquisition, signal processing and data communication. Its conceptual representation is presented in Fig. 17.2. Detailed explanations on this multichip microsystem along with experimental results obtained with it are presented in Sect. 5. The remaining sections of this chapter present efficient circuits and system-level strategies for use in such dense and power constrained implantable microsystems.

2.1 Low-Power Circuit Optimization

Low-noise amplifiers and filters are fundamental front-end building blocks in biomedical sensing devices. One such conditioning circuit is needed for each electrode. Therefore, it is important to optimize the circuits for very low-power operation through a dedicated methodology. A transconductance efficiency-based design approach can be used to achieve best performances [18]. This design strategy allows optimal sizing of transistors and for operation over the entire continuum of channel inversion levels, with respect to a pre-selected drain current and inversion coefficient (IC). IC is proportional to the level of channel inversion of a MOS transistor and is related to parameters such as transconductance, noise, intrinsic bandwidth, output conductance, etc. Therefore, the parameters of analog circuits like the input-referred noise or the gain-bandwidth product can be optimized through a careful selection of devices, sized according to a required IC and a minimum dc current. The inversion coefficient is defined as [22],

$$IC = \frac{I_D}{2n\mu C_{\text{ox}}(W/L)U_T}, \tag{1}$$

where n is the slope factor, μ is the carrier mobility, C_{ox} is the gate-oxide capacitance per unit area, W and L are the channel width and length, and U_{T} is the thermal voltage (kT/q). The MOS transconductance efficiency ($g_{\text{m}}/I_{\text{D}}$) is maximum for transistors operating in weak inversion ($IC < 0.1$), and reaches its minimum for transistors in strong inversion ($IC > 10$). Specifically, operation at very low IC results in optimally high transconductance for a given current, but requires large W/L ratios. Conversely, operating at high IC yields low transconductance, but offers larger bandwidth. We define the transconductance ratio valid for all levels of inversion as

$$g_m/I_D \cong \frac{1}{nU_T}\varphi(IC), \tag{2}$$

where

$$\varphi(IC) = \frac{1}{0.5 + \sqrt{0.25 + IC}} \tag{3}$$

is taken from the EKV model [23], and yields values between 0 and 1. The design method first consists of choosing transistor drain currents that meet the circuit specifications regarding the input-referred noise, the low-frequency gain, the output resistance and the 3-dB bandwidth. Then, power consumption is optimized through a careful selection of devices sized according to the best-suited channel inversion level. This avoids wasting power in over-designed circuits and enables for higher efficiency, which is very important in such sensitive implantable microsystems designs.

2.2 Low-Noise Amplification

A low-noise amplifier must be used in the first stage of a neural data acquisition circuit. Such an amplifier, often called a neural amplifier or bioamplifier, has two main purposes. First, it must retrieve the weak bioelectrical signals from the background neural activity with high SNR. Most neural amplifiers commonly provide at least 40 dB of gain. Then, it must present a band-pass characteristic such the

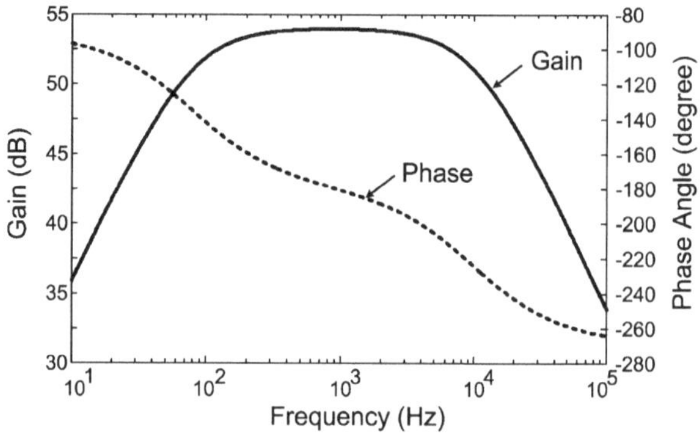

Fig. 17.3 A band-pass characteristic is typically required in the first rank amplifier to remove low frequency noise and dc voltages across differential electrodes

one depicted in Fig. 17.3. This particular type of response is needed to attenuate any low-frequency components such as electromagnetic noise pickup and dc potential mismatch across differential electrodes. If not removed by proper means, dc electrode voltage mismatch can attain hundreds of millivolts in some electrode materials, which can saturate the output of the amplifier.

Moreover, electrodes generate thermal noise. The first stage amplifier must be designed to exhibit an input referred-noise level below the thermal noise floor of the electrode, the amplitude of which depends on the impedance of the electrode and the bandwidth [24]. The two main well-known noise sources in CMOS amplifiers are the thermal noise and the $1/f$ noise. While the thermal noise can be related to MOS device transconductance, and hence their bias current, the $1/f$ noise can roughly be related to the total area circumscribed by the transistors [25]. Consequently, designing for low noise always yields a trade-off with power consumption and circuit size.

Besides, several amplifiers have been proposed for neural recording applications, but only a few meet the requirements for massive integration in a multi-channel device. A suitable amplifier must achieve an optimized noise-versus-power trade-off and allow small-size and massive integration, since one such amplifier per channel must be integrated. The noise efficiency factor (NEF) is commonly used to characterize the noise-versus-power trade-off achieved in a low-noise amplifier design [13, 17–19, 26]. The NEF is expressed as

$$NEF = v_{\text{ni(rms)}} \sqrt{\frac{2I_{\text{Total}}}{4\pi \cdot U_T kT \cdot BW}}, \tag{4}$$

where I_{Total} is the supply current of the amplifier and BW is its 3-dB bandwidth. The schematic of a compact low-noise amplifier featuring dc suppression is depicted in Fig. 17.4 (left hand side). This design uses a closed-loop low-frequency suppression scheme that removes unwanted low-frequency components, such as electrode potentials, while providing small implementation size and preserving the amplifier high-input impedance. The band-pass type transfer function of this amplifier topology is described by

$$\frac{v_1(s)}{v_{\text{in}}(s)} = \frac{-s\tau A_{v1}}{s\tau + A_{v1}}, \tag{5}$$

Table 17.2 Characteristics of reported integrated neural amplifiers

Amplifier	Power (μW)	Noise (μVrms)	NEF	Size (mm^2)	DC suppression
[13]	68	8.9	16.9	0.177	Passive
[17]	80.0	2.2	4.0	0.160	Passive
[14]	42.4	5.1	10.2	0.160	Passive
[24]	247.5[1]	3.0	10.4	0.076	Passive
[18]	7.6	3.1	2.67	0.160	Passive
This amplifier [12]	8.6	5.6	4.9	0.050	Active

[1] A supply voltage of 3.3 V is assumed for this 0.35-μm amplifier

where A_{v1} is the open loop gain of the main amplifier and τ is the time constant of the Miller integrator formed by the secondary amplifier, C_I and the equivalent resistor implemented by M_a and M_b. This transfer function yields a 3-dB high-pass cutoff frequency given by

$$f_{hp} = \frac{1}{2\pi}\frac{A_{v1}}{\tau}. \tag{6}$$

This low-noise amplifier design was fabricated and used in a dense multichannel neural recording microsystem [8]. It provides a gain of 50 dB and an input referred noise of 5.6 μVrms for power consumption below 9 μW. It achieves a measured NEF of 4.9 and presents a bandwidth ranging from 100 Hz to 9 kHz. For comparison, Table 17.2 summarizes the characteristics of reported integrated amplifiers suited for use in multi-channel microsystems. The topology presented in this chapter allows for the smallest reported sizes among similar purpose amplifiers using a differential input stage for better noise immunity.

2.3 Low-Power Signal Conditioning, Sampling and Digitization

After the first rank amplifier, a second amplification stage is needed to boost the neural signal to a sufficient value. This is to enable for high dynamic range in the data acquisition channel and to increase the SNR. The neural signal must also be low-pass filtered before being sampled and digitized, to attenuate out-of-band noise and avoid aliasing. The S/H circuit used in the chain must be optimized to enable for proper slew-rate and settling time, while minimizing power consumption. A sampling frequency of at least 3 to 5 times the neural signal bandwidth must be used to allow for proper signal integrity. For digitization, low-power medium-speed data converter topologies are naturally well suited for implantable applications in general. This is because operating at low frequency enables for a very low power budget in such data converters. Successive approximation (SA) analog-to-digital (A/D) converters dissipating a few microwatts have been demonstrated [27]. Such data

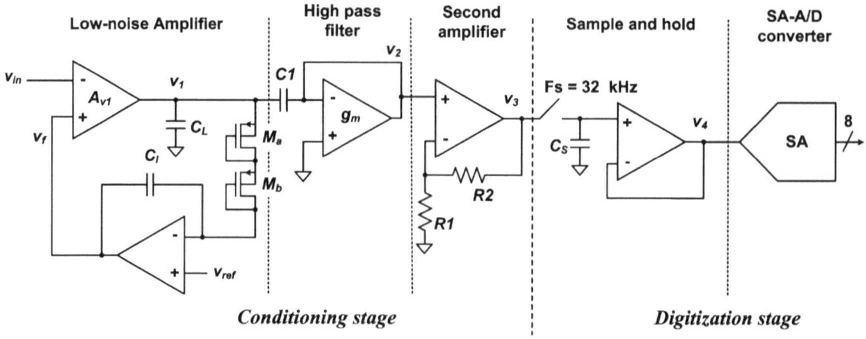

Fig. 17.4 Schematic of a conditioning and digitization circuit used in a 16-channel neural recording microsystem [8]. A low-noise amplifier, a highpass filter, and a second amplifier are cascaded for implementing the conditioning stage (*left hand side*). An S/H circuit and a SA-ADC are used for the digitization stage (*right hand side*)

converters benefit from very precise and compact capacitor arrays to perform voltage or charge scaling, which allows for a small implementation size.

The schematic of a low-noise low-power neural data acquisition circuit is presented in Fig. 17.4. This circuit was used in a 16-channel integrated neural recording interface [8]. It is composed of a low-noise amplifier, a high-pass filter, a second gain amplifier, an S/H circuit and an 8-bit SA-A/D converter. The first rank amplifier uses the low-noise amplifier topology described above. The high-pass filter, the transistors of which operate in the weak inversion regime, uses an operational transconductance amplifier (OTA)-C configuration. It is used to match dc levels with the remaining building blocks and to further attenuate the 60/50 Hz noise. A small transconductance is used in the OTA to set an additional high-pass cutoff frequency near 100 Hz. The transfer function of this filter is

$$\frac{v_2(s)}{v_1(s)} = \frac{sC_1 g_m^{-1}}{sC_1 g_m^{-1} + 1}. \tag{7}$$

Then, an additional gain stage of 20 dB is used in this chain to increase the dynamic range. It uses a two-stage amplifier topology in non-inverting configuration, and also operates in weak inversion. A RC network that employs a 1-pF capacitor and a 10-kΩ polysilicon resistor is used to compensate the amplifier. Its gain is set by two poly-resistors $R1 = 30$ kΩ, $R2 = 300$ kΩ. The optimized high-pass filter and the non-inverting amplifier consume 2.25 μW of power all together, when tested with a 1-kHz 50-$\mathrm{mV_{pp}}$ input sine wave. The whole conditioning stage, including the 1st rank amplifier, features a total harmonic distortion (THD) of 0.67% for a 160-μV 1-kHz input sine wave.

A sampling frequency of 32 kHz is used in the S/H circuit and in the A/D converter. The A/D converter uses a low-power SA topology detailed in [8] implementing a binary search algorithm based on a switched-capacitors voltage scaling circuit. The converted samples are stored in output registers which serve as output readout for subsequent digital building blocks in the integrated neural interface.

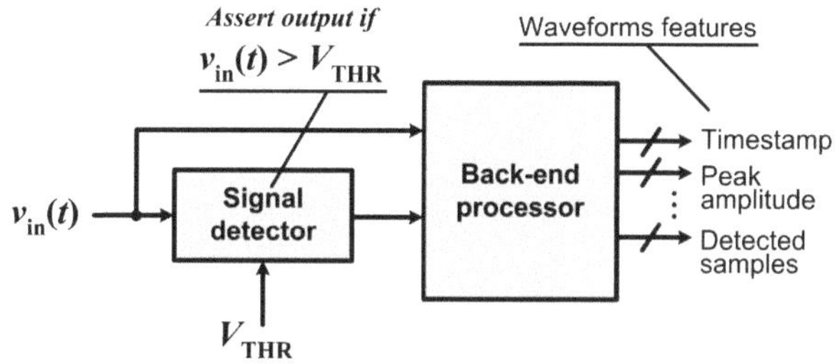

Fig. 17.5 A signal detector and its back-end processor used in a data management scheme. An event is detected when the amplitude of the input signal crosses the threshold (V_{THR})

3 Data Management

Data management is needed in neural recording microsystems to reduce power consumption and to transfer more channels over a same data link. An effective data reduction strategy must enable for high reduction factors, but preserve data integrity for further processing steps, such as waveform sorting. In such a scheme, automatic biopotential detectors made from integrated circuit blocks are required to discriminate neural events from noise (Fig. 17.5). The neural signal must be sufficiently amplified by means of a dedicated low-noise amplifier such as the first rank amplifier introduced above, before being assessed by the detector.

Integrated biopotential detectors have been proposed for real-time detection in multi-channel devices [14–16, 18, 26, 28–30]. However, most suggested designs are power-hungry and present poor data integrity. In fact, they are merely able to capture a limited number of AP features, such as the time of occurrence or the peak amplitude. Harrison has suggested an analog detector, the threshold of which is adjusted automatically according to the background noise characteristics [18]. Such a detector measures solely the portion of an AP that is higher than a positive threshold. Olsson has presented a digital threshold-based detector which sets the threshold value according to channel statistics [14]. It measures three distinct AP features to allow for further discrimination. Digital implementations of more intensive algorithms can also be used, but those using several multiply and accumulate operations, such as the *matched filter*, can exceed the circuit area and power budget allowed. Besides, detecting AP in an analog format enables for efficient channel and data management in further mixed-signal circuits (multiplexers, A/D converters, etc.), and digital back-end modules (controller, DSP, telemetry, etc.) as demonstrated in [16].

3.1 Digital Bandwidth Reduction

Digital data management schemes are employed to perform data reduction in neural interfaces. Such schemes are composed of a threshold detector and on-chip memory blocks to buffer the detected waveform prior to data transmission. A digital bandwidth reduction scheme that preserves full data integrity was implemented based on this principle [8]. It is composed of an absolute value threshold detector and on-chip SRAM blocks used for data buffering. The SRAM implements FIFOs and output memory buffers. The FIFOs continuously buffer 16 samples of incoming neural

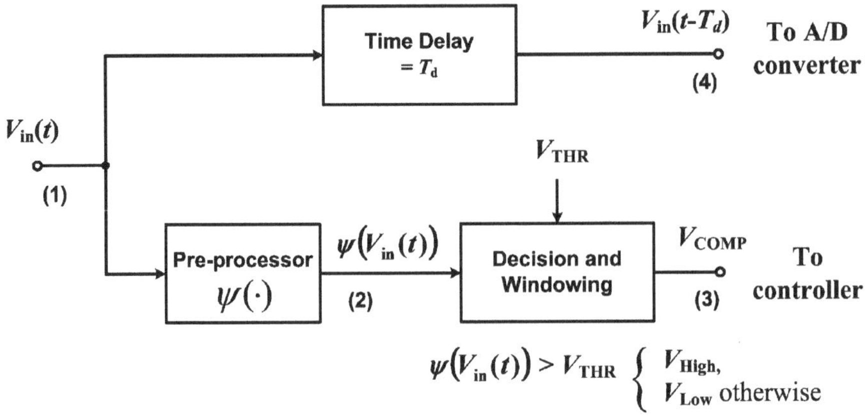

Fig. 17.6 Block diagram of an automatic detection strategy. It is composed of a pre-processor, a decision block and a time-delay element

waveforms on each channel to catch the first part of the detected events. A detector compares the absolute value of the amplitude of each incoming sample with a programmable threshold, and triggers a finite state machine (FSM) upon threshold crossing. This FSM manages data buffering of the remaining samples (up to 64 samples, or a 2-ms window in this case), and encloses the detected waveform samples in data packets along with a timestamp and the channel address. The implemented scheme uses 256 bytes for the 16 pre-detection FIFOs, and 1 Kbytes for the memory buffer that stores the detected events. This tested scheme preserves the integrity of the detected waveshapes entirely for further waveform sorting, whereas it achieves a typical bandwidth reduction factor of 8.

3.2 Analog Detection Schemes

Analog circuits can be used instead of digital building blocks to implement integrated detectors. Analog schemes imply less overhead than digital strategies since digitization is not required prior to signal detection. A sub-microwatt biopotential detector based on weakly inverted MOS devices was implemented for use in multi-channel neural recording microsystems [31]. The proposed analog detector, the block diagram of which is depicted in Fig. 17.6, enhances biopotentials for detectability and captures complete waveforms by means of a continuous-time linear-phase filter. It comprises three main building blocks; first, a pre-processor emphasizes APs shapes and attenuates out-of-band noise; secondly, a threshold function determines AP locations; finally, a delay element provides delayed copies of detected APs to preserve waveform integrity. As detailed in [31], the custom analog processor is based on an energy operator (the Teager energy operator), which exhibits excellent sensitivity to transients and good immunity to low-frequency artefacts such as hearth depolarisations, respiration, or chewing noises which can corrupt neural waveforms. The decision block is implemented by a low-power latched comparator and an asynchronous counter that determines the length of a detection window.

Compared to the digital scheme presented above, the delay filter is used instead of FIFOs to hold the waveform so the data acquisition begins ahead of the detection point and waveform integrity is preserved. The pre-processor and the filter are implemented using ultra-low-power OTA-C building blocks, the transistors of which are biased in the weak inversion regime. In addition to enable full waveform integrity, like the digital scheme presented above, this strategy alleviates the needs for consuming FIFOs. The automatic detector and the filter were both integrated in a TSMC 0.18-μm six-metal one-poly CMOS process, and fit in an area of 272 × 257 μm^2, which represents 44% of the 400 × 400 μm^2 per-channel-area allotment that is typically allocated to multi-channel interfacing circuits to match up with the pitch of a microelectrode array [15]. The fabricated detector and filter dissipate 420 nW and 360 nW, respectively.

The analog detector, used along with an appropriate back-end processor, can achieve data reduction factors equivalent to those obtained with a digital counterpart, while dissipating less power [32]. Moreover, an analog detector can be combined with an averaging filter to adaptively adjust its detection threshold automatically, according to the characteristics of a measured signal, for achieving a better detection rate [33].

3.3 Ultra-Low-Power Signal Processing Circuits

The analog automatic detection strategy presented above enables to preserve complete waveform integrity by means of an ultra-low-power continuous-time filter [31]. This filter provides delayed copies of APs so the data acquisition circuits can be turned on ahead of incoming waveforms for not losing any part while being off most of the time for saving power. Such a filter introduces a linear-phase shift over the bandwidth of interest with very low distortion. Implementing a suitable neural signal delay requires a high-order filter which unfortunately presents higher sensitivity to parasitics and circuit non-idealities. Moreover, achieving ultra-low-power in CMOS circuits requires operating the MOS devices in the weak inversion regime, which are thus likely to exhibit poor matching and higher sensitivity to process variation. Therefore, particular attention must be paid to minimize these effects in the design of a suitable delay filter.

Filter design theory shows that a 9th-order equiripple allpass transfer function presenting a constant group delay can implement a suitable time delay of about 600 μs in the 0–10 kHz range [31]. This delay represents approximately a third of a typical AP duration, which is enough to hold the part of the AP waveform which comes before the threshold crossing point. The ideal response of the filter is shown in Fig. 17.7. Low-power, small-g_m OTAs and small-size integrated capacitors can be employed to implement this delay filter response. The schematic of the

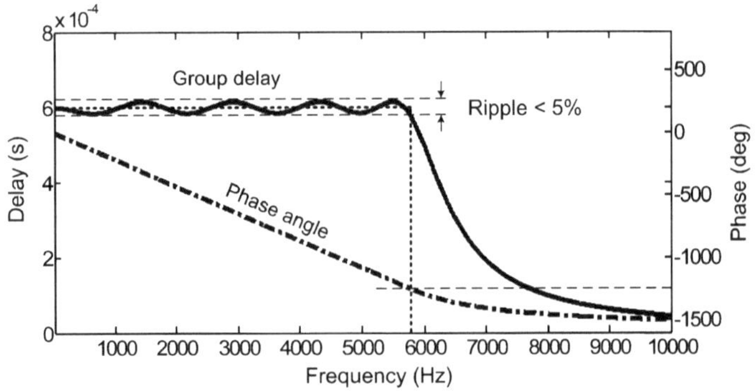

Fig. 17.7 Phase response and group delay of the allpass filter with equiripple delay. The filter exhibits a linear phase response and a constant delay within the prescribed bandwidth

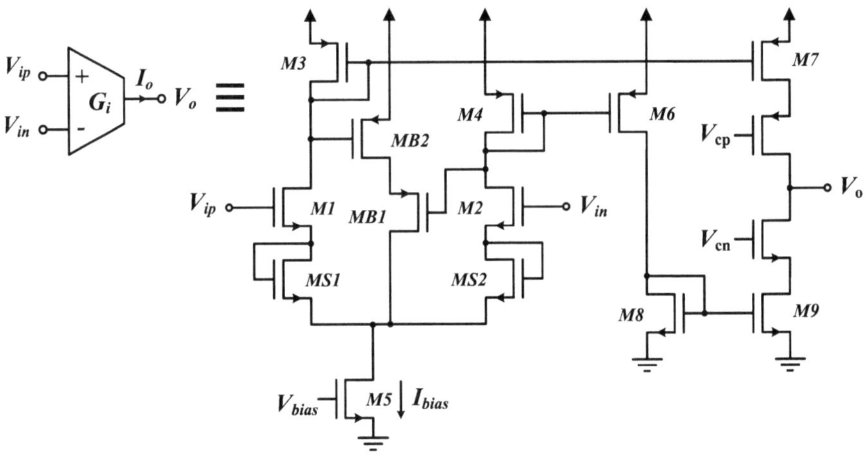

Fig. 17.8 A subthreshold biased OTA with source degeneration and bump linearization devices

current mirror OTA employed for this realization is depicted in Fig. 17.8. Its overall transconductance equals the transconductance of $M_{1,2}$, which is $G_m = g_{m1,2}$. The output current I_o of the OTA is then

$$I_o = g_{m1,2} \cdot (V_{ip} - V_{in}), \tag{8}$$

and its dc gain is

$$\frac{V_o}{V_{ip} - V_{in}} = G_m \cdot R_o \cong g_{m1,2} \cdot \frac{g_m r_o^2}{2}, \tag{9}$$

where R_o is the output impedance of the cascoded output stage. The OTAs use weakly inverted MOS devices featuring very low channel inversion levels (i.e. with an inversion coefficient $IC \ll 0.1$), the transconductances of which are linearly related to I_D with

$$g_m = \frac{I_D}{nU_T} \cdot \frac{1}{0.5 + \sqrt{0.25 + IC}} \cong \frac{I_D}{nU_T}. \tag{10}$$

This enables us to use currents of the order of one nA to a few tens of nA in the OTA building blocks employed both for the detector and for the delay filter, addressing low-power and small-scale transconductance issues. However, operation at very low current comes at the cost of a reduced dynamic range (*DR*). Therefore, the chosen OTA circuit (Fig. 17.8) incorporates *source degeneration* (SD) and *bump linearization* (BL) to simultaneously achieve a low overall transconductance, and a better linear range. SD is implemented using diode connected nMOS devices M_{S1}, M_{S2} in series with the sources of M_1, M_2, in the differential pair. The current flowing through the SD devices is converted into a voltage that is fed back to the source of M_1, M_2, to decrease their current, thus their transconductance (g_m). As a result, the overall transconductance of the OTA is also reduced. We can show that g_m of M_1, M_2 are decreased by a factor of $1 + N$ by SD, where $N = g_{m1,2} \cdot (g_{mS1,2})^{-1}$. Having a unitary ratio in the current mirrors formed by M_5, M_6 and M_7, M_8, the overall transconductance G_m of the OTA, including SD, is

$$G_m \cong \frac{g_{m1,2}}{1 + g_{m1,2} \cdot g_{mS1,2}^{-1}}. \tag{11}$$

This reduction of the overall G_{m} increases the linearity of the amplifier while it decreases the capacitor sizes needed for operation at low-frequency. BL transistors are also used to extend the linear operating range of the subthreshold OTA. The two series connected pMOS devices M_{B1}, M_{B2} in the central arm of the diff-pair are the *bump transistors*. These devices draw current from the two outer arms according to a bump-shaped function for linearizing the *tanh*-like characteristic of the subthreshold differential pair output current. The W/L ratios of the bump transistors must be chosen twice as large as one of the two outer arm devices (M_3, M_4) in order to achieve maximal linearity [34]. An interesting fact with BL is that this appropriate sizing of M_{B1}, M_{B2} can be shown to eliminate the 3rd-order distortion terms in the diff-pair transfer characteristic. Moreover, BL is particularly well suited for subthreshold OTAs because it increases the amplifier linear range without increasing the noise [34]. Cascode transistors (M_{CN}, M_{CP}) are used as well in the output stage to increase the amplifier output impedance and open-loop gain, and reduce non-ideal effects when cascading building blocks.

A cascaded topology is selected to implement the filter. It uses one first-order allpass section followed by four cascaded biquadratic allpass sections. The biquadratic sections are synthesized according to an allpass equal-capacitor-values topology, introduced in [35], and which yields a minimum number of components. Each biquad uses four OTAs and two equal-value capacitors to implement the required polynomial. The transfer function of one biquad section (Fig. 17.9) is

$$H(s) = \frac{G_{ma}}{G_{mb}} \frac{s^2C^2 - sCG_{m(2n+1)} + G_{m(2n+1)}G_{m(2n)}}{s^2C^2 + sCG_{m(2n+1)} + G_{m(2n+1)}G_{m(2n)}}, \tag{12}$$

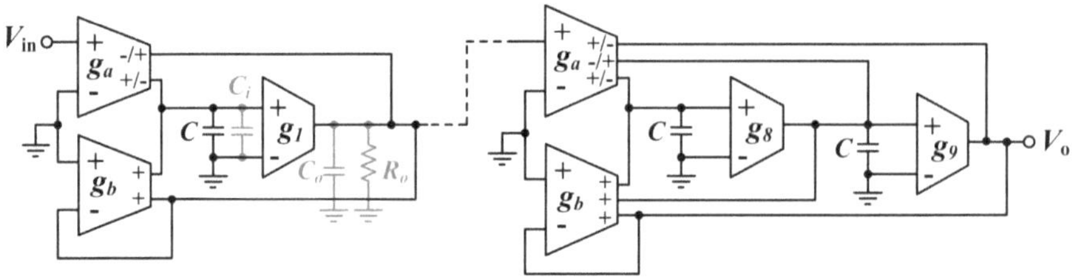

Fig. 17.9 Implementation of a 9th order OTA-C continuous-time delay-filter using a cascaded topology. The OTA parasitics are shown for g_1

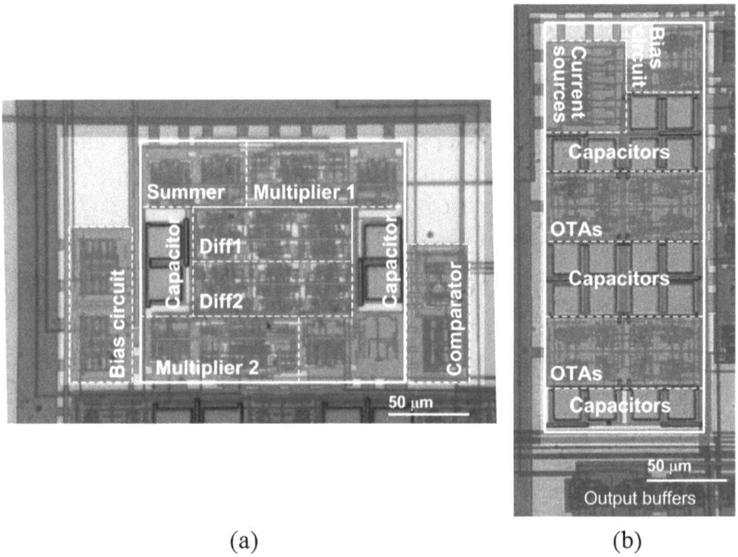

Fig. 17.10 Microphotograph of the fabricated analog detector chip. (**a**) The integrated TEO pre-processor and other constituting building blocks. (**b**) The integrated cascaded allpass delay filter

where integer $n = 1$ to 4. The filter coefficients are implemented by OTA G_{m} ratios, whereas the capacitors scale the *constant delay bandwidth* and the delay achieved by the filter. The biquads use grounded capacitors and single-ended OTAs to minimize the parasitic poles that cause excess phase shift, and to avoid the generation of parasitic zeros. Grounded capacitors absorb parasitic capacitances and need smaller area than floating ones.

In the implementation of the circuit layout, a better matching in OTAs can be achieved when realizing the filter by distributing currents instead of voltages and by using local mirroring of currents. Moreover, a tuning strategy is recommended in [36] to reduce mismatch and process variation effects. We have applied these strategies in our implementation of the delay filter. A microphotograph of the integrated filter is presented in Fig. 17.10.

Fig. 17.11 Measured output of the analog detector. (**a**) Synthetic input signal (*top*). TEO pre-processor output (*second trace*). Delayed waveforms at the filter output (*third trace*). Windows generated by the decision block (*bottom*). (**b**) A closer view on detected waveforms with a measured delay of ≈ 600 μs

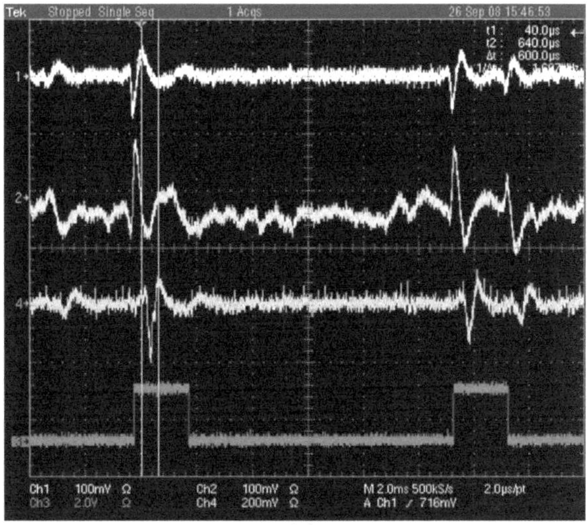

(a)

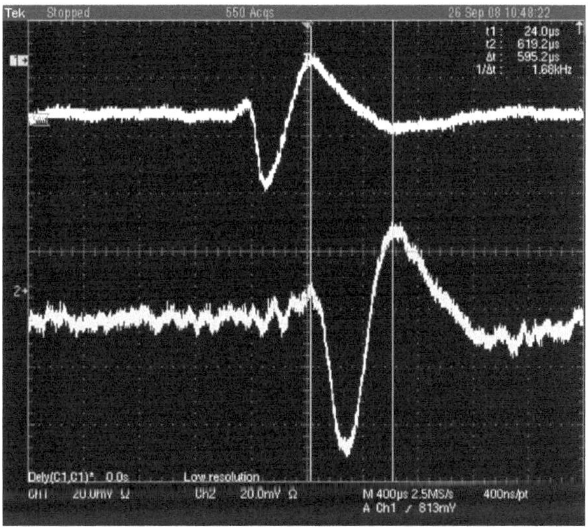

(b)

The fabricated analog detector along with the delay filter dissipates 776 nW, which corresponds to less than 10% of typical power consumption achieved by recently reported dedicated first-rank amplifiers (one such amplifier per channel is required). A synthetic neural signal presenting a realistic firing rate was constructed and used as an input signal for testing the detector and the filter. Figure 17.11a presents the detector output response to this synthetic signal. The multiple APs embedded in the background noise (top trace) are successfully detected and isolated by the detector when the TEO output (second trace) crosses the detection threshold. The linear-phase filter was adjusted by tuning its bias current until it yields a signal delay close to 600 μs (third trace). Discrimination windows of proper lengths are generated upon threshold crossing (bottom trace). Figure 17.11b shows a delayed waveform measured at the output of the equiripple delay filter. The measured delay for this chip was 592 μs. As seen, the waveform is slightly amplified by the filter, but not distorted. The filters typically achieve a THD smaller than 1.2% within an input range of 40 mVpp. The typical phase response measured within the integrated delay filters presents a total phase shift of around 1.4 kdeg near 6 kHz, which is in accordance with the ideal response presented in Fig. 17.7.

4 Power Management

Several low-power methods can be applied, from process-level to system-level, throughout microsystems design. While the process technology is usually not flexible, the circuit designer can achieve power reduction though dedicated circuit techniques and architectural design approaches. Techniques like dynamic voltage scaling, body biasing, frequency scaling, reduction of switching activity through optimization, clever encoding of sequential circuits, etc., are available to optimize such power. Further reduction of power in digital systems can be achieved through architectural design, using parallelism, pipelining, or distributed processing [37]. Dynamic power management strategies such as clock- and power-gating have also proven their efficacy for reducing dynamic and leakage power. But, while static power is only now becoming an issue in digital electronics at advanced technology nodes, it has long been a major source of power dissipation in analog and mixed-signal circuits due to the intrinsic need for biasing. The low-power design methodology introduced above can be applied to optimize analog circuits. However, dedicated power management techniques must be elaborated to reduce dissipation further.

4.1 Event-Driven Schemes

The power consumption of neural recording implants can be improved using dynamic power management strategies [38]. Power-gating is extensively used in digital electronics to reduce power leakage in deep submicron circuits [39]. This technique refers to gating, or cutting off, a circuit from its power supply rails during standby mode. This strategy can be applied to low-duty cycle analog and mixed-signal neural data acquisition circuits in order to reduce static power dissipation. Power gating is realized by placing a power switch, called a header, or a sleep transistor, in series with a circuit block (or a power island), as shown in Fig. 17.12. A footer, which is an nMOS switch placed between the circuit block and V_{SS}, can also be used in combination with the header. When no input activity is detected, the header is turned off by the power management unit (PMU) to disconnect the circuits from the power rail. When circuits are required, the PMU turns on the header

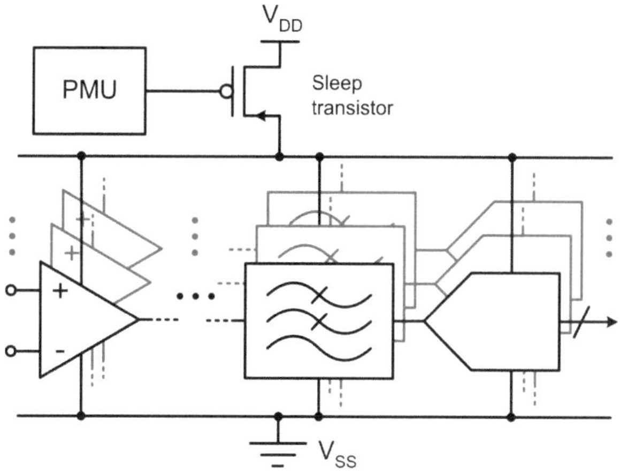

Fig. 17.12 A power-gating strategy for analog and mixed-signal circuit blocks

again so that the circuits are reconnected to the power rails. In order to achieve this strategy, the biophysical events must be detected in advance to activate the power-gated islands on the right time. The analog detector presented above can be used to achieve such synchronization. This detector dissipating very low power can remain continuously active to locate the biophysical events and trigger the PMU for powering up the gated islands. Since such an analog detector does not require much power compared to a full data acquisition chain, power is not wasted when no event occurs. In fact, the detector only drains about 5% of the total power per channel achieved in [31]. A complete data acquisition chain, including amplifiers, filters, an S/H, and a data converter can be controlled by one detector and dedicated headers. Subsequent digital building blocks (DSP, memory, etc.) in the data path, can also be included in associated digital power islands that are activated with a dedicated switch. In the case of a multi-channel device that records from multiple sites, one detector per channel is required to selectively activate the corresponding circuits upon event occurrences.

Besides, an efficient implementation of this management strategy requires a low error rate in the detector. Adaptive adjustment of the detector according to the characteristics encountered in each channel can improve the detection rate, as in [33, 40]. Also, a high power supply and common mode rejection ratios should be prioritized in the design of analog circuits when using this scheme to minimize ground bounces and switching noise. The distribution of control signals and circuit overhead must also be addressed in the design. In addition, wake-up transients can appear at the output of analog continuous-time devices when powered up, but simulation of a typical data acquisition building blocks using this scheme shows that the wake-up transients only represent a very small percentage of the duration of typical biophysical events (see [38] for more details).

The presented analog and mixed-signal power-gating approach is demonstrated with the simulation of a voltage follower, a building block that is extensively used in continuous-time and sampled-mode data acquisition circuits such as S/H, data converters, filters, etc. A pMOS sleep transistor was added between the voltage follower

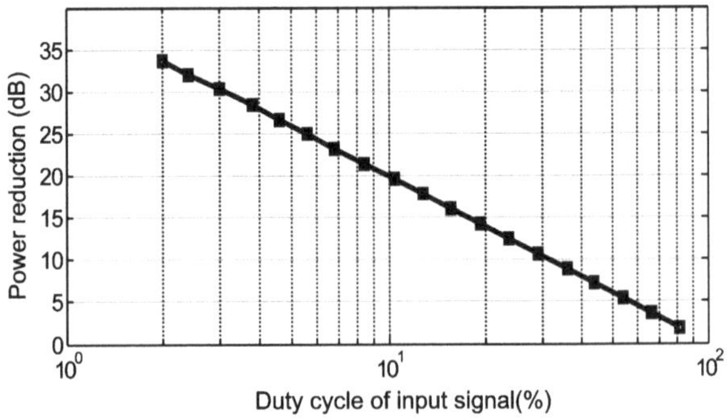

Fig. 17.13 Power dissipation reduction in a power-gated voltage follower versus the duty cycle of the input signal

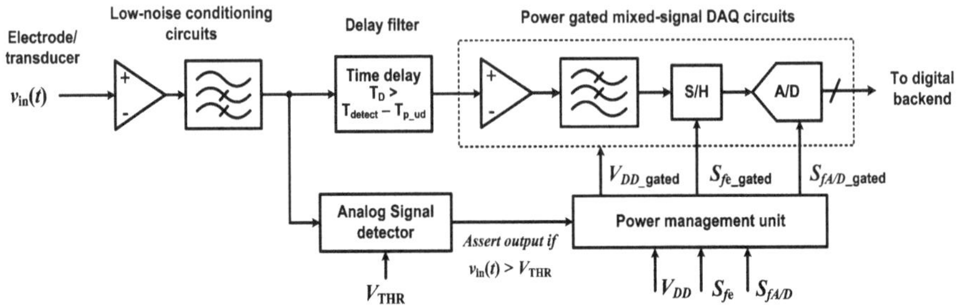

Fig. 17.14 A power gated data acquisition circuit

supply rail and the power source, and a 1-kHz, 200 mV_{pp} sine wave was used as input signal to the follower. The slew rate, the settling time and the total power consumption of the follower was optimized for operation in a 30-kHz sampled-mode circuit, using the low-power methodology introduced above. The sleep transistor was controlled with pulses of 2 ms of duration to model the measured biophysical events. Figure 17.13 presents the root mean square power dissipation of the voltage follower versus the duty cycle of the simulated input signal. As seen, the power reduction achieved is directly proportional to the duty cycle of the input signal. These results suggest that a neural recording channel using the power management strategy described in this section will dissipate 10 times less power than conventional configurations, for a typical input neural signal featuring 60 events/s, each lasting 2 ms, which represents a duty cycle of 12%. We report similar power reductions when applying the presented power gating strategy to other conditioning circuits such as filters and gain amplifiers.

The schematic of an event-driven power gated mixed-signal data acquisition circuit is presented in Fig. 17.14. The power gated circuits come after the 1st rank amplifier and a high-pass filter. The PMU is controlled by an analog detector. The power gated circuits are activated upon detection of an AP and stay on during the occurrence of the waveform. Otherwise, the circuits are off. A time delay element is used before the power gated circuits, so complete waveforms can be captured.

5 System Integration for High Density

5.1 Microsystem Integration Strategies

A vertical integration approach was developed to build a miniature multichannel neural recording system. The elaborated strategy consist in assembling several ICs with a microelectrode in 3D [8]. The resulting multichip structure is composed of a pile of stacked ASICs vertically mounted on the base of a stainless-steel microelectrode array. A metal layer is micromachined on the base of the array to route the electrodes and the circuits. The ICs and the electrodes are interconnected with wirebonds through this layer. We have assembled a 16-channel neural recording prototype employing this strategy (Fig. 17.2). It is composed of a mixed-signal IC and a digital IC. The mixed-signal IC implements 16 data-acquisition channels and features high parallelism. It employs one instance of the neural interfacing circuit previously introduced in Sect. 2.3 per channel. Such parallelism decreases the speed compared to an analog multiplexed architecture and it relaxes the requirements for bandwidth, slew rate, and settling time in the conditioning and digitization circuits. This allows operation at low frequency which significantly decreases inter-channel crosstalk and power consumption. This approach has also the practical advantage to be fully scalable to accommodate the design of any system topologies. Moreover, it allows for simultaneous sampling of neural signals, and decreases switching noise and crosstalk, which are typically associated with analog multiplexing.

The digital IC implements a digital management scheme (introduced in Sect. 3.1), performs control, and allows communications toward a remote controller. The assembled multi-chip microsystem transfers the detected neural waveforms on a 4-wire serial bus. Both the mixed-signal and the digital ICs were fabricated in a 0.18-μm six-metal one-poly CMOS process and subsequently mounted on a dedicated printed circuit board for testing.

The final microsystem is composed of the multiple ICs stacked on the base of a stainless steel microelectrode array. The ICs and the electrodes are interconnected using wirebonds and chromium-gold (Cr-Au) conductive traces micromachined on the base of the array using photolithography and chemical etching. The minimum distances achievable between the Cr-Au traces are on the order of a few μm. It worth mentioning that the size of the chip-on-board assembly used to test the assembled prototype and the associated interconnecting network do not give a suitable indication of the size of the final microsystem version (Fig. 17.15). Testing purposes required to use an independent interconnecting pad to connect each chip pad on a test board (a total of 80 pads and associated routing network with minimum distances of 127 μm is used to wirebond the ICs to the PCB). In contrast, one interconnecting pad is used to wirebond several chip pads in the final version, which contributes to further reducing the size of the interconnecting part compared with the chip-on-board version.

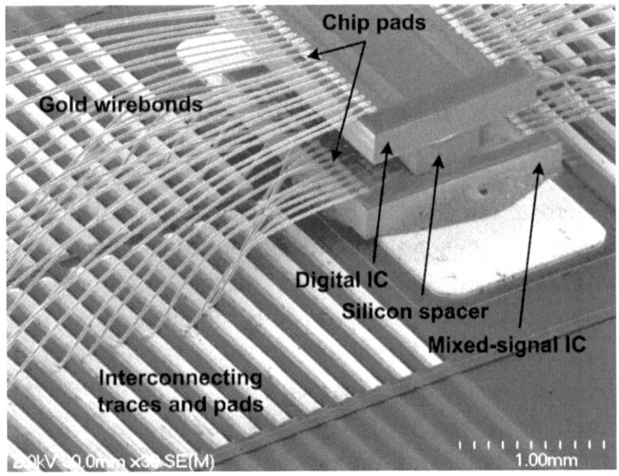

Fig. 17.15 SEM photograph of the stacked and wirebonded ICs on the test board. The two ICs are stacked and wire bonded on a test PCB. A silicon spacer is used between the stacked ICs

5.2 Microsystem Assembly and Test Platform

A chip-on-board configuration was used to test the neural interfacing ICs in accordance with the multi-chip integration strategy detailed in [8]. ICs were mounted and wirebonded directly onto the test board. The digital IC was stacked on the mixed-signal IC to implement the targeted multiple stage microsystem assembly structure. The two ICs are separated by a small silicon spacer (Fig. 17.15) stuck to the ICs using epoxy layers. A soft gold finish was used for the test board conductive traces and the wirebonded interconnections were realized with a wedge bonder and gold bonding wires of 25 μm diameter. Minimum distances and track width of 127 μm (5 mil) are used for the PCB traces and for the interconnecting pads employed to bond the ICs. Each minimum-width bonding trace is routed towards a via. Subsequent interconnections use minimum distances and employ traces of 8 mil width. A remote host controller was implemented with a Xilinx Spartan 3 FPGA prototyping board connected with the neural interface through the four-wire serial bus.

5.3 Test and Performances of the Neural Integrated Interface

The 16-channel multi-chip neural integrated interface described above was tested with synthetic neural waveforms. Synthetic signals featuring different firing rates were used to evaluate the bandwidth reduction factor achieved by the interface. The signals were played at the input of the interface and the resulting output data rates were measured. The measured data rates for input firing rates ranging from 10 spike/s to 120 spike/s are of 4.88 to 66.54 kbit/s per channel, respectively. This corresponds to data rates of 78 kbit/s to 1.06 Mbit/s for 16 channels. The resulting bandwidth reduction factors range from 3.5 to 48 for such inputs, which corresponds to a maximum bandwidth reduction of up to 98%. The multi-chip structure achieves a low-power consumption of 138 μW per channel, and a small size of 2.30 mm^2.

In addition, this multi-chip integrated neural recording interface allows complete acquisition of spike waveforms.

6 Conclusion

In this chapter, we have presented neural interfacing circuits and on-chip management strategies that reduce data rates and power consumption in multi-channel neural recording microsystems. We have detailed a transconductance-efficiency design approach to optimize analog and mixed-signal circuits used in sensitive applications such as neuroengineering microdevices and medical microsystems in general. The circuits covered were demonstrated with the implementation of a 16-channel multi-chip neural interface based on a parallel topology. We have explained two data reduction schemes based on digital and analog automatic detectors, and detailed power-gating approaches to control digital, analog and mixed-signal circuit blocks. An analog and mixed-signal power management strategy based on event-detection was also covered. This approach demonstrates its effectiveness to reduce power in a typical analog and mixed-signal building block by taking advantage on the low-duty cycle presented by extracellular APs and by other biophysical signals. The circuits and strategies presented in this chapter contribute to improve efficiency in neural recording implants and are also applicable to a broad variety of embedded medical microsystems in general.

References

1. Coulombe, J., Sylvain, C., Mohamad, S.: A power efficient electronic implant for a visual cortical neuroprosthesis. Artif. Organs **29**, 233–238 (2005)
2. Hu, Y., Sawan, M.: A fully integrated low-power BPSK demodulator for implantable medical devices. IEEE Trans. Circuits Syst. I, Regul. Pap. **52**, 2552–2562 (2005)
3. Hu, Y., Sawan, M., El-Gamal, M.N.: An integrated power recovery module dedicated to implantable electronic devices. Analog Integr. Circuits Signal Process. **43**, 171–181 (2005)
4. Sawan, M., Yamu, H., Coulombe, J.: Wireless smart implants dedicated to multichannel monitoring and microstimulation. IEEE Circuits Syst. Mag. **5**, 21–39 (2005)
5. Djemouai, A., Sawan, M.: Circuit techniques dedicated to effectively wireless transfer power and data to electronic implants. J. Circuits Syst. Comput. **16**, 801–818 (2007)
6. Harb, A., Hu, Y., Sawan, M., Abdelkerim, A., Elhilali, M.M.: Low-power CMOS interface for recording and processing very low amplitude signals. Analog Integr. Circuits Signal Process. **39**, 39–54 (2004)
7. Sawan, M., Mounaim, F., Lesbros, G.: Wireless monitoring of electrode-tissues interfaces for long term characterization. Analog Integr. Circuits Signal Process. **55**, 103–114 (2008)
8. Gosselin, B., Ayoub, A.E., Roy, J.F., Sawan, M., Lepore, F., Chaudhuri, A., Guitton, D.: A mixed-signal multichip neural recording interface with bandwidth reduction. IEEE Trans. Biomed. Circuits Syst. **3**, 129–141 (2009)
9. Ghafar-Zadeh, E., Sawan, M.: Charge based capacitive sensor array for CMOS based laboratory-on-chip applications. In: 5th IEEE Conference on Sensors, pp. 378–381 (2006)
10. Boyer, S., Sawan, M., Abdel-Gawad, M., Robin, S., Elhilali, M.M.: Implantable selective stimulator to improve bladder voiding: design and chronic experiments in dogs. IEEE Trans. Rehabil. Eng. **8**, 464–470 (2000)

11. Coulombe, J., Sawan, M., Gervais, J.F.: A highly flexible system for microstimulation of the visual cortex: design and implementation. IEEE Trans. Biomed. Circuits Syst. **1**, 258–269 (2007)
12. Mandal, S., Sarpeshkar, R.: Power-efficient impedance-modulation wireless data links for biomedical implants. IEEE Trans. Biomed. Circuits Syst. **2**, 301–315 (2008)
13. Gosselin, B., Sawan, M., Chapman, C.A.: A low-power integrated bioamplifier with active low-frequency suppression. IEEE Trans. Biomed. Circuits Syst. **1**, 184–192 (2007)
14. Olsson, R.H. III, Wise, K.D.: A three-dimensional neural recording microsystem with implantable data compression circuitry. IEEE J. Solid-State Circuits **40**, 2796–2804 (2005)
15. Harrison, R.R., Watkins, P.T., Kier, R.J., Lovejoy, R.O., Black, D.J., Greger, B., Solzbacher, F.: A low-power integrated circuit for a wireless 100-electrode neural recording system. IEEE J. Solid-State Circuits **42**, 123–133 (2007)
16. Sodagar, A.M., Wise, K.D., Najafi, K.: A fully integrated mixed-signal neural processor for implantable multichannel cortical recording. IEEE Trans. Biomed. Eng. **54**, 1075–1088 (2007)
17. Sarpeshkar, R., Wattanapanitch, W., Arfin, S.K., Rapoport, B.I., Mandal, S., Baker, M.W., Fee, M.S., Musallam, S., Andersen, R.A.: Low-power circuits for brain–machine interfaces. IEEE Trans. Biomed. Circuits Syst. **2**, 173–183 (2008)
18. Harrison, R.R., Charles, C.: A low-power low-noise CMOS amplifier for neural recording applications. IEEE J. Solid-State Circuits **38**, 958–965 (2003)
19. Wattanapanitch, W., Fee, M., Sarpeshkar, R.: An energy-efficient micropower neural recording amplifier. IEEE Trans. Biomed. Circuits Syst. **1**, 136–147 (2007)
20. Aziz, J.N.Y., Genov, R., Bardakjian, B.L., Derchansky, M., Carlen, P.L.: Brain/silicon interface for high-resolution in vitro neural recording. IEEE Trans. Biomed. Circuits Syst. **1**, 56–62 (2007)
21. Kandel, E.R., Schwartz, J.H., Jessell, T.M.: Principles of Neural Science. McGraw-Hill Medical, New York (2000)
22. Vittoz, E.A.: Micropower techniques. In: Franca, J.E., Tsividis, Y. (eds.) Design of Analog-Digital VLSI Circuits for Telecommunications and Signal Processing, pp. 53–96. Prentice-Hall, Upper Saddle River (1994)
23. Enz, C., Krummenacher, F., Vittoz, E.A.: An analytical MOS transistor model valid in all regions of operation and dedicated to low-voltage and low-current applications. Analog Integr. Circuits Signal Process. **8**, 83–114 (1995)
24. Mohseni, P., Najafi, K.: A fully integrated neural recording amplifier with DC input stabilization. IEEE Trans. Biomed. Eng. **51**, 832–837 (2004)
25. Johns, D., Martin, K.: Analog Integrated Circuit Design:. Wiley, New York (1996)
26. Perelman, Y., Ginosar, R.: An integrated system for multichannel neuronal recording with spike/LFP separation, integrated A/D conversion and threshold detection. IEEE Trans. Biomed. Eng. **54**, 130–137 (2007)
27. Azin, M., Mohseni, P.: A 94-μW 10-b neural recording front-end for an implantable brain-machine-brain interface device. In: The 2008 IEEE Biomedical Circuits and Systems Conference, pp. 221–224 (2008)
28. Horiuchi, T., Swindell, T., Sander, D., Abshier, P.: A low-power CMOS neural amplifier with amplitude measurements for spike sorting. In: The 2004 International Symposium on Circuits and Systems, vol. IV, pp. 29–32 (2004)
29. Rogers, C.L., Harris, J.G.: A low-power analog spike detector for extracellular neural recordings. In: The 2004 11th IEEE International Conference on Electronics, Circuits and Systems, pp. 290–293 (2004)
30. Haas, A.M., Cohen, M.H., Abshire, P.A.: Real-time variance based template matching spike sorting system. In: The 2007 IEEE/NIH Life Science Systems and Applications Workshop, pp. 104–107 (2007)
31. Gosselin, B., Sawan, M.: A low-power integrated neural interface with digital spike detection and extraction. Analog Integr. Circuits Signal Process. **64**, 3–11 (2010). doi:10.1007/s10470-009-9371-1

32. Gosselin, B., Zbrzeski, A., Sawan, M., Kerherve, E.: Low-power linear-phase delay filters for neural signal processing: comparison and synthesis. In: IEEE International Symposium on Circuits and Systems (ISCAS), pp. 1261–1264 (2009)
33. Gosselin, B., Sawan, M.: Adaptive detection of action potentials using ultra low-power CMOS circuits. In: The 2008 IEEE Biomedical Circuits and Systems Conference (BIOCAS), pp. 209–212 (2008)
34. Sarpeshkar, R., Lyon, R.F., Mead, C.: A low-power wide-linear-range transconductance amplifier. Analog Integr. Circuits Signal Process. **13**, 123–151 (1997)
35. Chun-Ming, C., Chun-Li, H., Wen-Yaw, C., Jiun-Wei, H., Chu-Kuei, T.: Analytical synthesis of high-order single-ended-input OTA-grounded C all-pass and band-reject filter structures. IEEE Trans. Circuits Syst. I **53**, 489–498 (2006)
36. Corbishley, P., Rodriguez-Villegas, E.: A nanopower bandpass filter for detection of an acoustic signal in a wearable breathing detector. IEEE Trans. Biomed. Circuits Syst. **1**, 163–171 (2007)
37. Bellaouar, A., Elmasry, M.: Low-Power Digital VLSI Design: Circuits and Systems. Springer, Berlin (1995)
38. Gosselin, B., Sawan, M.: Event-driven data and power management in high-density neural recording microsystems. In: Joint IEEE North-East Workshop on Circuits and Systems (NEWCAS'09) and TAISA Conference, pp. 1–4 (2009). Invited paper
39. Hyung-Ock, K., Youngsoo, S.: Semicustom design methodology of power gated circuits for low leakage applications. IEEE Trans. Circuits Syst. II **54**, 512–516 (2007)
40. Harrison, R.R.: A low-power integrated circuit for adaptive detection of action potentials in noisy signals. In: The 25th Annual International Conference of the IEEE Engineering in Medicine and Biology Society, pp. 3325–3328 (2003)

Chapter 18
Design Methods for Energy Harvesting

The Design of an Energy- and Data-Driven Platform

Cyril Condemine, Jérôme Willemin, Guy Waltisperger, and Jean-Frédéric Christmann

1 Introduction

Enhancing the battery life or even doing away with the battery altogether is today becoming a priority in autonomous microsystems. One possible solution is to harvest energy from the surrounding environment. The aim of this chapter is to introduce a solution maximizing the lifespan of autonomous communicating sensors. To reach this goal, an adaptable reconfigurable and robust microsystem including an energy harvesting system and power management will be presented. This work is part of a wider project aimed at developing an autonomous sensor node for smart buildings and medical implants. This node will include, on the same die: the power management platform, a multi-sensor unit (accelerometer, pressure, temperature, humidity, etc.) using a common ADC interface, a communication interface based on RFID and Zigbee solutions, and global digital signal processing resources.

Autonomous devices that are self-powered over a full lifetime, by extracting their energy from the environment, are crucial for applications such as ambient intelligence, active security or monitoring purposes. In such devices and applications, as the energy availability and power dissipation are not constant over time, energy management is a key function and determines the potential for information processing. Moreover, available average power and energy in the environment are low. All these challenging constraints have to be taken into account to develop an autonomous system.

Energy autonomy is not a new research topic. As an example, in 1956, Zenith put a TV set on the market with an autonomous, battery free remote control (the "Zenith Space Command TV", Fig. 18.1).

This remote control used no batteries, and instead worked by emitting ultrasonic noises that were created when a person pushed a button down, against a specific bar which created a specific frequency. The pushing of the buttons created a "click"

C. Condemine (✉) · J. Willemin · G. Waltisperger · J.-F. Christmann
CEA-LETI, MINATEC, 17 avenue des Martyrs, 38054 Grenoble Cedex 9, France
e-mail: cyril.condemine@cea.fr

G. Nicolescu et al. (eds.), *Design Technology for Heterogeneous Embedded Systems*,
DOI 10.1007/978-94-007-1125-9_18,

Fig. 18.1 Zenith space command TV

Fig. 18.2 Citizen clock

noise for the user and this, incidentally, is how television remotes became known as clickers [1].

One other example is the first thermoelectric watch (patent "Thermoelectric generators for wristwatches", CH 1975) developed in 1975 in Switzerland and used by Citizen for the thermal EcoDrive in 1999 (Fig. 18.2).

To achieve the goal of ultra low-power energy-autonomous microsystems, several issues must be solved.

The first concerns the focus on available energy instead of the power dissipation by increasing the conversion efficiency. This efficiency is divided into two parts: the source efficiency limited by physical principles and the electronic circuit efficiency needed to transform this energy into suitable electric energy (DC-DC converter, battery charger, linear regulator for example). There is often a significant difference between the voltage levels at the generator output (about 100 mV) and at the battery input (2 to 4 V), so an "up" converter also needs to scale the voltages as well as convert power. To increase the conversion efficiency, an optimal point must be reached: for example, a thermo-element generator gives a large amount of power with a low voltage output, but the up DC-DC converter used to increase this voltage to the battery level will have a very poor efficiency at this level of energy. An optimal point can be found with a higher generator output voltage (series to parallel implementation) and lower DC-DC conversion gain.

Another issue is the management of resources. The finite energy resources are often only temporarily available and one limitation is the capability of the system to store the available harvested energy in reliable storage units. This imposes a need to define concepts of energy-aware behavior for such systems. This includes overall energy optimization in nominal situations and fallback strategies in cases where energy is low.

The last major issue is the power consumption of applications, for example sensors (few μW to hundreds of μW) and communications (few mW to tens of mW). The sensors should be ultra low-power, should include wake-up functionalities and very low sleep-mode power consumption. The same constraints are applied to RF communication blocks: ultra low power, short range communication with low data rate to ensure the compatibility between communications and energy levels.

This chapter is organized as follows. In Sect. 2, the main issues and tradeoffs of autonomous systems will be presented and the architecture will be proposed. An overview of energy harvesting sources and the associated electronics are presented in Sect. 3. Section 4 introduces the power management strategy with power path optimization.

2 Global System Description

Energy Autonomous Systems can be roughly divided into three sections [2], as shown in Fig. 18.3:

- *Energy sources*: The harvester can be any device or system that can harvest energy from correlated or uncorrelated sources of energy. Different kinds of energy sources are available in typical surroundings: temperature differences, light radiation, electromagnetic fields, kinetic energy, etc. The harvesting technique could in fact take advantage of more than one source at a time. This part of the system includes a dedicated DC-DC converter to extract and up-convert the maximum energy from the sources using for example maximum power point tracking (MPPT). In this case, the objective is to adapt the DC-DC converter input impedance dynamically to the source output impedance, in order to transfer maximum energy from that available. We take into account any kind of energy storage element that could be used to both accumulate excess energy from the harvester and feed it into the system whenever the available energy from the surroundings is insufficient. Typically energy sources consist of electro-chemical elements such as batteries or fuel cells, or electrical storage systems such as capacitors or super-capacitors.
- *Energy and power management*: This functional block includes a battery charger, DC-DC converters to scale up or scale down the voltage level from batteries to loads, a linear regulator to generate low-noise power supply, and a low-power

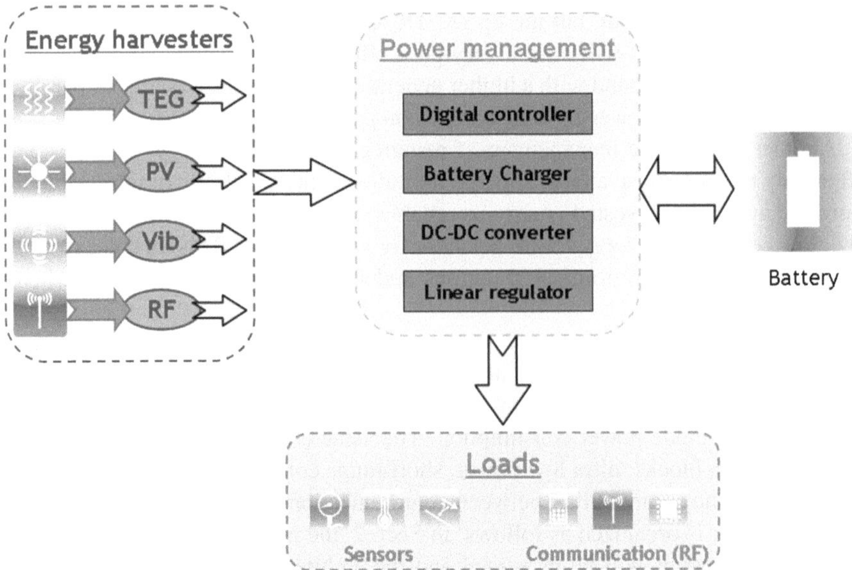

Fig. 18.3 Energy harvesting system synoptic and energy flow

digital platform to manage all the systems, and implement the energy- or data-driven management strategy. A global energy level measurement circuit can also be included, as a starting point for a quality of service strategy.

- *Energy consumption*: This is the part of the system that is related to data acquisition, storage and transmission. This is of fundamental importance since most of the opportunities for energy autonomy are related to the efficiency of this part.

The target application has a direct impact on the energy harvesting system architecture. But in every case, for self-powered capability, the following equation has to be fulfilled:

$$E_{\text{harvested}} > E_{\text{system}} + E_{\text{power-mgt}}$$

where $E_{\text{system}} = (1 - \theta)E_{\text{sleep}} + \theta E_{\text{active}}$, θ corresponding to the duty cycle (expressed as a fraction of time) that the system is active.

We can list 3 kinds of applications, where θ will differ significantly:

- *"always sleep" systems*: These systems measure physical parameters with a well-defined sampling period, for example %RH (relative humidity), T (temperature), or CO_2 concentration. The main application is long-term monitoring, for example in smart buildings. The system spends the main part of its time in sleep mode, and in that case, θ is very close to zero: e.g. one measurement of duration 50 ms every 10 minutes (or 8.3×10^{-3}% duty cycle). The main issue is the power consumption in sleep mode. All the other power consumptions are negligible due to a very low duty cycle, and only impact on the energy buffer size (for RF communication).

Origin	Type	Power Density (Max)
Photovoltaic	Indoor light	100µW/cm² (inside light)
Electromagnetic	RFiD 2.5MHz	10 µW @ 1m
Motion	Vibrations (wideband)	10µW per gram of proof mass
	Deformation	1.21J/g (Dielectric Polymer)
Heat	Thermo-electricity	50µW/cm², 0.5v @ ΔT=1K
Chemical	Thin-film AIC batteries	100 µA.h/cm²
	Thick-film batteries	2 mA.h/cm²

Fig. 18.4 Power density by energy source

- *"wake-up" systems*: These systems include a low performance, ultra-low power (ULP) constant monitoring functionality. This monitoring tracks a physical parameter and when an event occurs, wakes up the global system, with high performance and higher power consumption. The main applications are energy-driven applications such as shock tracking, cold chain monitoring, etc. In that case, θ is not constant and is application dependent. The main issue is the energy conversion efficiency and that direct paths from sources to load (without battery) can exist.
- *"monitoring" systems*: These systems measure a physical parameter at high frequency and with high performance. They are not energy-autonomous, and the energy harvester only increases the measurement time. The main applications are physiological monitoring, motion capture, etc. The main issue is the energy conversion efficiency from the sources to the battery and from the battery to the loads.

3 Energy Sources

Different kinds of sources can be used depending on the target application. To compare these sources, two parameters can be taken into account: the efficiency (η) and the energy density with the harvested power by area.

For small form-factor systems, the table shown in Fig. 18.4 shows the difficulties in using a single source. To allow autonomous functionality, a multi-source solution can be envisaged. In this section we describe the physical principles behind various energy harvesting approaches, and the key success factor for the associated energy converters.

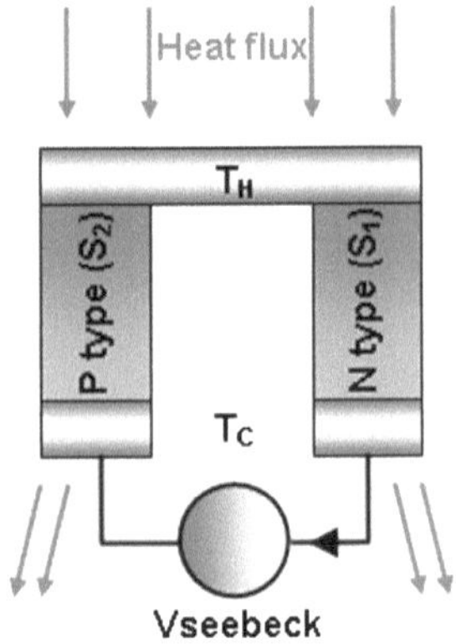

Fig. 18.5 Thermo-electric generator

3.1 Thermo-element Generator and Associated Electronics

3.1.1 The Seebeck Effect

The Seebeck effect [3] is the conversion of a temperature gradient into electricity. It occurs in the presence of a temperature difference between two different metals or semiconductors. The induced voltage, as shown in Fig. 18.5, is equal to the temperature difference ($T_H > T_C$) multiplied by the difference of the Seebeck coefficients (S_1, S_2) between the materials:

$$V_{\text{Seebeck}} \approx (T_H - T_C) \cdot (S_2 - S_1).$$

The approximation is due to the temperature dependence of the Seebeck coefficients. Several (N) junctions could be (electrically) serially connected to form a thermopile or thermoelectric generator (TEG):

$$V_{\text{TEG}} \approx N \cdot V_{\text{Seebeck}}.$$

3.1.2 TEG Efficiency

The efficiency of the thermoelectric generator is equal to:

$$\eta = \frac{\Delta T}{T_H} \cdot \frac{\sqrt{1 + Z \cdot T_{\text{AVG}}} - 1}{\sqrt{1 + Z \cdot T_{\text{AVG}}} + \frac{T_C}{T_H}}$$

where $\Delta T = T_H - T_C$ and $T_{\text{AVG}} = (T_H + T_C)/2$. $Z = \sigma ST/K$, where σ represents the electrical conductivity, S the Seebeck coefficient, and K the thermal conductivity. Those parameters are average quadratic values for P and N type material parameters.

This efficiency is composed of two parts: the Carnot efficiency $\eta_{\text{Carnot}} = \Delta T / T_H$ (which dramatically limits the efficiency for small temperature gradients), and the

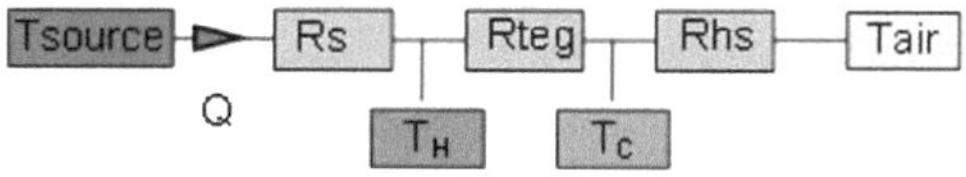

Fig. 18.6 Equivalent circuit of a thermal system

performance of the thermoelectric generator. For energy harvesting in the [200–450 K] ambient temperature range $Bi_2(Te,Se)_3$ is the most efficient material with Z for commercially available products in the range [0.5–1]. In this example, the product of the two terms will give an overall efficiency below 1% for $\Delta T = 10°C$–$20°C$, even with $Z = 1$.

The harvested power is proportional to the temperature difference between the pads of the thermoelectric generator. This temperature difference is due to heat (Q) flowing through a thermal circuit. Neglecting thermal capacitance and parallel parasitic thermal resistance, an equivalent circuit model between a source of heat, the TEG and a heatsink can be represented as in Fig. 18.6.

From this model, it is clear that the useful temperature difference for thermoelectric power generation is:

$$\Delta T = T_H - T_C = Rteg \cdot Q = (Tsource - Tair) \cdot \frac{Rteg}{Rs + Rteg + Rhs}.$$

As an example, the Micropelt MPG-D751 [4] thermal resistance is $Rteg =$ 12.5 K/W. Hence to maximize ΔT (in order to maximize the Carnot efficiency and output power), it is necessary to optimize the ratio between *Rteg* and the sum of thermal resistance.

3.1.3 Electronic Power Converter

Electronic systems require a supply voltage in the range of [1–5 V] in order to operate, so a direct connection to a TEG may not be possible due to its very low output voltage. In this case, a power up-converter should be connected between the TEG and the electronic charges. This up-converter can be of inductive [5] or of capacitive type [6–8].

Due to very low output power and voltage, capacitive up-conversion is very well adapted to providing the thermoelectric harvested power; indeed capacitive up-conversion can provide high efficiency, low input voltage, high gain conversion and full integration on an ASIC.

A capacitive up-converter (or "charge pump") delivers the following output voltage:

$$V_{out} = (N+1)(V_{in} - V_t) - \frac{N I_L}{f C},$$

where N represents the number of pump stages, V_{in} the input voltage, V_t the threshold of the diode/MOS, I_L the output current, f the pumping frequency and C the pump capacitance.

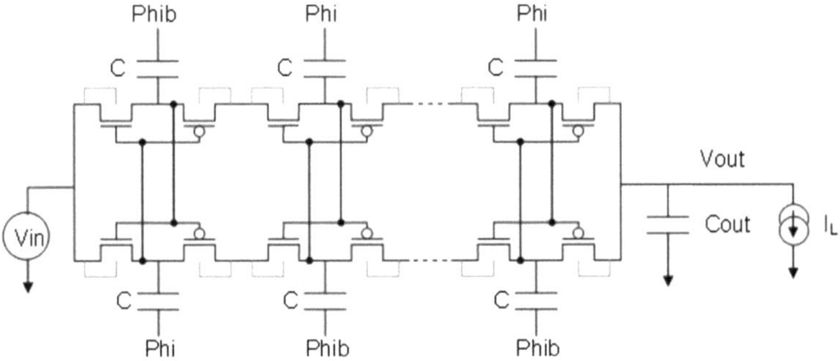

Fig. 18.7 Cross-coupled charge pump

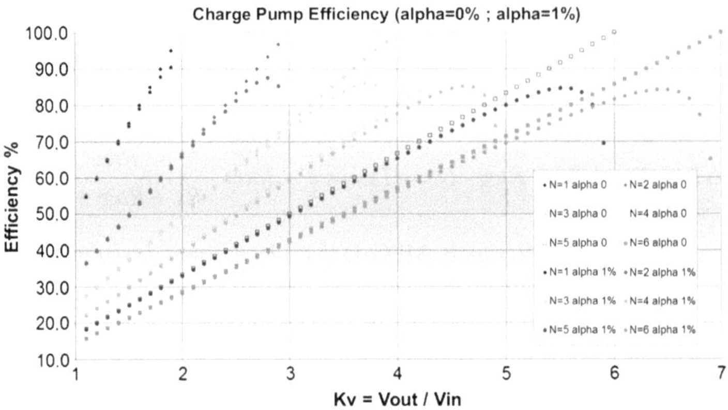

Fig. 18.8 Charge pump efficiency

In order to suppress the effect of V_t, which decreases converter gain and efficiency, a high-performance charge pump architecture with V_t cancellation can be used. A relevant example is that of a cross-coupled architecture, the performance characteristics of which are very close to a perfect pump despite a very straightforward structure (Fig. 18.7).

However, a triple-well technology is required to polarize the substrate of the NMOS and cancel the substrate effect.

The efficiency of a charge pump with zero V_t is equal to:

$$\eta = \frac{K_v}{N + 1 + \alpha \cdot \frac{N^2}{N+1-K_v}},$$

where $K_v = V_{\text{out}}/V_{\text{in}}$ and $\alpha = C_p/C$ (C_p is the parasitic capacitance). It is therefore clear that the efficiency depends on the number of stages N, the desired gain of the converter K_v and the technological parameter α.

The graph in Fig. 18.8 shows the efficiency for varying numbers of stages and values of α.

The optimal number of stages for a given value of α and K_v can be calculated as:

$$N_{op} = \left(1 + \sqrt{\frac{\alpha}{1+\alpha}}\right)(K_v - 1).$$

If the input voltage of the charge pump has to change (induced by a change of temperature difference for example), it can be advantageous to implement a dynamic optimization of the converter which will adapt the number of stages to maximize the efficiency. Of course the controller implementation has to be ULP in order to limit the impact on the energy budget.

3.1.4 Impedance Adaptation

Due to the thermoelectric generator input impedance, impedance matching must be carried out in order to maximize output power, i.e. the electronic up-converter should have the same impedance as the TEG. This can be achieved, for a given temperature difference, by choosing the appropriate fC product (f is generally chosen to be in the MHz range in order to make the capacitor C small and thereby save silicon area). The first step is to calculate the number of stages required (knowing that the input voltage will be the half of the TEG open-circuit voltage) and the efficiency of the pump. This will give access to the output current of the pump and finally to the calculation of the fC product.

Nevertheless, the up-converter has a minimum start-up voltage due to the threshold of the transistors (for example 400 mV in a standard 130 nm technology). Below this operating voltage, the converter does not operate. At impedance matching, the input voltage of the charge pump is half the open-circuit voltage of the TEG. So, if energy harvesting has to be done at a temperature difference where the open-circuit voltage is less than twice the threshold voltage, a matched impedance would cause the converter not to operate (because the input voltage would be less than the threshold voltage). Because of this, the impedance of the charge pump should vary so as to guarantee an input voltage equal to the maximum between the threshold voltage and half of the TEG open circuit voltage. In doing so, the impedance will be high for small temperature differences and equal to the TEG resistance at high temperature differences.

In practice, this dynamic impedance adaptation can be achieved for example by changing the pump frequency according to the input voltage. Figure 18.9 shows impedance adaptation of the charge pump input impedance (R_{in}) of MPG-D751 (140 mV/K with $R_{teg} = 300\ \Omega$) and with a threshold voltage of 400 mV.

By mixing adaptation strategies (both for input impedance and for the number of stages), a very high-efficiency (>60%) dedicated up-converter can be designed.

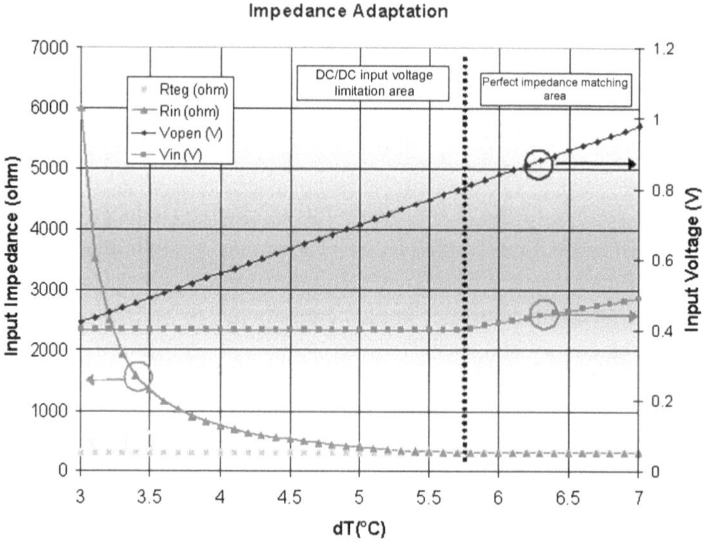

Fig. 18.9 Impedance adaptation

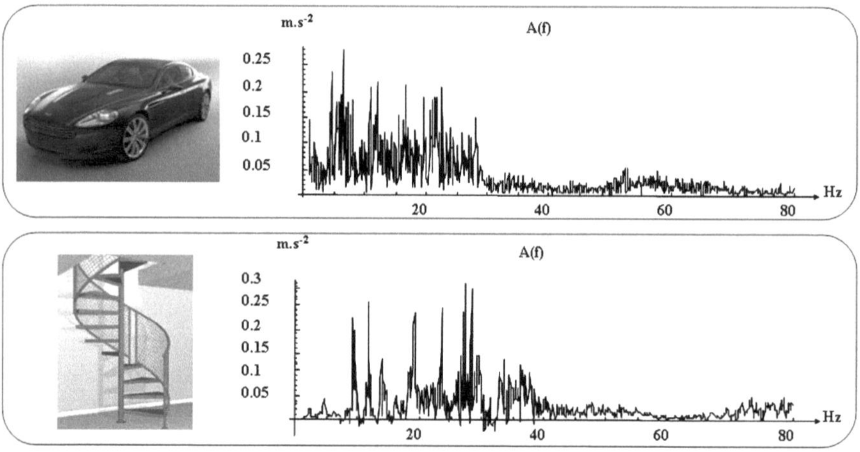

Fig. 18.10 Mechanical vibration spectral density

3.2 *Mechanical Vibration Harvester and Associated Electronics*

Another potential source of energy that can be harvested from the surroundings of autonomous sensor nodes, particularly in automotive and smart building applications, is that of mechanical vibrations. As shown in Fig. 18.10, mechanical vibration frequencies are mainly below 100 Hz and are fairly uniformly distributed [9].

Fig. 18.11 Flyback circuit

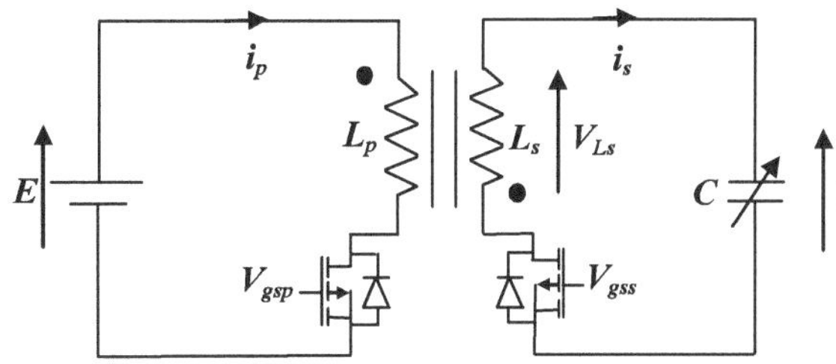

There are multiple techniques for converting vibrational energy to electrical energy. The two most commonly used techniques for low volume harvesters are electrostatic and piezoelectric conversion.

To convert most of these vibrations into electrical power, a movable proof mass is used either with piezoelectric materials or with MEMS capacitance solutions.

With the piezoelectric solution, larger deflection leads to more stress, strain, and consequently to a higher output voltage and power. In the case of the electrostatic solution, larger deflection leads to higher relative displacement and consequently to larger capacitance variations and gain in energy.

We chose to investigate conversion structures based on MEMS electrostatic transduction with high electrical damping. Many advantages are provided by electrostatic conversion: it is easy to integrate, and its power density is increased by size reduction. Moreover, high electrical dampings are easily achievable through this transduction principle. Thus, and contrary to most existing systems [10], these structures are able to recover power over a large spectrum below 100 Hz [11].

To transform mechanical vibrations into electrical energy, the proposed MEMS structures are included in an energy transfer circuit composed of one battery (as a power storage unit), an inductive transformer (flyback structure, Fig. 18.11) and 2 power MOS transistors, working in MEMS constant charge mode. Thus the energy is directly proportional to capacitance variations and the minimum and maximum voltages.

The energy is first up-converted from the battery to the MEMS capacitance when the capacitance is at its minimum value (charge injection part in the Fig. 18.12). The structure moves due to mechanical vibrations (one plate is "free") and the capacitance voltage increases (mechanical to electrical conversion part). When the capacitance is at its maximum point, the energy is transferred back from the MEMS to the battery (charge recovery part). The gain in energy is:

$$E_{\text{recup}} = \frac{1}{2} C_{\min} U_{\max}^2 - \frac{1}{2} C_{\max} U_{\min}^2.$$

To manage the charge transfer between the MEMS and the battery, the state of both MOS transistors must be controlled by the vibration frequency and amplitude. A minimum and maximum capacitance monitoring system manages a time control unit in order to shape a pulse signal controlling the power MOS states.

Depending on the level of integration (PCB or ASIC), the functions shown in Fig. 18.13 are realized with different blocks, and can be based on discrete operational amplifiers, comparators, inverters and RC delay lines for PCB solutions,

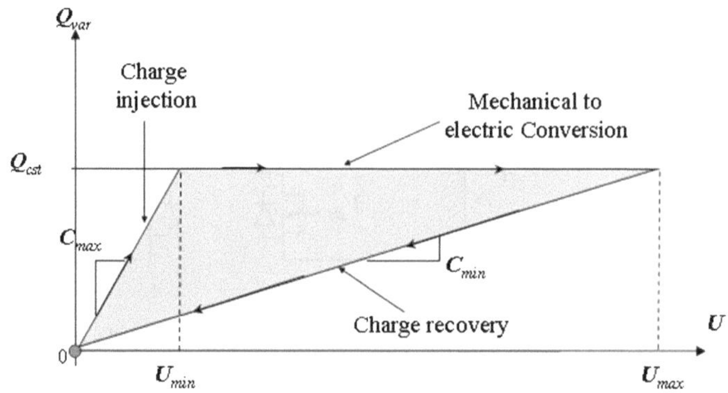

Fig. 18.12 Constant charge mode cycle

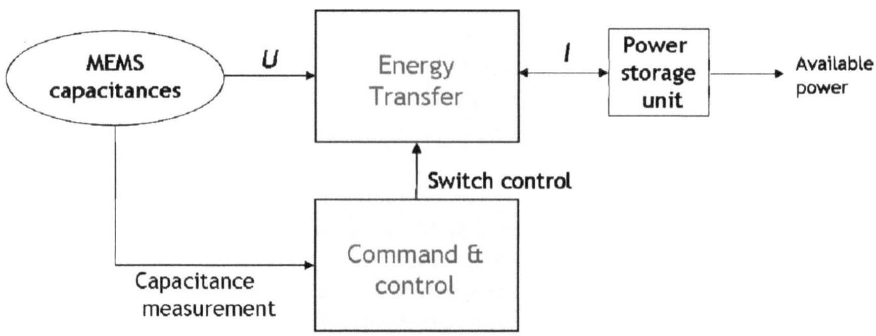

Fig. 18.13 Synoptic of functions in mechanical vibration energy harvester

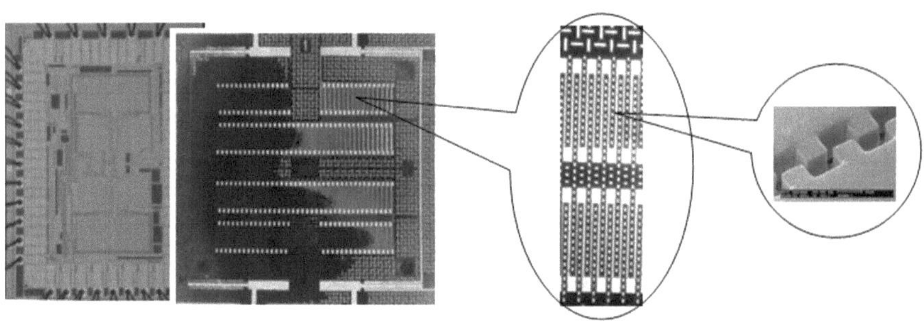

Fig. 18.14 System realization

or on an integrated transconductance amplifier for temporal differentiation [12] and a CMOS-based thyristor for timing control element [13] in ASIC solutions (Fig. 18.14).

This principle has been used for macroscopic proof mass (100 g) to microscopic proof mass (1 g) showing in all cases a positive energy balance: from 1 mW to 3 μW

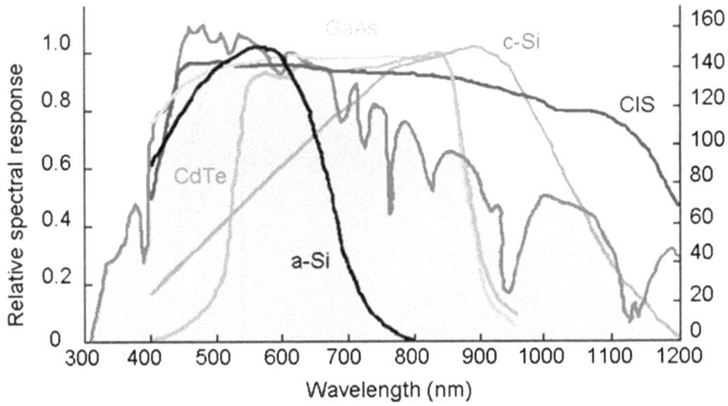

Fig. 18.15 PV cell efficiency vs. technology

depending on the proof mass weight, the voltage (50 to 300 V) and the acceleration (1 ms^{-2}).

3.3 Photovoltaic Cell and Associated Electronics

Ambient solar energy is one of the most abundantly available sources of energy and it can be easily harvested by using photovoltaic cells which can now be totally integrated in the device. The amount of energy and power harvested depends on the environment conditions and the capability of the device to adapt itself to the variation of the environmental conditions over time [14]. The ambient incident solar energy can vary from 100 mW/cm^2 (outdoor direct sun) to 100 μW/cm^2 (indoor at the surface of a desk). Silicon solar cells are characterized by an efficiency between 15–20%. For new thin film solar cells the efficiency is around 10% and technological progress should allow it to attain figures similar to those of silicon solar cells. Hence it is possible to estimate that the power available from photovoltaic cells varies from about 15 mW/cm^2 outdoors to 10 μW/cm^2 indoors.

A single solar cell has an open circuit voltage of about 0.6 volts. Photovoltaic systems are semiconductor devices and they have a current voltage characteristic which is affected by the radiation and the temperature. The power management unit has to be optimized to ensure an optimal and cost-effective energy management for these varying conditions.

Solar cells have efficiencies which are dependent on the spectral characteristic of the sources (Fig. 18.15), and therefore some cells are better adapted for a specific source [15].

Solar cells can be modeled thanks to a simple theoretical model based on a diode with a parallel shunt resistance (R_p) and series resistance (R_s). As a solar cell has a diode characteristic, the current follows an exponential function and the open-circuit

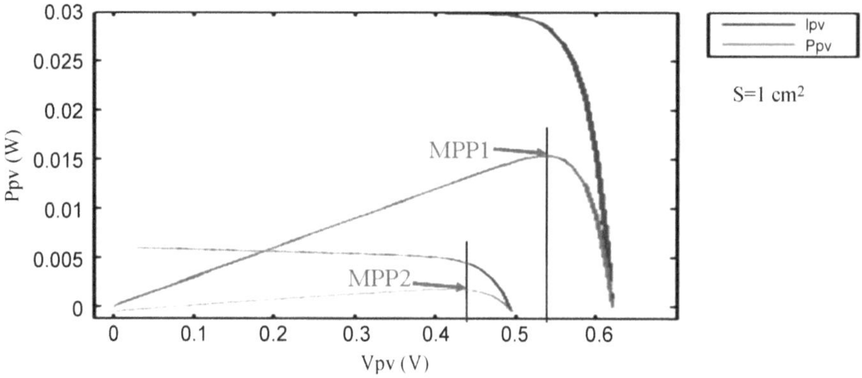

Fig. 18.16 Power and current vs. voltage, outdoor (Sun) and indoor (1% Sun)

voltage V_{co} is dependent on the short-circuit current I_{sc}:

$$V_{oc} = \frac{kT}{q} \ln\left(\frac{I_{sc}}{I_{sat}}\right),$$

$$I_{PV} = -I_{sat}\left[\exp\left\{\frac{q(V + I_0 \times R_S)}{kT}\right\} - 1\right] + \frac{V + I_0 \times R_S}{R_P} - I_{ph},$$

where I_{ph} is the generated photonic current, I_{sat} the saturation current, q the electric charge, k the Boltzman constant and T the temperature in Kelvins.

To obtain the maximum energy from the solar cell, the power converters must use MPPT since, as shown in Fig. 18.16, the Maximum Power Point (MPP) changes with light intensity and temperature.

There are a large number of algorithms used to reach this MPP, such as the "perturb and observe" algorithm, open circuit voltage sampling, short circuit current measurement or the incremental conductance algorithm [16]. However for microsystems applications, complex digital computation cannot be performed because of the energy consumption limitation due to the energy-limited node capacity. An interesting method (which is fully analog) exists, but one of the disadvantages is the need for a reference photovoltaic cell [17]. This reference cell will increase the surface needed and decrease the harvesting potential (active area) in the case of size-constrained applications.

For Microsystems, the harvesting module is small and the potential consumption part associated with the power management unit is reduced to a value which is challenging to achieve. A photovoltaic cell in this context needs a DC-DC converter, and the most suitable structure is a switched mode power supply (SMPS) architecture. By controlling the duty cycle, it is possible to reach the MPP. The efficiency of the conversion circuit is dependent on the photovoltaic cell (module). In such a module, photovoltaic cells can be associated in series and/or parallel to deliver a specific current and/or voltage (0.45 V, 15 μA/cm^2 at MPP for amorphous PV Cell under 200 lux, indoor use). This enables an increase in the conversion efficiency for specific

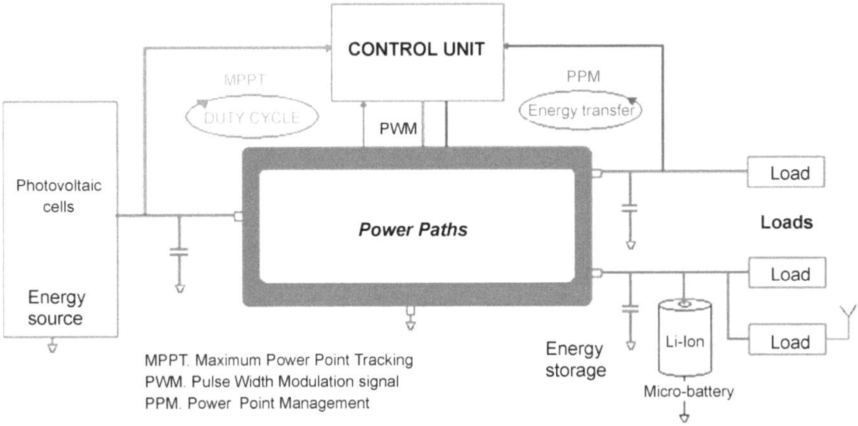

Fig. 18.17 Ultra low power MPPT implementation

illumination and temperature conditions, as long as the other constraints (such as size) are not violated.

If an up/down DC-DC converter is used for the MPPT, then the power can be delivered to the load with a higher voltage than the photovoltaic cell. The problem is to build a power converter that can efficiently charge a large capacitor at the optimal voltage and current of the photovoltaic cell (optimal power point). Unlike a typical voltage regulator that uses feedback from the output, a MPPT converter requires the photovoltaic cell input to be fed forward into the controlling circuit.

One idea to overcome this interface problem between, on one side, a variable maximum operating point (I, V) and, on the other side, a regulated output voltage, is to place a super-capacitor operating as a buffer. The MPPT charges the super-capacitor at the maximum power efficiency of the photovoltaic cells. The super-capacitor is used as input for the DC-DC converter. With this architecture, a minimal input voltage is set to the DC-DC converter which guarantees a good minimum level of transfer efficiency. This architecture will reduce significantly the loss due to ultra-low power input, as the DC-DC converter has an efficiency which sinks when the input power is ultra low. Another idea is to use a DC-DC converter with multiple outputs, enabling the asynchronous regulation of each output and storage of any excess energy to ensure the system delivers all the energy that is harvestable at any given time (Fig. 18.17).

4 Power Management

One application for harvester systems is in wireless autonomous sensors. However, some issues limit the development of such networks. On one hand, for the mechanical energy harvester (as for the other solutions such as the thermoelectric generator, or small PV cell), the maximum amount of available energy is unpredictable and not stable at high impedance. On the other hand, we need to provide circuits with low-impedance power sources and a regulated supply voltage value. As an example, the consumption of a digital sensor interface is in the range of 1 nJ per conversion, with a constant power consumption profile. For a transducer, the power consumption is in the range of 100 nJ per transmitted bits, with high current peaks (a few mA). So to be energy sufficient, these systems require high-efficiency multi-source harvesting systems, reliable power storage solutions and energy management.

Power management is needed to extract the energy at the maximum power point, with MPPT systems for each source, and to choose the best power path between sources, loads and battery in terms of efficiency.

The technical problem to solve can be expressed as "How to optimize in terms of efficiency, in real time, the consumption and the storage of energy, depending on the application power consumption and the available energy in the system?" The objective is to develop a generic power hub allowing optimization, in a dynamic way, of the power path between sources, loads and battery to reach the maximum efficiency. This implies the design and integration of the system architecture for the management of multiple power sources (harvesting and storage elements) aimed at powering multiple loads, in the context of small (<10 mW) power transfers.

To prove the advantages of this solution, several points have to be developed:

- Digital Voltage Scaling (DVS) and Digital Voltage and Frequency Scaling (DVFS) techniques for digital signal processing, embedded sensor interface and external digital loads.
- Energy consumption measurement or estimation.
- Power path optimization algorithm and its implementation using asynchronous digital signal processing.

This platform includes a large spread of functions such as algorithms to estimate the available energy, ULP digital circuit (dedicated asynchronous circuit and/or ULP microprocessor, DVFS) implementation.

4.1 Load Power Consumption

Figure 18.18 introduces a figure of merit (FOM) for sensor interface and RF communication. In both cases, the FOM is based on energy consumption relative to communication (bandwidth) or to conversion accuracy (resolution). It thus represents energy per conversion for sensor interfaces and energy per transmitted bit for RF circuits. Below several pJ/conv and several tens of nJ/bit, the power consumption can be considered to be low enough (especially in standby mode) for the design of an autonomous wireless sensor node.

To achieve this level of FOM, disruptive solutions should be studied such as asynchronous ADC for wake-up sensor, time domain ADC (analogue to time and time to digital) for Ultra Low Voltage capabilities. To be capable of handling measurement applications with a minimum of energy, a variety of techniques for wake-up

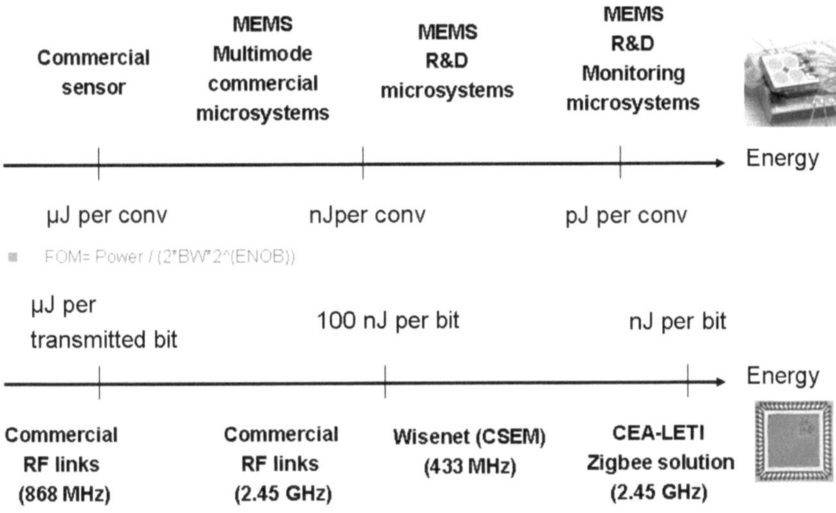

Fig. 18.18 Energy consumption figures of merit

modes (type, initiator, threshold, duration, etc.), sensor configurations (gain, offset, common mode, etc.) and DVFS have to be implemented.

As the RF parts have a very low duty cycle, ULP operating modes are essential (especially in sleep mode). A fast wake-up time is critical: the objective is to reduce the power consumption during the OFF to ON transition. A global system optimization is needed to avoid over-specification and to choose an energy efficient protocol (MAC etc.).

4.2 Power Path Optimization

At the present time, integrated circuits operate under 3.3 V to 0.5 V depending on the technology and the functionality (analogue or digital). The batteries are designed to store the maximum of energy, so the trend is to increase the battery voltage over 3 V. The harvester outputs are however below 1 V. The question is: how to avoid the power path from the sources (1 V) to the battery (3.3 V) through a first DC-DC converter ($\eta = 70\%$), and then from the battery to the loads (1.2 V for example) through a second DC-DC converter ($\eta = 70\%$) and an LDO regulator ($\eta = 90\%$)? In this case [18] with a 500 µW power consumption of the load, the total efficiency is no more than 40%: the total efficiency is the product of the efficiency of each converter ($70\% \times 70\% \times 90\%$) leading to 45% total efficiency (Fig. 18.19).

One solution is to manage the power path and to go directly from the source (1 V) to the loads (0.8 V) through the LDO regulator [19]. The total efficiency increases up to 70%! This solution implies designing very challenging functions with ULP techniques (in the range of tens of nW): harvested power level monitoring, battery SoC monitoring and charger, load power consumption monitoring, integrated

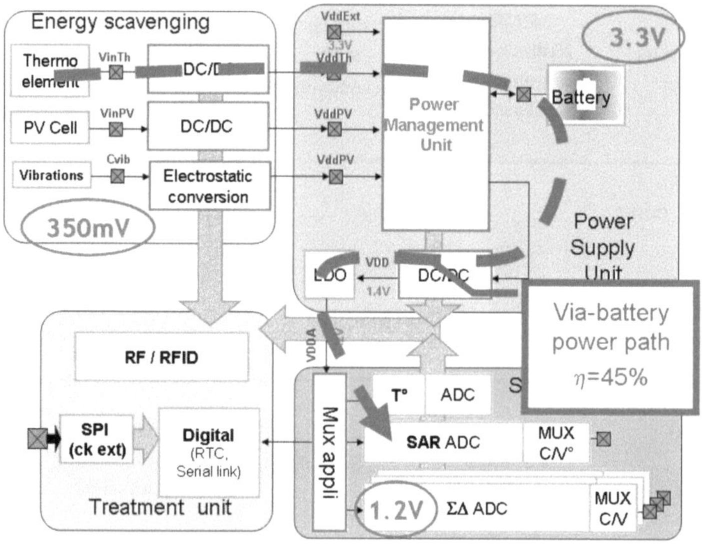

Fig. 18.19 Via-battery power path

DC-DC converters and LDO regulators. Dynamic power management would optimize energy extraction, thanks to a power path reconfiguration and a low-power environment-aware algorithm. Asynchronous solutions for ADC and ULP digital dedicated circuits appear to be a promising approach in this context. Asynchronous circuits are well-suited for the implementation of energy harvesting microsystem for several reasons. Regarding digital logic, intrinsic standby state and PVT (Process Voltage Temperature) robustness provided by asynchronous data-driven design techniques are promising. Asynchronous circuits can be easily supplied at very low voltage levels and their smooth current profile due to automatic speed regulation perfectly fits energy harvesters and battery requirements. Detecting and processing events is the fundamental behavior of asynchronous circuits and energy monitoring is among the main issues in energy harvesting based applications. An energy harvesting microsystem has indeed to be aware of both the available energy and the system activity. By detecting environmental energy changes, the microsystem can configure the optimal power path within the architecture in order to reach the best power efficiency trade-off.

To optimize the power path we propose an architecture with two power paths (Fig. 18.20). The first path is direct between the source and the microsystem thanks to a new DC-DC converter, the second path continues to use an energy buffer (battery or/and capacitance) to respond to high power requirements needed for example for data transmission. The optimal energy operating point that can be reached leads to two kinds of system optimization. The first approach allows the energy transfer optimization from the sources to the load. Whereas the second one optimizes the node activity level regarding the energy state. Optimal energy transfer involves the DC/DC converters and the battery charger optimization. In the proposed architec-

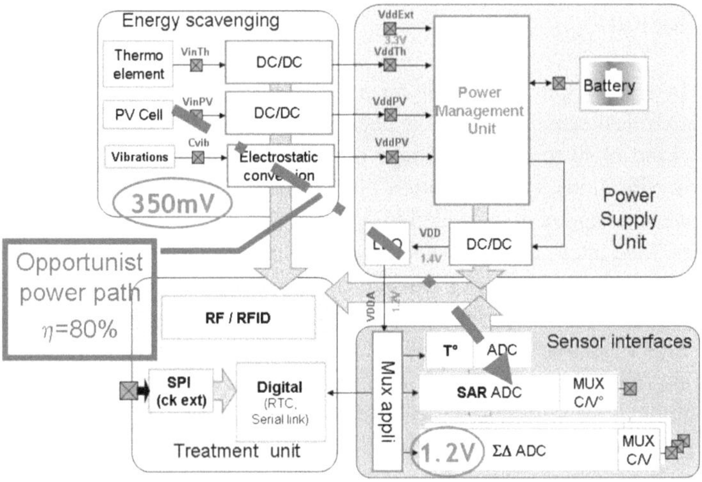

Fig. 18.20 Opportunistic power path

G	T_{batt}/T_{nrj}									
α_D	0	0.1	0.2	0.4	0.6	1	2	4	6	10
0%	0%	0%	0%	0%	0%	0%	0%	0%	0%	0%
10%	4%	4%	3%	3%	3%	2%	1%	1%	1%	0%
20%	8%	7%	7%	6%	5%	4%	3%	2%	1%	1%
30%	12%	11%	10%	9%	8%	6%	4%	2%	2%	1%
40%	16%	15%	13%	11%	10%	8%	5%	3%	2%	1%
50%	20%	18%	17%	14%	13%	10%	7%	4%	3%	2%
60%	24%	22%	20%	17%	15%	12%	8%	5%	3%	2%
70%	28%	25%	23%	20%	18%	14%	9%	6%	4%	3%
80%	32%	29%	27%	23%	20%	16%	11%	6%	5%	3%
90%	36%	33%	30%	26%	23%	18%	12%	7%	5%	3%
100%	40%	36%	33%	29%	25%	20%	13%	8%	6%	4%

Fig. 18.21 Gain of path reconfiguration architecture vs. indirect path only

ture, we further improve the total efficiency of the energy transfer through the use of the direct power path. This one is used as much as possible to benefit from its high inherent efficiency. As an example (Fig. 18.21, based on a realistic application), we consider efficiencies of the indirect and direct paths of respectively $\eta_i = 45\%$ and $\eta_d = 75\%$ and an application with a possible harvested energy duty cycle of 15 h/day (i.e. $T_{\mathrm{batt}}/T_{\mathrm{nrj}} = 0.6$). Further, 20% of power loads have to be supplied by the battery (i.e. $\alpha D = 80\%$), and the power requirements of the microsystems will be lowered by 20%. Gains up to 40% are possible if all the power loads are eligible for direct path (bottom line) and if energy is always available (Fig. 18.21).

5 Conclusion

Progress in energy efficiency will open new opportunities in the design of autonomous wireless sensors. To enable energy production in a large panel of applications, this kind of system must manage multiple sources such as thermal gradient, mechanical vibrations, solar radiation or magnetic fields. To maintain efficiency, global power management is needed to optimize the power path between the multiple sources, the battery and the loads. One key success factor is the load consumption and especially the control of leakage current in idle mode. Disruptive concepts are required in the roadmaps for sensor interfaces, RF blocks, and digital computation. The final objective is to propose an energy- and data-driven platform for WSN applications. To reach the objective of a small volume (several mm^3) autonomous and communicating sensor node, two major issues must now be considered: 3D packaging and software (at MAC and OS level).

References

1. http://en.wikipedia.org/wiki/Remote_control
2. CATRENE Working Group on Energy Autonomous Systems: Energy autonomous systems: future trends in devices, technology, and systems, white paper
3. Fournier, J.-M., Salvi, C., Stochkolm, J.: Thermoélectricité: le renouveau grâce aux nanotechnologies. Techniques de l'ingénieur: Thermoélectricité (2006)
4. http://www.micropelt.com
5. Lhermet, H., Condemine, C., Plissonnier, M., Salot, R., Audebert, P., Rosset, M.: Efficient power management circuit: from thermal energy harvesting to above-IC microbattery storage. In: ISSCC (2007)
6. Doms, I., Merken, P., Mertens, R.P., Van Hoof, C.: Capacitive power-management circuit for micropower thermoelectric generators with a 2.1 μW controller. In: ISSCC (2008)
7. Palumbo, G., Pappalardo, D., Gaibotti, M.: Charge-pump circuits: power-consumption optimization. IEEE Trans. Circuits Syst. **49**(11), 1535–1542 (2002)
8. Ker, M.-D., Chen, S.-L., Tsai, C.-S.: Design of charge pump circuit with consideration of gate-oxide reliability in low-voltage CMOS processes. IEEE J. Solid-State Circuits **41**(5), 1100–1107 (2006)
9. Despesse, G., Chaillout, J.J., Jager, T., Cardot, F., Hoogerwerf, A.: Innovative structure for mechanical energy scavenging. In: Solid-State Sensors, Actuators and Microsystems Conference, Transducers 2007, June 2007, pp. 895–898 (2007)
10. Roundy, S., Wright, P.K., Rabaey, J.: A study of low level vibrations as a power source for wireless sensor nodes. Comput. Commun. **26**, 1131–1144 (2003)
11. Despesse, G., Jager, T., Chaillout, J.-J., Leger, J.-M., Basrour, S.: Design and fabrication of a new system for vibration energy harvesting. In: Research in Microelectronics and Electronics, 25–28 July 2005, vol. 1, pp. 225–228 (2005)
12. Stocker, A.A.: Compact integrated transconductance amplifier circuit for temporal differentiation. In: Proceedings of the 2003 International Symposium on Circuits and Systems. ISCAS '03, 25–28 May 2003, vol. 1, pp. I-201–I-204 (2003)
13. Kim, G., Kim, M.-K., Chang, B.-S., Kim, W.: A low-voltage, low-power CMOS delay element. IEEE J. Solid-State Circuits **31**(7), 966–971 (1996)
14. Randall, F., Jacot, J.: The performance and modelling of 8 photovoltaic materials under variable light intensity and spectra. EPFL, Lausanne, Switzerland

15. Torres, E.O., Rincon-Mora, G.A.: Long-lasting, self-sustaining, and energy-harvesting system-in-package (SiP) wireless micro-sensor solution. In: International Conference on Energy, Environment and Disasters (INCEED), Charlotte, NC (2005)
16. Randall, J.F., Jacot, J.: Renew. Energy **28**, 1851–1864 (2003)
17. Brunelli, D., Benini, L., Moser, C., Thiele, L.: An efficient solar energy harvester for wireless sensor nodes. In: EDAA (2008)
18. Torres, E.O., Min, C., Forghani-zadeh, H.P., Gupta, V., Keskar, N., Milner, L.A., Hsuan-I, P., Rincon-Mora, G.A.: SiP integration of intelligent, adaptive, self-sustaining power management solutions for portable applications. In: Proceedings of IEEE International Symposium on Circuits and Systems. ISCAS 2006 (2006)
19. Amelifard, B., Pedram, M.: Optimal selection of voltage regulator modules in a power delivery network. In: Design Automation Conference. DAC '07, pp. 168–173 (2007)

Chapter 19
Power Models and Strategies for Multiprocessor Platforms

Cécile Belleudy and Sébastien Bilavarn

1 Introduction

Multiprocessor systems are a promising solution to answer the need for more processing power, memory and network bandwidth in future mobile applications. At the same time, this solution implies high power dissipation and one of the key challenges is to control the power consumption, especially for embedded systems where a critical constraint is to extend the battery lifetime of a device. There are other goals which may also need to be considered, some of which could be to limit the cooling requirements or to reduce the financial cost. But finding ways to reduce the energy consumption becomes crucial because of power-related environmental concerns. Each year, approximately, 160,000 tons of portable (consumer) batteries are sold in the European Union. The portable battery market consists of general purpose batteries, button cells and rechargeable batteries. These batteries contain metals, which may pollute the environment at the end of their life. Reducing the power consumption limits the pollution. The problem of managing the energy consumed by electronic systems can be addressed at different levels of integrated systems development, ranging from technological level (circuits, architectures) to application and system software capable of adapting to the available energy source. Many research and industrial efforts are currently underway to develop energy-aware scheduling as well as power services in operating systems. In the literature, many works describe the energy behavior from theoretical studies or from (essentially) monoprocessor measurement results [1]. In this chapter, we will experiment with multiprocessor platforms in various possible configurations.

C. Belleudy (✉) · S. Bilavarn
University of Nice-Sophia Antipolis, LEAT, CNRS, Bat. 4, 250 rue Albert Einstein,
06560 Valbonne, France
e-mail: belleudy@unice.fr

S. Bilavarn
e-mail: bilavarn@unice.fr

G. Nicolescu et al. (eds.), *Design Technology for Heterogeneous Embedded Systems*,
DOI 10.1007/978-94-007-1125-9_19, © Springer Science+Business Media B.V. 2012

Increasingly powerful SOC designs require ever-larger main memory, which leads to a heavy increase of the energy consumption. In particular this can be a limiting factor in many embedded system designs. For future technologies, the static power will become the dominant factor of power consumption for the off-chip memory. In order to solve this issue, an architectural solution is to adopt a multi-bank memory system instead of a monolithic (single-bank) memory system, such as the RAMBUS-DRAM (RDRAM) technology [2] or the Mobile-RAM (SDRAM) technology [3]. In this case, significant power reductions are expected with the help of an efficient policy to manage the low-power modes of the different banks (Standby, Nap, Power-Down). In these modes, the power supply of the different parts of line and row decoders are switched off progressively. In the last low-power mode, the self-refresh is disabled and data is lost.

In this chapter we propose to analyze the impacts on energy considering the application allocation in the multibank memory. We will also present experiments and power measurements obtained from executing a H264/AVC decoder on a multiprocessor platform in different configurations. We will discuss an efficient exploitation of the results for the definition of power-aware strategies controlled by the operating system. Two problems will be addressed: the power consumption of processors and the one of the main memory. The remainder of this chapter is organized as follows. We survey related work in Sect. 2. An overview of our power experimentation and detailed analysis of measures are provided in Sect. 3. A power model of the main memory is exposed in Sect. 4 with a quantitative study of the energy variation according to the memory mapping of the application. Section 5 then concludes the chapter.

2 Related Work

To date, energy management in a processor relies mainly on two mechanisms: DPM (Dynamic Power Management) to switch off the power supply of a part of the circuit and DVFS (Dynamic Voltage and Frequency Scaling) to tune a processor clock speed and its corresponding voltage according to requirements such as the workload (actual or expected) or the battery charge. The management of these mechanisms through an operating system requires software that must be able, first, to identify the various operating points of the processor(s) and second, to assess requirements derived from the application and its environmental context.

The longest-standing approaches were developed to extend the autonomic capacity of laptops. They rely on DPM (e.g. ACPI in Linux) to manage the power saving modes available efficiently. The number of modes can actually range from 1 to 6 according to the processor family, and power consumption in each mode can be divided by a factor from 3 to 1000. The corresponding wakeup latencies vary from microseconds to milliseconds. In these approaches [4, 5], the processor is modeled by a state graph where nodes represent the operating points (processor + peripherals) and edges are the transition conditions with their associated penalties (for example, time and energy overheads). A user can thus define its own energy management policy by monitoring application parameters on the processor like the number of instructions executed on a time window. The application can also force some execution requirements by constraints such as the bit rate or the resources needed. This is also applicable for environmental parameters like the battery level, which can impose limitations on the power consumption or on the set of minimum functionalities to maintain.

In embedded systems, reducing energy is one of the prime concerns. We can find several initiatives developed by processor manufacturers and dedicated to families of embedded architectures. For example, ARM [6] defined a technique called IEM (Intelligent Energy Manager), the role of which is to handle system configuration according to the actual and/or predicted workload. The IEM works together with a hardware unit called IEC (Intelligent Energy Controller) to retrieve information for a running period and select suited power modes using the two mechanisms DPM and DVFS. Intel [7] proposed a similar technology called SpeedStep that was integrated into the XScale family of Intel PXA27x [8]. These initiatives also have in common that they target monoprocessor platforms.

Another important field of investigation addresses strategies for low-power scheduling. The first techniques based on DPM proposed scheduling policies for sets of tasks (dependent or independent) given an application domain. The goal was to optimize CPU idle times and to reduce the number of processors returning to active modes. Many contributions exist for single processors; and some of them have been extended to multiprocessor systems. Most approaches trace the history of task executions to predict the idle periods and turn hardware to a low-power state. For example, in [9], the authors use a regressive analysis on the running tracks. Benini et al. [10] and Rong and Pedram [11] use Markov chains: processor execution is modeled by a state graph where the probabilities of mode transitions are revalued according to the actual execution of tasks. In practice, these approaches are difficult to implement in real time systems. To bring a solution, some approaches like [12] introduce the mobility of tasks (time separating the earliest and latest dates of execution) to obtain a better sequencing of the tasks on the same processor.

The majority of these approaches do not consider all the power modes available or the energy overhead due to the wakeup of processors. We can however note that such techniques based on DPM are increasingly used today to solve thermal dissipation problems [13]. Nowadays, the use of low-power or sleep modes is often associated with techniques for the dynamic control of processor speed (DVFS). Such techniques have a very significant potential for reducing power. Some previous works take place in a purely static context [14–17]: the number of processors in active mode, as well as the operating frequency of processors, are set once and before execution, on the basis of the worst case execution time (WCET). Dynamic approaches [18] start with a similar static analysis based on WCET, and carry out a re-evaluation of the frequency at runtime by reconsidering the effective execution time of the tasks. Finally, some other approaches are essentially defined in a dynamic (on-line) context without preliminary information on the system [19, 20].

A majority of these works have been developed for homogeneous platforms where the clock speed is controlled either globally for all processors (chip_wide)

[20], either locally by processor [14–16]. Heterogeneity in more realistic systems can be seen as a set of processors with DVFS or not [20], as a set of processors that can operate at different speeds, or as a set of different types of processors resulting in varying task execution time [21]. A power-aware strategy can thus have different objectives, for example minimizing the total energy consumption resulting from the execution of a task set [15, 17], or reducing the peak power consumption leading to important solicitations of the system battery [14, 22].

Among these strategies, some are said to be "global" [19]: all tasks can be allocated to different processors allowing a comprehensive management of energy consumption, which is optimal if the number of task migrations is not too high. The issue for real time applications and multiprocessor systems is that the global schedulability test is only proved for a workload of approximately $m/2$ for m processors [23]. This sub-optimality leads to the development of approaches said to be "local" [15]: task allocation is performed statically, and each processor controls its running speed or low-power mode according to its own workload. Finally, in mixed approaches [24], some of the tasks are allocated to processors, while the other tasks are distributed over all the processors depending on their workload. Only these latter tasks are allowed to migrate.

In many approaches, only the dynamic power consumption is considered. This assumption is very restrictive especially for technologies below 90 nm. The authors of [16, 21] introduce static power consumption in their model. The processor speed is then calculated considering the two sources of energy dissipation. The authors show that, because of the static power consumption, some operating points, depending on the allocated time slots, do not gain more energy. Moreover, it should be noticed that the penalties due to the switch of voltage-frequency couples are very rarely considered in the majority of these theoretical works. However, on some platforms like the IMX31 and PXA27x, transition times of some milliseconds are observed. This setting cannot be neglected. A set of non-experimental works have been presented so far. The majority of these works assess the power savings by using a simulation platform or by developing an *ad hoc* evaluation based on a theoretical model of the power consumption.

Another critical factor impacting power is the memory. Consumption of the main memory is becoming considerable and it is expected to increase for future hardware. Some approaches associate the power management of processors with the power modes of the main memory. Several techniques exploiting multibank memory architectures with low-power mode management were proposed. From access data pattern analysis, they try to determine when to power down and into which mode it is possible to switch the memory banks. These memory controller policies can be compiler-based data mapping [25–27], hardware-assisted [28], or operating system oriented [2, 29].

At the compiler level, [25] studied the impact of loop transformations on banked memory architectures. In [30], it was proposed to keep data blocks in the on-chip memory space in a compressed form. The compiler's job is to analyze the application code and extract information on data access patterns to determine the data blocks to compress/decompress every time. In [27], the authors proposed to perform extra computations if doing so makes it unnecessary to reactivate a bank which is in the low-power operating mode. Most of these approaches operate on data, exploit one low-power mode of the memory and predict the memory bank usage at compile time for the target application. In [26], the proposed approach automatically places data elements that are likely to be accessed simultaneously in the same memory bank by tuned data migration based on the temporal affinity of data. For hardware-assisted techniques [30], the self-monitored hardware transit automatically banks to low-power modes based on the information collected by the supporting hardware. These techniques showed better performance than the compiler based approach but they are not flexible and need extra hardware, which in itself consumes energy.

The operating system based approach has the advantage of a global view of the system, without introducing any performance or energy overhead. Lebeck, et al. [31] proposed a scheme for reducing DRAM energy by power aware page allocation algorithm. Delaluz, et al. [32] proposed a scheduler-based approach. They use a bank usage table (BUT) which is managed by the operating system, and reset all banks' power modes at the context switches. The BUT gives the bank usage information at the previous time a process was scheduled. This information helps in selecting which banks to set to a low-power mode when the scheduler picks this process the next time. This approach cannot take into account the energy-aware policy applied to the processor, such as the frequency variations.

To reduce the power consumption of the main memory, we propose a new approach based on the task-bank locality in SMP multiprocessor architectures and which is complementary to the energy-aware scheduling policy.

Firstly, in the following section, power characterization of a multiprocessor platform is described according to different operating points and a proposal for power management strategy is given.

3 Design of a Multiprocessor Power Management Strategy

3.1 Development Environment

We present in the following the design and evaluation of a strategy based on DVFS (Dynamic Voltage and Frequency Scaling) for multiprocessor power management. The application domain is in the field of video processing and the platform used for all developments is the CT11 MPCore from ARM. In a first step, we analyze in detail the performance, power and energy consumption of a representative video application, namely H.264/AVC decoding, in different execution conditions: platform configuration (number of CPUs activated, voltage/frequency), workload (number of threads of the decoder) and input data (video sequences). The next section describes the experimental procedure on which this characterization, and power measurement in particular, is based.

3.1.1 Platform Description

The multiprocessor development platform is composed of two parts: the multicore platform itself which is an ARM MPCore test chip, and an Emulation Baseboard on top of which it is connected. The Emulation Baseboard is in charge of the power supply, memory system, bus control (AMBA AXI) and peripherals. The MPCore test chip is composed of 4 ARM11 processors, each having 32 KB instruction memory and 32 KB data memory. The system includes hardware support for memory system coherence with the unified L2 cache (1 MB). For our needs, the MPCore is used in symmetric mode under the control of Linux SMP, but it can also operate in asymmetric mode (AMP). Adaptive shutdown of unused processors is supported as well as dynamic voltage scaling. But concerning frequency switching, it is only supported statically with the Emulation Baseboard. All these features are controlled by hardware resources that are detailed in the following.

Most development boards allow power measurements using standard probes which are a convenient way to provide fast global values, but in our case another solution can be used. Specific built-in registers labeled SYS_VOLTAGE$_x$ exist in the MPCore to perform direct sampling of the voltage and current of the chip or the PLL [33]. By writing into these registers, it is also possible to force the supply voltage of the cores.

Register SYS_VOLTAGE$_0$ has two fields DAC_DATA and ADC_DATA that are used respectively to force and read the voltage of the cores. DAC_DATA is an 8-bit field that allows specifying one of 256 possible values ranging from 0.95 to 1.45 V. This value is sent to the digital to analog converter and sets the supply voltage of the four cores.

The other ADC_DATA field, of 12 bits, returns the voltage of the cores. Register SYS_VOLTAGE2 operates like SYS_VOLTAGE0 except that it has read-only accessibility. Its purpose is to return a voltage measure from a resistor network of 0.025 Ohms in a way to indicate the supply current of the CPUs, and thus to derive the corresponding instantaneous power consumption of the MPCore. Finally, we also used another register called SYS_PLD_INIT, to process a static modification of the processor frequency. This register sets the initialization of two parameters: the Test Chip PLL Control register and the Test Chip Clock divider register. A reset of the emulation baseboard makes the change effective in SYS_PLD_INIT. The above registers let us set the voltage and frequency of the entire test chip (corresponding to Fig. 19.1), and there is no support for separate clock and voltage scaling for each processor. Also we must keep in mind in the following the fact that the Emulation Baseboard allows the voltage to be scaled dynamically, but frequency only statically.

To use these hardware resources, we developed Linux modules that could meet our specific needs of power measurements in different platform configurations (a module to set voltage and frequency statically, and a module for power and energy monitoring). In addition we developed shell scripts that automate other configurations, like the number of processors activated, and power measurements possibly selecting among different video sequences. During the execution of an application, the power module samples the monitoring registers. It is activated every 200 ms by

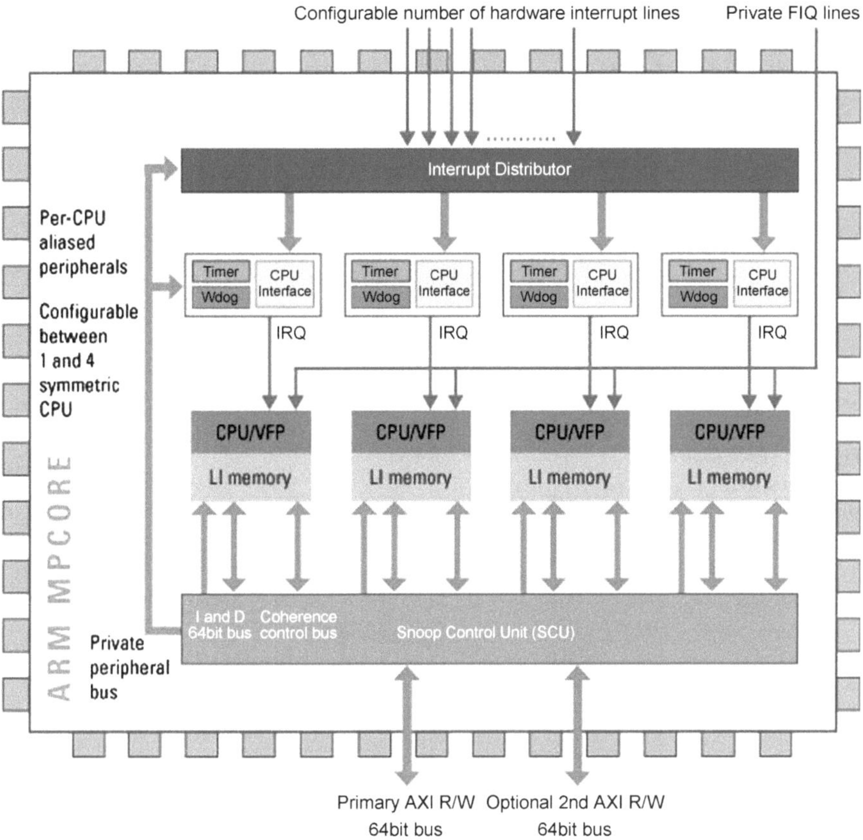

Fig. 19.1 ARM11 MPCore processor

an external timer in such a way as to minimize a possible time overhead on the application. With this method, we observed less than 1% execution time differences with and without using the power module. In the following, we present in detail the multithread H.264 decoder that has been used to benchmark the MPCore processor.

3.1.2 Application Mapping

The parallelization of the decoder exploits the possibility of slice decomposition of frames in the H264/AVC standard. Indeed a slice represents an independent zone of a frame: it can reference other slices of previous frames for decoding; therefore decoding one slice (of a frame) is independent from another (slice of the same frame). In our implementation, a slice is handled by a POSIX thread through the entire slice decoding process.

This way, the decoder can process different slices of a frame in parallel. There are very few data dependencies this way, and sequential execution remains only at

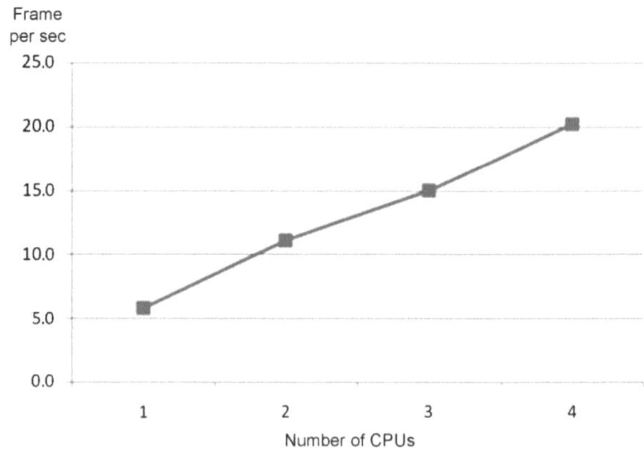

Fig. 19.2 Slice partitioning: frame rate of an 8 slices configuration using 1, 2, 3, 4 CPUs

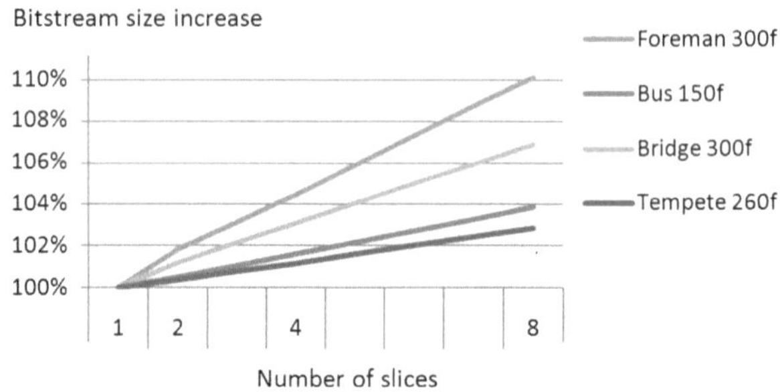

Fig. 19.3 Evolution of the bitstream size for 1, 2, 4, 8 slices and different video sequences

the input stream level because of the sequential nature of reading NAL packets (Network Abstraction Layer). So this parallelization is expected to scale performances according to the number of slices and processors. Given a number of slices, we proceed in creating as many threads that can run concurrently according to the number of CPUs available. The implied data locality of this solution also benefits greatly from the L1 cache of the core, provided the amount of slice data fits the cache size. The performance results of Fig. 19.2 emphasize the scalability of a decomposition of eight slices: the decoding speed increases from 6.3 to 20.6 fps with the number of CPUs (1, 2, 3, 4), and the maximum speedup corresponds to a value of 3.27 using four CPUs. Slice partitioning thus represents a good trade-off between performance and complexity, both in terms of implementation and processing, since there is very little dependence between slices to manage.

Another advantage of slice decomposition is its regularity and homogeneity which is suited to SMP implementations and makes balancing the workload between CPUs very easy (1 slice = 1 thread). Nevertheless, processing independent slices in a frame has a counterpart: it reduces the range of motion search within a frame, and compression efficiency is thus altered when increasing the number of slices [34].

Figure 19.3 shows the evolution of the bitstream size versus the number of slices. In a way to keep acceptable compression efficiency, we will not consider configurations of more than eight slices because it implies an increase of more than 10%

Fig. 19.4 Speedup of the multislice decoder for 1, 2, 4, 8 slices, different video sequences, using 4 CPUs (250 MHz)

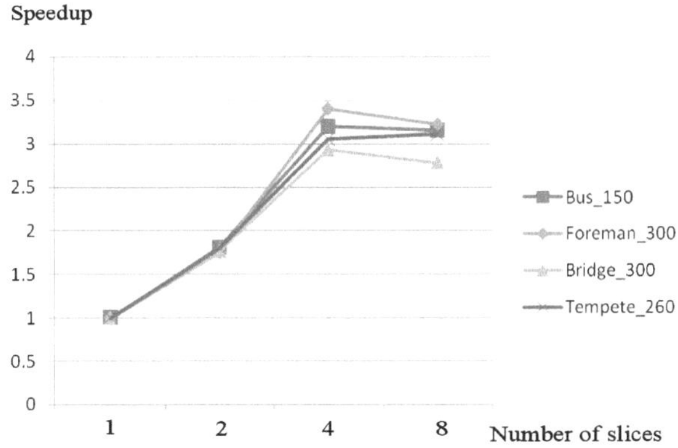

in the bitstream size (compared to a single slice configuration). Given this, we will consider the decoder only in the four following configurations 1, 2, 4, and 8 slices (or threads) in a way to provide the platform with different workloads.

3.2 Performance and Power Characterization

This section presents and discusses the performance and power measurements of the parallelized decoder on the MPCore platform in different configurations (1, 2, 4, 8 slices) and for several video sequences in 256×256 resolution (foreman, bus, bridge and tempete with 300, 150, 300, and 260 frames respectively). As explained previously, the parallelization associates one thread to each slice of a frame, so we can refer to the parallelism level in terms of the number of threads, which is exactly the number of slices per frame. The available configurations of 1, 2, 4 and 8 slices allow us to study the effect of different workloads on the platform. The next section addresses the performance analysis.

3.2.1 Performance Analysis

Figure 19.4 reports the speedup results of the decoder in configurations of 1, 2, 4, 8 threads when four CPUs are activated, compared to the execution of a monoslice version using a single CPU. The platform is configured in nominal conditions at 1.20 V/250 MHz. Figure 19.5 shows the corresponding performance in terms of number of frames decoded per second (fps).

The first observation is a linear acceleration growing from one to four slices. The execution speedup reaches a value of 3.19 (average on all sequences), which is close to the theoretical maximum of four. This denotes a good balance of the workload (i.e. threads) between the processors and shows the relevance of slice partitioning for SMP implementations. The loss in performance with respect to the theoretical maximum is due (i) to the sequential nature of processing the input stream (before

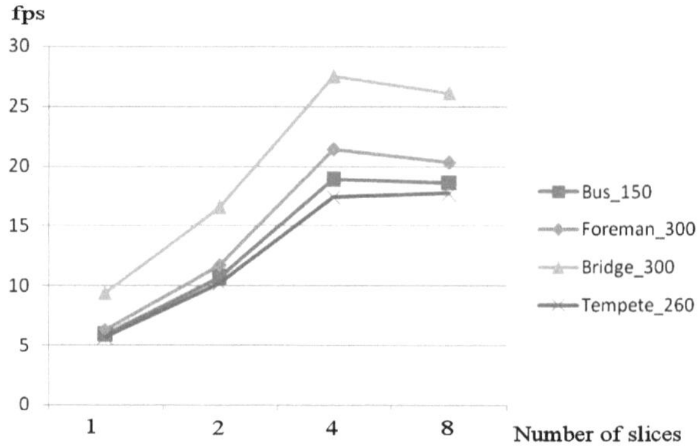

Fig. 19.5 Frame rate of the multislice decoder for 1, 2, 4, 8 slices, different video sequences, using 4 CPUs (250 MHz)

the creation of slice threads), (ii) to thread synchronizations and to a lesser extent (iii) to the influence of the operating system. These reasons are also responsible for the performance penalty in configurations of eight slices. In this case, the extra number of slices does not benefit from additional CPUs to speedup execution. Indeed, the number of threads per processor exceeds one, and therefore the effect of context switching and the associated L1 cache penalties may also have to be considered.

If we have a close look at the frame rates per video sequence (Fig. 19.5), we can notice important variations. Those variations are sensitive for instance between the bridge (27.2 fps) and the tempete sequence (18 fps). In the first one, small objects are moving in slow motion over a fixed background (low motion complexity). In the second one, the background is mobile and a group of objects are moving randomly and with fast motion (high motion complexity).

The frame rate appears to be very sensitive to motion properties in the video, with variations of about ±20%. This can be exploited to define a dynamic power strategy based on a frame rate adaptation using frequency scaling, which is discussed in Sect. 3.3.

3.2.2 Power Analysis

The following section analyzes the power consumption in different configurations of the decoder (workload) and the MPCore platform (voltage/frequency, number of CPUs). The test chip allows voltage variations ranging from 0.95 V to 1.45 V. To be compliant with ARM specifications ($V_{max} = 1.20 + 10\%$), we have set the following operating points: 0.95 V/150 MHz, 1.08 V/200 MHz, 1.20 V/250 MHz (nominal), 1.32 V/300 MHz. In these conditions, we have also considered various configurations of the platform (1, 2, 3, 4 CPUs) and workload (8, 4, 2, 1, 0 threads) to analyze the power consumption of the platform in different operational scenarios. Given the large quantity of results, we will first focus on the power consumption in nominal conditions presented in Fig. 19.6 (4 CPUS, 1.20 V/250 MHz). These measurements have shown very few differences in behavior between different video sequences, so we have represented only the power profiles of a foreman sequence. Indeed we

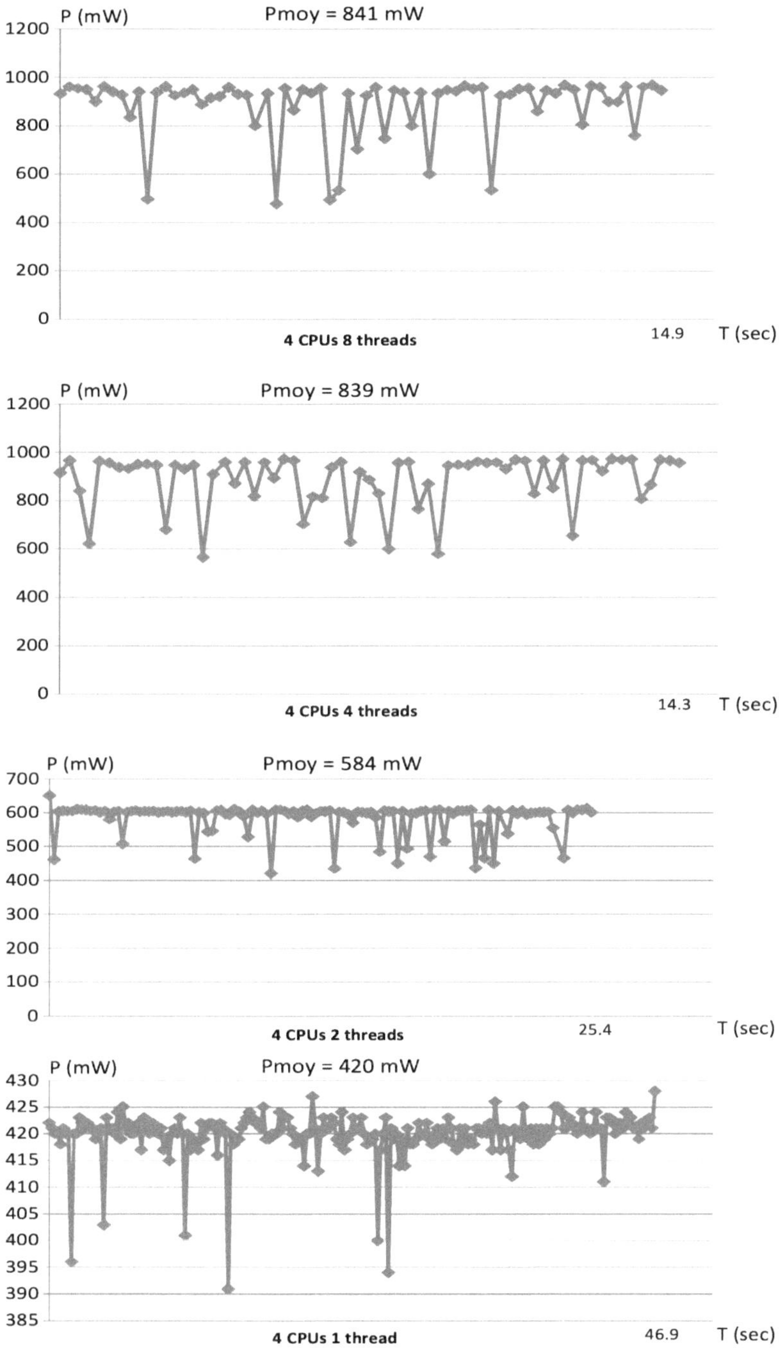

Fig. 19.6 Power profiles of 4 CPUs (1.20 V/250 MHz) in different workload configurations (8, 4, 2, 1 threads) using the foreman sequence

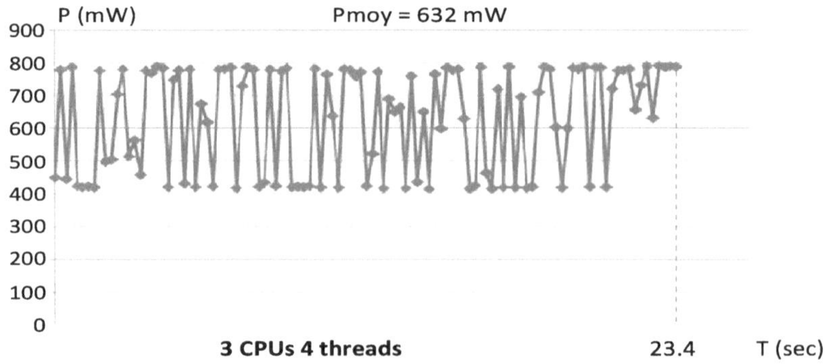

Fig. 19.7 Power profile of 3 CPUs (1.20 V/250 MHz) in a workload configuration of 4 threads using a foreman sequence

can observe the same trends of power consumption in each condition of workload: variations occur between a well-defined minimum and maximum value, and the distribution of points depends on the processor activity. For balanced CPU/workload configurations (e.g. 1, 2, 4, 8 thread/4 CPUs, Fig. 19.6), the power distribution is close to the maximum indicating a homogeneous and maximum load of processors. The average power consumption is well identified in this case. When the workload is unbalanced, for instance in a configuration of 4 threads/3 CPUs (Fig. 19.7), random variations occur between the minimum and maximum, indicating unequal demands on the processors resulting from a heterogeneous distribution of four threads on three CPUs. In this case, processing suffers from penalties caused by thread migrations and the associated L1 cache updates. As a consequence, these configurations are not energy efficient; the average power consumption is high and less predictable.

If we focus on the average power consumption for different configurations of CPU/workload (section 20 V/250 MHz of Table 19.1), we can see that it is easily predictable when CPU loads are balanced: 840 mW for 4 CPUs, 580 mW for 2 CPUs and 415 mW for 1 CPU. In the case of 3 CPUs, there are higher variations (typically in the 4- and 8-thread configurations) because of the non-homogeneous distribution of threads that leads to more unpredictable and irregular CPU activity. When there are fewer threads than active cores, we observe clearly that power consumption results from the number of active CPUs. For instance: 2, 4, 8 threads/2 CPUs, or 2 threads/2, 3, 4 CPUs exhibit very similar power consumption figures between 578 and 587 mW (1.20 V/250 MHz section of Table 19.1). This results from the operating system's ability to disable the unused cores using a sleeping mode called Wait For Interrupt (WFI). Finally, we have set other voltage-frequency couples in order to extend the power characterization of the platform.

The following values have been considered: 0.95 V/150 MHz, 1.08 V/200 MHz and 1.32 V/300 MHz. Regarding the previous analysis, we examine the average power consumption and execution time in these voltage/frequency configurations, and for different workloads including when no user thread is run (8, 4, 2, 1, 0 threads). In each case, power profiles follow the same trends as those observed in the nominal case (Figs. 19.6 and 19.7).

Table 19.1 Average power (mW) and decoding time (sec) of a foreman sequence (300 frames) in different configurations of platform (number of CPUs, voltage/frequency) and workload (number of threads)

MPCore config	Decoder config				
	8 threads	4 threads	2 threads	1 thread	0 thread
1.32 V/300 MHz					
1 CPU	633/44.0	634/43.1	636/41.7	636/41.5	361/–
2 CPUs	885/22.6	878/22.0	880/22.2	637/40.3	364/–
3 CPUs	1034/17.2	932/20.5	882/21.6	637/40.1	363/–
4 CPUs	1256/13.0	1187/12.2	867/21.5	642/40.7	355/–
1.20 V/250 MHz					
1 CPU	415/50.7	415/50.8	416/49.3	416/48.9	241/–
2 CPUs	587/26.6	579/25.9	581/25.6	418/47.0	241/–
3 CPUs	701/19.8	632/23.4	578/25.6	418/57.6	238/–
4 CPUs	841/14.9	839/14.3	584/25.4	420/46.9	238/–
1.08 V/200 MHz					
1 CPU	270/62.1	270/61.9	270/60.3	271/59.8	156/–
2 CPUs	381/32.2	376/31.5	375/31.5	271/58.2	147/–
3 CPUs	462/23.7	408/28.7	381/30.6	272/57.5	153/–
4 CPUs	561/17.9	568/17.0	380/31.1	273/57.5	154/–
0.96 V/250 MHz					
1 CPU	160/82.9	160/80.9	160/79.2	160/78.4	93/–
2 CPUs	225/42.0	224/41.2	225/41.5	160/75.7	89/–
3 CPUs	276/30.9	241/36.5	224/41.1	161/76.2	89/–
4 CPUs	335/23.0	335/22.3	223/40.3	162/75.9	92/–

These results show that it is possible to derive a first simple yet accurate power model, provided we exclude sub-optimal configurations where the workload is unequally balanced between processors. Such a basic model is able to provide reliable power predictions from the number of threads and the voltage/frequency configuration of the platform. We address other key results that are discussed in terms of development of a power management strategy in the following section.

3.3 Power Management Design and Implementation

3.3.1 Impact of Results

An interesting point from a DVFS perspective is the variation of about 40% in decoding speed resulting only from the motion properties of the video. Indeed this

Table 19.2 Differences of configurations (number of CPUs, workload, voltage/frequency) and performances for an equivalent level of energy consumption

Performance variations					
Nb CPUs	Nb threads	V/F (V/MHz)	*P* (mW)	Perf. (fps)	*E* (mJ)
1	8	0.95/150	160	**3.6**	**13**
4	4	1.32/300	1187	**24.6**	**14.4**

Table 19.3 Energy efficiency: effect of increasing the number of CPUs vs. frequency (at an equivalent performance level of 13.5 fps)

Energy efficiency				
Nb CPUs	V/F (V/MHz)	Perf. (img/sec)	*P* (mW)	*E* (mJ)
4	**0.95/150**	13.4	335	**7.51**
2	**1.32/300**	13.5	880	**18.80**

can be exploited to smooth the frame rate by a dynamic adaptation of processor frequency. Other results can also be exploited. Table 19.2 reports total performance variations ranging from 3.6 to 24.6 fps (extracted from Table 19.1) when considering all the possible configurations of execution (platform, decoder). This gives room for adapting the performance to different video qualities: 15, 20 or 25 fps for example.

We can also notice that the energy cost is comparable in both performance extrema (3.6 to 24.6 fps), and indicates that there are suboptimal configurations in terms of energy efficiency among all possibilities. We can also verify the intuitive assumption that a combination of a high number of active cores with low operating frequency is better from an energy point of view. To confirm this, we can consider a decoding constraint of 13.5 fps that can be satisfied in two configurations: four CPUs at 0.95 V/150 MHz and two CPUs at 1.32 V/300 MHz. The difference reaches a factor of two between the corresponding energy consumptions (Table 19.3). This shows that it is more interesting to increase the number of CPUs while minimizing the frequency, for a given level of performance (13.5 fps in this case).

Another aspect concerning energy efficiency is illustrated in Fig. 19.8, which relates the energy consumption measured in a configuration of four tasks for different number of cores and voltage/frequency. These results show that energy efficiency increases when using more CPUs, by a factor of 1.8 from 1 to 4 processors.

We can also notice that energy efficiency is very sensitive to the distribution of the workload with a clear inflection point in each configuration between 2 and 3 CPUs, showing that energy efficiency is altered. In these conditions, it is more energy efficient to maximize the number of CPUs while minimizing frequency, but also to pay attention to workload balancing in order to minimize the energy waste.

From these considerations, we have derived the following strategy: first, a static adaptation of the number of active CPUs to the workload according to a quality constraint, and second, a finer adjustment of performance variations using frequency scaling. In this case, we chose to use four CPUs to benefit from the maximum video

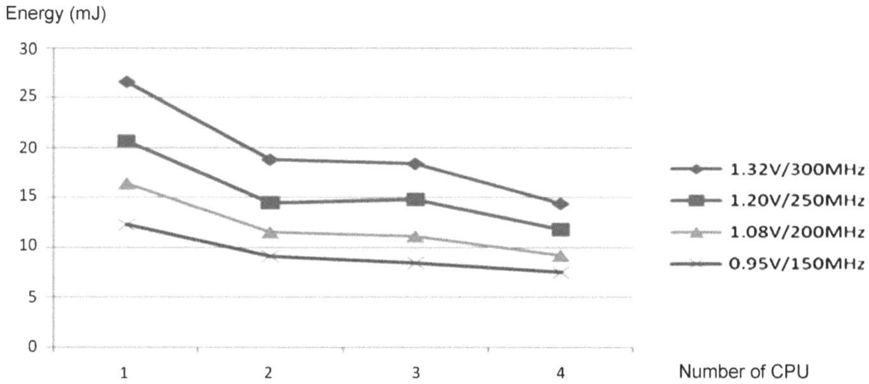

Fig. 19.8 Impact of load balancing on energy efficiency (configuration of 4 threads using 1, 2, 3, 4 CPUs in 4 voltage/frequency settings)

quality (25 fps), and then adjust the decoding speed by dynamic frequency scaling, around 20 or 15 fps, which will result in lowering the power consumption. It could also be possible to use 2 CPUs (12 fps at 250 MHz) and adapt the decoder speed at a lower value, such as 8 fps for example.

3.3.2 Implementation of the Adaptation Strategy

The principle of the adaptation strategy is to control the decoder speed around a frame rate constraint slightly lower than the average performance in nominal conditions (to decrease the operating frequency). The adaptation is based on changing the frequency (thus voltage) when the frame rate is between two defined thresholds. The number and values of these thresholds depend on the operating points (voltage-frequency couples) and on the performance constraint to satisfy. To implement this adaptation on the MPCore, we used seven fps thresholds that were derived from the following operating points: 0.96 V/150 MHz, 1.02 V/175 MHz, 1.08 V/200 MHz, 1.14 V/225 MHz, 1.20 V/250 MHz, 1.26 V/275 MHz, 1.32 V/300 MHz. Each threshold $tresh_i$ is associated with a given frequency f_i which is computed as follows:

$$tresh_i = adaptation\_const * f_{\mathrm{nom}}/f_i$$

where f_{nom} is the nominal frequency (250 MHz).

For an adaptation constraint of 20 fps, we have the following thresholds: 33.3, 28.6, 25, 22.2, 20, 18.2, 16.7 fps defined respectively for 150, 175, 200, 225, 250, 275, 300 MHz. When the decoder speed remains in a zone delimited by two consecutive thresholds during a sufficient amount of time (to minimize the number of clock switching), the processor frequency is switched to the value associated with this zone. As an illustration, if the frame rate is between 28.6 and 33.3 fps during 250 frames, the operating point is set to 0.96/150 MHz.

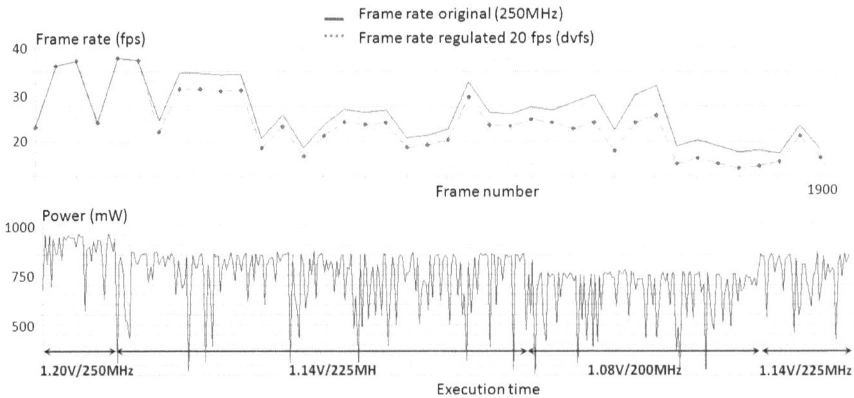

Fig. 19.9 Frame rate and power profile of a 20 fps regulation (9.7% energy gains)

Changing the supply voltage is done as explained in Sect. 3.1.1. Concerning frequency, it can not be set dynamically, but only statically after a reset of the Emulation Baseboard. Whereas dynamic clock switching is not supported on the platform, we have nevertheless implemented an adapted version of the DVFS strategy defined above in order to check the possible energy gains. The energy consumption could be measured this way using the procedure of Sect. 3.1.1, but considering dynamic voltage scaling only. Frequency scaling could not be included in the strategy we developed for evaluation, but we have tried to implement a strategy as close as possible to the DVFS strategy described above.

We have thus developed a DV(F)S driver for Linux with the following characteristics. It handles the switching of operating points with respect to the decoding speed. It samples the frame rate at regular time intervals every T_{monitor} images, and makes the decision of switching or not the operating point every $T_{\text{eval}} \cdot T_{\text{monitor}}$ images. A T_{switch} parameter delay is also simulated to take the effect of PLL setting times into account. To analyze the adapted performances of the decoder, the driver computes a trace of the frame rate which is extrapolated from the actual performance at 250 MHz and the frequency that should have been changed to by the driver. Power and energy are measured using the procedure of Sect. 3.1.1; results are reported in the next section.

3.3.3 Results

The power and performance measures for two regulations at 20 and 15 fps of a 1 minute 20 second video sequence are given in Figs. 19.9 and 19.10. In nominal conditions (1.20 V/250 MHz), the sequence requires 62 Joules to be decoded at an average speed of 24.8 fps. Both figures show traces of the original and regulated frame rates as well as the trace of power consumption. On the power profiles, we can clearly observe different voltage/frequency domains, thus the DVS switches that can be identified by distinct maximum values.

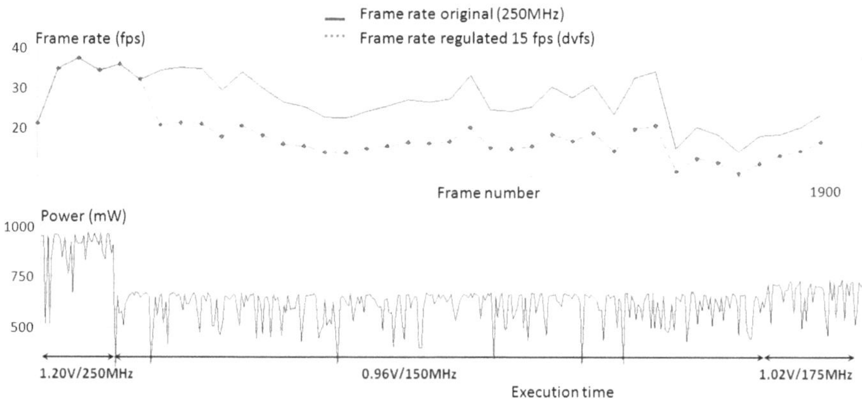

Fig. 19.10 Frame rate and power profile of a 15 fps regulation (30.6% energy gains)

Energy gains are good in both cases with respectively 9.7 and 30.6% for 20 and 15 fps. Because the platform remains at an operating frequency of 250 MHz, we must emphasize that the energy reported does not consider two effects that would result from an effective frequency scaling: the energy consumption should be higher because the execution time increases when frequency decreases. At the same time, the energy consumption decreases with frequency and voltage because dynamic power consumption is proportional to $\alpha C V^2 F$ [35]. These opposing effects will result in compensation, so we can reasonably assume that the energy gains measured are close to the results of an effective DVFS technique.

The performance profiles of Figs. 19.9 and 19.10 allow comparing the evolution of the decoder performances at 250 MHz (full line) versus scaling frequency (dotted line). Each point is evaluated every T_{monitor} images and a decision of changing the frequency or not is made every $T_{\text{eval}} \cdot T_{\text{monitor}}$ images.

In these conditions, three configuration switches are operated for a 20 fps adaptation (1,14 V/225 MHz–1,08 V/200 MHz–1,14V/225 MHz) and only two switches but at lower frequency (0,96 V/150 MHz–1,02 V/175 MHz) for a 15 fps adaptation, which results in better energy gains.

Obviously the potential energy gains are linked to the performance. This is due to the fact that performance improvement allows proportional frequency (and voltage) reduction. As a consequence, there is still room for optimizing the decoder using SIMD instructions in particular (1.5/2× speedups), that could permit to run the platform at lower frequency (150 MHz, 335 mW), or even to meet fps constraints using less than four CPUs.

3.4 Conclusion

The work exposed in this section provides many results on power and performance that have stressed the impact of load balancing on energy efficiency of multiprocessor platforms. They have led us to propose a management strategy for video processing that has been implemented with reported energy savings of up to 30.6%. Whereas the achievable gains depend on the level of application performance optimization and video quality required, the exploitation of strategies that has been tuned for domain specific applications lead us to anticipate high energy savings, compared to general purpose strategies. It is highly probable that power management in the future will have to adapt strategies to different application domains in order to achieve important energy gains. Another potential source of high power reduction concerns the memory system that has to be controlled under similar constraints. This is also a promising field of investigation which is the purpose of the following section.

4 Power Memory Model

4.1 Memory Architecture

Processor power management using techniques such as DVFS tends to increase application run time and therefore to stretch the active time and the energy consumption of the main memory. As an example, for a given processing rate, we can operate with 4 CPUs at a frequency of 300 MHz (1.32 V) for a duration of $T/2$ and afterwards go into WFI mode; or alternatively for a duration T with a frequency of 150 MHz (0.95 V). In the latter case, we notice that the power consumption of the MPcore is reduced by a factor of 2.32. If we add the memory consumption, for instance, let us consider one bank of RDRAM memory [2] operating in its first low-power mode, such that the power saving factor falls to 1.53. If a deeper low-power mode is used, this factor decreases again, consequently reducing the benefit that could be achieved by frequency tuning.

Moreover, for future technologies, static power will increase to become the dominant contribution to energy consumption in off-chip memory. To solve this issue, one solution consists of using a multi-bank memory system with an efficient management policy of low-power modes of the different banks. A methodology for allocation of tasks into the multi-bank memory could minimize the energy consumption by extending the time spent in low-power modes and/or by reducing the number of bank wake-ups. To service a memory request (read or write), a bank must be in active mode which consumes most of the power. When a bank is inactive, it can be put in any low-power mode (for instance standby, nap, power-down mode). Each mode is characterized by its power consumption and the time that it takes to transit back to the active mode (resynchronization time). The lower the energy consumption of the low-power mode, the higher the resynchronization time becomes. The addressed SMP architecture (like the MPcore) has a two level cache arrangement with a bus-based communication model. A shared multi-bank memory is linked to the processors as shown in Fig. 19.11. Each bank can be controlled independently and placed into one of the available low-power modes. Each low-power mode is characterized by the number of components being disabled to save energy. In the

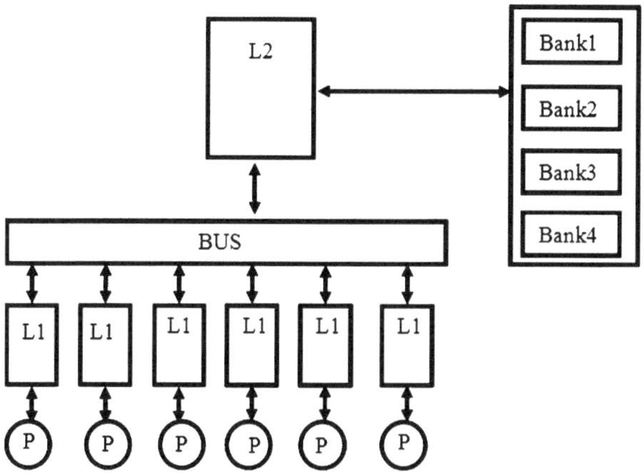

Fig. 19.11 Architecture of the memory system

following, our aim is to show that an appropriate memory allocation can enable significant power savings and to propose a strategy in relation to the energy-aware scheduling policy.

4.2 Multi-bank Memory and System Model

The memory is described by architectural parameters: the number of banks (NB_{mem}), the bank size (SB_{mem}) and the number of low-power mode (N_{LP}). In active mode, a bank is characterized by its power consumption ($P_{\text{active\_mem}}$) and the access time for read/write operation (T_{access}). In each low-power mode l, the memory features to consider are: power consumption ($P_{\text{lp\_mode-}l}$), wakeup time ($T_{\text{lp\_mode-}l}$) and an energy overhead when the bank goes to the active mode ($E_{\text{mode\_switch-}l}$). A dynamic system will be able to model the different modes of the memory as in the ACPI [4] and to assess the power savings in order to apply an efficient power strategy.

The application is composed of a set of tasks. Each task t_i is characterized by its worst case execution time $WCET_{t_i}$ (or the average execution time). In order to evaluate the memory power consumption, some additional parameters must be taken into account: the task memory size S_{t_j} (code and data) and the number of main memory accesses $L2\_M_{t_j}$. These last parameters correspond to the number of L2 cache misses and can be collected by profiling. Usually, this is the average number of cache misses derived from simulations or previous executions.We assume that the task size is less than or equal to the size of a memory bank.

4.3 Energy Model

The energy consumption of a multi-bank memory depends on the power dissipated in the different modes and the time spent in these modes. In the following equation, the energy model is given for the worst case execution time ($WCET_{t_i}$) of the task but it can be replaced by the actual execution time. Usually the worst case execution time is specified for the maximum speed of the processor and need to be adjusted when the frequency is scaled down: $WCET_{t_i}(f_{\text{CPU}})$. This time is divided into two time intervals: the first one, for read and write operation (bank is accessed), and the second one, when the bank is not accessed but still active: $T_{t_i} = T_{\text{access-}t_i} + T_{\text{noaccess-}t_i}$.

The access time is the result of the multiplication of the number of memory accesses and the read/write time: $T_{\text{access-}t_i} = L2\_M_{ti} \cdot T_{\text{access}}$. $T_{\text{access-}t_i}$ is independent of the processor frequency and therefore $E_{\text{access-}t_i}$ is a constant as regards the selected processor frequency. The time spent when no memory access occurs ($T_{\text{noaccess-}t_i}$) is the difference between the access time and the worst case execution time and consequently is a function of the processor frequency: $T_{\text{noaccess-}t_i}(f_{\text{CPU}}) = WCET_{t_i}(f_{\text{CPU}}) - T_{\text{noaccess-}t_i}$.

An allocation function noted ϕ has to be defined which associates each task t_i belonging to a set of N tasks to a bank b_j belonging to a set of k banks.

$$\varphi : \{t_1, t_2, \ldots, t_N\} \rightarrow \{b_1, b_2, \ldots, b_k\}, \quad \varphi(t_i) = b_j$$

The energy consumption of a memory of k banks and a given allocation of N tasks to these banks is evaluated by the sum of the energies consumed in the active mode (E_{active}), in the low-power modes ($E_{\text{mode\_switch}}$) and to switch between the modes ($E_{\text{lp\_mode}}$): $E_{\text{memory}} = E_{\text{active}} + E_{\text{mode\_switch}} + E_{\text{lp\_mode}}$. We now describe the individual contributions in detail.

4.3.1 E_{active}

The memory bank is assumed in an active mode during the whole duration of execution of a task. The time spent in active mode is divided into two time intervals: the first one for the read/write operation (the bank is accessed) and the second one, when the bank is not accessed but still active: $E_{\text{active}} = E_{\text{access}} + E_{\text{noaccess}}$. The equation for E_{access} is given by:

$$E_{\text{access}} = \sum_{b_j : j=1}^{k} \left(\sum_{t_i : \varphi(t_i) = b_j}^{N} L2\_M_{t_i} \cdot E_{\text{mem\_access}} \right)$$

E_{noaccess} is the energy due to the co-activation of the different banks of the memory when these banks are active but not servicing any read or write operation. The time spent in this configuration is called, for a bank b_j, $T_{\text{no\_access-}b_j}$.

$$E_{\text{noaccess}} = \sum_{b_j : j=1}^{k} (T_{\text{noaccess-}b_j} \cdot P_{\text{noaccess}})$$

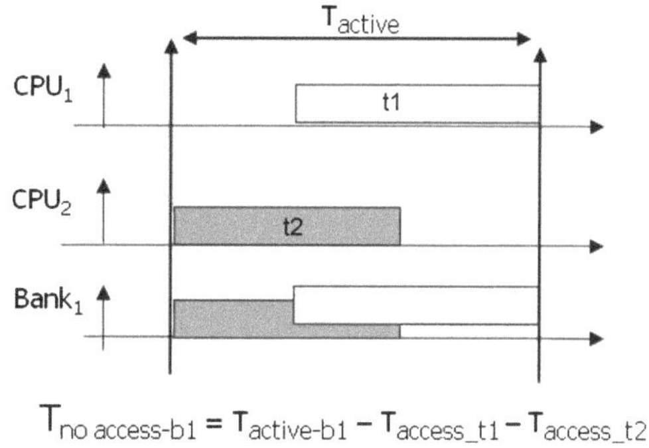

Fig. 19.12 Example of memory mapping and active time

When two tasks, stored in a same bank, are running in parallel, the no access time is the interval between the start time and the end time of the two tasks minus the access times of the two tasks. As an example let us consider two tasks running according to the scheme in Fig. 19.12.

The no access time is dependent on the processor frequency and also increases the no access energy. For instance, if a processor is slowed down by a factor S, the no access energy of the memory is multiplied by the same factor. This drawback can be mitigated by grouping tasks running in parallel.

4.3.2 $E_{\text{mode\_switch}}$

This energy is dissipated during the transition of the memory bank from a low-power mode to the active mode:

$$E_{\text{mode\_switch}} = \sum_{j=1}^{k} \left(\sum_{l=1}^{N_{\text{LP}}} N_{\text{mode\_switch-}l}(b_j) \cdot E_{\text{mode\_switch-}l} \right)$$

with $N_{\text{mode\_switch-}l}(b_j)$, the number of switches per bank.

4.3.3 $E_{\text{LP\_mode}}$

This energy is consumed in the memory banks in low-power modes:

$$E_{\text{lp\_mode}} = \sum_{j=1}^{k} \left(\sum_{l=1}^{N_{\text{LP}}} T_{\text{lp\_mode-}l}(b_j) \cdot P_{\text{lp mode-}l} \right).$$

$T_{\text{lp\_mode-}l}(b_j)$ is the time spent by the memory bank b_j in low-power mode l.

4.4 Energy-Aware Memory Allocation

4.4.1 Energy-Aware Memory Allocation: Principle

In most cases, the energy in the active mode is the major part of the memory energy consumption. In order to minimize this part of the energy and also to decrease the effect of the processor slowdown, our strategy is to try to allocate tasks running in parallel to the same bank. We can consider two kinds of DVFS in a multiprocessor platform: chip-wide (all the processors run with the same frequency) and per-core (the processors can run at different frequencies). Our approach is implemented with two queue structures consisting in the running task queue (RUQ) and ready task queue (REQ) and the knowledge of the memory requirements. Each bank is described by the list of running tasks which are allocated to it, the end time of these tasks (based on the WCET or the average actual execution time) and the remaining memory size. When a task begins its execution, the attributes of bank is updated. The scheduler in charge of each task selects a processor, a frequency and a memory bank. In the case of a chip-wide DVFS system, all the tasks are slowed down. The end time of all the running tasks must be updated. For a per-core DVFS system, only the considered task is affected by the slowdown factor. The recovery time between two tasks is defined as the time for which these two tasks are running in parallel. A task is allocated to the active bank that has the highest recovery time. In the reverse case, no active bank has enough memory to store the task, two choices are envisaged: the task execution is delayed or a bank in low-power mode is woken up. In order to assess the power saving of these two ways, the memory needs of the task in REQ are analyzed. The memory size of the tasks belonging to the REQ is compared to the memory size freed by the task that will finish their execution on a window time equal to the wakeup time of the bank. If the memory needs are inferior or equal to the freed memory size, the tasks are delayed. Otherwise, a bank is woken up and a time overhead is added

4.4.2 Results

In order to illustrate the benefit of our approach, we consider a simple example, with four tasks t_1, t_2, t_3, t_4 with respectively an execution time of $T/2$, T, $T/2$ and T at the maximum frequency (300 MHz). The memory size of these tasks is 3 Mb. The access time is ten percent of the execution time (average value for a H264 decoder). The memory is a RDRAM described in [2] and is composed of four banks of 8 Mb. The low-power mode used for this example is standby. When a processor is not busy, it goes into WFI mode. Three configurations of the MPcore were analyzed in Table 19.4 that gives the overall energy consumption of the memory system and the processors (MPcore).

This example shows that the memory consumption is significantly reduced with an adapted power management and it is complementary to processor power management. Other examples show that some operating points don't enable power savings

Table 19.4 Normalized energy for three configurations of the MPcore

Normalized energy	Without energy-aware allocation	With energy-aware allocation
4 CPUs (300 MHz, 1.32 V)	1	0.92
4 CPUs (150 MHz, 0.95 V),	0.78	0.65
2 CPUs (300 MHz, 1.32 V)	0.85	0.82

if we take into account the memory consumption. This fact leads us to believe that the energy benefit of an operating point must be assessed on the overall system.

5 Conclusion

Existing approaches of power management are mostly generic. For example, *cpu-freq* implemented in Linux is based on monitoring the CPU workload. When the workload decreases, processor frequency is reduced, and so is the power consumption. In case of video processing for example, this approach is inefficient because the application represents most of the time a maximum load which results in the processor operating at the maximum clock frequency. The results we have described demonstrate the need for domain specific strategies (for example multimedia or networking) in addition to general purpose strategies to enable higher levels of energy gains. This means that the software infrastructure of power management and the operating system must be flexible and able to handle different policies. Another point concerns the possibility of other techniques alternative to DPM and DVFS. For instance, interesting extensions to investigate should include memory and/or dedicated (possibly dynamically reconfigurable) accelerators, within power management strategies. For a simple example, we have shown the benefit of including a memory energy-aware strategy that is complementary to a scheduling policy that exploits the DVFS technique.

References

1. Contrebas, G., Martonosi, M.: Power prediction for Intel XScale processors using performance monitoring unit events. In: ISPLED (2005)
2. Fan, X., Ellis, C., Lebeck, A.: Modeling of DRAM power control policies using deterministic and stochastic Petri nets. In: Lecture Notes in Computer Science, vol. 2325, pp. 37–41. Springer, Berlin (2003)
3. Infineon Inc, Mobile-RAM data sheet (2004)
4. ACPI, Advanced configuration and power interface. http://www.acpi.info/ (2006)
5. IBM, MontaVista, Dynamic power management for embedded systems (2002)
6. ARM, Intelligent energy controller technical reference manual. ARM Limited. http://infocenter.arm.com (2008)
7. Intel, Wireless Intel speedstep power manager. White paper (2004)

8. Intel, Intel PXA270 processor, electrical, mechanical, and thermal specification. http://www.intel.com (2005)
9. Hwang, C.-H., Wu, A.: A predictive system shutdown method for energy saving of event-driven computation. In: International Conference on Computer-Aided Design, November 1997, pp. 28–32 (1997)
10. Benini, L., Bogliolo, A., Micheli, G.D.: A survey of design techniques for system-level dynamic power management. IEEE Trans. Very Large Scale Integr. Syst. **8**, 299–316 (2000)
11. Rong, P., Pedram, M.: Determining the optimal timeout values for a power-managed system based on the theory of Markovian processes: offline and online algorithms. In: Proc. of Design Automation and Test in Europe (2006)
12. Qiu, Q., Liu, S., Wu, Q.: Task merging for dynamic power management of cyclic applications in real-time multiprocessors systems. In: ICCD 04 (2004)
13. Merkel, A., Bellosa, F.: Energy power consumption in multiprocessor systems. In: EuroSys2006 (2006)
14. Luo, J., Jha, N.K.: Static and dynamic variable voltage scheduling algorithms for real-time heterogeneous distributed embedded systems. In: International Conference on VLSI Design, January 2002
15. Chen, J.-J., Kuo, T.W.: Energy efficient scheduling of periodic real-time tasks over homogeneous multiprocessors. In: PARC05 (2005)
16. De Langen, P., Jurlink, B., Vassiliadis, S.: Multiprocessor scheduling to reduce leakage power. In: 17th International Conference on Parallel and Distributed Symposium (2006)
17. Benini, L., Bertozzi, D., Guerri, A., Milano, M.: Allocation, scheduling and voltage scaling on energy aware MPSoCs. In: Integration of AI and OR Techniques in Constraint Programming for Combinatorial Optimization Problems. Lecture Notes in Computer Science, vol. 3990, pp. 44–58. Springer, Berlin (2006)
18. Choudhury, P., Chakrabarti, C., Kumar, R.: Online dynamic voltage scaling using task graph mapping analysis for multiprocessors. In: VLSI Design (2007)
19. Zhu, D., Melhem, R., Childers, B.: Scheduling with dynamic voltage/speed adjustment using slack reclamation in multi-processor real-time systems. IEEE Trans. Parallel Distrib. Syst. **14**(7), 686–700 (2003)
20. Chen, J., Yang, C.Y., Kuo, T., Shih, C.-S.: Energy efficient real-time system task scheduling in multiprocessor DVS systems. In: ASP-DAC, January 2007
21. Yang, C., Chen, J.-J., Kuo, T.-W., Thiel, L.: An approximation scheme for energy-efficient scheduling of real-time tasks in heterogeneous multiprocessor systems. In: DATE 09, April 2009
22. Chou, P.H., Liu, J., Li, D., Bagherzadeh, N.: IMPACCT: Methodology and tools for power aware embedded systems. Design Automation for Embedded Systems, Special Issue on Design Methodologies and Tools for Real-Time Embedded Systems, 205–232 (2002)
23. Srivastava, M., Chandrakasan, A., Brodersen, R.: Predictive system shutdown and other architectural techniques for energy efficient programmable computation. IEEE Trans. Very Large Scale Integr. (VLSI) Syst. **4**, 42–55 (1996)
24. Bhatti, M.K., Muhammad, F., Belleudy, C., Auguin, M.: Improving resource utilization under EDF-based mixed scheduling in multiprocessor real-time systems. In: 16th IFIP/IEEE Int. Conf. on Very Large Scale Integration, VLSI-SOC'2008, Rhodes, Greece (2008)
25. Kandemir, M., Kolcu, I., Kadayif, I.: Influence of loop optimizations on energy consumption of multi-bank memory systems. In: Proc. Compiler Construction, April 2002
26. Ozturk, O., Kandemir, M., Irwin, M.J.: Increasing onchip memory space utilization for embedded chip multiprocessors through data compression. In: CODES05 (2005)
27. Koc, H., Ozturk, O., Kandemir, M., Ercanli, E.: Minimizing energy consumption of banked memories using data recomputation. In: ISLPED'06 (2006)
28. Kandemir, M., Ozturk, O.: Nonuniform banking for reducing memory energy consumption. In: DATE'05, Munich, Germany (2005)
29. BenFradj, H., Belleudy, C., Auguin, M.: Multi-bank main memory architecture with dynamic voltage frequency scaling for system energy optimization. In: 9th EUROMICRO Conference on Digital System Design, September 2006

30. Delaluz, V., Kandemir, M., Vijaykrishnan, N., Sivasubramaniam, A., Irwin, M.J.: DRAM energy management using software and hardware directed power mode control. In: International Symposium on High Performance Computer Architecture, pp. 159–170 (2001)
31. Lebeck, A., Fan, X., Ellis, C.: Memory controller policies for DRAM power management. In: International Symposium on Low Power Electronics and Design (2001)
32. Delaluz, V., Sivasubramaniam, A., Kandemir, M., Vijaykrishnan, N., Irwin, M.J.: Scheduler based DRAM energy management. In: DAC (2002)
33. ARM, Core Tile for ARM11 MPCore user guide. ARM Limited. http://infocenter.arm.com (2005)
34. Roitzsch, M., Slice-balancing, H.: 264 video encoding for improved scalability of multicore decoding. In: 7th ACM & IEEE International Conference on Embedded Software (EMSOFT), September 2007, pp. 269–278 (2007)
35. Mudge, T.: Power: a first-class architectural design constraint. Computer **34**(4), 53–58 (2001)

Chapter 20
Dynamically Reconfigurable Architectures for Software-Defined Radio in Professional Electronic Applications

Bertrand Rousseau, Philippe Manet, Thibault Delavallée, Igor Loiselle, and Jean-Didier Legat

1 Introduction

Embedded electronic systems are mainly driven by the consumer market. Consumer embedded systems are generally produced in high volumes, which can sometimes reach several hundred millions of units. This is the case for instance for mobile phones or game consoles [13]. Such a large amount of produced systems makes viable the development of dedicated platforms.

Most of those consumer-oriented embedded systems use heterogeneous platforms, integrated into system-on-chips. This approach allows to reach the required performances within the power budget by associating different architectural solutions, each one for a specific functionality, at the cost of an increased silicon area. Developments for consumer electronics are also constrained by short time-to-markets and short product lifetimes. Consumer applications must generally not face strong requirements in terms of validation and qualifications.

At the opposite, professional embedded systems are very different. Their production volume is low since they target specific applications. However, although volumes are low, there is a great number of different designs, since many designs only fit precise applications. Those systems must comply to very specific constraints.

B. Rousseau (✉) · P. Manet · T. Delavallée · I. Loiselle · J.-D. Legat
Université catholique de Louvain (UCL), Louvain-la-Neuve, Belgium
e-mail: bertrand.rousseau@uclouvain.be

P. Manet
e-mail: philippe.manet@uclouvain.be

T. Delavallée
e-mail: thibault.delavallee@uclouvain.be

I. Loiselle
e-mail: igor.loiselle@uclouvain.be

J.-D. Legat
e-mail: jean-didier.legat@uclouvain.be

G. Nicolescu et al. (eds.), *Design Technology for Heterogeneous Embedded Systems*,
DOI 10.1007/978-94-007-1125-9_20, © Springer Science+Business Media B.V. 2012

Table 20.1 Comparison between consumer electronics and professional electronics

Consumer electronics	Professional electronics
High volumes	Low volumes
Standard products	High design count
Loosely defined requirements	Very precise requirements
Loosely coupled with other systems	Often integrated in very complex systems
Short development time & lifetime	Long development time, lifetime & support
Use dedicated SoC platforms	Use programmable or reconfigurable platforms
Weak or inexistent qualification & validation processes	Strict qualification & validation processes

They must sometimes stay operational during a long period of time and be maintained during several decades. It is not rare for such designs to be upgraded several times during their use. They are often integrated in complex systems. For instance, a radar subsystem in an airplane is integrated into the onboard avionics. Development times for professional electronics are long since they must meet very strict validation and qualification constraints. DO-254 and the DO-178B norms are two examples of such qualification constraints for avionics [29, 30]. DO-254 is applied to the hardware part of the design and DO-178B is applied to the software part of the design. These two norms impose to respect a strict development process with strict traceability requirements. The development of embedded systems for professional applications is quite different than the development of consumer electronics. Table 20.1 summarizes the differences between both domains.

Professional applications have a higher diversity of wireless communication standards than in consumer electronics since many user groups maintain their own standards. Moreover, some advanced communication systems, like in military telecommunication applications, have to support a higher level of interoperability. Supporting this diversity while providing professional communication systems with a higher functionality level is increasingly difficult and costly. The software-defined radio allows to define completely and dynamically a communication standard by software. This approach provides a versatile programmable solution that can be used to implement complete wireless communication applications. Furthermore, using a programmable platform simplifies the application development since it is done using software. SDR is thus a very promising solution for professional electronics since it allows to face the high standard count and allows to easily build complex and interoperable communication applications. SDR is also a generic solution that can be reused on a larger set of applications compared to conventional dedicated professional communication systems. This makes SDR a cost-attractive solution since non recurring engineering costs can be amortized on larger volumes.

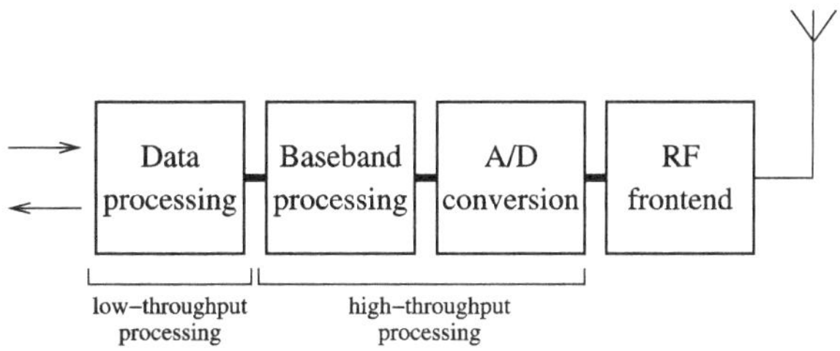

Fig. 20.1 Functional layers of a wireless communication system

1.1 Wireless Communication Systems

A wireless communication system is composed of four functional layers: the data processing layer, the baseband modulator, the analog/digital converter and the RF frontend. These layers are illustrated in Fig. 20.1.

- The data processing module handles operations carried on the transmitted data, like error correcting codes, cryptography, interleaving, symbol formation, etc. These operations generally require a low throughput.
- The baseband modulator performs the signal modulation. This is where the transmitted symbols get encoded in the waveform. It also handles the interpolation/decimation operations. This module typically works at a high data throughput.
- The analog/digital conversion translates modulated digital data in an analog signal and conversely. This module also works at a high throughput. It could require very high performance analog/digital converters capable of working at a very high frequency depending on the data rate.
- The radio frequency stage is in charge of the transmission and reception of the radio waves. It is composed of filters, mixers, power amplifiers, and an antenna. This layer has a very high data throughput and is implemented using analog circuits.

1.2 Evolution of Wireless Communication Systems

Many different communication standards have emerged and nowadays many wireless communication systems have to handle several of those. Moreover, those standards are evolving and are regularly updated to include more functionalities or to improve performances. In order to be able to implement those waveforms, wireless communication systems typically use dedicated solutions, which have been specifically designed for a targeted standard. This lessens greatly the ability to upgrade the system and the adaptability of those solutions.

Future standards will be even more versatile and flexible than they are today. In this context, development of dedicated solutions will become too complex to be viable. For instance, telecommunication roadmaps foresee new types of communication systems like the cognitive radio and the intelligent radio. The cognitive radio is

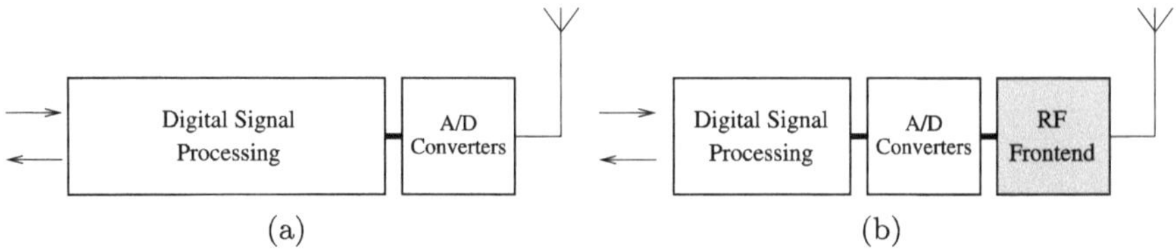

Fig. 20.2 (**a**) Ideal SDR modulation chain. (**b**) Actual SDR modulation chain

a communication system able to observe its environment and to adapt itself in order to reach better performances. The intelligent radio is a cognitive radio which is able to guess itself the best adaptation to realize to optimize the communication quality. This kind of communication system could use high-level mechanisms like artificial intelligence to perform these optimization choices. These future applications motivate research and development of new solutions in order to fulfill the requirements in flexibility and performances of future wireless communication systems.

In radios, RF frontends are designed for a single radio band. Building RF frontends capable of working on different bands is a challenge since the performances quickly degrade beyond a given frequency range. Reconfigurable RF frontends are a promising solution. They use tunable components to enable adaptation of a single RF transmission/reception chain to several bands. Some of them use MEMS [5, 6].

2 The Software-Defined Radio

A software-defined radio (SDR) is a wireless communication system where most or all the functional layers are defined by software [22]. The ideal SDR would have digital/analog converters right after the antenna in the modulation chain (as shown in Fig. 20.2(a)), so that all signal transformations could be performed digitally. Although this approach could be adopted for some low-frequency radio signals (e.g.: HF or VHF), it cannot be used for high frequency carrier due to analog/digital converters bandwidth or power consumption limitations. An intermediate stage is thus generally placed between the converters and the antenna (as in Fig. 20.2(b)) to move the modulated signal between a high carrier frequency and a lower baseband frequency where analog/digital and digital/analog conversions are feasible. This solution reduces the flexibility of the SDR approach however.

The SDR approach allows to implement different communication standards on a single programmable platform. The platform allows to build a generic and reusable wireless communication system, capable of implementing several communication standards, and moreover, capable of supporting their evolutions. Using software also allows to decrease development time of new communication applications. A generic solution allows to produce the platform in higher volume since it can be reused for several applications. As such, software-defined radio is an anticipated solution, both in the professional and the consumer electronics domains. Nevertheless, since it uses a programmable solution, it has to face a power efficiency challenge due to the power consumption overhead introduced by the software-based computation compared to a dedicated hardware implementation.

2.1 SDR for Professional Electronics

The SDR is a solution with a real interest for consumer and professional electronics. Both markets are however very different, and SDR solutions for each market do not have to meet the same requirements.

In the consumer market, telecommunication applications are defined by a set of standards published by normalization organizations such as IEEE or ISO. The challenge in this field consists in producing ready-to-use solutions integrating several of these standards in one platform. They are currently built by integrating several dedicated IPs in a system-on-chip. This approach requires a large amount of silicon area. In this context, SDR is a very interesting approach to face the growing amount of standards since those platforms makes a better utilization of the silicon area, which reduces costs. SDR also allows to reduce development time since it uses software.

Regarding professional embedded electronics, there are many waveforms in use since each group of users uses and maintains its own standards. Some standards are even kept secret for strategic reasons, as it is the case for military waveforms for instance. This makes the development of ready-to-use multi-standard solutions as it is done for the consumer market non-viable or impossible. The objective for SDR solutions targeting professional electronics is thus different. Professional SDR systems must provide highly flexible platforms that allow customers or a third party to build their custom telecommunication applications while still providing good power efficiency. This makes the development of SDR solutions for the professional embedded market considerably more difficult, and even more when considering strict validation of its functionality.

2.2 The Software Communication Architecture

In telecommunication applications, the waveform refers to how a signal is shaped but is generally extended to all the physical modulation chain [42]. It is specified by several characteristics, including carrier frequency, symbol generation, channel equalization, interpolation/decimation, frequency-domain transformations and filtering. Sometimes, error correction codes and encryption are also included in this specification. Communication systems are initially designed to use a particular waveform. For instance, military tactical terrestrial, air-to-ground and air-to-air radios are implemented by dedicated communication systems, each of them using a different waveforms. This allows to fit their particular functional requirements but does not allow direct interoperability between those systems. For example, the tactical terrestrial radio cannot directly communicate with an airplane. To improve the

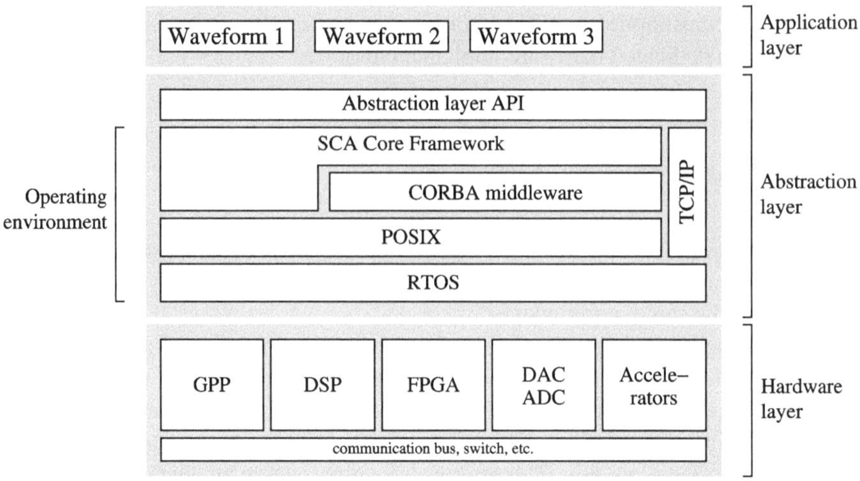

Fig. 20.3 SCA framework layers

coordination between actors in the field, there is an important need for a better interoperability. Regarding telecommunication platforms, interoperability leads to the following requirements:

- a single platform must handle several waveforms;
- a given waveform has to run on all interoperable platforms.

An advantage of SDR is to decouple the waveform definition from its hardware implementation by defining it in software. This paradigm eases the interoperability amongst software-defined communication systems. However, as they generally use different hardware solutions, a common infrastructure must be used on different hardware configurations. It is a common interface between the hardware and software parts of the radio. This common infrastructure also allows to take benefit from third-party telecommunication software providers to implement custom waveforms.

An influential architecture framework in the SDR domain targeting those issues is the Software Communications Architecture (SCA), which is a part of the Joint Tactical Radio System (JTRS) project [9]. The JTRS aims to define the next generation U.S. military radio system. The SCA is nevertheless not limited to military use. The objective of the SCA is to create an architecture framework allowing to perform a separation between the definition of a waveform and the radio platform implementing it. The SCA aims to enable implementation of software-defined waveforms on heterogeneous platforms, composed of several different components like GPPs, DSPs, or FPGAs.

SCA is mainly composed of 3 layers (as shown in Fig. 20.3):

- the application layer is where the communication applications live. Waveforms are described here. Higher application functionalities like protocols and communication system services are also described at this level;

- the abstraction layer defines a standard operating system for the high-level applications of the application layer. This is where most platform-level features (like interprocess communications, real-time tasks) are defined. The abstraction layer is the interface between the waveform application and the actual platform implementation;
- the hardware layer is the hardware platform on which the whole system is based. It could be composed of one or several processors, which have different architectures, and communicate through a set of communication structures like busses and network-on-chips.

The SCA defines an operating environment composed of several components. The SCA Core Framework is one of them, it provides tools to handle waveforms. It allows to perform control operations on the distributed objects composing the communication application. The operating environment also contains a CORBA middleware. This middleware allows communications and interactions between heterogeneous application objects by decoupling them from their actual implementation. Finally, the operating environment also provides a POSIX compatible real-time operating system, for task execution scheduling and handling.

The SCA Core framework is strongly influential to other initiatives in the domain, and holds a reference position for other frameworks dedicated to software-defined radio systems.

2.3 Hardware Platforms for SDR

SDR hardware platforms for the professional market raise several challenges. They must have a very high flexibility while still providing very high power efficiency and being affordable at low volumes. Existing SDR platforms for professional electronic applications typically use a combination of a general purpose processor (GPP) and a digital signal processor (DSP). The GPP is then in charge of the applicative part of the service and the DSP handles the high computational part. However those solutions built upon standalone components have a low energy efficiency due to the lack of integration. Computation throughput of standalone DSPs cannot meet future requirements which foresee high data rate. Furthermore the implementation of more functional layers in software raises new challenges.

Recently proposed platforms in consumer electronics use a combination of a GPP and a programmable accelerator, e.g.: ADRES [4], Montium [27], Pact XPP [23]. Nevertheless, there are no components directly available for professional electronic applications and their specific constraints. Another approach is to use a combination of a GPP and a reconfigurable platform, like an FPGA. This approach has several advantages: FPGA designs can reach a high throughput, they have a good flexibility thanks to their reconfigurability, are available for low volumes and for extended operation conditions ranges. Professional applications generally need to specialize certain parts of a modulation to implement their specific waveform. A reconfigurable platform has several advantages for such requirements since it allows to implement the specific custom functionalities efficiently in the reconfigurable logic of the FPGA.

However FPGAs are also known to show a very weak abstraction of their programmable logic and typically require long development times. Large FPGA components have also a high power consumption, which can prevent their use in applications where power consumption must be kept very low.

3 Dynamic Reconfiguration of FPGA

Dynamic partial reconfiguration (DPR) consists in modifying dynamically a part of a design configured in an FPGA without service interruption. In a design containing several functional blocks, DPR enables to replace one or several functional blocks without stopping the functions running in the other blocks. This technique allows to modify the functionalities implemented in a system during its use.

DPR can reduce constraints on resources since unused hardware can be reassigned to other tasks dynamically. It also allows a system to reach a higher flexibility and adaptability level since the functionality of the platform can be modified dynamically to match more closely the desired performances or functions. DPR has been widely studied in academia during the past years, and it has been shown that it could offer interesting opportunities for FPGA designs [2, 3, 25, 33, 35, 36].

At present, the only supplier of FPGAs capable of DPR with a large amount of reconfigurable logic is Xilinx with its Virtex serie (in versions Virtex II and above) [40]. Most of the concepts presented here are therefore related to this serie of devices. Other FPGA models are known to support dynamic reconfiguration as well, for instance the Atmel AT40K can also perform DPR [1], but has far less resources.

3.1 FPGA Configuration

FPGA reconfigurable logic elements are configured by the data stored in a configuration memory. The content of this memory defines the content of the lookup tables and memory blocks, and configures routing resources in the device. Other components, like DSP blocks, high speed IOs controllers, or digital clock managers are also configured with the content of this memory. The configuration memory is modified through a configuration chain, which is illustrated in Fig. 20.4 for Virtex FPGAs.

The first element of this chain is the configuration bitstream. This bitstream is a binary file composed of configuration data. For dynamically reconfigurable FPGAs from Xilinx, the configuration bitstream is organized in packets. Each packet contains a header and a payload. The header specifies the target address in the configuration memory for the configuration data stored in the payload. This payload is composed of configuration frames, which are the smallest amount of configuration data that can be written in the FPGA configuration memory.

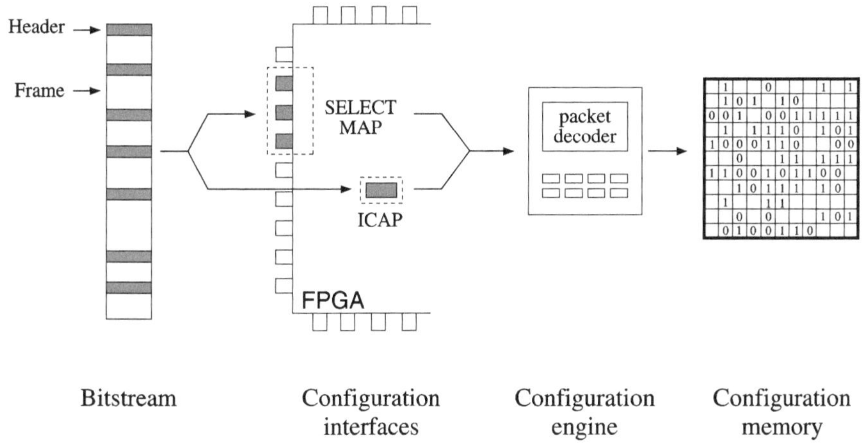

Fig. 20.4 Reconfiguration chain for a Virtex FPGA

During configuration of the FPGA, a bitstream is written to a configuration port using its specific protocol. These configuration ports could be external or internal. In Xilinx FPGAs, the internal configuration port is called ICAP, for Internal Configuration Access Port. The bitstream written in the configuration port is then interpreted by a reconfiguration engine which has access to the configuration memory and rewrites it. There is therefore no direct access to the configuration memory from the user logic.

A dynamic partial reconfiguration of an FPGA is performed by sending a bitstream containing only some configuration frames to a configuration port. This kind of bitstream is called a partial bitstream. Only part of the configuration memory is then rewritten and therefore, only a part of the design is modified. The unmodified hardware can continue to work without any interruption during reconfiguration, which allows to maintain platform services.

3.2 *Design for DPR*

To use dynamic partial reconfiguration, an FPGA design must follow specific design rules. First, the design must be partitioned in reconfigurable regions and non-reconfigurable regions. In Xilinx terminology, the non reconfigurable region is called the static region. This part of the design will not be altered by a partial reconfiguration of the FPGA. The other regions are called partial reconfigurable regions (PRR). There must be a separated region for each part of the design that will be reconfigured individually. Regions generally match the functional block partitioning.

Special constraints must be given to the placement and routing tools to tell them to keep resources from a block strictly inside their region. Dividing the design in PRRs and a static region allows to identify the configuration frames needed to compose a partial bitstream.

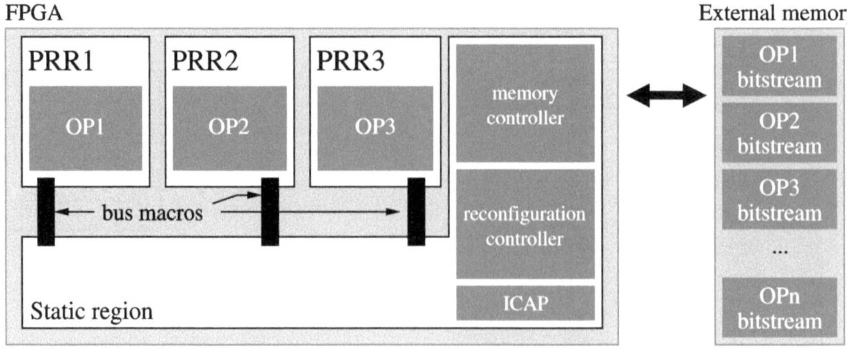

Fig. 20.5 A DPR-enabled design with a static region, several partial reconfigurable regions, and access to partial bitstreams in memory

The second rule to observe to enable DPR in an FPGA design is to use specific communication structures to communicate with PRRs. Those communication structures are called bus macros. Bus macros are pre-routed structures of the FPGA that implements a communication link. Since they are pre-routed, bus macros must be generated once for each bus configuration. They must be instantiated in the design for each link between a PRR and another module. Using a pre-routed structure allows to assure that, even if the logic of a PRR is modified, the logic composing the link will stay the same after reconfiguration, and its functionality will be preserved.

Figure 20.5 illustrates a design enabled for DPR: the design is partitioned into a static region and three PRRs. Communications between the PRR and the static region are made through bus macros. All the necessary hardware to perform reconfiguration is also present: a reconfiguration controller connected to the internal configuration port, and a memory controller to access bitstreams on an external memory.

3.3 Evolution of DPR in the Latest Components

DPR was introduced in the Virtex family with the Virtex II component. Its layout was organized in columns, defining an entire column of the component as the smallest configuration granularity. Therefore, the DPR had to cope with severe hardware constraints that lessened its use.

Since the Virtex 4 component, several improvements facilitate the use of DPR, making it a viable solution for some specific applications like SDR. The improvements for DPR have been made on the layout architecture, the clock tree and the ICAP. Virtex 4 components and the next generations are still organized in columns but the partial reconfiguration granularity is reduced here to only a part of a column. Therefore, PRRs can be almost any height and width. Since clock regions are now also rectangular, it allows to match them with a PRR.

Moreover, the output frequency and phase shift of the digital clock managers can also be modified using DPR. Finally, the width of the ICAP port has been extended to 32 bits and its speed has been increased up to 100 MHz. This significantly speeds up the reconfiguration time, which is a main concern when using DPR since the hardware in the PRR cannot be used during the reconfiguration process. In the next generations of Virtex FPGAs (namely, the Virtex-5 and the Virtex-6), the same structures regarding reconfiguration have been preserved, allowing to easily port applications to new devices.

3.4 Advantages of DPR Usage in Professional Electronics Applications

The DPR allows using more hardware over time than physically present in the FPGA. This can be used to reduce the required size of the FPGA and its power consumption. This also allows to execute an algorithm with an optimized implementation depending on its parameters and data set. Furthermore, upon the usual speed and power research goals, DPR offers system-level advantages for professional electronics [19]. Such advantages are:

- Task speed: execution time can be reduced by adopting faster hardware. More functions can be implemented in hardware with less constraints on the size of the component.
- Power reduction: power consumption could be reduced by having less hardware instantiated and running, and by using a smaller FPGA to reduce leakages. However, FPGA reconfiguration introduces additional energy consumption depending on the reconfiguration frequency and the bitstream size.
- Survivability: by allowing reconfiguration of the system in a degraded but safe mode when a part of it is damaged. It is necessary for applications running in harsh environments where environmental conditions can exceed the normal operating range.
- Mission change: DPR enables reconfiguration of an application for an entire mission without interrupting services. This is crucial for system where the real-time issue is not critical, but the interruption issue is. DPR provides an easy and safe way to strongly modify an entire system without having the complexity of implementing all the functionalities in one design.
- Environment change: during operation, the application can be developed specifically for several environments and switch dynamically.
- Adaptive algorithm change: DPR allows to adapt dynamically an algorithm depending on the external conditions. Compared to environmental changes, this aspect has a lower granularity, since it is performed at algorithm level.
- On-line system test: a system in harsh environment can be damaged. DPR can be used here to temporarily instantiate a testing module to evaluate system functionality level.

- Hardware virtualization: DPR allows to have more hardware available than physically present in the FPGA. It allows to manage a set of hardware modules as a component library. This is a key advantage for SDR applications.

4 Dynamic Partial Reconfiguration for SDR

DPR is a promising solution for professional SDR since this application requires high performances and high flexibility. SDR naturally benefits from the general advantages of DPR: reduction on hardware constraints, optimized hardware functions, and higher flexibility.

Hardware virtualization is a very important advantage that DPR brings to software-defined communication systems. Thanks to this feature, SDR is foreseen by Xilinx as the main target applications for DPR [16]. Being able to manage hardware implementations of specific functions in the same way as a software library opens the perspective for efficient, yet flexible, SDR platforms implemented on FPGAs [7, 41]. Dedicated hardware implementing a specific operator could be instantiated dynamically in an FPGA to provide the functionality with the targeted performance. These implementations can be stored in memory as a library of hardware operators and fetched when needed by the reconfiguration manager.

Several FPGA-based platforms capable of DPR have been developed to build a SDR platform. It is the case for instance of a platform from ISR technologies [15], which used a Virtex-II FPGA [34]. Another platform, from Thales, is also leveraging DPR. This solution has been shown to be able to reach the requirements of SDR applications, and to provide interesting system-level advantages [21, 31].

5 Impacts of DPR

Hardware platform in a design has traditionally been considered as being statically defined before runtime and kept constant for the whole duration of the service. Dynamic hardware is thus a new concept, which is transverse to the design abstraction layers since its usage has repercussions from the bit level up to the system level. As a consequence, usage of DPR in an FPGA design has many impacts on several levels. In this section, some of them are described.

5.1 Impacts on the Design

In order to use DPR, usual FPGA design techniques must be adapted. DPR introduces new design rules, new interfaces and protocols, and new constraints. This makes FPGA design for DPR more complex than traditional FPGA designs.

The use of DPR requires to follow specific design rules. The design must be partitioned into regions, and those regions must communicate between each other using pre-routed design structures called bus macros. Partitioning is realized by giving placement constraints to the placement and routing tools. Design partitioning can actually be compatible with validation and qualification since testing is easier when functional modules are clearly separated.

The reconfiguration hardware is vendor-specific since there is no standard reconfiguration module and no standard protocol to interact with it. As such, a specific module must be developed for each version of the reconfiguration port. This is an important issue since it implies that different versions of a reconfiguration module must be maintained. Another issue comes from the fact that different versions sometimes present very different performances. For instance, the first version of the Xilinx ICAP (in Virtex-II) were only capable of writing 8-bit words at 66 MHz, while latest ICAP version can write 32-bit words at 100 MHz.

Storage and access to configuration bitstreams also implies to adopt a specific memory hierarchy. Bitstreams can be large: a bitstream containing the complete configuration of an FPGA can go up to several megabytes. Typical partial bitstreams weight several hundreds of kilobytes. Large memories are thus needed to store them, and those must be external since internal memory blocks are too small. Access to those memories must be carefully designed in order to reach real-time and power consumption requirements.

DPR introduces new needs in terms of development: a configuration port controller and a configuration scheduler are needed. The configuration port controller allows bitstream transfers from a memory to the configuration port at full speed (for instance, by using DMA transfers). A scheduler must also be implemented to manage reconfigurations and perform bitstream transfers from memory to the configuration port controller. It can be implemented in software, running on an embedded soft-core or hard-core CPU.

5.2 Impacts on the Development Flow

FPGA development flows are composed of several steps which allows to progress in a top-down approach from the specifications to the actual implementation of the design. For each step of this flow, traditional FPGA development has mature tools providing efficient solutions. DPR has potentially an influence on every level of a system. When it comes to use DPR in a design, those development flow steps miss some important elements. Existing development flows from the vendors [38] do not provide all the necessary tools. There is no standardized toolset for DPR development [17].

In particular, there is at present no standard tools allowing to model dynamic reconfiguration in a design. However, modeling tools, such as SystemC [12, 26] could provide a solution for these needs. Some works have already studied the possibility of using it [10, 11, 14, 32]. There is also no behavioral model of DPR, capable of modeling delays and latencies of a reconfiguration. This prevents the possibility to simulate correctly the behavior and functionality of a design using DPR with an integrated toolchain. This makes also difficult to test a design using DPR before it is fully implemented on the device. As a consequence, developers using DPR must complete the implementation and integration of the system before being able to test it. This forces developers to jump from development to integration without having processed to complete validation. The lack of validation is a major issue which can make development with DPR infeasible or inefficient and difficult for professional uses where a design must be strictly validated at each step of the work [28].

5.3 Impacts on Power Consumption

DPR allows to adapt the hardware in order to provide the best implementation of a required function to perform a task. This brings additional degrees of freedom allowing to reduce power consumption. Typically, static power could be reduced by adopting a smaller FPGA, something that DPR makes possible since it allows to have virtually more hardware than in reality. Some works have already studied this aspect of DPR, and have shown that DPR could provide new perspectives for dynamic and static power reduction [24].

However power consumption reduction is worsened by the need to transfer large bitstream files from memories. Since bitstreams are large, those memories are generally external and accessing them consumes a significant amount of power. Figure 20.6(a) shows the repartition of energy during the reconfiguration of an FPGA. Measures have been made on a Virtex 4 FPGA, manufactured in 90 nm technology, with a supply voltage of 1.2 V [20]. Reconfigurations were performed with a partial bitstream of 500 KB stored in external memories: a 16-bit DDR2 SDRAM and a 16-bit flash memory. The leading part of the total energy consumed by the FPGA during reconfiguration is due to the static part of the design (shown in dark grey), which continues to operate while a part of the device is reconfigured. It is however not involved in the reconfiguration process. Regarding the energy related to the reconfiguration itself, we can see that the energy due to the reading of the bitstream in the external memories is more important than the energy consumed by the reconfiguration logic itself. Even for DDR SDRAM where it is lower, the consumed energy is about 8.95$\times$ higher. Note that when using external DDR memories, a bitstream should still be transfered from flash and stored in DDR since this memory can only be used as a temporary cache memory.

For applications with strong real-time constraints, the problem is even worse, since high-speed memories are then necessary to meet real-time requirements. A specific memory hierarchy would then be implemented in order to make sure that the bitstream is transferred with the full available bandwidth. For this, a dedicated memory bus is required for reconfiguration. Specific memories such as zero bus turnaround (ZBT) memory are used to reach high data transfer rates. Nevertheless, since those memories are relatively small, they can only store one or few

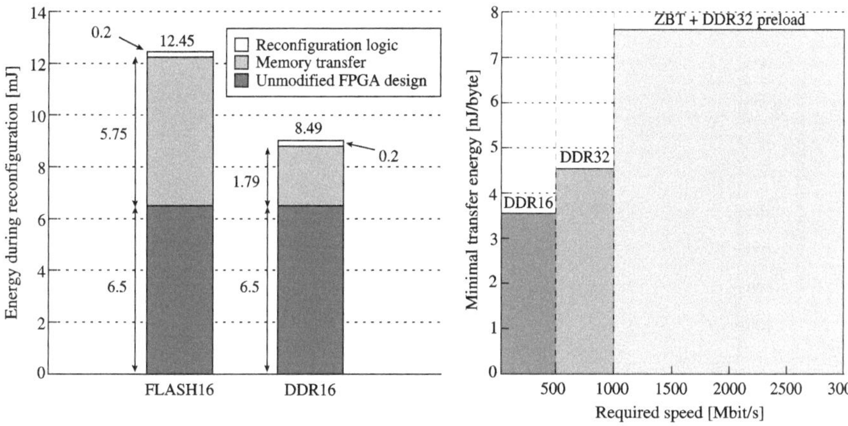

Fig. 20.6 (**a**) Energy consumed by the FPGA during reconfiguration. (**b**) Cumulative energy of the memory hierarchy during bitstream transfers

bitstreams. They require to transfer previously a bitstream from a slower but bigger memory to the ZBT memory. As a consequence, more power is consumed in transfers. Figure 20.6(b) shows an evaluation of the cumulative power consumption for different memory hierarchy configurations [8]. Each configuration allows to reach a specific data rate range. Faster reconfigurations consume more power due to memory transfers.

5.4 Impacts on Virtualization

DPR provides true hardware virtualization since it enables to store a set of hardware functions in memory. This allows to use hardware functions in the same way as a software library. However handling abstracted reconfigurable hardware is still a complicated task because of the transversal nature of dynamic reconfiguration. Some functions could possibly not be ready when the system needs it, and the system would then need to handle the reconfiguration of its hardware. A dynamic hardware manager is thus required to handle reconfiguration and provide information about the availability of a specific function to the reconfiguration scheduler situated in the middleware or operating system. It has to provide information on reconfiguration delay and latencies as well. There is at present no standard allowing to take care of this issue. This lack of tools to manage and model dynamic hardware makes this task difficult.

A solution consists in using a programmable controller in front of a reconfigurable region in order to handle the low-level layers of the virtualization. Such a controller locally checks the status of its reconfigurable region and request for a reconfiguration if needed. Figure 20.7 compares the hardware complexity of a microcontroller with two hardware functional blocks used in OFDM modulation [18].

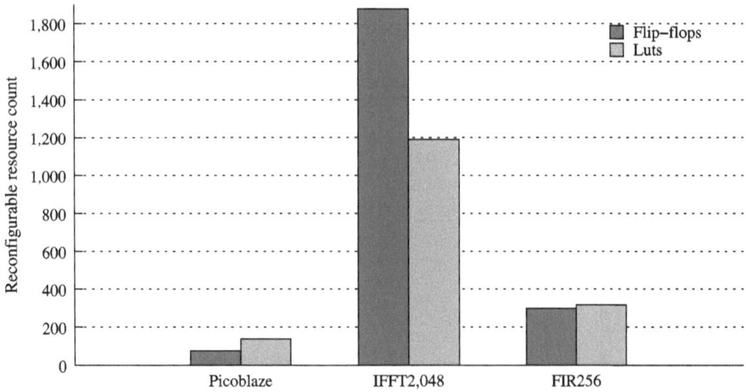

Fig. 20.7 Comparison of reconfigurable resources needed for the implementation of a Picoblaze microcontroller and hardware OFDM functional blocks in an FPGA

The microcontroller used for the comparison is a Picoblaze soft-core microcontroller from Xilinx [39]. The functional blocks have been generated using Xilinx CoreGen hardware block generator [37]. In the figure, one can see that the resource count needed to implement a microcontroller in an FPGA is small compared to the resources needed to implement typical functional blocks. Using a microcontroller to virtualize a reconfigurable region causes thus only a small resource consumption overhead in comparison to the complexity of the functional blocks implemented in this region. Such as solution is thus interesting to provide an intermediate level of virtualization.

6 Conclusions

Professional electronic applications have many specificities compared to consumer electronics. They target specific applications and their low production volumes do not allow to build dedicated platforms. Many of them must comply to strict validation and qualification rules. Moreover, highly flexible and yet efficient platforms are required to reach the requirements of their challenging applications.

Software-defined radio is the evolution of the traditional single-standard radio. It allows to define communication waveforms entirely in software. SDR brings a higher adaptability to communication systems and allows them to be used for different standards. Moreover, SDR allows to address interoperability challenges. The great diversity of professional electronics telecommunication standards requires to provide a very high level of customization of those SDR solutions. It makes the development of SDR platforms a challenge even more complex than for the consumer market.

Current professional SDR platforms are based on standalone components and are therefore inefficient. Future requirements cannot be reached by those traditional solutions associating GPP for the high-level part of the application and a DSP for the high-throughput part. New solutions must be found that enables to keep a sufficient flexibility while presenting improved performances and being available at low volumes.

FPGAs are interesting platforms to fulfill these needs: they can reach a high throughput, they offer configurable logic for custom functionalities required by advanced applications, and they are accessible even for low production volumes. Moreover, a new technique called dynamic partial reconfiguration allows to reconfigure a part of a design without interrupting the services. Beside the possibility to better manage logic resources, DPR brings many advantages at the system level. It allows to implement complete and adaptive systems in an FPGA. Regarding SDR, DPR allows to virtualize hardware and to manage hardware resources like a software library, a feature which is of real interest for a highly versatile and demanding application as SDR. However DPR is a solution that has many impacts. Using DPR on an FPGA has consequences on design, development flows, power and virtualization.

This chapter has provided an overview of the possibilities that DPR can bring to professional electronic applications like SDR, and has pointed at the different aspects that are affected when a system designer decides to make use of DPR. It has shown that DPR is a promising solution that brings potentially very interesting advantages. However this technique still suffers from a lack of model and tools. In the future, efforts should be done in those fields in order to make a mature solution from DPR.

Acknowledgements Bertrand Rousseau holds a F.R.S.-FNRS fellowship (Belgian Fund for Scientific Research). Philippe Manet, Thibault Delavallée and Igor Loiselle are funded by the Walloon region of Belgium.

References

1. Atmel AT40K data sheet, http://www.atmel.com
2. Alderighi, M., Casini, F., D'Angelo, S., Mancini, M., Pastore, S., Sechi, G.R.: Evaluation of single event upset mitigation schemes for SRAM based FPGAs using the FLIPPER fault injection platform. In: 22nd IEEE International Symposium on Defect and Fault-Tolerance in VLSI Systems, DFT '07, pp. 105–113 (2007)
3. Becker, J., Donlin, A., Huebner, M.: New tool support and architectures in adaptive reconfigurable computing. In: IFIP International Conference on Very Large Scale Integration, VLSI–SoC 2007, pp. 134–139 (2007)
4. Bougard, B., De Sutter, B., Verkest, D., Van der Perre, L., Lauwereins, R.: A coarse-grained array accelerator for software-defined radio baseband processing. IEEE MICRO **28**(4), 41–50 (2008)
5. Brown, E.R.: RF-MEMS switches for reconfigurable integrated circuits. IEEE Trans. Microw. Theory Tech. **46**(11), 1868–1880 (2008)
6. Craninckx, J., Liu, M., Hauspie, D., Giannini, V., Kim, T., Lee, J., Libois, M., Debaillie, D., Soens, C., Ingels, M., Baschirotto, A., Van Driessche, J., Van der Perre, L., Vanbekbergen, P.: A fully reconfigurable software-defined radio transceiver in 0.13 μm CMOS. In: IEEE International Solid-State Circuits Conference, Digest of Technical Papers, ISSCC 2007, 11–15 Feb. 2007, pp. 346–607 (2007)

7. Delahaye, J.-P., Palicot, J., Moy, C., Leray, P.: Partial reconfiguration of FPGAs for dynamical reconfiguration of a software radio platform. In: Mobile and Wireless Communications Summit (2007)
8. Delavallée, T., Rousseau, B., Manet, P., Vandierendonck, H., Legat, J.D.: Modeling the impact of bitstreams transfer in dynamically reconfigurable platforms. In: Faible Tension Faible Consommation (FTFC08) (2008)
9. Joint Program Executive Office (JPEO) JTRS, Software Communication Architecture specifications v2.2.2. http://sca.jpeojtrs.mil
10. Gailliard, G., Mercier, B., Sarlotte, M., Candaele, B., Verdier, F.: Towards a SystemC TLM based methodology for platform design and IP reuse: application to software defined radio. In: Proc. of the Second European Workshop on Reconfigurable Communication-Centric SoCs, RECOSOC (2006)
11. Gailliard, G., Nicollet, E., Sarlotte, M.: Transaction level modelling of SCA compliant software defined radio waveforms and platforms PIM/PSM. In: DATE (2007)
12. Ghenassia, F.: Transaction-Level Modeling with SystemC. Springer, New York (2005)
13. Hammes, M., Kranz, C., Seippel, D., Kissing, J., Leyk, A.: Evolution on SoC integration: GSM baseband-radio in 0.13 μm CMOS extended by fully integrated power management unit. IEEE J. Solid-State Circuits **43**(1), 236–245 (2008)
14. Herrholz, A., Oppenheimer, F., Hartmann, P.A., et al.: The ANDRES project: analysis and design of run-time reconfigurable, heterogeneous systems. In: Proceedings of the 17th International Conference on Field Programmable Logic and Applications (FPL'07), Amsterdam, The Netherlands, pp. 396–401 (2007)
15. ISR Technologies, Inc. http://www.isr-technologies.com/
16. Kao, C.: Benefits of partial reconfiguration. Xilinx Xcell J. **2005**(55), 65–67 (2005)
17. Lysaght, P., Blodget, B., Mason, J., Young, J., Bridgford, B.: Enhanced architectures, design methodologies and CAD tools for dynamic reconfiguration of Xilinx FPGAs. In: Proceedings of International Conference on Field Programmable Logic and Applications (FPL '06), Madrid, Spain, pp. 1–6 (2006)
18. Loiselle, I., Rousseau, B., Manet, P., Vandierendonck, H., Legat, J.D.: Virtualization analysis of a reconfigurable region for adaptive DSP and multimedia applications. In: Faible Tension Faible Consommation (FTFC08) (2008)
19. Manet, P., et al.: RECOPS: reconfiguring programmable devices for military hardware electronics. In: DATE (2007)
20. Manet, P., Rousseau, B., Kuti Lusala, A., Legat, J.D.: Opportunité d'utiliser la reconfiguration dynamique pour optimiser la consommation d'applications implémentées en FPGA. In: Faible Tension Faible Consommation (FTFC07) (2007)
21. Manet, P., Gailliard, G., Maufroid, D., Tosi, L., Mulertt, O., Di Ciano, M., Legat, J.-D., Aulagnier, D., Gamrat, C., Liberati, R., La Barba, V., Cuvelier, P., Gelineau, P., Rousseau, B.: An evaluation of dynamic partial reconfiguration for signal and image processing in professional electronics applications. EURASIP J. Embed. Syst. (2009). doi:10.1155/2008/367860
22. Mitola, J. III: Software radios-survey, critical evaluation and future directions. In: National Telesystems Conference, NTC-92, 19–20 May 1992, pp. 13/15–13/23 (1992)
23. PACT Inc: XPP-III processor overview (2006)
24. Paulsson, K., Hubner, M., Becker, J.: Exploitation of dynamic and partial hardware reconfiguration for on-line power/performance optimization. In: International Conference on Field Programmable Logic and Applications, FPL 2008, pp. 699–700 (2008)
25. Pionteck, T., Albrecht, C., Koch, R.: A dynamically reconfigurable packet-switched network-on-chip. In: Design, Automation and Test in Europe, Proceedings of DATE '06, vol. 1, p. 8 (2006)
26. Qu, Y., Tiensyrja, T., Soininen, J.-P.: SystemC-based design methodology for reconfigurable system-on-chip. In: Proceedings of the 8th Euromicro Conference on Digital System Design (DSD '05), Porto, Portugal, pp. 364–371 (2005)
27. Rauwerda, Gerard K., Heysters, Paul M., Smit, Gerard J.M.: An OFDM Receiver Implemented on the Coarse-grain Reconfigurable Montium Processor. In: 9th International OFDM-Workshop, InOWo, Dresden, Germany, 15–16 September (2004)

28. Rousseau, B., Manet, P., Galerin, D., Merkenbreack, D., Legat, J.D., Dedeken, F., Gabriel, Y.: Enabling certification for dynamic partial reconfiguration using a minimal flow. In: Proceedings of the Conference on Design, Automation and Test in Europe, DATE 2007 (2007)
29. RTCA DO-178B/EUROCAE ED-12B, Software considerations in airborne systems and equipment certification (1992)
30. RTCA DO-254/EUROCAE ED-80, Design assurance guidance for airborne electronic hardware (2000)
31. Sarlotte, M., Counil, B., Gelineau, P., Chau, R., Maufroid, D.: Partial reconfiguration concept in a SCA approach. In: SDR Forum Technical Conference (2007)
32. Schallenberg, A., Nebel, A., Oppenheimer, F.: OSSS+R: modelling and simulating self-reconfigurable systems. In: Proceedings of the 16th International Conference on Field Programmable Logic and Applications (FPL '06), Madrid, Spain, pp. 1–6 (2006)
33. Sedcole, P., Blodget, B., Anderson, J., Lysaghi, P., Becker, T.: Modular partial reconfigurable in Virtex FPGAs. In: International Conference on Field Programmable Logic and Applications (2005)
34. Uhm, M., Bezile, J.: Meeting software defined radio cost and power targets: making SDR feasible. In: Military Embedded Systems, pp. 7–8 (2005)
35. Wang, H., Delahaye, J.-P., Leray, P., Palicot, J.: Managing dynamic reconfiguration on MIMO Decoder. In: IEEE International Symposium on Parallel and Distributed Processing, IPDPS 2007 (2007)
36. Wu, K., Madsen, J.: Run-time dynamic reconfiguration: a reality check based on FPGA architectures from Xilinx. In: 23rd NORCHIP Conference, pp. 192–195 (2005)
37. Xilinx CoreGen, CORE generator 10.1 user guide (2008)
38. Xilinx XAPP290, Two flows for partial reconfiguration: module based or difference based (2004)
39. Xilinx XAPP627, PicoBlaze 8-bit microcontroller for Virtex-II series devices (2003)
40. Xilinx, Inc., http://www.xilinx.com
41. Xilinx SDR platform, http://www.xilinx.com/prs_rls/dsp/0626_sdr.htm
42. Zhang, Y., Dyer, S., Bulat, N.: Strategies and insights into SCA-compliant waveform application development. In: MILCOM 2006, pp. 1–7 (2006)

Chapter 21
Methods for the Design of Ultra-low Power Wireless Sensor Network Nodes

Jan Haase and Christoph Grimm

1 Introduction

Wireless Sensor Networks (WSN) are a very active research field. However, there is still a wide gap between the power consumption requirements of most applications and the actual power consumption of platforms and nodes. Depending on the WSN application, power consumption requirements may differ with respect to lifetime, the energy available, the kind of communication, security needs and other factors.

A major challenge is the architecture level optimization, including partitioning HW/SW or even HW/SW/analog. For power optimization at architecture level, both accurate and fast system simulation is needed, augmented with means for power estimation.

1.1 Related Work

In [22] SystemC AMS was used to simulate nodes of a WSN at architecture level, including analog components. In [3] SystemC was successfully used to co-simulate the hardware and software parts of a wireless sensor network. Some preliminary tests were made with a pure SystemC model with excellent results. However, the lack of support for modeling upper layer network communication in SystemC finally led to the usage of NS-2 [13] to model the network.

A quite natural approach to model communication in an abstract and therefore efficient way are the SystemC TLM extensions. TLM has already been used successfully for network simulation. In [24], a protocol for cooperative MIMO mobile

J. Haase (✉) · C. Grimm
Institute of Computer Technology, Vienna University of Technology, Gußhausstraße 27-29/E384, 1040 Wien, Austria
e-mail: haase@ict.tuwien.ac.at

C. Grimm
e-mail: grimm@ict.tuwien.ac.at

G. Nicolescu et al. (eds.), *Design Technology for Heterogeneous Embedded Systems*,
DOI 10.1007/978-94-007-1125-9_21,

sensor networks was proposed. To model such a large and complex network, TLM and SystemC was used, resulting in a very fast and efficient simulator that permitted to evaluate the network performance.

However, to compare power consumption of different node architectures and communication protocols, at least qualitative power estimation is required. Power estimation is an established technique for digital systems. Some estimation techniques have been proposed in [15]. Today, there are even commercial solutions to estimate power at the system level, e.g. PowerOpt (ORINOCO) [18].

WSN also include analog components and must consider network activity while running a communication protocol. This requires very efficient power estimation at functional and architecture level, including analog/RF components and software running on an MCU.

Power optimization of WSNs has been previously analyzed in the Power Aware Wireless Sensor (PAWiS) project [5, 23]. It is based on the OMNeT++ Simulation Environment [21] and abstracts both intra-node and inter-node communication as a number of modules that exchange messages among them. In fact, many of the approaches used in the PAWiS simulation framework are reused for the SNOPS framework, therefore we will describe it in more detail in Sect. 3.

The remainder of the chapter is organized as follows: In Sect. 2, we sketch typical challenges faced during optimization of WSN. After introducing the PAWiS simulation framework in Sect. 3, we describe a new sensor network simulation approach based on TLM in Sect. 4. The chapter then concludes in Sect. 5.

2 Challenges for WSN Power Optimization

The application scenario described in this chapter is an automotive WSN. The main constraint is energy consumption, as battery replacement is often not feasible, e.g. in tire pressure sensors. Another feature of this application scenario is that certain nodes may be connected to the car's internal bus (e.g. a CAN bus), such that we actually face a mixed wireless/wired sensor network.

In Fig. 21.1, a possible topology of a sensor network in a car with wired and wireless connections is shown. Since a multi-hop protocol is used here, some sensors would act as transition nodes. Thus the network perspective must also be taken into account.

Evaluation of architectures, techniques and protocols designed for energy saving is frequently done by virtual prototyping and simulation. Nevertheless, modeling wireless sensor networks and nodes involves the development of hardware, software, and network models. Co-simulation becomes necessary at the expense of simulation performance and interoperability issues.

In this chapter, we will use the ongoing SNOPS project as an example [2]. The core idea of SNOPS is the development of ultra low power reconfigurable sensor node building blocks, designed to relieve the Micro Controller Unit of some of its tasks. This strategy makes the hardware/software co-simulation even more important, since tasks are shifted between the hardware and the software part. There-

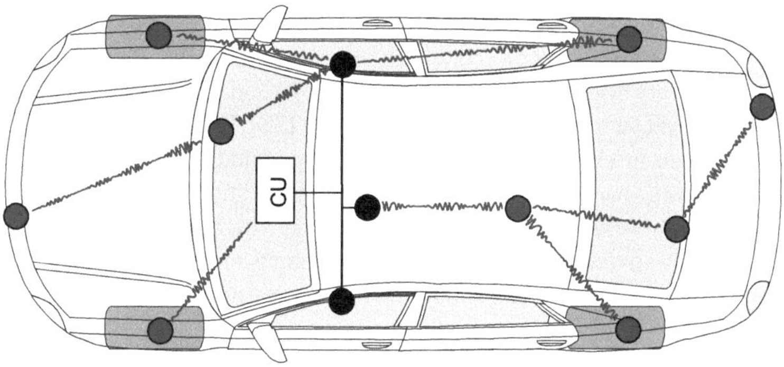

Fig. 21.1 A mixed wireless/wired sensor network in an automotive scenario with a central unit (CU) and an additional field bus

fore, new ways for simulation are explored utilizing SystemC and Transaction Level Modeling (TLM).

2.1 Requirements for WSN Power Optimization

The MCU is the most complex subsystem within the sensor node architecture. Therefore, it is the most difficult to model. It is usually modeled as a finite state machine. However, estimating the time the MCU requires for executing a certain task is far from trivial. In order to be able to compare the efficiency of different algorithms, a cycle-accurate model is preferable. An Instruction Set Simulator (ISS) provides this level of abstraction. This level of accuracy is achieved at the expense of reducing simulation speed.

The sensor subsystem and the transceiver can be modeled very accurately by a finite state machine. In this case, the time spent at each state is usually fixed by the software executed on the MCU and can be easily recorded. The same holds for the remaining subsystems, in which the power consumption is specified and the time they are active is easy to estimate.

Network behavior is another key point of the simulation which adds an order of complexity to the whole system. Nodes may not be equally important depending on the role they play within the network. Apart from the power consumed by its activity, some nodes may have to route messages or provide information about the network, depending on the implemented protocols. Also, due to our automotive application scenario, we might see the wireless network interacting with some bus based network, e.g. a CAN bus system.

Signal propagation is subjected to several effects: delay, noise and distortion. These propagation effects have to be taken into account in order to estimate when a signal is detected by a receiver and whether the message is correctly received or not.

Some parts of these effects are linear and time invariant and can be easily estimated. However, there are also other effects which are very difficult to model accurately.

Hence, there are four main simulation levels, according to power simulation. The software level, which requires the use of an ISS. The hardware level, which requires hardware models which can be usually modeled as finite state machines. The network level, that requires a topology and traffic model, in order to estimate which nodes are more likely to run out of energy and which protocols distribute load in an optimal way so that the overall network lifetime is maximized. And finally, the radio model, which models the channel and estimates the signal received on each node.

Summarizing, the simulation framework has to meet the following requirements:

- The simulation of the single sensor nodes has to provide a detail level that is sufficient for determining meaningful information on power consumption.
- It must be possible to simulate a mixed wireless/wired sensor network, where the wired part would be a field bus in the targeted application scenario.
- The simulation should also give the designer information on how a chosen protocol or certain protocol parameters might influence the overall power consumption as well as the power consumption of a chosen node.

3 The PAWiS Framework

In the PAWiS project [11] three items were addressed. For the optimization of wireless sensor network (WSN) nodes a simulation framework was developed. As an actual part of a WSN node a wakeup receiver (WuR) was designed and implemented [8, 19]. As a second module of the WSN node an energy management unit including a harvester was designed [6].

One of the main design goals of WSN nodes is an extremely low energy consumption. The optimization does not only require to reduce the consumption of every single part of the node but also to employ global optimization of the whole node including cross-layer optimization techniques to the network stack and optimization at a system or even network level [9]. To achieve a fast design-to-product cycle this optimization is performed with simulation of a virtual prototype of the WSN node. Since available simulation environments did not offer features like power simulation, modularization and the modeling of a wireless network, the PAWiS Simulation Framework [4, 5, 23] was developed.

The PAWiS Simulation Framework is based on the OMNeT++ [20] network simulation framework. To simulate and optimize WSN nodes, the node is split into modules which are modeled separately, one C++ class per module. These modules communicate via so called "*functional interfaces*" which are basically remote procedure calls and implemented as C++ methods. Every module can offer and implement multiple functional interfaces, which themselves can invoke functional interfaces of other modules. To provide this functionality the OMNeT++ classes are wrapped by PAWiS classes which are then the parent class for the modules.

WSN nodes typically consist of a microcontroller with an integrated CPU, memory and peripherals like an analog-to-digital converter (ADC), a real time clock (RTC), bus interfaces (I^2C, SPI, etc.) and digital IO. Additionally it requires a power supply, sensors, and an RF interface. All these components are connected physically and controlled by the firmware executed by the CPU. For an appropriate model of such nodes the PAWiS Simulation Framework offers to implement modules as hardware modules, software modules or mixed modules. The distinction is not really based on modules but on functional interfaces: these are either software or hardware.

Functional interfaces which model a software implementation have to consider two things. Firstly, they have to implement the functionality, simply as C++ code. Secondly, the surrounding effects, namely the execution delay and the power consumption, have to be considered. This is delegated to the central CPU model by using the `requireCpu(...,`*duration*`)` function. The *duration* parameter references the total execution time of the current functionality referred to the so called "norm CPU" [5] which is then converted to the execution time of the actually employed CPU. There is also a function `requireCpuUntil(...,`*predicate*`)` which allows to specify the end of processing with a predicate function.

Hardware functional interfaces have to implement their own timing and power consumption simulation. The functionality is still achieved with C++ code, while the timing is modeled with the `wait(...)`, `waitUntil(...)`, and `waitOrUntil(...)` functions.

The power consumption is implemented by specifying *reporters* which handle the V/I characteristic of the module. There are constant power ($I = I_{\text{const}}$), resistive power ($I = \frac{V}{R}$) and linear power ($I = I_{\text{const}} + \frac{V}{R}$) reporters available and the user can define his own custom reporters. These also consider the efficiency of the supply modules as well as the voltage drop across their output resistance.

To simulate wireless communication, the PAWiS Framework provides a model of the *environment*. A component of this environment, called *Air*, is also provided, containing typical parameters of wireless communication channels (attenuation, noise, etc.). It calculates the (received) RF power at all instantiated nodes depending on the distance from the transmitter, obstacles, etc. The effects of the obstacles are manually introduced, e.g. by specifying an additional attenuation factor between some pairs of nodes.

Wireless sensor networks are usually mobile. Furthermore, nodes may run out of energy and some conditions may change during the network operation. In order to simulate this dynamic behavior of the network, the PAWiS framework supports scripting with the embedded scripting engine Lua. Two main ways of binding Lua are supported. A first kind of scripts are executed during initialization time, called from the network configuration file. The second type of scripts have *module context* (they are assigned to a module) and are executed for every instance of the module type. Most important functionalities of Lua scripts are moving, grouping and creating sensor nodes, besides accessing and providing functional interfaces.

In the following paragraphs, a simple example of the use of the PAWiS framework is provided. The example in an implementation of a simple application about a blinking LED, as it is represented in Fig. 21.2.

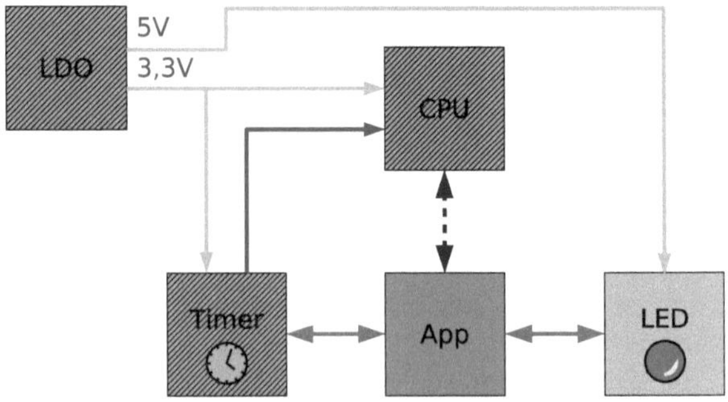

Fig. 21.2 Diagram of the LED example simulated with PAWiS

Through two of the modules (LED and application), whose implementations can be read in Listings 21.1 and 21.2, the main techniques and mechanisms provided by the PAWiS framework are introduced.

Module activities are separated in tasks. Every task is implemented as one class method. The execution within one task is sequential but different tasks may run concurrently. A task can be started in two different ways. Tasks that should "just run" from the beginning are started from the `onInit()` method (see Listing 21.1) using the `startTask(`*name*, *task method*, *params in*, *params out*`)` call. Note that this function can be used anywhere and anytime to create new tasks. Its first parameter is the method (using a macro to cast to the proper type). The second and third parameters are pointers to `ParameterList` objects for input and output parameters, respectively. A call to `startTask()` is always non-blocking which means that the started task runs in parallel and independently. The starter does not wait for any condition (e.g. finishing) of it.

The second kind of task creation is using functional interfaces. In the example shown in Listing 21.2 there are three methods. In the method `onStartup()` a functional interface named *set* is registered (note that it is not done in `onInit()`) with `registerFunctionalInterface(`*name*, *method*, *multiInvokable*`)`. The first parameter is a string with the name of the interface (referred to by other modules). The second parameter is a casted pointer to the task method. The third parameter tells the framework, if this functional interface is intended to be invoked multiple times in parallel. When set to `false` the framework does additional sanity checks to warn the user of multiple invocations.

Other modules can then invoke the functional interface with the method `invoke(`*module*, *interface*, *params in*, *params out*`)`.

During the `onInit()` method, a `PowerSourceAdapter` object is created. This is necessary to connect a `PowerReporter` to a `PowerSource`. This `PowerReporter` is then assigned to the `PowerSourceAdapter` using `setReporter()`.

Finally, the `set()` method is the implementation of the already registered *set* functional interface. It updates the LED current consumption value. If it is switched on, the forward voltage U_f is set to 2.2 V and the series resistor to 280 Ω. To switch off the LED its series resistor is set to ∞. This is a trick because we don't want to

Listing 21.1 Implementation of the PAWiS module "App" from `app.cpp` showing a task

```
#include "app.h"

Define_Module(App);

void App::onInit() {
  // init member variables
  // ...
  // register ISRs
  registerIsr(CAST_TASK(App::onTimer), 17, "timer_isr");
  // schedule initial task
  startTask("main",CAST_TASK(App::myMain),0,0);
}

void App::onTimer(TaskControl &pa_oControl) {
  // do our internal work
  m_bLedState = !m_bLedState;

  // set the LED
  pa_oControl.requireCpu(0.5, 0.0, 0.2, 0.3, 0.001f);
  pa_oControl.invoke("Led", "set", m_bLedState);

  // setup new timer event
  pa_oControl.requireCpu(0.5, 0.0, 0.3, 0.2, 0.005f);
  pa_oControl.invoke("Timer", "start", (unsigned int)(20));

  ++m_nCounter;
}

// predicate function for requireCpuOrUntil()
bool App::done() {
  // test an external condition,
  // e.g. ADC has finished conversion
  if (m_nCounter > 2) {
    return true;
  }
  return false;
}

void App::myMain(TaskControl &pa_oControl) {
  // ...
  // start the LED toggling
  onTimer(pa_oControl);
  // Before Start waiting 50ms or the predicate
  pa_oControl.requireCpuOrUntil(0.1f,0.0f,0.0f, 0.9f, 0.05,
    CAST_PREDICATE(App::done));
  // main program running infinitely
  while (true) {
    pa_oControl.setCpuState(
      new SimpleCpu::CpuState(SimpleCpu::CpuState::pmSleep));
    pa_oControl.requireCpu(/* ... */, 1.0e-6);
  }
}
```

Listing 21.2 Class implementation of the PAWiS module "Led" from `led.cpp`

```
/* ... */

void Led::onStartup() {
  registerFunctionalInterface("set", CAST_TASK(Led::set), false);
}

void Led::onInit() {
  // Create power source adapter and reporter
  m_pPowerAdapter.reset(
    new PowerSourceAdapter(*getContext().findPowerSource("ldo"),
      LdoPowerSource::src5V, 0, true));
  m_pPowerAdapter->setReporter( new LedReporter(*this));
  // Initial consumption
  m_pPowerAdapter->
    getReporter<LedReporter>()->set(0.0f, /* infinity */);
}

void Led::set(TaskControl &pa_oControl) {
  m_bState = pa_oControl.inParams()->
    get<bool>(0);
  if (m_bState) {
    m_pPowerAdapter->
      getReporter<LedReporter>()->set(2.2f, 280.0f);
    setFillColor("#40ff40");
  } else {
    m_pPowerAdapter->
      getReporter<LedReporter>()->set(0.0f, /* infinity */);
    setFillColor("#206020");
  }
}
```

simulate a switch. Note that the method `set()` immediately exits and assumes that the power consumption continues.

Module Library Within the PAWiS project, the framework has been supplemented with an interface specification and a module library, which contains the implementation of simple and common protocols used in WSNs. The interface specification defines the network layers and their interfaces. It proposes a cross-layer communication model [7] in order to optimize and enhance the ordinary network protocol stack. As it can be seen in Fig. 21.3, besides the typical network layers (physical, MAC, routing, transport and application), additional cross-layer planes are defined: the energy management plane, the security plane, the node management and the cross-layer management plane. Specifying the interfaces permits to easily develop a module library in which each module can be easily added, replaced or removed.

According to the interface specification, a module library has been created, which implements this model with simple protocols, protocols specifically developed for WSNs (EADV [12], CSMA-MPS [10]) and common implementations used in practice (CC2400, Zigbee, S-MAC).

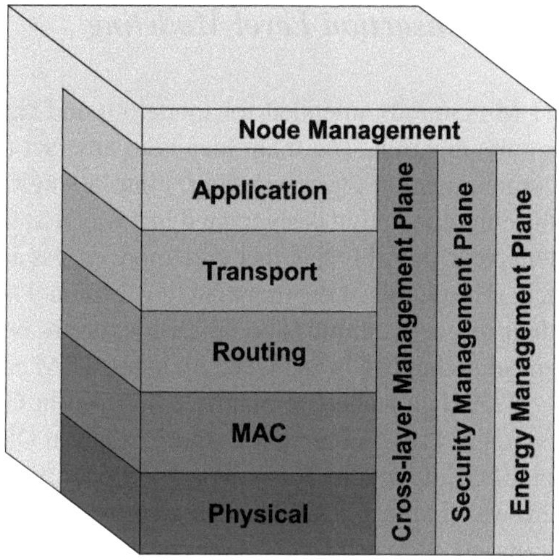

Fig. 21.3 Cross-layer protocol model proposed in PAWiS

Data Post Processing All the output information is logged in a text file. A data processing tool is provided to analyze this log file and to graphically represent the power consumption.

The PAWiS project (framework, module library and data processing tool) is free and open source software and distributed under the GNU General Public License (GPL) [17].

4 Using TLM for Wireless Sensor Network Simulation

This section describes a prospective WSN simulation approach currently under investigation in the course of the SNOPS project. A current implementation, using the basic ideas and algorithms from the PAWiS simulation framework, suggests that the approach is feasible, but comparative figures with respect to other simulation approaches and frameworks cannot be given yet.

Power simulation on the node level within the simulation of a whole sensor network can turn out to be very expensive with respect to simulation performance. To assess the power consumption benefits of power saving techniques, modeling and simulation on a high detail level is required. It would include an instruction set simulator (ISS) for the processor at hand, together with a power simulator, such that the power consumption of the node can be computed to get comparative figures.

To make up for this costly detailed node level simulation, it is desirable to get a network level simulation approach which is as lightweight as possible. When looking at simulation speedup approaches in similar modeling areas, Transaction Level Modeling (TLM, see [1]) is a very promising technique.

4.1 Transaction Level Modeling

TLM is mainly intended for modeling and simulation of systems using busses for communication. The main idea is to abstract the low-level events occurring in bus communication ("pin wiggling") into a single element called *transaction*. That is, the communication is abstracted in a way that focuses on data transfers between system parts, possibly together with more or less accurate timing estimates for the transfers. The details of the transfers (e.g. protocol issues like address and data phase, and their respective timing) are left aside, or are only modeled coarsely. The simulation speedup that can be achieved by using TLM can reach up to a factor of 1000.

TLM is a technique mainly employed in C-based design, and especially within SystemC [16]. For SystemC, there exists an OSCI standard: TLM 1.0 was presented in 2005, and mainly focuses on the modeling of the communication channels. TLM 2.0 was finalized in 2008, with a focus on model interoperability with standardized transaction objects and abstract protocol phases, and can be used as an extension library to SystemC. Models which are conformal to the TLM 2.0 standard can be used together in one model "out of the box", even if they were developed independently.

The standard transaction class in TLM 2.0 is called *generic payload*, and has attributes like a command (read or write), a target address, a data section and the data length, which basically sets a focus on memory mapped busses. The generic payload can also be extended with additional attributes.

Transactions are then passed by reference via method calls among the different components of the system model. For example, a read request from a CPU to a Memory module together with the resulting data transfer can be captured by one transaction. In TLM, every component in the system model basically assumes one of three roles:

- *Initiators* create new transactions (e.g. a CPU model).
- *Interconnects* do forward transactions (e.g. busses and routers). They might also modify these transactions, e.g. by mapping virtual addresses to local addresses.
- *Targets* are the final destinations of the transactions (e.g. memories and I/O components), where the command associated to the transaction is actually executed.

Note that even if several interconnects are involved in forwarding a transaction, all operations are performed using *one* single transaction object whose reference is passed along. The TLM 2.0 standard foresees several method interfaces for passing transactions, with the most important ones being the blocking and the nonblocking interface. The blocking interface has just one method implemented by targets, namely `b_transport(`*`transaction`*`, `*`time-offset`*`)`, with no return value. The name "blocking" results from the fact that an implementation of this method is allowed to call a SystemC `wait(`*`time/event`*`)`. This interrupts the current process until a certain (simulation) time span has passed or a certain event occurs, thus effectively blocking the initiator process which generated the transaction. The time offset parameter passed with the transaction indicates when the transaction is valid with respect to the current simulation time. For example, if a transaction is send to a target with an offset of 5 ms, the target might choose to call `wait(5, SC_MS)`, and then proceed with processing the transaction. Note, however, that there are different ways to act on the offset value. For details, consult [1].

The nonblocking interface has two similar methods, one for the initiator side and one for the target side, which are not allowed to call `wait(...)`. Instead, if there is the need to delay the transaction processing, the transaction is stored and an event is registered for the time when to proceed. Therefore, there is a method on the side of the initiator such that targets can respond actively. Again, see [1] for details.

The lifetime of a transaction can have various granularities. If it is a single time span (which might be zero), this is usually modeled using the so called *loosely timed coding style*. This coding style typically involves using only the blocking interface, and also has the additional feature that *temporal decoupling* can be used, which means that initiators can run ahead of the global simulation time by maintaining a local timing offset, and synchronize with the global simulation time either periodically or depending on certain circumstances. This technique allows for a further simulation speedup, since it reduces context switches.

It is also possible to subdivide the lifetime of a transaction into different phases (accompanied by respective method calls), e.g. capturing address phase, data phase and latencies. This is known as the *approximately timed coding style*, and involves using the nonblocking interface. While a transaction is finished with the return of the blocking transport in the loosely timed coding style, the (reference to the) transaction might get passed back and forth several times in the approximately timed coding style.

4.2 TLM and WSN

At a first glance, the TLM approach seems not to be very useful regarding wireless sensor network simulation at the network level, where no dedicated busses, routers or arbiters are used. But from a simulation point of view, an environment model (e.g. as it is implemented in the PAWiS framework [23]) behaves not much different from a bus model in TLM (see Fig. 21.4): it gets transmitted data packets (which correspond to transactions), and forwards them. However, it is important to note the differences:

- A TLM bus model will be part of a later implementation, while the environment model is an estimate of the operating conditions the implementation will face.
- In WSNs, no memory map is used, therefore using READ and WRITE commands also makes no sense.
- The smallest information unit in the TLM standard is a byte, whereas in WSNs, the data granularity refers to bits.
- A WSN node might be the source *as well* as the destination of data packets, while it might also be used to route data packets. Therefore, from the TLM point of view, it acts as initiator, target and interconnect.
- While a transaction in a usual TLM model is passed along a straight path through the system, the transmission of a data packet in a WSN network might fork.

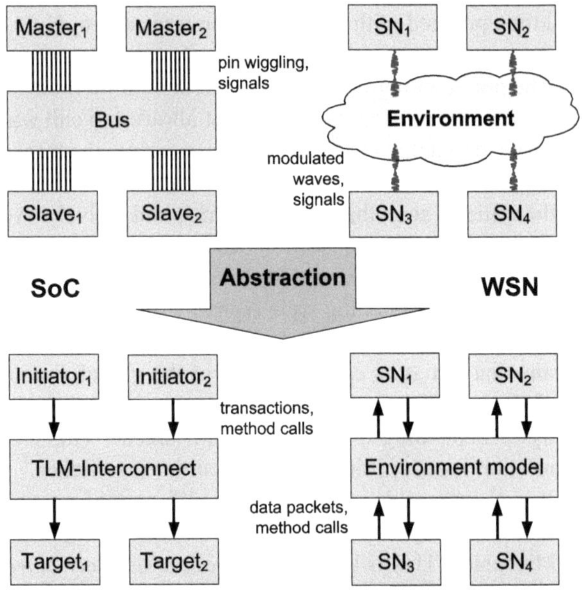

Fig. 21.4 Similar abstractions for SoCs and WSNs (SN: Sensor Node)

The last point requires some explanation. In a WSN, a data packet transmitted by one node can be received by all nodes close enough. Therefore the environment model, which gets the packet from the node, has to determine all nodes which are able to receive the packet, and forward the packet to them (see Sect. 3). A node then decides by itself if it receives this packet completely: If the packet contains a target ID, it might stop receiving upon decoding a target ID not matching its own ID (the same holds if message IDs similar to the operation of a CAN bus are used). Note that it is important to model these canceled receiving attempts, since they influence the power consumption in the network.

For simple functional verification, it could make sense to *not* model these effects. That is, the environment could be modeled such that it only forwards the data packet to the node with the matching target address. But even in this case, forks can occur: if a node routes a packet, it might route it to more than one node, depending on the routing strategy used. Also, broadcast messages might be used which have to be distributed through the whole (or a part of the) network, e.g. to exchange routing information. In contrast to this, a transaction in a TLM model is always passed along a straight path (see Fig. 21.5), where the accesses to the transaction are non-concurrent. In turn, if we choose to represent the transmission of a data packet through a WSN by a single transaction object, we might have concurrent accesses to it.

Obviously, the TLM approach has to be altered to be useful for wireless sensor network simulation. Fortunately, the TLM 2.0 standard is very flexible. It is possible to extend the generic payload by additional attributes (e.g. signal strength or bit error rate for the case at hand), but the TLM 2.0 interfaces can also be used with an entirely custom transaction class. Also, an extended or entirely custom set of

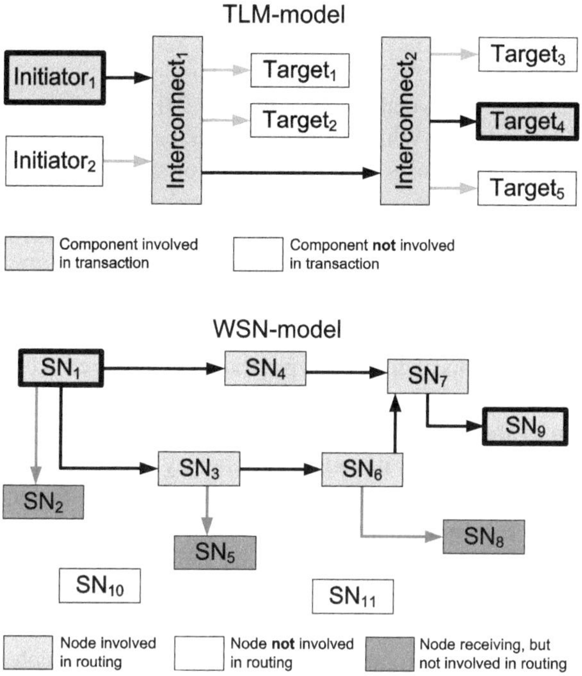

Fig. 21.5 TLM Transaction path versus a WSN packet path

phases can be introduced, which could be used to capture typical wireless communication steps like carrier sensing. A successful adaptation of the TLM 2.0 approach to wireless sensor networks can yield multiple benefits:

- With regard to the mixed wireless/wired sensor network scenario, we stay within a modeling paradigm where bus modeling is straightforward. This is an issue not addressed by the PAWiS framework properly.
- We can benefit from the general simulation performance enhancement prospects of TLM.
- If TLM is also used to model the sensor nodes internals, it might be possible to achieve seamless integration of node and network simulation.

Moreover, by associating power consumption information to transactions, we can get a whole new view on power consumption in a wireless sensor network by introducing concepts like the power consumption of a transaction. We can determine in how far certain transactions influence the whole system's (or certain critical node's) power consumption, and use this information to optimize WSN protocols and routing strategies.

4.3 Implementation

Since the transaction is a pivotal concept in TLM 2.0, it is vital to find a suitable adaption to WSN modeling. While some of generic payload attributes of TLM 2.0 are not suitable for our purposes (e.g. byte enable or streaming width), we have to add others (e.g. signal strength, transmission duration). Fortunately, with the generic payload extension mechanism there is an appropriate tool available for this purpose. Informally, a generic payload extensions can be thought of as a "Post-it" attached to the generic payload. Technically, it is an arbitrary C++ structure implementing the TLM 2.0 generic payload extension interface (see [1]). It is possible to attach arbitrarily many extensions of different types (but only one of each type) to a generic payload.

Our adoption of the TLM approach to WSNs works as follows: When a node transmits a data packet, it generates a new generic payload, with the data going into its data section and the ID of the final target node going into the address parameter. This ID is only for simulation purposes and does not represent any kind of real address, which would depend on the protocols and implementations selected depending on the application. It then attaches a extension with additional information, e.g. to which node to route to, or the signal strength, and sends the transaction to the environment model. The environment model now determines which nodes are able to receive the data packet associated to the transaction, and sends the transaction to them.

Keeping the transaction point of view is the most demanding part of this implementation. In a message implementation, like the one in PAWiS, each node owned a message and they were duplicated when sending to several nodes. However, in the transaction approach, a single transaction must cover the whole communication process. As the transaction is likely used by several nodes at the same time, data modifications in the transaction may lead to inconsistencies. Also, some of the data is different for every node, e.g. the signal strength.

Therefore, we use a second extension for the transactions sent from environment to the nodes. We refer to this extension as *node extension*, while the other is called *environment extension*. The basic data contained in the node extension is the same, but now organized using the map data structure from the standard template library, where the key is the ID of the node the transaction is sent to. Since a transaction might pass a node more than once, we keep the data within a dynamic array, and keep a map of indexes (again with the node ID, which is supposed to be unique and unchangeable, as a key) to indicate to the respective node which array entry holds the actual data. However, each node can in principle also access data not intended for it. Therefore, transaction data must be interfaced so that only permitted operations can be carried out and the integrity of the transaction is preserved.

Figure 21.6 shows an example (without vectors). As the transaction is sent to the environment the first time in step 1, it only contains the environment extension. Before the environment passes it to the nodes 2 and 3 in step 2, it adds the node extension with entries for these two nodes. Note that the environment extension is

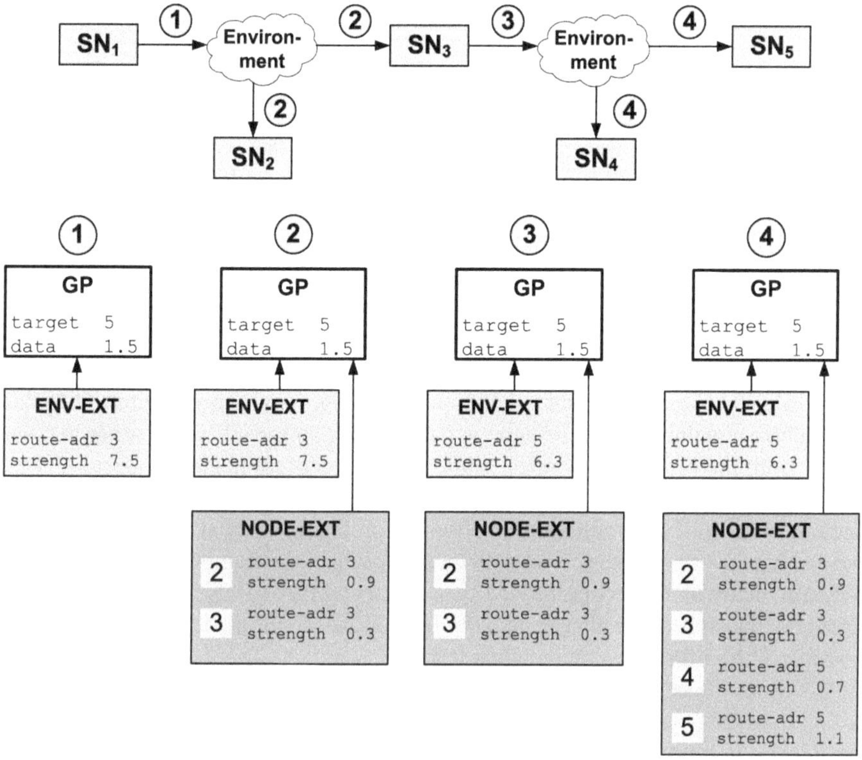

Fig. 21.6 A generic payload with WSN extensions in the course of a packet transmission

still present, but is not accessed by the nodes. In step 3, node 3 reuses the environment extension before it passes the transaction to the environment, which then adds information for node 4 and 5 in step 4.

Maintaining a dynamic array with all the indexes is of course costly in terms of memory, but avoids replicating a message whenever it is forwarded, which not only involves memory usage but also creating and destroying these messages.

Apart from the improvement in simulation performance, this TLM approach has been developed with ultra-low power in mind. One of the greatest benefits of keeping the transaction point of view is that a complete history of the data transfer associated to the transaction can be kept. Since the correlation of all single transmission and receiving events caused by this data transfer is kept, we can use this information to e.g. optimize routing strategies with respect to power consumption, if we add the consumed power by each node to the node extension. Hence, TLM is not only very useful to improve simulation performance, but it becomes a very powerful tool for energy profiling. This is explained in detail in [14].

It is also conceivable to add an additional extension for power consumption or other kinds of "simulation meta-data" and, regarding the network architecture, to add extensions for each network layer. In general, adding any kind of user defined extension depending on the protocols used is possible.

5 Conclusion and Outlook

This chapter dealt with Wireless Sensor Network simulation supporting ultra-low power design of wireless sensor nodes. After summarizing the requirements for WSN simulation for power optimization, we presented the existing PAWiS framework, which was specifically designed for this endeavor. Finally, a new TLM-inspired approach regarding the communication level was presented, which is currently explored and integrated into the PAWiS framework with the intention to enhance simulation performance and to allow for new views on power optimization.

The prospects of using TLM within a WSN simulation framework is twofold. Apart from yielding a fast simulation framework with new capabilities, it is also a way to explore in how far the TLM paradigm has to be altered to be useful for the area of WSN modeling and simulation. Currently, some aspects of the TLM 2.0 mechanisms (mainly within the generic payload) are "misused" in the sense that we depart from the original semantics, while other aspects are neglected because they are of no apparent use. Since this is done within a single framework, which does not need to be TLM 2.0 compliant, this poses no problem, especially considering that the end user does not have to deal with these aspects of the framework directly. Note however, that with respect to *timing* (i.e. considering the loosely and approximately timed coding styles), the altered network level TLM should work well together with possible TLM 2.0 models on the node level.

However, the experience gained could show the way for an approach for general wireless network modeling and simulation which might play a similar role like the TLM 2.0 approach for bus oriented modeling and simulation. That is, an approach for SystemC based modeling by providing a standardized extension library for model interoperability in wireless networks.

Acknowledgements This work is conducted as part of the Sensor Network Optimization by Power Simulation (SNOPS) project which is funded by the Austrian government via FIT-IT (grant number 815069/13511) within the European ITEA2 project GEODES (grant number 07013).

References

1. Aynsley, J.: OSCI TLM2 User Manual. Technical report, Open SystemC Initiative (2008)
2. Damm, M., Moreno, J., Haase, J., Grimm, C.: Using transaction level modeling techniques for wireless sensor network simulation. In: Design, Automation Test in Europe Conference Exhibition (DATE 2010), pp. 1047–1052 (2010)
3. Fummi, F., Quaglia, D., Ricciato, F., Turolla, M.: Modeling and simulation of mobile gateways interacting with wireless sensor networks. In: DATE '06: Proceedings of the Conference on Design, Automation and Test in Europe, pp. 106–111. European Design and Automation Association, Leuven (2006)
4. Glaser, J., Weber, D.: Simulation framework for power aware wireless sensors. In: Langendoen, K., Voigt, T. (eds.) 4th European Conference, EWSN 2007, Adjunct Poster/Demo Proceedings, Delft, The Netherlands. Parallel and Distributed Systems Report Series, report number PDS-2007-001, pp. 19–20 (2007)
5. Glaser, J., Weber, D., Madani, S.A., Mahlknecht, S.: Power aware simulation framework for wireless sensor networks and nodes. In: EURASIP JES (2008)

6. Hambeck, C., Mahlknecht, S., Herndl, T., Halvorsen, E.: An energy harvesting system for in-tire TPMS. In: 1st International Workshop on Power Supply on Chip, Cork, Irland (2008)
7. Madani, S.A., Mahlknecht, S., Glaser, J.: CLAMP: Cross Layer management plane for low power wireless sensor networks. In: Fifth International Workshop on Frontiers of Information Technology (2007)
8. Mahlknecht, S., Spinola Durante, M.: WUR-MAC: energy efficient wakeup receiver based MAC protocol. In: Proceedings of the 8th IFAC International Conference on Fieldbuses & Networks in Industrial & Embedded Systems, FET 2009, Hanyang University, Ansan, Korea (2009)
9. Mahlknecht, S.: Energy-self-sufficient wireless sensor networks for home and building environment. PhD thesis, Institute of Computer Technology, Vienna University of Technology (2004)
10. Mahlknecht, S., Böck, M.: CSMA-MPS: a minimum preamble sampling MAC protocol for low power wireless sensor networks. In: Proceedings of the IEEE International Workshop on Factory Communication Systems. WFCS (2004)
11. Mahlknecht, S., Glaser, J., Herndl, T.: PAWiS: towards a power aware system architecture for a SoC/SiP wireless sensor and actor node implementation. In: Proceedings of 6th IFAC International Conference on Fieldbus Systems and Their Applications, Puebla, Mexico, pp. 129–134 (2005)
12. Mahlknecht, S., Madani, S.A., Rötzer, M.: Energy aware distance vector routing scheme for data centric low power wireless sensor networks. In: 4th International IEEE Conference on Industrial Informatics INDIN'06 (2006)
13. McCanne, S., Floyd, S.: NS Network Simulator—version 2. http://www.isi.edu/nsnam/ns (1996)
14. Moreno, J., Haase, J., Grimm, C.: Energy consumption estimation and profiling in wireless sensor networks. In: Workshop Proceedings of International Conference on Architecture of Computing Systems (ARCS 2010), pp. 259–264 (2010)
15. Nebel, W.: System-level power optimization. In: Proceedings of Digital System Design, Euromicro Symposium, DSD 2004, pp. 27–34 (2004)
16. OSCI. SystemC™. Open SystemC Initiative. http://www.systemc.org
17. PAWiS. PAWiS Simulation Framework. http://pawis.sourceforge.net/ (2009)
18. Poweropt, http://www.chipvision.com
19. Spinola-Durante, M., Mahlknecht, S.: An ultra low power wakeup receiver for wireless sensor nodes. In: Proceedings of the Third International Conference on Sensor Technologies and Applications, SENSORCOMM 2009, Athens/Glyfada, Greece (2009)
20. Varga, A.: The OMNeT++ discrete event simulation system. In: European Simulation Multiconference (ESM'2001), Prague, Czech Republic (2001)
21. Varga, A.: OMNeT++ Discrete Event Simulation System User Manual (2005)
22. Vasilevski, M., Pecheux, F., Beilleau, N., Aboushady, H., Einwich, K.: Modeling and refining heterogeneous systems with SystemC-AMS: application to WSN. In: Design, Automation and Test in Europe Conference and Exhibition, pp. 134–139 (2008)
23. Weber, D., Glaser, J., Mahlknecht, S.: Discrete event simulation framework for power aware wireless sensor networks. In: Proceedings of the 5th International Conference on Industrial Informatics, INDIN 2007, Vienna, Austria, vol. 1, pp. 335–340 (2007)
24. Zhang, Q., Cho, W., Sobelman, G.E., Yang, L., Voyles, R.: TwinsNet: a cooperative MIMO mobile sensor network. In: Ubiquitous Intelligence and Computing, pp. 508–516. Springer, Berlin (2006). ISBN 978-3-540-38091-7

Index

G. Nicolescu et al. (eds.), *Design Technology for Heterogeneous Embedded Systems*, DOI 10.1007/978-94-007-1125-9, © Springer Science+Business Media B.V. 2012

MIX
Papier aus verantwortungsvollen Quellen
Paper from responsible sources
FSC® C105338
FSC
www.fsc.org

If you have any concerns about our products,
you can contact us on
ProductSafety@springernature.com

In case Publisher is established outside the EU,
the EU authorized representative is:
Springer Nature Customer Service Center GmbH
Europaplatz 3, 69115 Heidelberg, Germany

Printed by Libri Plureos GmbH
in Hamburg, Germany